Topics in Applied Physics   Volume 33

Topics in Applied Physics Founded by Helmut K. V. Lotsch

Volumes 1–56 are listed on the back inside cover

Electrets

Edited by G. M. Sessler

With Contributions by
M. G. Broadhurst   G. T. Davis
R. Gerhard-Multhaupt   B. Gross   S. Mascarenhas
G. M. Sessler   J. van Turnhout   J. E. West

Second Enlarged Edition

With 227 Figures

Springer-Verlag Berlin Heidelberg GmbH

Professor Dr. *Gerhard M. Sessler*

Fachgebiet Elektroakustik, Universität Darmstadt, Merckstraße 25,
D-6100 Darmstadt, Fed. Rep. of Germany

Library of Congress Cataloging in Publication Data. Electrets. (Topics in applied physics; v. 33) Includes bibliographies and index. 1. Electrets. I. Sessler, G.M. (Gerhard Martin), 1931–. II. Broadhurst, M.G. (Martin Gilbert), 1932–. III. Series. QC585.8.E4E43 1987 537'.24 86-31636

ISBN 978-3-540-17335-9   ISBN 978-3-540-70750-9 (eBook)
DOI 10.1007/978-3-540-70750-9

© Springer-Verlag Berlin Heidelberg 1980 and 1987

Ursprünglich erschienen bei Springer-Verlag Berlin Heidelberg New York 1987.

2153/3150-543210

Preface to the Second Edition

The first edition of this volume has been well received by readers and reviewers. In addition to the original English version published in 1980, MIR in Moscow issued a Russian edition in 1983. Since copies of the first edition are now exhausted while interest in the material continues, Springer-Verlag has asked the editor to prepare a new edition.

The present second edition contains the seven chapters of the original book and one additional chapter outlining recent progress in the field. The older chapters are essentially unchanged, except for the correction of misprints that came to the attention of the authors. Since the literature on electrets has significantly increased in the interim period, the discussion in the new chapter had to be much more concise than in the existing parts of the book. Even so, many of the new papers could, for reasons of space, not be included. A listing of recent literature concludes the book.

The editor expresses his gratitude to his fellow contributors for providing valuable suggestions concerning the new edition and to Springer-Verlag, especially to Dr. H. K. V. Lotsch, for a most gratifying collaboration.

Darmstadt, January 1987   *Gerhard M. Sessler*

Preface to the First Edition

Electrets have, over the past decade, emerged as invaluable components in an ever increasing number of applications. Their usefulness is responsible for the recent impressive growth of research work in a field which had been actively investigated since about 1920.

This volume aims to present the fundamental aspects of electret research as well as a detailed review of recent work in this area. The book is broad in scope, extending from the physical principles of the field to isothermal and thermally stimulated processes, radiation effects, piezoelectric and pyroelectric phenomena, bioelectret behavior, and, last, but not least, to applications of electrets. The emphasis of the experimental work discussed is on polymer electrets, but work performed on other organic substances, notably biomaterials, and on inorganic materials, such as ionic crystals or metal oxides, is also reviewed.

The interest in polymer electrets is due to the fact that these show extremely good charge-storage capabilities and are available as flexible thin films. In the 1960s attention focussed on highly insulating polymers, such as polytetrafluoroethylene, which have deep traps that store charges for extremely long periods of time. Around 1970, discovery of the strong piezoelectric properties of polyvinylidenefluoride attracted the imagination of many researchers and an enormous amount of work was devoted to the investigation of the physical and chemical properties of this and similar materials. Today, very active research is underway on charge-storage properties of both classes of polymers.

The chapters of this book are generally self-contained in the sense that each can be understood on its own. There are, however, many cross-references between chapters which will help to guide the reader to related or supplemental material in other parts of the volume. Uniform symbols and abbreviations are employed for the most-frequently used quantities and polymer names. A list of polymer names will be found in Chapter 1, a partial list of symbols at the end of the volume.

Although there have been a few monographs on specific topics of electret research and a number of conference proceedings, a cohesive treatment of the entire field of electrets has so far been lacking. The present volume, by covering many aspects of the field in a relatively small space, is an attempt in this direction. We realize, however, that a number of important questions are not, or not sufficiently, discussed, and that the views held by the different contributors are not always congruent.

It is with great pleasure that the editor expresses his gratitude to his fellow contributors, each being a renowned authority in his field, for their collaboration. The preparation and updating of the manuscripts placed a considerable burden on these colleagues, which they carried with understanding.

The book is dedicated to Professor *Bernhard Gross*, himself a contributor, by his fellow contributors. *Bernhard Gross* is the nestor of electret research, both theoretical and experimental. Apart from this, he has enhanced the knowledge in many other parts of physics. Without his contributions, electret research would not be what it is today. It is with admiration and gratitude that his coauthors devote this book to him.

Darmstadt, September 1979   *Gerhard M. Sessler*

Contents

3. Thermally Stimulated Discharge of Electrets

5. Piezo- and Pyroelectric Properties

6. Bioelectrets: Electrets in Biomaterials and Biopolymers

7. Applications

Contributors

Broadhurst, Martin G.
  United States Department of Commerce, National Bureau of Standards,
  Washington DC 20234, USA

Davis, G. Thomas
  United States Department of Commerce, National Bureau of Standards,
  Washington DC 20234, USA

Gerhard-Multhaupt, Reimund
  Heinrich-Hertz-Institut für Nachrichtentechnik Berlin GmbH,
  Abteilung S + E, Einsteinufer 37, D-1000 Berlin 10, Germany

Gross, Bernhard
  Institute of Physics and Chemistry of São Carlos-University of São Paulo,
  Caixa Postal 369, 13560 São Carlos, S. P., Brazil

Mascarenhas, Sérgio
  Institute of Physics and Chemistry of São Carlos-University of São Paulo,
  Caixa Postal 369, 13560 São Carlos, S. P., Brazil

Sessler, Gerhard M.
  Fachgebiet Elektroakustik, Universität Darmstadt
  Merckstraße 25, D-6100 Darmstadt, Fed. Rep. of Germany

van Turnhout, Jan
  Plastics and Rubber Research Institute TNO, PO Box 71,
  2600 AB Delft, The Netherlands

West, James Edward
  AT & T Bell Laboratories, 600 Mountain Avenue,
  Murray Hill, NJ 07974, USA

1. Introduction

G. M. Sessler

With 2 Figures

An electret is a piece of dielectric material exhibiting a *quasi-permanent electrical charge*. The term "quasi-permanent" means that the time constants characteristic for the decay of the charge are much longer than the time periods over which studies are performed with the electret.

The electret charge may consist of "real" charges, such as surface-charge layers or space charges; it may be a "true" polarization; or it may be a combination of these. This is shown schematically in Fig. 1.1 for a dielectric plate. While the true polarization is usually a frozen-in alignment of dipoles, the real charges comprise layers of trapped positive and negative carriers, often positioned at or near the two surfaces of the dielectric, respectively. The electret charges may also consist of carriers displaced within molecular or domain structures throughout the solid, resembling a true dipole polarization. If the charges are displaced to domain boundaries they are referred to as Maxwell–Wagner polarization. On metallized electrets, a compensation charge may reside on the electrode, unable to cross the energy barrier between metal and dielectric. Mostly, the net charge on an electret is zero or close to zero and its fields are due to charge separation and not caused by a net charge.

An electret not covered by metal electrodes produces an *external electrostatic field* if its polarization and real charges do not compensate each other everywhere in the dielectric. Such an electret is thus in a sense the electrostatic analogue of a permanent magnet, although electret properties may be caused by dipolar *and* monopolar charges while magnetic properties are only due to magnetic dipoles. The existence of an external field and the corresponding analogy with a magnet has often been used to define the electret.

However, as *Heaviside* already realized in 1892 [1.1], the fields of an electret may be compensated within a short time period by the relative motion of real charges and dipoles. This is observed in many piezoelectric substances. If one prefers to include some of these materials in the electret category, as *Heaviside* did, it is necessary to use the broader definition of the electret introduced above based on the *permanency of at least one of its charge components* and waive the necessity of an external field. However, such a broad definition includes the entire class of piezoelectric substances. We shall in the following adhere to this definition but include from the group of piezoelectric materials only the polymeric substances in the discussions in this book (see also Sect. 1.3).

While the classical electrets were made of *thick plates of carnauba wax* or similar substances, present electret research frequently deals with *thin-film*

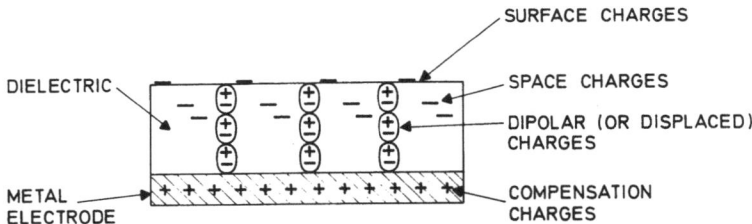

Fig. 1.1. Schematic cross section of a one-sided metallized electret having deposited surface charges, injected space charges, aligned dipolar charges (or microscopically displaced charges), and compensation charges

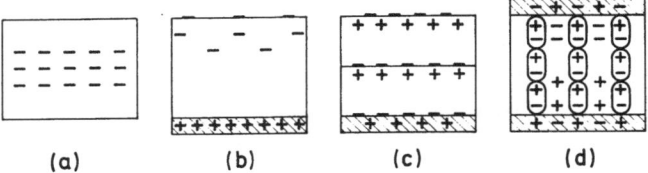

Fig. 1.2a–d. Schematic cross section of some typical electrets without electrodes or with grounded electrodes: (a) nonmetallized monocharge electret, (b) one-sided metallized electret with surface and space charges, (c) one-sided metallized electret with surface charges and charges displaced within domains (Maxwell-Wagner effect) and (d) two-sided metallized electret with dipolar and space charges

polymers such as the Teflon materials polyfluoroethylene propylene (FEP) and polytetrafluoroethylene (PTFE) or polyvinylidene fluoride (PVDF). Typical electrets in present use are 10–50 μm thick films of a few cm^2 area, often coated on one or both surfaces with evaporated metal layers. The materials are polarized to charge densities of 10^{-8}–10^{-6} C cm^{-2}, the charges being trapped real charges in Teflon and predominantly dipolar charges (but additionally trapped charges) in PVDF. In the latter case, the dipolar charges are statically (but not dynamically) compensated by space charges due to conduction in the material.

Examples of various nonmetallized and metallized electrets are shown in Fig. 1.2a–d. The charged dielectrics in these figures are either real-charge or dipolar electrets or a combination of these. While the nonmetallized and one-sided metallized samples usually exhibit external and internal electric fields, the fields of the two-sided metallized samples are completely contained within the dielectric.

1.1 Remarks on the History of Electret Research

Electret properties were already described by *Gray* in 1732 [1.2] when he mentioned the "perpetual attractive power" of a number of dielectrics, in particular waxes, rosins, and sulphur. He had generated the static electricity of these materials by contact electrification caused by cooling of the melts in iron ladles. More than a century later, in 1839, *Faraday* [1.3] theorized about electret properties due to application of an external electric field when he referred to a "dielectric which retains an electric moment after the externally-applied field has been reduced to zero". The word "electret" was then coined in 1892 by *Heaviside* [1.1].

Systematic research into electret properties began in 1919 when the Japanese physicist *Eguchi* formed electrets from essentially the same materials used by *Gray* by a *thermal method* consisting of the application of an electrical field to the cooling melt [1.4, 5]. He found that the dielectrics exhibited charges on their two surfaces which changed sign after a few days from a polarity opposite to one equal to that of the adjacent forming electrode. The charges were later named "heterocharges" and "homocharges", respectively, indicating their relation to the forming electrodes.

In the following decades, electrets from wax materials and a number of other substances were produced by *charging techniques* different from *Eguchi*'s thermal method. One such process, pioneered by *Selenyi* [1.6] in 1928, depends on the *injection of electrons or ions* into insulators. Subsequent extension of this work toward development of electrostatic recordings with powders [1.7] and later investigations of photoconductive image formation by *Carlson* [1.8] culminated in the development of xerography in the 1940s.

Related investigations of the *effect of light* on photoelectric layers by *Nadjakoff* [1.9] in 1938 remained largely unrecognized. *Nadjakoff* studied the charge separation occurring in such layers by simultaneous application of light and an electric field, thus discovering the photoelectret. The effect of photoconductivity on internal polarization was later extensively studied by *Kallmann* and co-workers [1.10, 11] and by *Fridkin* and *Zheludev* [1.12].

In the 1950s, a number of charging methods depending on the application of *high-energy ionizing radiation* were developed. Simplest among these is the bombardment of a dielectric with an electron beam of range smaller than the thickness of the dielectric [1.13, 14].

While *Selenyi*'s work with electron beams was mainly intended for recording, these investigations were performed to study breakdown and thermal charge release. Other similar methods are based on irradiating suitable dielectrics with ionizing radiation, such as penetrating gamma rays or electron beams, and applying an electric field to separate charge carriers generated by the radiation [1.15]. Even without an electric field, charging with gamma irradiation is possible due to Compton currents in the dielectric, as has been shown by *Gross* [1.16].

During the development of xerography, a simple charging method related to *Selenyi*'s ion-beam technique but depending on the application of a *corona discharge* was used [1.8, 17] and later extended to thin films [1.18]. Dielectrics were also charged by application of a *magnetic field* and heat to a dielectric [1.19]. More recently, the charging of thin films with low-energy *electron beams* [1.20] and *liquid contacts* [1.21] was reported.

While the early period produced a significant amount of experimental data, an understanding of charge storage and charge decay was initially not achieved. The first step toward an *explanation* of these phenomena was *Mikola*'s observation [1.22] in 1925 that two charges of different nature and of opposite polarity are found on nonmetallized dielectrics charged by application of an external voltage. One of the charges was assumed to be an internal polarization caused by ionic displacement while the other charge was ascribed to a deposition of ions onto the surfaces. *Adams* [1.23] in 1927, in an attempt to explain the charge decay theoretically, postulated again the presence of two charges, both being due to internal phenomena. These were assumed to be a volume polarization and a "compensating" charge at the surface, the latter being induced by the field of the former. The charge reversal and eventual decay was then explained as being due to the slow decay of the volume polarization and the time lag in the corresponding decay of the compensating charge. Further investigations were made by *Gemant* [1.24] in 1935 who confirmed the polarity reversals found by *Eguchi* and introduced the terms "heterocharge" for a charge of polarity opposite to that of the adjacent forming electrode and "homocharge" for a charge of equal sign.

Convincing evidence for the presence of *two different types of charge* and identification of these charges was eventually offered in a series of fundamental studies by *Gross* [1.25, 26] in the 1940s. He extended the experimental knowledge by measurements of polarization and depolarization currents and charges at varying temperatures and by sectioning experiments. It was demonstrated that the heterocharge is linked to dielectric absorption involving dipoles in polar substances or ionic charges in other materials while the homocharge is due to interfacial charging between electrode and dielectric. *Gross* also derived many of the fundamental relationships governing the fields and charges in a dielectric and the adjacent air gaps and electrodes. The two-charge theory was subsequently used by *Swann* [1.27] as the basis of a phenomenological theory of open-circuit charge decay, extended later on by *Gubkin* [1.28] to short-circuit conditions.

In 1937, *Gross* [1.29] first used the *Boltzmann–Hopkinson superposition principle*, valid for linear dielectrics, to develop a mathematical formalism for the discharge of electrode charges over the internal resistance of the dielectric and for the decay of a polarization in the dielectric. Later, *Wiseman* and *Feaster* [1.30] obtained further extensive confirmation of the validity of this principle by demonstrating that the response of a dielectric to a series of polarization steps is a linear superposition of the individual responses. *Perlman* and *Meunier* [1.31] then applied the principle to the open-circuit decay of unshielded dielectrics and thus explained the charge decay in carnauba wax.

Further insight into the nature of charge retention was achieved from the studies of *Gerson* and *Rohrbaugh* [1.32] which indicated that *carrier trapping* could play an important role in electrets.

Dramatic progress was made by the introduction of the *thermal depolarization method*, first used by *Randall* and *Wilkins* [1.33] for the investigation of phosphorescence, into electret research. In 1964, this method was applied to the case of dipole polarization by *Bucci* and co-workers [1.34, 35] who suggested the name ionic thermal conductivity (ITC). It allows determination of the activation energy and dipolar relaxation time from a measurement of the depolarization currents obtained upon linear heating of the dielectric. A host of recent work has been devoted to the application of ITC methods or related thermally stimulated current techniques for investigating dipolar and space-charge phenomena respectively [1.36–41]. These studies have culminated in a very comprehensive treatment by *van Turnhout* [1.42].

Of similar importance are recent studies of *isothermal charge transport* in insulators, taking into consideration the effect of trapping on excess-charge currents [1.43–49]. This work has also been extended to irradiated materials [1.50] and materials with non-uniform conductivity [1.46]. Additional information has been gainel from dc and ac conductivity measurements on poorly conducting materials and from the interpretation of such data [1.47, 48, 51, 52]. Recently, charge transport in amorphous solids has been discussed in terms of a stochastic transport model [1.53]. This model describes the dynamics of a group of carriers executing hopping processes in an electric field with a wide distribution of hopping times. Such analyses with and without consideration of trapping phenomena yield a number of important conclusions about charge transport.

Much of the newer experimental work was done on *polymer electrets* which, although chemically not as well defined as many inorganic materials, have gained importance due to their usefulness in practical applications.

Piezoelectric properties of biological and polymeric materials were already investigated by *Fukada* and others in the 1950s and 1960s [1.54]. Of particular importance was the discovery by *Kawai* in 1969 of a strong piezoelectric effect in the polymer PVDF [1.55]. Due to a host of actual and potential applications of piezoelectric polymer materials, this subject has now developed into a very active area of research [1.56–60].

Apart from the use of charge-storage phenomena in xerography and a few other areas, where extreme permanence of the polarization is not required, *applications of electrets* came only recently. Actually, long-lived electrets were generally considered a scientific curiosity until the late 1960s when, a few years after the description of the first polymer-film electret microphone in 1962 [1.61], such transducers were introduced commerically on a large scale. Presently, work on the use of electrets in dosimeters, transducers, pyroelectric detectors, prosthetic and switching devices, gas filters and other instruments is being performed by a number of laboratories [1.62–67] and a series of commercial applications have been reported (see Chap. 7). This has stimulated a considerable increase in very recent electret research.

The early *literature on electrets* was reviewed by several authors (e.g., [1.68, 69]), notably by *Fridkin* and *Zheludev* [1.12] and *Gross* [1.70], while the more recent work is discussed in a number of articles of different scope (e.g.,

[1.71, 72]), in the proceedings of several conferences [1.73–75], and in *van Turnhout*'s monograph on thermally stimulated currents [1.42].

1.2 Survey of Physical Properties

Physical properties of particular interest in electret research are conductivity, mobility, dipolar relaxation frequency, and piezo- and pyroelectric constants. To explain the differences between various classes of electrets, and distinguish them from other dielectrics, a brief discussion of these properties is necessary.

If a dielectric has mobile intrinsic charge carriers, it exhibits an *intrinsic conductivity g*. This conductivity is responsible for the decay of excess real charges within the material. In the simplest case, if the excess charges are located on the surfaces of an open-circuited sample (e.g., a nonmetallized dielectric) and decay through the dielectric without giving rise to space-charge effects, the time constant τ of the decay is given by $\tau_1 = \varepsilon\varepsilon_0/g$.

Even in nonconductive materials (devoid of intrinsic carriers), an excess charge decays due to the *carrier mobility*. A charge located initially on the nonmetallized surface of a one-sided metallized sample of thickness s spreads under its field through the dielectric within the so-called transit time given by $t_{0\lambda} = s^2/\mu V_0$, where μ is the mobility and V_0 the initial voltage generated across the sample by the charge. The mobility may be affected by trapping events.

The decay of a uniform dipole alignment is, in the simplest case of a single *dipole relaxation frequency* and in the absence of an internal field (i.e., for a short-circuited dielectric), determined by a relaxation time $\tau_2 = 1/\alpha$. Mostly, an entire spectrum of relaxation frequencies is present.

Piezoelectric properties are found in noncentrosymmetrical crystals which account for 21 of the 32 crystal classes. Ten of these belong to the polar classes and are also *pyroelectric*. Piezoelectricity (and in some cases pyroelectricity) occurs in single crystals (such as quartz), in polycrystalline materials (such as PZT), and in materials with crystalline and amorphous regions (such as PVDF). While piezoelectric crystals show spontaneous polarization, the polycrystalline or partially crystalline materials have to be poled at an elevated temperature by application of an electric field. Piezo- and pyroelectric effects can also occur in space-charged dielectrics, but only in cases where either a mechanical anisotropy (e.g., due to a domain structure or to stiff electrodes deposited on the material) or inhomogeneous strain is present [1.76]. Whether this yields a sizeable contribution to the observed piezo- and pyroelectric effects remains in question [1.57, 77]. Some piezoelectric polymers show also cooperative interactions within groups of molecules and thus resemble ferroelectric materials. Evidence of cooperative phenomena is, for example, found in the poling hysteresis of PVDF [1.60]. X-ray diffraction studies on PVDF also give results consistent with the assumption that the material is ferroelectric [1.78].

Typical properties of some electret materials (including piezoelectric substances) are shown in Table 1.1. The piezoelectric materials are all characterized by a very slow dipolar relaxation. Due to their relatively high conductivity,

Table 1.1. Selected properties of some electret materials at room temperature, in order of increasing conductivity g; τ_1 = dielectric relaxation time, τ_2 = dipole relaxation time, d_{31} = piezoelectric constant, p = pyroelectric coefficient

Class	Material (Example)	Structure	g [$\Omega^{-1}\,cm^{-1}$]	$\tau_1 = \varepsilon\varepsilon_0/g$ [s]	$\tau_2 = 1/\alpha$ [s]	d_{31} [cm V^{-1}]	p [nC cm^{-2} K^{-1}]
Nonpolar polymer electret	PTFE FEP	crystalline and amorphous regions	$\approx 10^{-22}$ [a]	$\approx 10^9$	–	0	0
Classical electret	Carnauba wax		10^{-18} [b]	2×10^5	10^3–10^5 [c]	0	0
Photo-electret	Anthracene Naphthalene	crystalline or polycrystalline	10^{-15} [d]	2×10^2	–	10^{-11} [d]	
Piezo-electric polymer	PVDF	crystalline and amorphous regions	10^{-14} [e]	10^2	$\approx 10^{9}$ [f]	2×10^{-9} [e]	4 [g]
Piezo-electric crystal	Quartz	crystalline	$<10^{-14}$ [h]	$>10^2$	no exponential decay measurable	2×10^{-10} [i] (d_{11})	
Piezo-electric ceramic	PZT-5	polycrystalline	$<10^{-13}$ [i]	$>10^3$		2×10^{-8} [i]	5 [k]

[a] G.M. Sessler, J.E. West: J. Electrostat. **1**, 111 (1975).

[b] M.M. Perlman, J.L. Meunier: J. Appl. Phys. **36**, 420 (1965).

[c] M.M. Perlman: J. Appl. Phys. **42**, 2645 (1971).

[d] [1.12].

[e] N. Murayama, K. Nakamura, H. Obara, M. Segawa: Ultrasonics **14**, 15 (1976).

[f] Estimated from decay of piezoelectric constants.

[g] K. Takahashi, R.E. Salomon, M.M. Labes: Bull. Am. Phys. Soc. **23**, 378 (1978)

[h] R.E. Bolz, G.L. Tuve (eds.): *Handbook of Tables for Applied Engineering Science* (Chemical Rubber Co, Cleveland 1970).

[i] D.A. Berlincourt, D.R. Curran, H. Jaffe: In *Physical Acoustics* Vol. IA, ed. by W.P. Mason (Academic Press, New York 1964).

[k] S.B. Lang, F. Streckel: Rev. Sci. Instr. **36**, 929 (1965)

their polarization is statically screened by intrinsic charges. This screening occurs in some materials during the poling process, in others (those with a strong pyroelectric effect) after cooling. The classical electret materials show relatively rapid dipolar and space-charge decay in open circuit. Their use as electrets depends traditionally on short-circuit storage, where the space-charge decay is slowed down due to the absence of internal fields while compensation by atmospheric charges is also avoided. Piezoelectricity is absent in carnauba wax, although a weak effect $(d \approx 10^{-11} \, \text{cm V}^{-1})$ exists in some crystalline photoelectrets. The smallest conductivities are observed in nonpolar polymers. Values range from 10^{-18} to $10^{-23} \, \Omega^{-1} \, \text{cm}^{-1}$ where the lowest values are probably limited by ambient radiation and can only be estimated from extrapolated time constants of charge decays. In these materials, charge transport frequently occurs by excess-charge currents. In some polymers with extremely small trap-modulated mobilities $(10^{-17} \, \text{cm}^2 \, \text{V}^{-1} \, \text{s}^{-1}$, see Sect. 2.6.6), the corresponding transit times are very long. Storage of real charges is therefore quite permanent with time constants of the decay often of the order of years or centuries under open-circuit and short-circuit conditions. These materials assume no dipolar polarization and show therefore no piezoelectric properties under normal conditions.

1.3 Organization of the Book

Chapter 2 reviews some of the fundamental aspects of charge storage. It includes an analysis of the fields, forces, and currents due to charged dielectrics and a survey of the charging and charge-measuring methods. These are the basic analytical tools and experimental methods for the study of electrets. Also covered in this chapter are isothermal decay processes which belong to the most elemental phenomena in such dielectrics.

The investigation of electret properties by thermally stimulated discharge (TSD) is the topic of Chap. 3. This method has great significance in electret research since it is capable of resolving processes characterized by very different activation energies or preexponential factors in a single, relatively short experiment. As demonstrated in this chapter, the TSD method, applied in a variety of ways, has provided extensive information on dipolar and trapped-charge effects in dielectrics.

Chapter 4 presents a review of radiation-induced effects in dielectrics, mostly in the context of charging phenomena and changes in electrical properties of the irradiated materials. Particularly one kind of radiation experiment, namely charging with electron beams, offers a great deal of control with respect to charge uniformity and charge depth; this makes this method one of today's favorite charging procedures in research and production. Apart from this, radiation methods have become extremely useful as diagnostic tools, where they have already revealed a wealth of information on electret properties.

The investigation of piezoelectric and pyroelectric properties of some polymer electrets is discussed in Chap. 5. Since the discovery of a strong

Table 1.2. Structures of polymers and abbreviations of polymer names (prepared by J. van Turnhout)

Name	Abbreviation	Structure $\begin{array}{cc} H & x \\ \mid & \mid \\ -C-C- \\ \mid & \mid \\ H & y \end{array}$
Polycyclohexyl methacrylate	PCHMA	$x = CH_3, \; y = COO$—⟨H⟩
Polyethyl methacrylate	PEMA	$x = CH_3, \; y = COOC_2H_5$
Polymethyl methacrylate	PMMA	$x = CH_3, \; y = COOCH_3$
Polyphenyl methacrylate	PPhMA	$x = CH_3, \; y = COO$—◯
Polyethylene	PE	$x = H, \; y = H$
Polypropylene	PP	$x = H, \; y = CH_3$
Polyvinyl chloride	PVC	$x = H, \; y = Cl$
Polyvinylidene chloride	PVDC	$x = Cl, \; y = Cl$
Polyvinyl fluoride	PVF	$x = H, \; y = F$
Polyvinylidene fluoride	PVDF	$x = F, \; y = F$

Name	Abbreviation	Structure
Polybisphenol A carbonate	PC-n	$\overset{\displaystyle O}{\overset{\|}{-C}}-O-◯-\overset{\displaystyle CH_3}{\underset{\displaystyle CH_3}{\overset{\|}{\underset{\|}{C}}}}-◯-O-$
Polyethylene terephthalate (Mylar)	PET	$\overset{\displaystyle O}{\overset{\|}{-C}}-◯-\overset{\displaystyle O}{\overset{\|}{C}}-O-(CH_2)_2-O-$
Polytetrafluoroethylene	PTFE	$\begin{array}{cc} F & F \\ \mid & \mid \\ -C-C- \\ \mid & \mid \\ F & F \end{array}$
Tetrafluoroethylene-hexa-fluoropropylene copolymer	Teflon-FEP	$\begin{array}{cccc} F & F & F & F \\ \mid & \mid & \mid & \mid \\ -C-C-C-C- \\ \mid & \mid & \mid & \mid \\ F & F & CF_3 & F \end{array}$
Tetrafluoroethylene-per-fluoromethoxyethylene copolymer	Teflon-PFA	$\begin{array}{cccc} F & F & F & F \\ \mid & \mid & \mid & \mid \\ -C-C-C-C- \\ \mid & \mid & \mid & \mid \\ F & F & O & F \\ & & \mid & \\ & & CF_3 & \end{array}$

piezoelectric effect in polyvinylidenefluoride in 1969, this field has seen an extremely rapid development. Piezoelectric polymers are morphologically and electrically related to conventional polymer electrets and are therefore generally referred to as electrets. Thus, a book on electrets would be incomplete without a discussion of piezoelectric polymers. The same could be said, of course, about conventional piezoelectric materials which are, in a strict sense, also electrets. However, to keep the present volume focussed on those electret materials which have not received wide attention in the past, we have excluded nonpolymeric piezoelectric substances from our considerations.

Chapter 6 is concerned with electret effects in biomaterials. Although most of the discussion deals with genuine biological substances, mostly biopolymers, reference is also made to a few artificial polymers, such as Teflon, which are useful in biomedical applications. This chapter is truely interdisciplinary as it analyzes the biological and medical effects of charged materials. Bioelectrets do play an important role in life processes and added knowledge in this field will no doubt open up new approaches in medicine.

Applications are generally not covered in Chaps. 2–6. Exceptions are the discussion of dosimetry in Sect. 4.7 and the analysis of biomedical phenomena in Chap. 6. Since dosimetry is intimately linked to radiation effects in general and biomedical phenomena are tightly interwoven with the basic biological processes it appeared reasonable to present these topics in the respective chapters.

Most other applications of electrets are described in Chap. 7. Although applications are subject to relatively rapid change, it was felt that a review of the present state of the art will not only document the wide usefulness of today's electrets, which is perhaps still unrecognized by many interested readers, but will also show how a simple concept can have a prominent place in the generally complex technology of today.

Table 1.2 lists the structures and abbreviations of the polymers referred to in the book.

References

1.1   O. Heaviside: *Electrical Papers* (Chelsea, New York, 1892) pp. 488–493
1.2   S. Gray: Philos. Trans. R. Soc. London, Ser. A: **37**, 285 (1732)
1.3   M. Faraday: *Experimental Researches in Electricity* (Richard and John Edward Taylor, London 1839)
1.4   M. Eguchi: Proc. Phys. Math. Soc. Jpn. **1**, 326 (1919)
1.5   M. Eguchi: Philos. Mag. **49**, 178 (1925)
1.6   P. Selenyi: Z. Tech. Phys. **9**, 451 (1928)
1.7   P. Selenyi: J. Appl. Phys. **9**, 637 (1938)
1.8   C. F. Carlson: "History of Electrostatic Recording", in *Xerography and Related Processes*, ed. by J. H. Dessauer, H. E. Clark (Focal Press, London 1965) pp. 15–49
1.9   G. Nadjakoff: C. R. Acad. Sci. **204**, 1865 (1937)
1.10 H. Kallmann, B. Rosenberg: Phys. Rev. **97**, 1596 (1955)
1.11 J. R. Freeman, H. P. Kallman, M. Silver: Rev. Mod. Phys. **33**, 553 (1961)
1.12 V. M. Fridkin, I. S. Zheludev: *Photoelectrets and the Electrophotographic Process* (Consultants Bureau, New York 1961)
1.13 L. G. Brazier: Engineer **196**, 637 (1953)
1.14 B. Gross: J. Polym. Sci. **27**, 135 (1958)
1.15 P. V. Murphy, S. C. Ribeira, F. Milanez, R. J. de Moraes: J. Chem. Phys. **38**, 2400 (1963)
1.16 B. Gross: Z. Phys. **155**, 479 (1959); J. Appl. Phys. **36**, 1635 (1965); IEEE Trans. NS-**25**, 1048 (1978)
1.17 R. W. Tyler, J. H. Webb, W. C. York: J. Appl. Phys. **26**, 61 (1955)

1.18 R. A. Creswell, M. M. Perlman: J. Appl. Phys. **41**, 2365 (1970)
1.19 C. S. Bhatnagar: Indian J. Pure Appl. Phys. **2**, 331 (1964)
1.20 G. M. Sessler, J. E. West: Appl. Phys. Lett. **17**, 507 (1970)
1.21 P. W. Chudleigh: Appl. Phys. Lett. **21**, 547 (1972); J. Appl. Phys. **47**, 4475 (1976)
1.22 S. Mikola: Z. Phys. **32**, 476 (1925)
1.23 E. P. Adams: J. Franklin Inst. **204**, 469 (1927)
1.24 A. Gemant: Philos. Mag. **20**, 929 (1935)
1.25 B. Gross: An. Acad. Bras. **17**, 219 (1945); Phys. Rev. **66**, 26 (1944)
1.26 B. Gross: J. Chem. Phys. **17**, 866 (1949)
1.27 W. F. G. Swann: J. Franklin Inst. **255**, 513 (1955)
1.28 A. N. Gubkin: Sov. Phys. Tech. Phys. **2**, 1813 (1958)
1.29 B. Gross: Z. Phys. **107**, 217 (1937)
1.30 G. G. Wiseman, G. R. Feaster: J. Chem. Phys. **26**, 521 (1957)
1.31 M. M. Perlman, J. L. Meunier: J. Appl. Phys. **36**, 420 (1965)
1.32 R. Gerson, J. H. Rohrbaugh: J. Chem. Phys. **23**, 2381 (1955)
1.33 T. J. Randall, M. H. F. Wilkins: Proc. R. Soc. London A**184**, 347, 366, 390 (1945)
1.34 C. Bucci, R. Fieschi: Phys. Rev. Lett. **12**, 16 (1964)
1.35 C. Bucci, R. Fieschi, G. Guidi: Phys. Rev. **148**, 816 (1966)
1.36 N. Januzzi, S. Mascarenhas: J. Electrochem. Soc. **115**, 382 (1968)
1.37 R. A. Creswell, M. M. Perlman, M. A. Kabayama: In *Dielectric Properties of Solids*, ed. by F. E. Karasz (Plenum New York 1972) pp. 295–312
1.38 J. van Turnhout: Polym. J. **2**, 173 (1971)
1.39 B. Gross: J. Electrochem. Soc. **119**, 855 (1972)
1.40 E. B. Podgorsak, P. R. Moran: Appl. Phys. Lett. **24**, 580 (1974)
1.41 G. M. Sessler, J. E. West: Phys. Rev. **10**, 4488 (1974)
1.42 J. van Turnhout: *Thermally Stimulated Discharge of Polymer Electrets* (Elsevier, Amsterdam 1975)
1.43 A. K. Jonscher: Thin Solid Films **1**, 213 (1967)
1.44 M. A. Lampert, P. Mark: *Current Injection in Solids* (Academic Press, New York 1970)
1.45 H. Seki, I. P. Batra: J. Appl. Phys. **42**, 2407 (1971)
1.46 H. J. Wintle: J. Appl. Phys. **43**, 2927 (1972); Thin Solid Films **21**, 83 (1974)
1.47 R. M. Hill: Thin Solid Films **15**, 369 (1973)
1.48 R. M. Hill: J. Phys. C: **8**, 2488 (1975)
1.49 L. N. de Oliveira, G. F. L. Ferreira: J. Electrostat. **1**, 371 (1975)
1.50 B. Gross, L. N. de Oliveira: J. Appl. Phys. **45**, 4724 (1974); J. Appl. Phys. **46**, 3132 (1975)
1.51 T. J. Lewis: in *1976 Annu. Rep. Conf. Electr. Insul. Dielectr. Phenom.* (NAS, Washington, D.C. 1977); IEE Conf. Publ. **129**, 261 (1975)
1.52 A. K. Jonscher: Thin Solid Films **36**, 1 (1976); *The Universal Dielectric Response* (Chelsea, London 1978)
1.53 H. Scher, E. W. Montroll: Phys. Rev. B**12**, 2455 (1975)
       G. Pfister, H. Scher: Phys. Rev. B**15**, 2062 (1977)
1.54 E. Fukada: J. Phys. Soc. Jpn. **10**, 149 (1955); Ultrasonics **6**, 229 (1968)
1.55 H. Kawai: Jpn. J. Appl. Phys. **8**, 975 (1969)
1.56 S. Mascarenhas: In *Electrets, Charge Storage and Transport in Dielectrics*, ed. by M. M. Perlman (Electrochemical Society, Princeton, N.J., 1973) pp. 650–656
1.57 G. Pfister, M. Abkovitz, R. G. Crystal: J. Appl. Phys. **44**, 2064 (1973);
       G. Pfister, M. A. Abkovitz: J. Appl. Phys. **45**, 1001 (1974)
       H. Burkhard, G. Pfister: J. Appl. Phys. **45**, 3360 (1974)
1.58 R. Hayakawa, Y. Wada: Adv. Polym. Sci. **11**, 1 (1973); Jpn. J. Appl. Phys. **15**, 2041 (1976)
1.59 E. Fukada: Adv. Biophys. **6**, 121 (1974); Jpn. J. Appl. Phys. **15**, 43 (1976)
       M. Oshiki, E. Fukada: Jpn. J. Appl. Phys. **15**, 43 (1976)
       M. Tamura, S. Hagiwara, S. Matsumata, N. Ono: J. Appl. Phys. **48**, 513 (1977)
1.60 F. I. Mopsik, M. G. Broadhurst: J. Appl. Phys. **46**, 4204 (1975)
       M. G. Broadhurst, G. T. Davis, J. E. McKinney: J. Appl. Phys. **49**, 4992 (1978)
       G. T. Davis, J. E. McKinney, M. G. Broadhurst, S. C. Roth: J. Appl. Phys. **49**, 4998 (1978)

1.61 G.M.Sessler, J.E.West: J. Acoust. Soc. Am. **34**, 1787 (1962)
1.62 G.W.Fabel, H.K.Henisch: Phys. Stat. Sol. a **6**, 535 (1971)
      M.W.Harper, B.Thomas: Phys. Med. Biol. **18**, 409 (1973)
      H.Bauser, W.Ronge: Health Phys. **34**, 97 (1978)
1.63 J.H.McFee, J.G.Bergman, G.R.Crane: IEEE Trans. SU-**19**, 305 (1972)
1.64 M.Tamura, T.Yamaguchi, T. Oyaba, T.Yoshimi: J. Audio Eng. Soc. **23**, 21 (1975)
1.65 G.M.Sessler, J.E.West: J. Acoust. Soc. Am. **53**, 1589 (1973)
      P. V. Murphy, S.Merchant: In *Electrets, Charge Storage, and Transport in Dielectrics*, ed. by
      M.M.Perlman (Electrochemical Society, Princeton, N.J. 1973) pp. 627–649
1.66 J.L.Bruneel, F.Micheron: Appl. Phys. Lett. **30**, 382 (1977)
      D.Perino, J.Lewiner, G.Dreyfus: L'onde electrique **57**, 688 (1977)
1.67 J.van Turnhout: J. Electrostat. **1**, 147 (1975)
      J.van Turnhout, C.van Bochove, G.J.van Veldhuizen: Staub-Reinhalt. Luft **36**, 36 (1976)
1.68 F.Gutmann: Rev. Mod. Phys. **20**, 457 (1948)
1.69 V. A.Johnson: *Electrets*, Parts I and II (US Dept. of Commerce, Washington D.C. 1962)
1.70 B.Gross: *Charge Storage in Solid Dielectrics* (Elsevier, Amsterdam 1964)
1.71 H.Kiess: RCA Rev. **36**, 667 (1975)
1.72 K.J.Euler: J. Electrostat. **2**, 1 (1976)
1.73 L.M.Baxt, M.M.Perlman (eds.): *Electrets and Related Electrostatic Charge Storage
      Phenomena* (Electrochemical Society, New York 1968)
1.74 M.M.Perlman (ed.): *Electrets, Charge Storage and Transport in Dielectrics* (Electrochemical
      Society Princeton, N.J. 1973)
1.75 M.S.de Campos (ed.): *International Symposium on Electrets and Dielectrics* (Academia Brasil
      Ciencas, Rio de Janeiro 1977)
1.76 K.Nakamura, Y.Wada: J. Polym. Sci. Part A: 2, **9**, 161 (1971)
1.77 N.Murayama, T.Oikawa, T.Katto, N.Nakamura: J. Polym. Sci. Phys. **13**, 1033 (1975)
      R.E.Salomon, H.Lee, C.S.Bak, M.M.Labes: J. Appl. Phys. **47**, 4206 (1976)
1.78 R.G.Kepler, R. A.Anderson: J. Appl. Phys. **49**, 1232 (1978)

2. Physical Principles of Electrets

G. M. Sessler

With 28 Figures

2.1 Electric Fields, Forces, and Currents

Knowledge of the electric fields, forces, and currents due to a given charge distribution and its variations in a dielectric is of great importance for the understanding of electret properties. In the following, these quantities will be derived under the assumption that the charge-carrying dielectric is in the form of a sheet bounded by two other nonchargeable dielectric sheets (gaps) as shown in Fig. 2.1. All lateral dimensions (i.e., dimensions parallel to the electrodes) are assumed to be much greater than the thickness.

Within the center layer, quasi-permanent real charges of volume density $\varrho_r(x)$ or of planar density $\sigma_r(x)$ and of uniform lateral distribution are to be present. In addition to the real charges, a laterally uniform polarization $P(x)$ is assumed. This polarization can be resolved into two components

$$P = P_i + P_p . \tag{2.1}$$

Here, $P_i(x)$ is the component which follows instantaneously an applied external field $E(x)$ and which is given by

$$P_i = \varepsilon_0(\varepsilon - 1) E , \tag{2.2}$$

while $P_p(x)$ is the quasi-permanent, frozen-in component due to dipole polarization or microscopic charge displacement. This component produces charges of local density

$$\varrho_p = -dP_p/dx \tag{2.3}$$

in the volume. If the polarization $P_p(x)$ changes by $\Delta P_p(x)$ at a certain interface, this can be treated like a planar charge of density

$$\sigma_p = -\Delta P_p \tag{2.4}$$

at that interface.

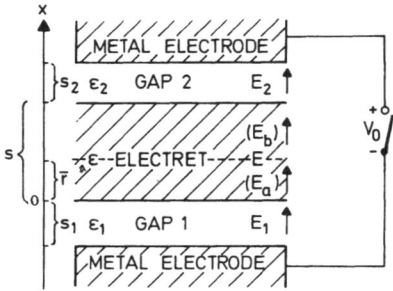

Fig. 2.1. Arrangement of charged dielectric (electret), gaps, and electrodes; s, s_1 and s_2 are the thickness dimensions, ε, ε_1, and ε_2 the dielectric constants, E, E_a, E_b, E_1, and E_2 the electric fields. A voltage V_0 is applied or generated across the electrodes

The total quasi-permanent volume and planar charge densities $\varrho(x)$ and $\sigma(x)$ are given by, respectively,

$$\begin{aligned} \varrho &= \varrho_r + \varrho_p, \\ \sigma &= \sigma_r + \sigma_p. \end{aligned} \qquad (2.5)$$

These charge distributions characterize the electret.

A voltage V_0 can be applied or generated across the electrodes if the open-circuit position of the switch in Fig. 2.1 is chosen. Under external short-circuit conditions, V_0 equals zero. The voltage, charges, fields, and geometrical dimensions are assumed to be either constant or to vary slowly with time. The time dependence will be explicitly invoked in Sect. 2.1.4.

2.1.1 Fields Due to Charge Layers

Assuming first the presence of a *single charge layer* of planar density σ at a location $x = \bar{r}$, one can write Gauss's law for the electric displacement applied to the interfaces $x=0$, $x=\bar{r}$, and $x=s$, respectively:

$$\begin{aligned} -\varepsilon_1 E_1 + \varepsilon E_a &= 0, \\ -\varepsilon E_a + \varepsilon E_b &= \sigma/\varepsilon_0, \\ -\varepsilon E_b + \varepsilon_2 E_2 &= 0, \end{aligned} \qquad (2.6)$$

where the fields E_1, E_a, E_b, and E_2 are defined in Fig. 2.1.

According to Kirchhoff's second law one has

$$V_0 + s_1 E_1 + \bar{r} E_a + (s - \bar{r}) E_b + s_2 E_2 = 0. \qquad (2.7)$$

One obtains from (2.6) and (2.7) the following results for E_1 and E_a:

$$SE_1 = -(V_0/\varepsilon_1) - (\sigma/\varepsilon_0 \varepsilon_1 \varepsilon \varepsilon_2)[\varepsilon_2(s - \bar{r}) + \varepsilon s_2], \qquad (2.8)$$

$$SE_a = -(V_0/\varepsilon) - (\sigma/\varepsilon_0 \varepsilon^2 \varepsilon_2)[\varepsilon_2(s - \bar{r}) + \varepsilon s_2], \qquad (2.9)$$

with

$$S = (s_1/\varepsilon_1) + (s/\varepsilon) + (s_2/\varepsilon_2).$$

Corresponding equations hold for E_b and E_2.

The fields on the two sides of a charge layer in a homogeneous sheet of dielectric, covered with metal electrodes on both sides, follow from (2.9) for $V_0 = 0$ as

$$E_a = -(\sigma/\varepsilon_0\varepsilon)(1 - \bar{r}/s), \tag{2.10}$$

$$E_b = (\sigma/\varepsilon_0\varepsilon)(\bar{r}/s). \tag{2.11}$$

For example, a charge layer of density $\sigma = 10^{-8}\,\mathrm{C\,cm^{-2}}$ in the center of a dielectric with $\varepsilon = 2$ generates fields of $\pm 28\,\mathrm{kV\,cm^{-1}}$ on its two sides, respectively.

Assume now the presence of *two charge layers* of planar density σ_1 and σ_2 on the surfaces $x = 0$ and $x = s$, respectively. One obtains from (2.8, 9) and the corresponding equations for E_b and E_2 by linear superposition [2.1]

$$SE_1 = -(V_0/\varepsilon_1) - (s/\varepsilon_0\varepsilon_1\varepsilon)\sigma_1 - (s_2/\varepsilon_0\varepsilon_1\varepsilon_2)(\sigma_1 + \sigma_2) \tag{2.12}$$

$$SE = -(V_0/\varepsilon) + (s_1/\varepsilon_0\varepsilon_1\varepsilon)\sigma_1 - (s_2/\varepsilon_0\varepsilon\varepsilon_2)\sigma_2 \tag{2.13}$$

$$SE_2 = -(V_0/\varepsilon_2) + (s_1/\varepsilon_0\varepsilon_1\varepsilon_2)(\sigma_1 + \sigma_2) + (s/\varepsilon_0\varepsilon\varepsilon_2)\sigma_2. \tag{2.14}$$

These equations are often used for field calculations of electrets. Equation (2.12) shows that an electret with a single-polarity charge layer has a much larger external field than an electret with two charge layers of vanishing net charge ($\sigma_1 + \sigma_2 = 0$) if $s_2 \gg s$, as is often the case. The larger external field of such electrets causes compensation effects (e.g., by attraction of atmospheric ions) and thus leads to reduced charge stability.

The induction charges σ_{i1} and σ_{i2} on the metal electrodes are given by

$$\sigma_{i1} = \varepsilon_0\varepsilon_1 E_1, \tag{2.15}$$

$$\sigma_{i2} = -\varepsilon_0\varepsilon_2 E_2, \tag{2.16}$$

with

$$\sigma_{i1} + \sigma_{i2} = -(\sigma_1 + \sigma_2). \tag{2.17}$$

2.1.2 Fields Due to Volume-Charge Distributions

The field E_1 of the volume-charge distribution $\varrho(x)$ defined in (2.5) can be derived by calculating from (2.8) the field contribution of an infinitesimal charge layer $\varrho(x)\,dx$ and integrating over all such charge layers [2.2]. This yields

$$SE_1 = -(V_0/\varepsilon_1) - \int_0^s \{[(s-x)/\varepsilon_0\varepsilon_1\varepsilon] + [s_2/\varepsilon_0\varepsilon_1\varepsilon_2]\}\varrho(x)\,dx. \tag{2.18}$$

It is convenient to define the following abbreviations for the integral over the charge distribution and its first moments:

$$\hat{\sigma} = \int_0^s \varrho(x)\,dx, \tag{2.19}$$

$$\hat{\sigma}_1 = \frac{1}{s}\int_0^s (s-x)\varrho(x)\,dx, \tag{2.20}$$

$$\hat{\sigma}_2 = \frac{1}{s}\int_0^s x\varrho(x)\,dx, \tag{2.21}$$

where $\hat{\sigma} = \hat{\sigma}_1 + \hat{\sigma}_2$. Using (2.20) and (2.21), (2.18) can be rewritten as

$$SE_1 = -(V_0/\varepsilon_1) - (s/\varepsilon_0\varepsilon_1\varepsilon)\,\hat{\sigma}_1 - (s_2/\varepsilon_0\varepsilon_1\varepsilon_2)(\hat{\sigma}_1 + \hat{\sigma}_2), \tag{2.22}$$

which corresponds to (2.12). Similarly, the field E_2 is obtained as

$$SE_2 = -(V_0/\varepsilon_2) + (s_1/\varepsilon_0\varepsilon_1\varepsilon_2)(\hat{\sigma}_1 + \hat{\sigma}_2) + (s/\varepsilon_0\varepsilon\varepsilon_2)\,\hat{\sigma}_2, \tag{2.23}$$

which corresponds to (2.14).

Comparison of (2.22, 23) with (2.12, 14) demonstrates that a volume-charged electret behaves, as far as its external field is concerned, as if it had surface-charge densities $\hat{\sigma}_1$ and $\hat{\sigma}_2$. These quantities are therefore often referred to as "projected" or "effective" surface-charge densities. According to (2.20), $\hat{\sigma}_1$ is obtained by projecting all the charges, weighted by their respective ratios $(s-x)/s$, onto the surface $x=0$.

Equations (2.15) and (2.22) yield for the induction charge σ_{i1}, under the assumption $s_2 = 0$ and $V_0 = 0$,

$$\sigma_{i1} = -\hat{\sigma}_1/(1 + s_1\varepsilon/s\varepsilon_1). \tag{2.24}$$

A corresponding relation holds for σ_{i2}. For $s_1 = 0$, (2.24) can be written as

$$\sigma_{i1} = -\hat{\sigma}_1. \tag{2.25}$$

The charge induced on a contacting electrode thus equals $-\hat{\sigma}_1$, as expected.

The field E within the electret depends, of course, upon location. For $x = x'$, where $0 < x' < s$, $E(x')$ may be calculated from Gauss's law, applied to the volume bounded by $x=0$ and $x=x'$:

$$\varepsilon E(x') - \varepsilon_1 E_1 = \frac{1}{\varepsilon_0} \int_0^{x'} \varrho(x)\,dx. \tag{2.26}$$

Substituting E_1 from (2.22), this yields

$$SE(x') = -(V_0/\varepsilon) + (s_1/\varepsilon_0 \varepsilon_1 \varepsilon)\,\hat{\sigma}_1 - (s_2/\varepsilon_0 \varepsilon \varepsilon_2)\,\hat{\sigma}_2$$
$$- (S/\varepsilon_0 \varepsilon)\left(\hat{\sigma}_1 - \int_0^{x'} \varrho(x)\,dx\right). \tag{2.27}$$

This equation corresponds to (2.13).

The above equations for the internal and external fields of a charge distribution can also be derived from an equivalent-circuit model where the static quantities are replaced by dynamic quantities [2.3].

It is often convenient to use the differential form of (2.26),

$$\varepsilon_0 \varepsilon \, dE/dx = \varrho(x), \tag{2.28}$$

which is Poisson's equation.

The potential $V = -s_1 E_1$ of the nonmetallized surface $x=0$ of a one-sided metallized electret ($s_2 = 0$) between short-circuited electrodes ($V_0 = 0$) follows from (2.22):

$$V = \frac{\hat{\sigma}_1}{\varepsilon_0(\varepsilon_1/s_1 + \varepsilon/s)}. \tag{2.29}$$

If the electrode at $x = -s_1$ is at a large distance $s_1 \gg s$ so that it does not appreciably affect the field within the electret, this yields

$$V \approx s\hat{\sigma}_1/\varepsilon_0 \varepsilon. \tag{2.30}$$

When a floating electrode is approached to the nonmetallized surface of the electret this surface remains at the potential given by (2.30).

2.1.3 Electric Forces

For the force calculations, it is assumed that the material occupying the gaps (Fig. 2.1) is not electrostrictive, and further that any motions of the electret relative to the electrodes can be neglected. The forces on the electret due to a dc or ac voltage V_0 and due to the electret charges may then be calculated as follows.

The electric forces per unit area on the lower and upper electrodes are, respectively, for laterally uniform fields under the above conditions

$$F_1 = \tfrac{1}{2}\varepsilon_0\varepsilon_1 E_1^2,$$
$$F_2 = -\tfrac{1}{2}\varepsilon_0\varepsilon_2 E_2^2.$$

This yields with Newton's third law for the force per unit area on the electret

$$F = -(F_1 + F_2) = \tfrac{1}{2}\varepsilon_0(-\varepsilon_1 E_1^2 + \varepsilon_2 E_2^2). \tag{2.31}$$

Assuming now $\varepsilon_1 = \varepsilon_2$, one obtains with (2.12) and (2.14) for an electret with surface-charge layers only [2.2]

$$F = (1/2S)(\sigma_1 + \sigma_2)$$
$$\cdot [-2(V_0/\varepsilon_1) - (s/\varepsilon_0\varepsilon_1\varepsilon)(\sigma_1 - \sigma_2) + (s_1 - s_2)(\sigma_1 + \sigma_2)/\varepsilon_0\varepsilon_1^2]. \tag{2.32}$$

For an electret with a volume-charge distribution, the force is obtained by replacing σ_1 and σ_2 in this equation by $\hat{\sigma}_1$ and $\hat{\sigma}_2$, respectively. Equation (2.32) shows that F depends linearly on V_0. The force is equal to zero if either the net charge on the electret is equal to zero ($\sigma_1 = -\sigma_2$) or if the geometry and the charge arrangement is symmetrical and no external voltage is applied ($\sigma_1 = \sigma_2$, $s_1 = s_2$, $V_0 = 0$).

Equation (2.32) does, however, in general not represent mechanically stable conditions. Stability may only be obtained if the electret is in contact with one of the electrodes and if furthermore V_0 meets certain conditions. Criteria also exist for achieving bistable arrangements, allowing stable contacts with either of the electrodes [2.4].

An electret metallized on one side ($s_2 = 0$) is subject to a force from the other electrode

$$F = -\tfrac{1}{2}\varepsilon_0\varepsilon_1 E_1^2. \tag{2.33}$$

This yields with (2.12) for a single charge layer (σ_1 finite, $\sigma_2 = 0$)

$$F = -\frac{\varepsilon_0}{2\varepsilon_1 S^2}\left(V_0 + \frac{s\sigma_1}{\varepsilon_0\varepsilon}\right)^2. \tag{2.34}$$

For a volume charge, σ_1 in this equation is replaced by $\hat{\sigma}_1$.

The forces acting on charged dielectrics have also been derived from the energy relation $F dx = dW - dU$, where dx represents an infinitesimal motion of the electret, dW is the energy supplied by the external circuit and dU is the change in internal energy of the system [2.4, 5]. This and other methods have been used to calculate forces for a variety of electret arrangements [2.6].

2.1.4 Currents

Currents in an electret may occur for a variety of reasons, such as the presence of an electric field, temporal variations of such a field, or a time dependence of the electret charges. Generally, the currents are composed of a conductive and a displacement component where the former is due to physical motion of charges through a given cross section of the dielectric while the latter involves inductive effects. To calculate these currents, it is necessary to invoke the time dependence of some of the variables introduced above.

The conduction-current density $i_c(x, t)$ is related to the real-charge density $\varrho_r(x, t)$ by the equation of continuity

$$\partial \varrho_r(x, t)/\partial t = -\partial i_c(x, t)/\partial x. \tag{2.35}$$

From the Poisson equation (2.28), ϱ_r can be eliminated with (2.35) by considering (2.5). Integration over x yields with (2.3) a space-independent quantity, namely the total-current density, $i(t)$,

$$i(t) = \varepsilon_0 \varepsilon \frac{\partial E(x, t)}{\partial t} + \frac{\partial P_p(x, t)}{\partial t} + i_c(x, t). \tag{2.36}$$

Here the terms on the right represent, respectively, the densities of displacement current, depolarization current, and conduction current. The latter is often further resolved into its physical components,

$$i_c(x, t) = [g + \mu_+ \varrho_{r+}(x, t) + \mu_- \varrho_{r-}(x, t)] E(x, t), \tag{2.37}$$

where $g = e(n_+ \mu_+ + n_- \mu_-)$ is the conductivity of the dielectric and μ_+ and μ_- are the trap-modulated mobilities of the positive and negative charges whose densities $en_+ + \varrho_{r+}$ and $en_- + \varrho_{r-}$ are composed of space-independent intrinsic components (en_+ and en_-) and generally space-dependent excess components (ϱ_{r+} and ϱ_{r-}). In (2.37), the terms within the brackets represent ohmic current and excess-charge currents of positive and negative carriers.

Apart from being independent of location anywhere in the electret, the total current $i(t)$ is also the same in the air gaps and in the external circuit. The air-gap current is a pure displacement current, so that

$$i(t) = \varepsilon_0 \varepsilon_1 \frac{dE_1(t)}{dt}. \tag{2.38}$$

It is useful to apply (2.36) to one of the electret–air gap interfaces. Gauss's law may be written for the interface at $x = 0$ as

$$\varepsilon_0 \varepsilon E(0, t) - \varepsilon_0 \varepsilon_1 E_1(t) = \sigma_r(0, t) - P_p(0, t), \tag{2.39}$$

where it has been assumed that a real surface charge σ_r and a polarization charge P_p is present at the surface. Substituting E from (2.39) into (2.36) with $x=0$ yields

$$i(t) = \varepsilon_0 \varepsilon_1 \frac{dE_1(t)}{dt} + \frac{d\sigma_r(0,t)}{dt} + i_c(0,t). \qquad (2.40)$$

With (2.38) one has thus

$$i_c(0,t) = -d\sigma_r(0,t)/dt, \qquad (2.41)$$

which means that the current from the surface layer is due to a decay of the surface charge.

For the case of vanishing gaps ($s_1 = s_2 = 0$ in Fig. 2.1), E follows from (2.27) for $x' = 0 + \Delta x$ with

$$\int_0^{\Delta x} \varrho(x,t)dx = -P_p(0,t)$$

(due to the presence of an electrode, a real surface charge does not exist at $x=0$):

$$\varepsilon_0 \varepsilon E(0,t) = -\hat{\sigma}_1(t) - P_p(0,t).$$

Substituting this into (2.36) with $x=0$ yields [2.7]

$$i(t) = \frac{d\sigma_{i1}(t)}{dt} + i_c(0,t), \qquad (2.42)$$

where $\hat{\sigma}_1(t)$ has been replaced by $-\sigma_{i1}(t)$ because of (2.25). This equation is useful for investigating the nature of electret charges (see Sect. 2.4.1).

2.2 Charging and Polarizing Methods (Forming Methods)

Methods for forming space-charge electrets and dipolar electrets differ frequently. The charging of *space-charge* (or *surface-charge*) *electrets* is mostly achieved by injecting (or depositing) charge carriers by discharges, particle beams, contact electrification, or other techniques through (or onto) a non-metallized surface. Injection from a deposited metal layer is also possible at relatively high fields. Other methods consist in the generation of carriers within the dielectric by light, radiation or heat and simultaneous charge separation by a field.

Dipolar electrets, on the other hand, are generally polarized by application of an electric field to the material at room temperature or temperatures decreasing from a properly selected higher to a lower value. Dipolar orientation has also been achieved by corona charging [2.8]. In this case, the orientation is caused by the field of the deposited charges (see Chap. 5).

Electret charging is limited by internal and external breakdown whenever the charges produce fields of sufficient magnitude. The occurrence of *internal breakdown effects* depends on the dielectric strength of the material. For polymers, the dielectric strength is of the order of a few $MV \cdot cm^{-1}$, e.g., 2.2×10^6 and $1.3 \times 10^6 \, V \cdot cm^{-1}$ for 12.5 and $100 \, \mu m$ Teflon. Thus, according to Gauss's law [second expression in (2.6) with $E_a = 0$], charge densities of up to $4 \times 10^{-7} \, C \cdot cm^{-2}$ and $2.4 \times 10^{-7} \, C \cdot cm^{-2}$ can be stored on the nonmetallized surfaces of such Teflon samples, respectively, without danger of breakdown. The maximum storage capabilities are thus only weakly dependent on sample thickness and are sufficiently high for most applications.

The situation is different with respect to *external breakdown*. Such a breakdown occurs when the voltage across the gap between the nonmetallized face of an electret and the closest electrode reaches a value determined by the sample-electrode geometry and the gas composition and pressure. This phenomenon, usually referred to as Paschen breakdown, is of importance in a number of charging and charge-measuring techniques and will therefore be discussed in the following.

The voltage $V_1 = -E_1 s_1$ between an electret metallized on one surface and a planar electrode placed at a distance s_1 from its nonmetallized surface follows from (2.29) as

$$V_1 = s s_1 \hat{\sigma}_1 / \varepsilon_0 (\varepsilon s_1 + \varepsilon_1 s) . \tag{2.43}$$

A plot of V_1 as function of s_1 with $\hat{\sigma}_1$ and s as parameters is shown in Fig. 2.2, assuming $\varepsilon_1 = 1$ (air) and $\varepsilon = 2$. Also plotted is the breakdown voltage between parallel electrodes in air at atmospheric pressure (Paschen curve). Breakdown occurs if the air-gap voltage exceeds the Paschen curve [2.9]. This happens upon approach or removal of an electrode to or from an electret for a gap thickness at which the voltage and Paschen curves in Fig. 2.2 cross. The charge of the electret is then lowered to a value for which the voltage curve is lower than the Paschen curve. Figure 2.2 also indicates that the threshold charge density causing breakdown increases with decreasing electret thickness.

This is brought out more clearly in Fig. 2.3, where the maximum charge density not resulting in breakdown at any gap spacing is plotted against electret thickness. The plot is obtained from (2.43) and the analytical expression for the Paschen curve [2.10]. It shows that charge densities on classical plate electrets (thickness about 1 cm) are limited to a few $nC \cdot cm^{-2}$ while those on thin-film electrets (thickness $10 \, \mu m$) can reach about $10^{-7} \, C \cdot cm^{-2}$ without danger of external breakdowns. However, charge densities up to and exceeding $10^{-6} \, C \cdot cm^{-2}$ can be achieved by using dielectric inserts during the charging process

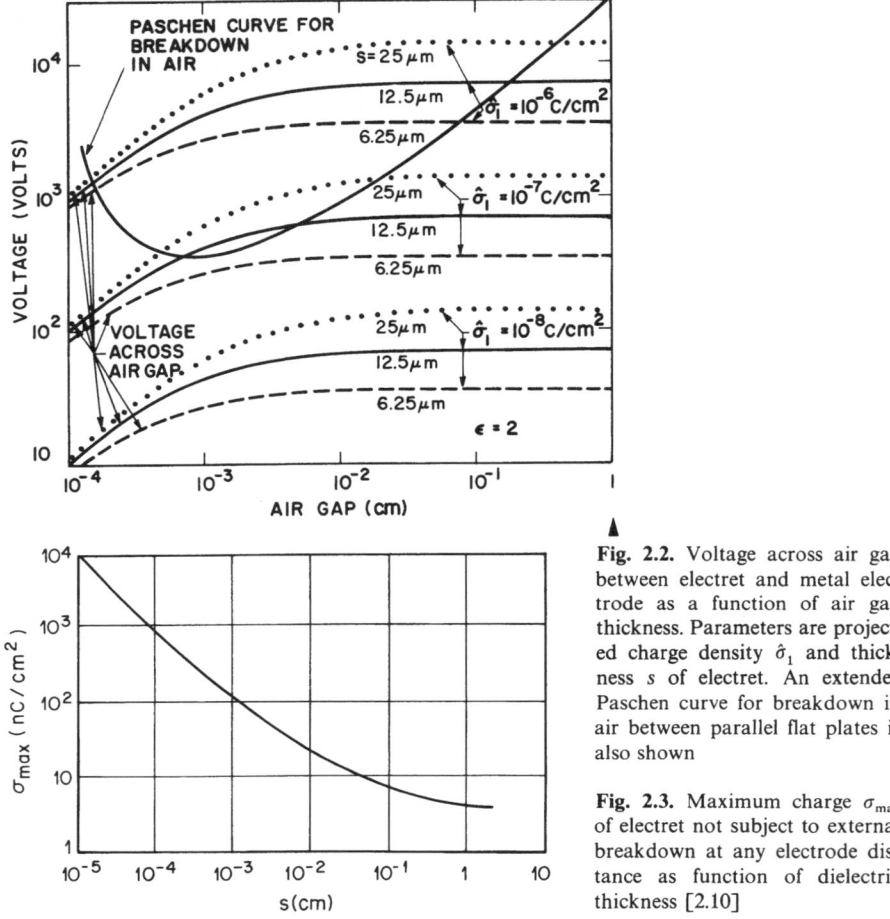

Fig. 2.2. Voltage across air gap between electret and metal electrode as a function of air gap thickness. Parameters are projected charge density $\hat{\sigma}_1$ and thickness s of electret. An extended Paschen curve for breakdown in air between parallel flat plates is also shown

Fig. 2.3. Maximum charge σ_{max} of electret not subject to external breakdown at any electrode distance as function of dielectric thickness [2.10]

(see Sect. 2.2.3). To avoid breakdown effects in this case, metal electrodes have to be kept away from the nonmetallized surface(s) of the electret.

2.2.1 Triboelectricity: Contact Electrification

Triboelectricity, defined as the charging of two dielectrics in contact with each other, is a superposition of a kinetic effect and an equilibrium effect [2.11, 12]. Generally, the *kinetic effect* is caused by asymmetric rubbing of two pieces of the same material. An example is a dynamic rod sliding over another static rod like a violin bow over a string. Electrification due to the kinetic effect has been attributed to the fact that the rubbed section of the static rod is heated more than the dynamic rod. Of greater importance for electret studies is the *equilibrium effect*, also known as *contact electrification*. This effect originates from static contacts between different materials and was observed as early as 1732 by Gray. Recent reviews of the field of contact electrification have been given by *Seanor* [2.13], *Inculet* [2.14], *Bauser* [2.15], and *Fuhrmann* [2.16].

Studies of contact electrification in vacuum, where atmospheric effects such as humidity are excluded, indicate that the charging is due to electron transfer from or to the insulator. It can be described by introducing the concept of work functions, widely used in the theory of metals and semiconductors, for the dielectric materials [2.17–19].

Assuming that the positive or negative charge in the insulator reaches a uniform volume distribution of density ϱ to depth $\bar{r}$, corresponding to a planar charge density $\sigma = \varrho\bar{r}$, an average field $\sigma/2\varepsilon\varepsilon_0$ is generated. Equating this to the field due to the difference $\Delta\phi$ in work functions between the two surfaces, $\Delta\phi/e\bar{r}$, one obtains for the injected charge density [2.18]

$$\sigma = 2\varepsilon\varepsilon_0\Delta\phi/e\bar{r}. \tag{2.44}$$

Depending upon the assumptions about charge trapping, this equation predicts different relations between σ and $\Delta\phi$. For example, a linear relationship follows as long as the penetration depth $\bar{r}$ is independent of $\Delta\phi$. The latter condition implies a dependence of the volume density ϱ on $\Delta\phi$. If, however, one discrete trapping level is present, ϱ has to be assumed constant and (2.44) together with $\bar{r} = \sigma/\varrho$ yields [2.20]

$$\sigma = \sqrt{2\varepsilon\varepsilon_0\varrho\Delta\phi/e} \tag{2.45}$$

similar to a relation derived by *Harper* [2.17]. For surface charging one obtains always a linear dependence [2.21].

If one considers the effect of band bending due to the metal–insulator contact, the following picture emerges [2.20]: For a uniform trap distribution over the band gap, σ is proportional to $\Delta\phi$. However, for traps at discrete energy levels, the relation between σ and $\Delta\phi$ ranges from exponential for a trap-free insulator to quadratic for complete ionization of the traps.

Experimental data for two polymers are shown in Fig. 2.4 [2.18, 22]. While the polycarbonate data show a proportionality between σ and $\Delta\phi$, the PTFE data exhibit a nonlinear relation between these quantities. As discussed above, the linear data can be explained by a constant penetration depth or by an energetically uniform trap distribution. However, uniform trap distributions are generally not found in polymers (see also Sect. 2.6.2), while a constant penetration depth cannot be reconciled with physical models for bulk traps [2.23, 24]. The nonlinear data on PTFE has been interpreted [2.22] with (2.44) as well as with a two-trap model giving a better fit [2.20]. This latter model is also open to question since it predicts an electron affinity of about 3 eV, believed to be too large for PTFE [2.23]. The interpretation of contact charging remains therefore in doubt.

As an electret-charging method, contact electrification has not been used extensively, mostly because of a lack of accurate reproducibility. Contact

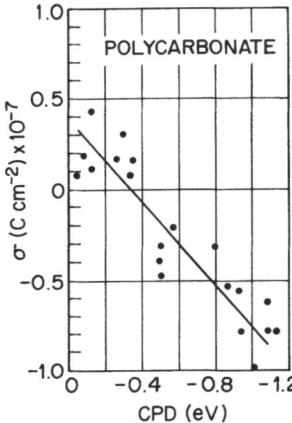 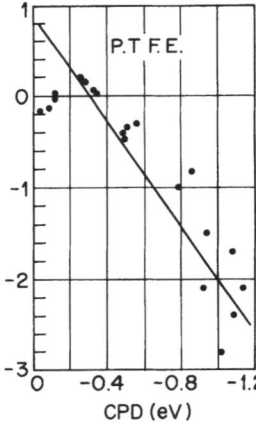

Fig. 2.4. Plot of charge density σ due to contact electrification between various metals and two polymers. Abscissa is the contact potential difference (CPD) of the metals against gold [2.18,22]

charging is, however, important in situations where dielectrics get into contact with metal or other dielectric surfaces, leading to generally undesirable charging effects.

2.2.2 Thermal Charging Methods

Thermal methods of electret charging or polarization consist in the application of an electric field to a dielectric at an elevated temperature and subsequent cooling while the field is still applied. For waxes, the elevated temperature is often chosen to be the melting point, while for polymers a temperature above the glass transition temperature but well below the melting point is selected. Application of the field may be through deposited (e.g., evaporated) or external (e.g., laid-on or distant) electrodes. If external electrodes are used, the presence of air gaps (which are, for laid-on electrodes, of microscopic dimensions) results in more complicated charging phenomena. In the present definition, thermal methods will comprise charging methods operated in the absence or in the presence of air gaps. Electrets formed by thermal methods are referred to as thermoelectrets. The literature on thermoelectret formation has been reviewed extensively [2.25–28].

Analysis of Thermal-Charging Phenomena

Basically, three kinds of phenomena can occur in thermal charging, namely (I) internal polarization with the sign of a heterocharge due to dipole alignment or charge separation within the dielectric (dielectric absorption), (II) homocharge deposition due to spark discharges in the air gaps, and (III) homocharge injection through contacting electrodes. The relative importance of these phenomena depends, among others, on the geometry of the experiment, on temperature and field strength prevailing during the charging process, and on the physics of the electrode–dielectric interface.

Dipole alignment occurs at elevated temperatures where the molecules or molecular chains are sufficiently mobile. Upon cooling, the aligned dipoles are frozen in, giving the dielectric a permanent polarization. The alignment is controlled by the Debye equation relating the change in polarization to the already existing polarization P_p, the electric field E and the distribution of dipole-relaxation frequencies. Assuming first a single dipole-relaxation frequency $\alpha(T)$, one has [2.29]

$$dP_p(t)/dt + \alpha(T) P_p(t) = \varepsilon_0 (\varepsilon_s - \varepsilon_\infty) \alpha(T) E . \qquad (2.46)$$

Here, ε_s and ε_∞ are the (weakly temperature dependent) static and optical dielectric constants, respectively.

If the dielectric is cooled during a time period sufficiently short compared to the time it is kept at the forming temperature, the dipoles align predominantly during the elevated-temperature phase and an isothermal solution of (2.46) is reasonably accurate. In this case, setting $P_p(0) = 0$, the time dependence of the polarization follows as

$$P_p(t) = \varepsilon_0 (\varepsilon_s - \varepsilon_\infty) E [1 - e^{-\alpha(T)t}] , \qquad (2.47)$$

which yields a maximum value of $\varepsilon_0 (\varepsilon_s - \varepsilon_\infty) E$. For polymers, the temperature dependence of a dipole relaxation frequency below the glass transition point is generally given by

$$\alpha(T) = \alpha_r \exp(-U_d/kT) , \qquad (2.48)$$

where U_d is a dipolar activation energy. A formally similar relationship (with T replaced by $T - T_\infty$) holds at higher temperatures. Thus, for small t, the polarization in polymers with a single $\alpha(T)$ increases as

$$P_p(t) = \varepsilon_0 (\varepsilon_s - \varepsilon_\infty) E \alpha_r t \exp(-U_d/kT) . \qquad (2.49)$$

The buildup of the polarization is therefore weakly field dependent and strongly temperature dependent in the sense that it is favored by higher temperatures. For a distribution of relaxation frequencies, which may be due to a distribution of activation energies U_d or a distribution of pre-exponential factors α_r (see Sect. 3.6.3), this dependence of P_p on field and temperature is essentially preserved [2.29].

The internal polarization due to *charge separation* within the dielectric is caused by conduction either between interfaces, such as domain boundaries (Maxwell–Wagner effect), or over the entire thickness of the electret. For simplicity, a nonpolar electret with a single adjacent air gap is assumed. The charge variation $-d\sigma_r/dt$ at the interface due to an applied voltage V_0 is, by means of (2.41), given by the conduction current $i_c = g(T)E$. Thus, one obtains from (2.13) with $s_2 = 0$

$$d\sigma_r(t)/dt = [g(T)/S][(V_0/\varepsilon) - (s_1\sigma_r/\varepsilon_0\varepsilon_1\varepsilon)]. \tag{2.50}$$

For $\sigma_r(0) = 0$, this yields under isothermal conditions

$$\sigma_r(t) = \varepsilon_0\varepsilon_1 V_0/s_1 \{1 - \exp[-g(T)s_1 t/\varepsilon_0\varepsilon_1\varepsilon S]\}, \tag{2.51}$$

which assumes a maximum value of $\varepsilon_0\varepsilon_1 V_0/s_1$. Then the entire voltage V_0 drops across the gap s_1 and the charging ceases.

The temperature dependence of the conductivity g is given by

$$g(T) = g_0 \exp(-U_c/kT), \tag{2.52}$$

where U_c is the activation energy for conduction. Thus, (2.51) exhibits formally the same field, time, and temperature dependence as (2.47). For small t, the interfacial charge density increases as

$$\sigma_r(t) = (g_0 V_0 t/\varepsilon S) \exp(-U_c/kT). \tag{2.53}$$

Charge deposition by spark discharges in an air gap adjacent to the electret requires a gap voltage V_1 in excess of a threshold value V_T, given for plane-parallel geometry by the Paschen curve (see Fig. 2.2). After the discharge is initiated the current density in the simplest case (namely for a corona discharge) follows the relation [2.30]

$$i = \gamma V_1(V_1 - V_T), \tag{2.54}$$

where γ is a constant determined by the geometry of the setup and the ion mobility. For voltages V_1 just in excess of V_T, (2.54) can be approximated by $i = \gamma V_T(V_1 - V_T)$.

Assuming now the existence of a homogeneous field E_1 in the air gap, the gap voltage $V_1 = s_1 E_1$ can be calculated from (2.12) with $s_2 = 0$ as function of the deposited surface charge σ_r and the externally applied voltage $-V_0$ (the negative sign is chosen so that for positive V_0, a positive V_1 is obtained). Substituting this value of V_1 into the above equation for i, and setting $i = d\sigma_r/dt$ [cf. (2.41) with $i_c = -i$], yields

$$\frac{d\sigma_r}{dt} = -\gamma V_T\left(V_T - \frac{V_0 s_1}{\varepsilon_1 S} + \sigma_r \frac{ss_1}{\varepsilon_0\varepsilon_1\varepsilon S}\right), \tag{2.55}$$

which is of the form of (2.50). It has the solution

$$\sigma_r(t) = \sigma_m(1 - e^{-t/\tau}) \tag{2.56}$$

with the final value

$$\sigma_m = (\varepsilon_0 \varepsilon / s)(V_0 - \varepsilon_1 V_T S / s_1)$$

and a time constant

$$\tau = \varepsilon_0 \varepsilon_1 \varepsilon S / \gamma s s_1 V_T.$$

· This equation shows that charge deposition exhibits a time dependence similar to dipole alignment and internal charge separation. The voltage dependence, however, is controlled by a threshold value $\varepsilon_1 V_T S / s_1$ of the externally applied voltage while the temperature dependence, which is due to γ and V_T, is very weak. At the elevated temperatures used in thermal charging, charge penetration from the surface into the bulk occasionally takes place.

Charge injection through deposited electrodes may occur if the contacts are not blocking. It is, under steady-state conditions, controlled by Schottky emission which yields a current density of

$$i = i_0 \exp\left(-\frac{\Delta\Phi - \beta E^{1/2}}{kT}\right), \tag{2.57}$$

where E is the field at the injecting electrode, $\Delta\Phi$ is the difference in the work functions between the electrode and the dielectric and β is the Schottky coefficient which equals $(e^3/4\pi\varepsilon\varepsilon_0)^{1/2}$.

However, in highly insulating materials with strong trapping, a time dependence of the injection current is observed. In this case the injection is essentially limited by trap filling, either at the interface or in the bulk (see Sect. 2.6.5), and the injection current is often described by [2.31]

$$i \propto [E(t)]^m t^{-n}. \tag{2.58}$$

While the exponent n is generally below unity but approaches this value as the trapping becomes increasingly effective, m is expected to exceed 1. For polymers, typical values of n are between 0.1 and 1 and for m between 1 and 3, with the latter values depending on temperature.

In the case of strong trapping, the charging leads to the formation of a space-charge layer (Schottky layer) in the vicinity of the injecting electrode. For negligible conduction currents, the stored charge grows initially proportional to $[E(0)]^m$, where $E(0)$ is the initial field at the injecting electrode. Thus, the initial field dependence of the stored charge is stronger than for internal polarization or charge deposition; charge injection is therefore expected to be important at high fields. The final charge follows from the condition that the field at the injecting electrode due to the trapped charge is equal in magnitude and opposite in sign to the applied field. This yields $\sigma_m = -\varepsilon_0 \varepsilon V_B / \bar{r}$, where V_B is the applied voltage and $\bar{r}$ the mean depth of the trapped charge.

Thermoelectret formation is therefore dependent on a number of phenomena. In many processes of this kind, however, the electrode–dielectric interface (or interfaces) can be considered a blocking contact so that charge injection is absent. The frequently used aluminum electrodes on polymers belong into this category. In such cases, the forming parameters voltage and temperature have the following effect on the charging process: At voltages below the threshold for air-gap breakdown, merely an internal heterocharge polarization is developed. At higher applied voltages and in the presence of an air gap, homocharge deposition occurs and increases more than linearly with voltage. Higher voltages thus favor charge deposition over internal polarization, while higher temperatures have the opposite effect. The composition of the resulting charge is, of course, also dependent on material properties and dimensions.

Experimental Results

The classical *wax electrets* are formed in the presence of air gaps by application of fields of 1 to $50 \, \text{kV} \cdot \text{cm}^{-1}$ or more to the substance while the temperature is lowered from above the melting point to room temperature. Investigation of the charging process is possible by measuring separately the charging currents and the induction charges on the electrodes [2.7]. It was determined that at relatively low forming fields (a few $\text{kV} \cdot \text{cm}^{-1}$) an absorption current with a t^{-1} time dependence flows, which is due to internal charge separation or dipole alignment. This causes a relatively short-lived heterocharge. At intermediate fields (about $10 \, \text{kV} \cdot \text{cm}^{-1}$), carrier deposition by air breakdown occurs, resulting in a small but stable homocharge in addition to a heterocharge. The electret thus exhibits a charge changing its sign from hetero to homo within periods of the order of hours. At higher forming fields, a large homocharge, which masks the smaller heterocharge, is observed [2.32].

Polymer-film thermoelectrets (with the exception of piezoelectric substances) are usually formed in the presence of an air gap. Generally, a maximum forming temperature somewhat above the glass transition temperature but well below the melting point is chosen (about 150–200 °C for Teflon and 100–120 °C for Mylar). Heterocharge polarization is achieved for relatively small forming fields of up to about $10–50 \, \text{kV} \cdot \text{cm}^{-1}$, depending on sample thickness. As predicted by (2.47) and (2.51), the charge is found to be proportional to the forming voltage [2.33, 34]. If predominantly surface-charge deposition is to be achieved, the polymer sample is exposed to a field of $100–800 \, \text{kV} \cdot \text{cm}^{-1}$ where breakdown in the air gap occurs [2.35–37] (for isothermal charging of this kind see Sect. 2.2.3). At the high formation temperatures, the charges are deeply trapped and thus thermally stable.

The voltage (or field) dependence of the stored charge on Mylar films is shown in Fig. 2.5. According to these results, a heterocharge is formed at small fields, while a sudden onset of homocharge formation occurs at a threshold

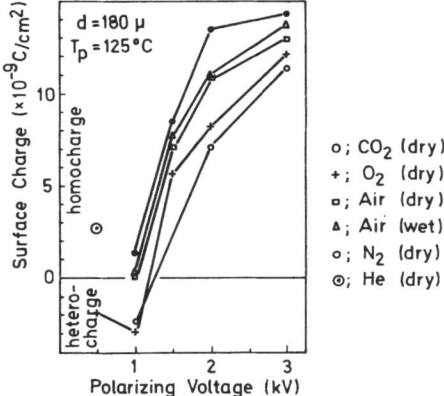

o; CO_2 (dry)
+; O_2 (dry)
□; Air (dry)
▲; Air (wet)
o; N_2 (dry)
⊙; He (dry)

Fig. 2.5. Initial charge density on 180 μm thick Mylar PET as function of voltage applied during charging in different atmospheres at 500 Torr [2.34]

voltage. The homocharge density rises steeply upon further increase of the applied voltage, as predicted by (2.56).

Piezoelectric polymer films are obtained by application of fields of up to $800 \, kV \cdot cm^{-1}$ through contacting (evaporated) electrodes at elevated temperatures or at fields up to $4 \, MV \cdot cm^{-1}$ at room temperature. Dipolar orientation takes place and can be improved in certain materials by uniaxial stretching prior to thermal charging [2.38]. Details of various "poling processes" for piezoelectric polymers (e.g., PVDF) will be given in Sect. 5.6.

Thermal charging methods have also been applied to *other solids*. Experiments on naphthalene showed evidence of the formation of a Schottky barrier during polarization [2.39], while charging of CaF_2 was attributed to traces of charge-compensated impurities [2.40]. The significance of the latter experiments is in the generation of an internal field in the sample which is potentially useful in radiation dosimetry (see Sect. 2.2.6).

In the absence of surface-charge deposition or charge injection, the polarization of dielectrics exhibits generally a reversible character in the sense that later depolarization without field yields currents opposite in sign and equal in magnitude to the charging currents [2.41, 42] (see also Fig. 3.61).

Thermal charging in the presence of an air gap can result in cyclic *discharge phenomena* through the gap [2.7]. Initially, the applied field causes an increase in the heterocharge and thus an increase of the field in the gap. When the field reaches breakdown strength, spray discharges occur, resulting in homocharge injection. Since these charges compensate the heterocharge, the field decreases and the discharge is extinguished. Thereupon the heterocharge starts rising again, repeating the cycle. It has been shown that the discharges are much more frequent and the final homocharge is larger if a dielectric insert is placed in the air gap [2.36].

A thermal charging method depending on the simultaneous application of heat and a *magnetic field* has also been described [2.43]. In this procedure, magnetic fields of several kG are applied to the cooling melt of waxes and other substances. The resulting polarization is of the order of $10^{-9}\,\mathrm{C\cdot cm^{-2}}$. A full explanation of this "magnetoelectret" effect is presently not available.

The *advantage* of all thermal charging methods is the great stability of the surface- and space-charge polarizations achieved on certain nonpolar materials, such as Teflon (see Sect. 2.6.5). This makes it, together with corona and electron beam charging, one of the preferred industrial charging procedures. The method is also ideally suited for polarizing dipolar electrets for piezoelectric applications (see Chap. 5). Drawbacks of this charging method are the nonuniformity of the lateral charge distribution on surface and space-charge electrets, and the slowness of the charging process.

2.2.3 Isothermal Charge-Deposition Methods

The isothermal charge-deposition methods discussed here depend on charge transfer due to a discharge in an air gap. Since heating is not applied in these cases, heterocharge effects due to dielectric absorption are generally absent. Charge-deposition methods have recently gained in importance due to the ease and speed at which they allow polymer films to be charged.

Most widely applied is the *corona-charging technique*, depending on the use of an inhomogeneous field to produce a discharge in air at atmospheric pressure. The field is generated by application of a voltage between a point-shaped or knife-shaped upper electrode placed at a certain distance from (or in contact with) one side of the dielectric and a planar back electrode on the other side [2.44, 45]. For voltages in excess of a threshold value, a current given by (2.54) is observed.

If the upper electrode is negatively biased, as is customary for the charging of Teflon, negative carriers flow toward the dielectric. In air at atmospheric pressure, these carriers are primarily CO_3^- ions of thermal energy [2.46]. Due to their small energy, the ions are deposited in the surface layer and do not penetrate into the material. It is likely that they transfer their charge to surface traps (as has been observed on amorphous selenium surfaces [2.47]) and reenter the air. Charge penetration into the bulk depends on charge polarity, charge density and surface characteristics. On Teflon, for example, negative charge penetrates little or not at all even at densities exceeding $10^{-8}\,\mathrm{C\cdot cm^{-2}}$, while a substantial penetration is observed for positive charge (see Sect. 2.6.3).

A corona arrangement offering a great deal of control over the charging process consists of a sharp needle electrode biased at a few kV against the dielectric and a wire mesh at a potential of at least a few hundred Volts [2.48]. The mesh controls the current to the sample which has initially a bell-shaped distribution modulated by the "shadows" of the wires (R. Gerhard, personal communication). However, the eventual distribution of the deposited charge is generally uniform if the charging is carried to the point where the entire sample surface has assumed a saturation potential which equals the grid potential. If the surface potential of the sample and the current into the rear electrode are

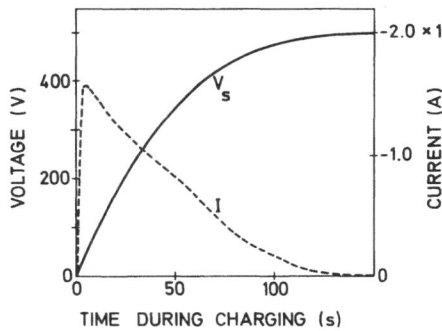

Fig. 2.6. Voltage V_s across, and current through, 25 μm Teflon PTFE as function of time during corona charging (sample area 25 cm², corona voltage −5.5 kV). Final charge on sample is about 3.6×10^{-8} C cm⁻² [2.49]

monitored, the equivalent surface-charge density and the conduction current through the sample can be determined during the charging process [2.49]. Figure 2.6 shows results for negative charging of Teflon, where, due to the absence of conduction phenomena, the entire current is a displacement current $I = CdV_s/dt$, with V_s representing the voltage across the sample.

In another modification of such a method [2.50], used for continuously charging long strips of film, the dielectric is moved from a supply reel through a corona discharge to a charge-measuring setup and onto a takeup reel. The corona discharge is generated in this case by a knife-edge electrode.

The *advantages* of corona charging are simplicity of the setup required and speed of charging. A drawback in setups not using a grid electrode is the relatively large lateral nonuniformity of the charge distribution. Also, an irreversible lowering of the short-circuit TSC peak temperature [2.51] indicates somewhat inferior charge stability under certain conditions. However, the advantages of this method have led to its extensive use in the large-scale production of film electrets for electret microphones. Corona charging is also widely used in xerography [2.52].

Spark-discharge methods, apart from their use for depositing homocharges on thermoelectrets (see Sect. 2.2.2), can also be used to charge dielectric films isothermally. To obtain satisfactory charge densities without destroying portions of the film by arcing, a sandwich consisting of the film and a much thicker dielectric insert of lower resistivity, acting as a protective series resistance, may be used in the gap between two parallel plate electrodes [2.53, 54].

Application of a voltage to such a setup will result in a gradual charge transfer by means of controlled spark breakdowns in the minute air gap between the dielectric insert and the film. Charge deposition is again due to ionic species, as in corona charging. The charging, which is formally given by (2.56), ceases when the voltage in the gap falls below an extinction value V_E lower than the threshold value V_T for inception of the discharge.

Apart from preventing destructive arcing through the film, the dielectric insert also permits removal of the film electret from the charging setup without loss of charge due to breakdown through the air gap. This can be seen by calculating from the generalized equation (2.12) the voltage $V_1 = s_1 E_1$ across the air gap in the presence of an insert of thickness s_1' and permittivity ε_1'. One obtains for $V_0 = 0$ and $\varepsilon_1 = 1$, assuming a charge layer of density σ_1 at $x = s$,

$$V_1 = -ss_1\sigma_1/\varepsilon_0 [\varepsilon s_1 + s + (\varepsilon/\varepsilon_1')s_1'].$$

Comparison with (2.43) indicates a decrease in V_1 due to the presence of the insert. This is particularly evident for an insert thickness much in excess of the electret and air gap thicknesses. For example, an electret with $\sigma = 10^{-6}\,C \cdot cm^{-2}$ will generate V_1 values below the Paschen curve in Fig. 2.2 for all air gap thicknesses if a dielectric insert of a few mm thickness is used. Consequently, no breakdowns in the air gap will occur. Measurements of such stored-charge densities require, of course, noncontacting methods (see Sect. 2.3).

The initial charge densities on 25 µm Mylar films charged with this method are depicted in Fig. 2.7. Also shown are the limits imposed by the applied voltage [σ_m in (2.56)] and by dielectric strength, the value of the latter being somewhat uncertain. The largest charge densities and full-trap densities achieved with this method are of the order of $10^{-6}\,C \cdot cm^{-2}$ and $10^{17}\,cm^{-3}$, respectively (see also Sect. 2.6.2).

At ambient pressures of about 0.1 atm and less the disruptive spark discharge converts into a uniform and well behaved *Townsend discharge* which can also be employed for isothermal charging. This method has been used to deposit charge densities of up to $6 \times 10^{-7}\,C \cdot cm^{-2}$ on Mylar films [2.55].

2.2.4 Charging with Liquid Contact

Another charging method depends on the use of a small amount of liquid to furnish intimate contact between an electrode and the dielectric surface [2.56, 57]. A dielectric, e.g., a polymer film metallized on one surface, is contacted on its nonmetallized side by a wet electrode such that a thin liquid layer resides between that contact and the dielectric (see Fig. 2.8a). Liquids such as water and ethyl alcohol have been used for this purpose. If a potential is applied between the electrode and the rear metallization, charge double layers form at both solid–liquid interfaces. The interaction of electrostatic and molecular forces causes a charge transfer to the polymer film. A compensation charge of opposite sign but equal magnitude flows into the rear metallization. The potential V_0 of the polymer surface assumes a value close to the applied voltage V_c. The difference $V_c - V_0$ is plotted in Fig. 2.8b as function of V_c for charging with an HCl solution.

By moving the electrode over the surface, large areas of the dielectric may be charged as desired. To ensure charge retention on the dielectric, the electrode has to be withdrawn (or the liquid evaporated) before removal of the voltage. Recently, this method has also been used with nonwetting liquid–insulator

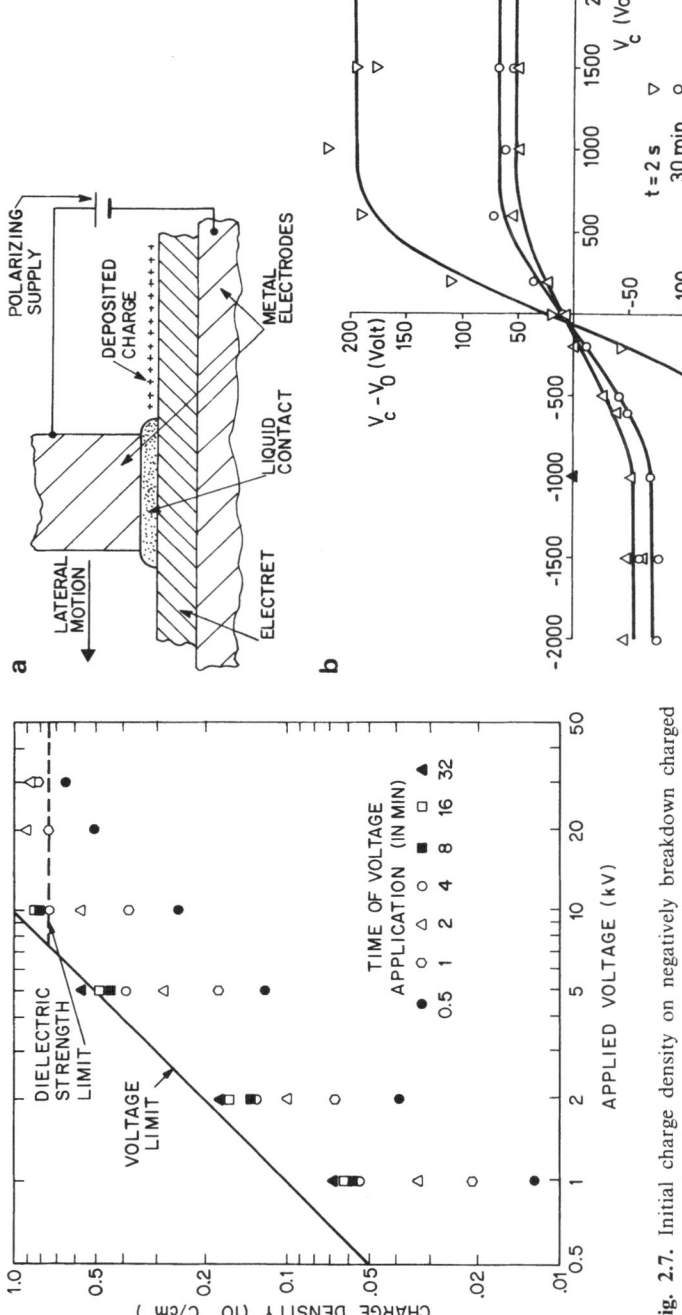

Fig. 2.7. Initial charge density on negatively breakdown charged 25 μm PET as function of charging voltage with charging time as parameter. During charging, one sheet of 0.15 cm soda-lime glass is used as resistive spacer between PET film and negative electrode. [2.53]. For calculation of the voltage limit, $V_T = 0$ has been assumed

Fig. 2.8. (a) Schematic view of setup for charging with liquid contact [2.56] **(b)** Charge-transfer curves for 25 μm FEP. V_c applied voltage, V_0 electret surface potential, t contact time. [2.57]

contacts, which permit the recording of high-resolution (10 μm) charge patterns [2.58].

The liquid-contact method also allows one to produce monocharge electrets, i.e., electrets possessing only a single-polarity charge (see Fig. 1.2a). This can be achieved by placing a nonmetallized dielectric between two electrodes, leaving two air gaps, and filling one of the gaps with a liquid. If a voltage is applied to the electrodes while the liquid evaporates, charge of one polarity is transferred through the liquid to one side of the dielectric and remains there. The compensation charge resides on the other electrode.

The *advantages* of the liquid-contact method are simplicity, control of initial charge density by the applied voltage, and uniform lateral charge distribution.

2.2.5 Partially Penetrating Electron and Ion Beams

Methods for injecting monoenergetic particle beams of range smaller than the thickness of the dielectric, dating back about 50 years (see Sect. 1.1), have recently been developed into extremely controllable and versatile tools for charging dielectrics. Most of the injection methods are based on the use of monoenergetic electrons which require much less energy (and thus impart less damage to the dielectric) than ions of equal range [2.59]. We shall therefore primarily discuss electron-beam charging and only briefly comment on the use of ion beams.

The *practical range of electrons* as function of their energy is shown in Fig. 2.9 for aluminum and a number of dielectrics used in electret research (see also Figs. 2.17, 4.2 below). The figure illustrates that beams of 0.5–1 MeV, which can be applied under atmospheric conditions, are only useful for the charging of dielectrics of 0.1 cm thickness or more. For thinner dielectrics, lower-energy beams have to be used and the injection can only be performed in vacuum.

The *physical picture* of the charging process for a sample metallized and grounded on the side facing away from the beam (rear side) is as follows: When striking the surface, the electrons release some secondaries, leaving a positively charged surface layer. The secondary emission yield, defined as the ratio of the numbers of emitted and primary electrons, is dependent on electron energy and surface properties. For most polymers, it reaches a maximum between 2 and 5 at primary energies V of 150–300 eV and decreases as V^{-1} at higher energies [2.60]. At $V \geqq 10$ keV, yields of less than 0.2 are expected. As the electrons penetrate into the dielectric, they generate secondary carrier pairs of relatively small energy which are quickly trapped. The secondaries cause a radiation-induced conductivity (RIC) orders of magnitude greater than the intrinsic conductivity of the material (see Sect. 4.1). For Teflon, for example, these conductivities are 10^{-13} and 10^{-22} $\Omega^{-1} \cdot$ cm^{-1}, respectively [2.61]. Due to the collisions, the primary electrons are eventually slowed down enough to be trapped, forming initially a distribution of negative charge around the average

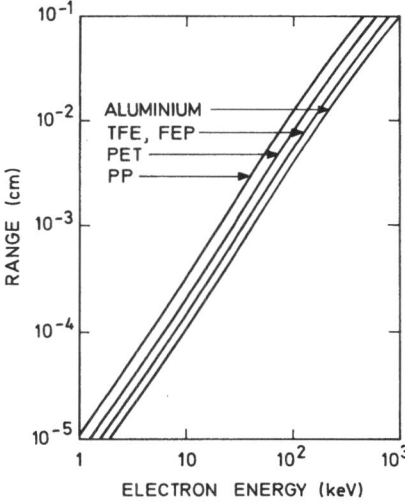

RANGE (cm)

ELECTRON ENERGY (keV)

Fig. 2.9. Practical or extrapolated electron range for aluminum; PTFE, FEP (Teflon); PET (Mylar); and PP (polypropylene); (the data for the polymers are derived from the aluminum range by dividing through the density ratio)

range, which is about 2/3 of the practical range shown in Fig. 2.9. Depending upon the field direction, additional charge motion takes place. In one-sided metallized samples the field is essentially directed toward the electrode (a small field toward the positive surface layer also exists). Thus, most electrons penetrate beyond the average range into regions where the RIC is still large enough to allow inward motion of electrons or outward motion of holes. This process is completed within minutes after termination of charging (see also Sect. 3.13). More details of the charge-penetration process will be discussed in Sect. 2.6.3. For Teflon, the resulting charge arrangement is stable over periods of the order of tens of years at room temperature [2.62].

Practical implementations of electron-beam setups use glow cathodes or rf discharges for electron generation, electrostatic beam focusing, and an acceleration voltage of 5–50 kV. To ensure a uniform current density over the irradiated cross section, scanned beams are occasionally used [2.59, 63, 64]. Alternatively, scanning electron microscopes have been employed as electron beam sources. Because of their beam uniformity, such instruments can be used in the static or scanned mode. In the target chamber dielectric samples, generally metallized on the rear surface, are exposed to the electron beam for periods of 1 ms up to a few s by opening and closing a mechanical shutter. During exposure the current to the rear electrode, corresponding in magnitude to the net injection current, is monitored. Thus the dielectric can be charged to a predetermined value. The method has also been adapted to charging long strips of foil utilizing a roller arrangement [2.64].

The stored charge has great lateral uniformity if the electron beam is uniform. Measurements of the lateral charge distribution on typical electron-beam charged samples show nonuniformities of less than ±5% which are probably due to spurious charges present on the dielectric before irradiation.

Ion-beam injection, having the above-mentioned drawback, is not widely used. However, charging of selenium by bombardment with a variety of 5–400 eV ion species [2.47] has demonstrated the feasibility of this technique.

The *advantages* of electron-beam charging are the complete control over charge depth, lateral charge distribution, and charge density possible with this method. These features have made electron-beam injection an important tool for the charging of electrets used in research and for the study of solid-state properties relating to charge trapping and charge decay [2.29, 59, 61, 62, 64, 65]. In addition, electron-beam charging is now being used extensively for charging membranes of electret microphones.

2.2.6 Penetrating Radiation

Dielectrics can be charged by internal displacement of carriers generated by various kinds of completely penetrating radiation. The carrier displacement may be accomplished by externally applied or internal fields. A number of experiments with γ rays, X-rays, β rays, and monoenergetic electron beams were performed along these lines yielding electrets which are generally less stable than those produced by other methods (see below). Correspondingly, most of these studies were motivated by an interest in the effect of radiation on various properties of solids (related, among others, to radiation dosimetry) rather than by the desire to produce stable electrets [2.66]. A brief account of these methods will be given here and further discussion deferred to Chap. 4.

In the classical charging processes of this kind the dielectric, with a *voltage applied* across its thickness, is subjected to penetrating radiation from a particle accelerator, an X-ray machine or a radioactive source. After exposure to a total dose of $1–10^6$ rads, the radiation is terminated and the voltage removed. The sample now exhibits electret properties and is often called a "radioelectret."

The charging has been explained in terms of molecular ionization followed by carrier drift and carrier trapping [2.67]. The irradiation produces electron–hole pairs which drift in the applied field toward the electrodes, resulting in charge separation. Eventually, the carriers are trapped, giving the dielectric a heterocharge.

More recently, thermoluminescent materials such as LiF and CaF_2 have been charged by similar methods [2.68, 69]. In these experiments, relatively small doses ($1–10^3$ rads) of X-rays are used to irradiate samples subject to external or internal electrical fields, the latter being due to prior thermoelectret polarization. The irradiation generates again secondary electrons and holes which are trapped, forming a space-charge distribution detectable in thermally stimulated current experiments. In CaF_2 the charge generated per rad of irradiation is five orders of magnitude greater than in Teflon-disk radioelectrets and about one order of magnitude greater than in Teflon film electrets [2.70]. The effect is thus of interest in radiation dosimetry (see Sect. 4.7).

Charging of dielectrics with penetrating radiation is also possible in the *absence of an applied electric field*. One such method depends on the use of Compton electrons produced by high-energy X-rays and γ rays [2.71]. These electrons are scattered preferentially in the forward direction and thus give rise to a space-charge polarization of the dielectric. Since the angular distribution of the scattered electrons becomes increasingly isotropic with decreasing energy, the effect is more pronounced at higher energies of the primary photon radiation (around 1 MeV). Thus, rather thick layers are needed for optimal polarization.

Another method not requiring an applied field utilizes an electron beam that penetrates the sample to be charged before being absorbed by a backup insulator. Due to the charge buildup in the insulator and the related electrical fields, electrons drift from the insulator to the sample (possibly by spark discharges in the gap) causing a charging effect [2.72]. The method has been used for the charging of polymer-film electrets.

All charging methods depending on penetrating radiation generate a radiation-induced conductivity throughout the dielectric which decays slowly after irradiation [2.61]. Electrets produced by such methods are therefore inherently less stable than electrets charged by other methods. However, as has been pointed out above, the charging effects due to penetrating radiation can be utilized in radiation dosimetry.

2.2.7 Photoelectret Process

The photoelectret process depends on the use of light rather than high-energy radiation but is otherwise very similar to the above radioelectret procedure. The materials, generally photoconductors, are coated with one or two transparent electrodes and irradiated with ultraviolet or visible light under an applied field. After termination of the irradiation and removal of the field, a permanent polarization (often referred to as "persistent internal polarization") is found in the dielectric, now called a "photoelectret". The classical experiments on ZnCdS, phosphors, anthracene, etc., have been thoroughly reviewed in the literature [2.25, 73]. More recently, photoelectret charging was also performed on thin films of cadmium sulfide [2.74] and on amorphous semiconductors [2.75].

The "polarization" is attributed to carrier generation by the light, displacement by the applied field, and eventual trapping; dipole polarization has been ruled out [2.73, 76]. The carrier generation can be readily caused by light of wavelengths shorter than the absorption edge of the photoconductor. The displacement of carriers due to the applied field is dependent upon the carrier mobilities in the bulk of the material as well as the height of the dielectric–electrode barrier. If the bulk mobilities of both carrier types are large and the barriers are high for at least one type of carrier, charges will accumulate at one or both dielectric–electrode interfaces. One obtains a "barrier polariza-

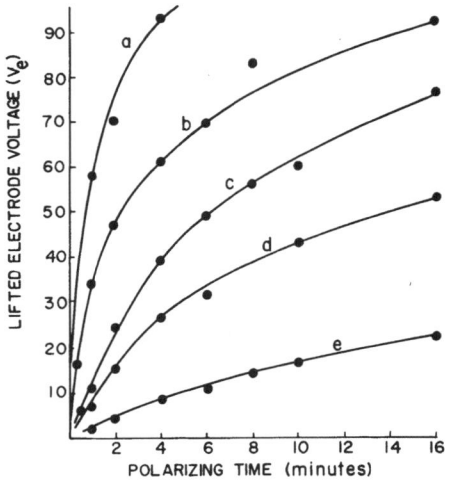

Fig. 2.10. Photoelectric polarization in anthracene as function of polarizing time for various ultraviolet intensities: (a) 2.2×10^{-8} W cm^{-2}; (b) 4×10^{-9} W cm^{-2}; (c) 7.3×10^{-10} W cm^{-2}; (d) 1.3×10^{-10} W cm^{-2}; (e) 1.4×10^{-11} W cm^{-2} (polarizing voltage ~ 300 V)

tion". If, however, the mobilities are such that only one carrier type is significantly displaced, or if the irradiation is nonuniform over the thickness of the sample, "bulk polarization" is produced. This case can obviously occur irrespective of the barrier heights. Trapping, detrapping, and retrapping may occur during irradiation; after irradiation, the charge distribution is greatly immobilized, but still subject to a dark decay (typically over periods of the order of days).

Results demonstrating the effect of polarizing time and light intensity on the electret charge of anthracene are shown in Fig. 2.10 [2.76]. The figure indicates that for very low intensities the charge increases almost linearly with time but appears to saturate at higher intensities. Other experiments yielded a time dependence given by an expression of the form of (2.56), where the time constant τ relates to light intensity while, for a given material, σ_m is only a function of voltage (generally proportional to voltage). This means that the curves in Fig. 2.10, obtained for the same polarizing voltage, will eventually all reach the same saturation charge. It was also shown that the charge is a weak (logarithmic) function of the light intensity.

Certain materials show a considerable "dark polarizability" (the ability to produce a polarization by field alone) if preirradiated in the absence of a field. This dark polarizability decreases more slowly with time than the dark decay mentioned above [2.73]. Such different behavior has been attributed to the fact that the dark decay occurs due to drift under the internal field while the dark polarizability decreases because of the recombination of radiation-produced positive and negative carriers in the field-free sample. The second process, taking place in the absence of a field, is slower than the first. Similar differences in field-dominated and field-free decays were recently found in gamma-irradiated Teflon samples [2.70].

The photoelectret process is of little practical use as a charging method. The reverse effect, however, namely the discharge of a polarization on photoconductors (particularly selenium) by irradiation with light is of great practical importance in xerography. In this application, the charging is achieved by a corona process (for details on xerography see Chap. 7).

2.3 Methods for Measuring Charge Density

A measurement of the net charge $\hat{\sigma} = \int_0^s \varrho(x)dx$ [see (2.19)] in a dielectric can easily be performed by dropping the dielectric into a Faraday cup (a metal container) and measuring the induction charge flowing from ground into the cup. It follows from simple electrostatics that this induction charge is opposite in sign but otherwise equal to the net charge in the dielectric.

In most cases, however, one is more interested in determining the moment of a charge distribution rather than the often small (or vanishing) net charge. For *nonmetallized* or *one-sided metallized* electrets, this can be achieved by measuring the induction charge on a planar electrode parallel to the electret surface. As has been shown in (2.24), the induction charge σ_{i1} generated by an electret backed up by a metal plate is proportional to the projected or effective surface-charge density $\hat{\sigma}_1$. For a contacting electrode, σ_{i1} actually equals $-\hat{\sigma}_1$. As has been further demonstrated, a surface charge $\hat{\sigma}_1$ generates the same external field as the actual charge distribution $\varrho(x)$. Thus $\hat{\sigma}_1$ is the only quantity that can be determined by a single induction measurement.

In spite of the limited information obtainable from induction measurements, such methods are widely used because of their simplicity. They will be discussed in Sects. 2.3.1–4 (see also reviews in [2.77, 78]). In Sect. 2.3.5 a method to determine the total charge $\hat{\sigma}$ will be reviewed. This method consists of two separate measurements.

Data on charge density of *two-sided metallized* electrets can be obtained in certain cases from measurements of polarization or depolarization currents, as shall be shown in Sect. 2.3.6. Methods for determining the polarization charge of piezoelectric electrets are based upon the actual evaluation of the piezoelectric and pyroelectric properties [2.79–81].

Other techniques of measuring charge density, depending on the deflection of particle beams [2.82, 83] and the force exerted on electrodes (ponderomotive meters [2.77, 84]) are less convenient to instrument than the purely electrical procedures and often affect the charges to be measured. They are therefore not widely used. Optical methods based on a measurement of birefringence or the amount of second-harmonic generation have also been described [2.85, 86].

For a discussion of these methods, the reader is referred to the literature.

2.3.1 Dissectible Capacitor

The dissectible capacitor, going back to Benjamin Franklin (see [2.87]), is the classical method of determining the effective surface-charge density on dielectrics. In its simplest implementation, it consists of two electrode plates connected by a ballistic galvanometer or by a capacitively shunted electrometer. The charged dielectric is placed between, and in contact with, the electrodes generating induction charges on them. If one of the electrodes is lifted from the dielectric, its induction charges will flow into the galvanometer or capacitor. They can be measured if the time constant of the galvanometer or that of the capacitor–electrometer combination is much larger than the time required to remove the electrode.

Due to possible air gaps between the electrodes and the dielectric, the induced charge σ_{i1} on the electrode may be smaller than the effective surface-charge density $\hat{\sigma}_1$ on the electret. The relation between σ_{i1} and $\hat{\sigma}_1$ for a one-sided metallized electret has been given in (2.24). For plate electrets, s_1/s is generally small so that σ_{i1} equals $-\hat{\sigma}_1$. For thin-film electrets, however, s_1 and s are often comparable, and s_1 has to be accurately determined, e.g., from a capacitance measurement.

The dissectible-capacitor method requires only simple instrumentation. Its drawbacks are the occurrence of contact electrification and breakdown in the air gap, affecting the measurement of small and large charge densities, respectively. A further problem with the study of thin-film electrets is the effect of the air gaps discussed above and their nonuniformity. These drawbacks have largely eliminated dissectible-capacitor methods in modern electret research.

2.3.2 Capacitive Probe

The capacitive probe is related to the dissectible capacitor but avoids its drawbacks by using a well-defined and relatively large air gap of thickness s_1 between probe and electret [2.77, 88–91]. If the probe, which may be stationary, is exposed to the field of the electret, a charge $a\sigma_{i1}$ flows into it from a parallel capacitance C. This charge is given by $a\sigma_{i1} = -CV$, where V is the voltage generated across C. As long as C is much larger than the capacitance of the probe, σ_{i1} equals the short-circuit induction charge and one has therefore for a one-sided metallized electret from (2.24)

$$\hat{\sigma}_1 = \left(1 + \frac{s_1\varepsilon}{s\varepsilon_1}\right)\frac{CV}{a}, \tag{2.59}$$

where all the quantities on the right are accurately measurable.

A schematic view of such a setup is given in Fig. 2.11. For measurements on thin-film electrets the probe is typically a few mm above the electret surface permitting charge-density measurements up to $10^{-6}\,\mathrm{C\cdot cm^{-2}}$. Use of a shutter

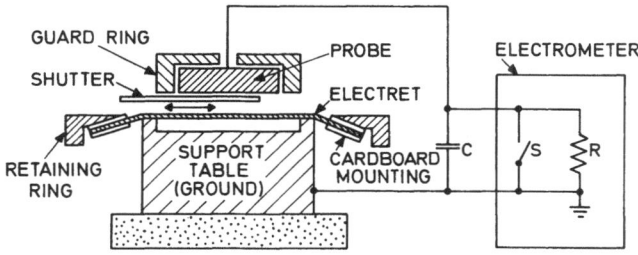

Fig. 2.11. Schematic drawing of a capacitive-probe setup for measuring projected-charge densities. C = external capacitor, R = input resistance of electrometer, S = switch. [2.91]

allows one to shield the probe from, and later expose it to, the field of the electret. The voltage across C is then measured with an electrometer (input resistance R) and the charge density determined with (2.59) within a time much less than $\tau = RC$. Calibration of such a setup is possible by replacing the electret by a two-sided metallized sample of equal thickness whose top surface is charged to a known voltage. Various probe sizes yield different resolutions, while a micropositioner allows use of the instrument as a "local-charge meter" for mapping charge distributions. Fully automatic local charge meters of similar design have also been described [2.77].

Advantages of capacitive-probe methods are the great accurancy typical for static techniques with well-defined geometry and the noncontacting measuring procedure. Such methods can be easily used to determine average charge density as well as lateral charge distribution. The measuring range extends from 5×10^{-11} to above 10^{-6} C·cm^{-2}.

2.3.3 Dynamic Capacitor

The dynamic-capacitor method utilizes a measurement of the ac voltage induced on an electrode near an electret [2.92]. The voltage is due to field variations which are either caused by mechanical excitation of the electrode (or the electret) or by periodic shielding of the electrode by a rotating shutter. In all cases, the relation between charge density and generated voltage depends on circuit details and will not be given here. The measuring accuracy may be improved by filtering, amplifying, and automatically registering the generated ac voltage.

As compared to the capacitive probe, the dynamic-capacitor method is mechanically somewhat more complicated and not as easy to use. On the other hand, it allows the application of signal-processing techniques and the performance of measurements with small probes at operating frequencies ω sufficiently high so that $RC \gg 1/\omega$ holds for the RC constant of the circuit.

2.3.4 Compensation Method

Like the dynamic capacitor, this method depends on vibrating the electret or the opposing electrode. However, instead of measuring the induced ac voltage, one compensates the field of the electret in the air gap with a dc voltage V_0 until the ac voltage (and thus the field E_1) vanishes [2.29, 35, 93, 94]. The effective surface-charge density for a one-sided metallized electret then follows from (2.22) with $E_1 = 0$ and $s_2 = 0$ as

$$\hat{\sigma}_1 = -\varepsilon_0 \varepsilon V_0 / s. \tag{2.60}$$

This method is also a noncontacting procedure. It has the additional advantage that a measurement of the air-gap thickness is not necessary to determine $\hat{\sigma}_1$. Highly automated setups of this kind have been developed in several laboratories [2.29, 95].

2.3.5 Thermal-Pulse Method

An elegant method to determine the total net charge stored in one-sided metalized electrets is the thermal-pulse technique, which is in wide use for the study of charge centroids (for details of this method, see Sect. 2.4.5). If the centroid location $\bar{r}$ of the charge distribution is determined with this method, the total charge follows from

$$\hat{\sigma} = \frac{\varepsilon \varepsilon_0 V}{\bar{r}},$$

where V is the surface potential of the electret. Thus, measurements of V and $\bar{r}$ have to be performed.

2.3.6 Polarization and Depolarization Currents

The induction methods discussed in Sects. 2.3.1–5 can only be used on samples having at least one nonmetallized surface. On samples *metallized on both surfaces* charge densities may be determined *in certain cases* by measuring the polarization or depolarization currents from the two electrodes. One example is real-charge injection, e. g., by electron-beam charging with a partially penetrating beam. In this case the injected charge is equal in magnitude to the integral over the compensation current into the nonirradiated (rear) electrode if the front electrode is floating. Another example is the decay of a uniform volume polarization. Since every polarization charge in the dielectric induces an image charge on the electrode, the polarization is equal in magnitude to the integral over the externally measured polarization (or depolarization) current [2.8].

The situation is more complicated for real-charge release within the dielectric, where drift of charges to the electrodes generally does not release an equal number of image charges. According to (2.20) and (2.25) the image charge induced on a contacting electrode changes very little if a space charge initially trapped close to an electrode drifts to this electrode. For example, a charge layer of density σ_0 at a depth x_0 in the dielectric releases a charge $\sigma_0 x_0/s$ from the far electrode if the drifts are unidirectional and less if they are bidirectional. Since the latter predominates for deep charge layers in homogeneous dielectrics, the released charge is actually always less than 15% of the stored charge [2.29]. However, in electrets with regions of relatively high conductivity, such as samples charged by injection methods, drifts may be unidirectional and the amount of released charge can approach the amount of stored charge. An example is an electret in which carriers injected through one surface and trapped close to the other surface generate a high radiation-induced conductivity throughout the penetrated volume. Upon depolarization, the carriers drift to the surface of injection, thus releasing a large electrode charge. In cases of unidirectional drift (near-surface charge or conductive region) measurement of the charge released by depolarization will therefore give information about the density of the stored charge if its average depth is known.

2.4 Methods for Measuring Charge Distributions

Sectioning methods and potential probe techniques are the classical procedures of measuring the distribution of polarization charges and real charges [2.96, 97]. Sectioning techniques in particular have been used extensively on thick electrets and have yielded a wealth of information. Because of limitations of the slicing procedures, these methods are of little use for thin-film electrets. For such dielectrics, techniques depending on the measurement of charging and charge-release currents or on the evaluation of voltage changes due to heat, ultrasonic, or optical excitation have been developed. These methods yield the *centroid location* (mean depth) of the charge distribution or even the *actual distribution* in the thickness direction of the electret.

2.4.1 Sectioning and Planing Methods

On electrets of sufficient thickness, sectioning and planing methods permit the measurement of the *distribution* of polarization and real charges. Such experiments have been originally performed on wax electrets which are soft enough to allow cutting and shaving operations, but have been extended more recently to polymer electrets. Typically, the electret is divided into a number of thin disks by cutting along planes perpendicular to the direction of polarization. To avoid loss of charge, the cutting is performed at low temperatures. Also, care has to be taken to prevent contact charging due to the cutting. The sections may now be studied by charge and depolarization-current measurements.

The net charge of an electret section of thickness s' can be determined by putting the slice into a Faraday cup. Since polarization charges (dipoles and microscopically displaced charges) have zero net charge, this experiment determines the *net real charge*, $\hat{\sigma}_r = \int_0^{s'} \varrho_r(x)dx$. If these charges are of one polarity within the section, as is to be expected for $s' \ll s$, $\hat{\sigma}_r$ equals the *total real charge*.

A measurement of the short-circuit depolarization current of a thin electret section having a uniform polarization $P_p(t)$ and a uniform real charge $\varrho_r(t)$ yields the *polarization* P_p [2.98]. This is seen by substituting (2.20) with (2.3) and (2.5) into (2.42). One obtains

$$i(t) = \frac{d}{dt} P_p(t) - \frac{d}{dt} \varrho_r(t) \int_0^{s'} \left(1 - \frac{x}{s'}\right) dx + i_c(0, t). \qquad (2.61)$$

Integration over time yields the charge released over the electret area a,

$$Q_\infty = aP_p(0) - \tfrac{1}{2} as' \varrho_r(0) + a \int_0^\infty i_c(0, t)dt. \qquad (2.62)$$

On thick sections, Q_∞ includes polarization and space-charge contributions. However, with decreasing s', the second term on the right goes to zero. The same is true for the third term since i_c is proportional to $E(0, t)$ [see (2.37)] which decreases with decreasing real charge within the section. Progressively thinner slices of an electret will therefore give a depolarization current which tends toward a thickness-independent value determined only by the polarization P_p. Measurements on such sections thus yield the *polarization charge* of an electret.

Alternatively, planing experiments can be performed by shaving thin layers off the dielectric and measuring either the charge on the shaved-off sections or the charge on the remainder of the sample [2.99].

Present slicing techniques permit the removal of layers down to about 5 µm thickness [2.100]. This determines the current resolution of the sectioning method.

2.4.2 Split Faraday Cup

The split Faraday cup method [2.61, 101] allows one to determine the *centroid location* of real charges for a two-sided metallized electret charged by electron injection. Measurements of the induction charges on the two electrodes ("split Faraday cup") during and immediately after charging are required. The spatial depth on electrets with only one electrode can also be determined by a modification of this method. These procedures yield accurate results only if dipole or other volume polarizations are absent.

The method is illustrated in Fig. 4.16 and discussed in Sect. 4.4.6. Briefly, the dielectric is irradiated with floating front electrode (open circuit) and a buildup of the induction charge on the rear electrode is measured. Thereafter, the irradiation is terminated and the front electrode connected to ground.

The induction charges are now distributed between front and rear electrodes and have the values Q_1 and Q_2, respectively. Measurement of these quantities allows determination of the mean charge depth $\bar{r}$ by means of [2.61]

$$\bar{r}/s = Q_2/(Q_1 + Q_2). \tag{2.63}$$

Often, the spatial depth changes with time after short circuit. This change can also be detected from the time dependence of Q_1 and Q_2.

On samples metallized on the rear side only, a measurement of the rear-electrode charge Q_m at the end of beam charging and of the charge Q_i induced on a contacting front electrode after charging yields the mean depth [2.62]

$$\bar{r}/s = (Q_m - Q_i)/Q_m. \tag{2.64}$$

The Faraday-cup method depends on measurements taken during the charging of the electret. If such measurements are not available, the combined induction-depolarization method or the thermal-pulse method can be used.

A method developed for Townsend discharge electrets utilizes bridge measurements to determine the charge density on the sample and the potential difference across it. These measurements also allow one to determine the space-charge depth [2.55, 102].

2.4.3 Combined Induction-Depolarization Method

The *centroid location* of real charges on previously charged electrets can be determined by a method utilizing an induction-charge measurement and a measurement of the integrated depolarization current [2.103]. It is applicable to nonmetallized or one-sided metallized electrets with either a single-polarity charge distribution or a distribution consisting of a positive and a negative cloud of charges. Dipole or other volume polarizations are to be absent.

The first step of the procedure consists of a measurement of the effective surface-charge density σ_1 on the top (nonmetallized) surface by an appropriate method (see Sect. 2.3). Thereafter, a complete depolarization of the sample by application of heat, radiation, light, or other means is performed while the integral over the depolarization-current density $q_\infty = \int_0^\infty i \cdot dt$ through contacting electrodes is determined. For a single-polarity distribution, the mean charge depth $\bar{r}$, measured from one of the electret surfaces, is then given by

$$\frac{\bar{r}}{s} = \frac{1}{1+|\sigma_1/q_\infty|},\tag{2.65}$$

where the assumption is made that all charges drift to that surface (see below). For a distribution consisting of positive and negative charge clouds, the expression on the right-hand side of (2.65) can also be used to measure charge depth if all the positive carriers drift to one surface while all the negative carriers drift to the other surface. In this case, $\bar{r}$ is the sum of the mean positive and negative drift distances.

Under the same conditions, the total trapped charge in the dielectric can be determined from

$$\int_0^s |\varrho_r(x)|dx = |q_\infty| + |\sigma_1|.\tag{2.66}$$

The requirement of unidirectional drift for carriers of one polarity is approximately met in a number of cases of which two are particularly important. Such drifts are always encountered if the charges are stored close to one surface (approximately within 20% of the sample thickness) and if retrapping of the carriers is fast [2.29, 103]. These conditions are met in some polymer materials, such as Teflon, surface-charged with thermal, liquid-contact, or other methods. Unidirectional drifts occur also in electrets charged to any depth with electron-beam methods. In this case the drift is back to the surface of incidence through the volume modified by the radiation-induced conductivity. This holds even for samples where the electrons have been injected to a depth beyond the center.

2.4.4 Light-Radiation-Release Method

The *centroid location* of charges in photoconducting electrets metallized on both surfaces with semitransparent electrodes can be determined by a light-radiation-release method [2.104]. The sample is first irradiated through one of its electrodes with light that is sufficiently absorbed so that it does not penetrate beyond the charge layer. Due to detrapping of the stored charge Q_t, or drift of secondary carriers generated by the light, the stored charge is completely neutralized. Thus, an external charge Q_A equalling $Q_t(s-\bar{r})/s$, where $\bar{r}$ is the depth measured from the nonirradiated surface, is released. A similar experiment (on a different portion of the sample) performed with irradiation through the other electrode yields a released charge $Q_B = Q_t\bar{r}/s$. From these relations, $\bar{r}/s$ can be determined as

$$\bar{r}/s = Q_B/(Q_A + Q_B).\tag{2.67}$$

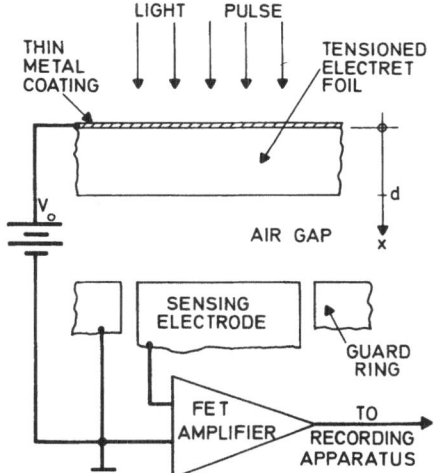

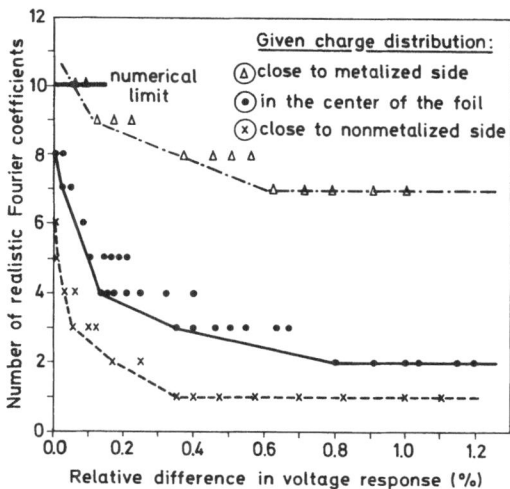

Fig. 2.12. Experimental setup for measuring charge distribution in x-direction by thermal-pulse method [2.106]

Fig. 2.13. Number of Fourier coefficients of charge distribution that can be determined with an accuracy of 20% or better as function of the accuracy to which the voltage response in a thermal-pulse experiment is measured [2.108]. Each symbol represents a charge distribution

The main features of the charge distribution in a relatively thick electret can also be determined by an optical method. The distribution follows from the charge released by irradiation with a light beam propagating through the sample normal to the direction of polarization [2.105].

2.4.5 Thermal-Pulse Method

The *centroid location* of real and polarization charges in electrets can also be determined from measurements of the potential change across the dielectric upon absorption of a light pulse by the sample surface and diffusion of the corresponding heat into the sample [2.106]. A schematic illustration of the method is given in Fig. 2.12. The potential of the nonmetallized surface of a one-sided metallized electret relative to the electrode is by virtue of (2.30) $V = s\hat{\sigma}_1/\varepsilon\varepsilon_0$. As the heat diffuses through the dielectric, changes in s and $\hat{\sigma}_1$ due to thermal expansion and changes in ε cause corresponding changes ΔV of V. If measurements of ΔV are taken at time t_1 when the thermal pulse has just entered the electret and at time t_2 when it is uniformly distributed over the dielectric, one has

$$\bar{r}/s = \Delta V(t_2)/\Delta V(t_1). \tag{2.68}$$

Thus, a measurement of the voltage transient allows one to determine $\bar{r}$. Experiments of this kind are nondestructive since the thermal pulse results only in a minor temperature increase of the sample.

It has also been suggested to use this method to detect more details of the *spatial distribution* of the charge [2.106]. Since the voltage change at time t as function of the temperature increase $\Delta T(x', t)$ is, for a real charge, given by

$$\Delta V(t) \propto \int_0^s \left[\varrho_{\rm r}(x) \int_0^x \Delta T(x', t) dx' \right] dx \tag{2.69}$$

it appears possible to calculate $\varrho_{\rm r}(x)$ by deconvolution of the integral if ΔT is known. Yet, because of inaccuracies of the deconvolution procedure, a unique solution for $\varrho_{\rm r}(x)$ is generally not obtained. All that can be determined under ideal conditions are 5–10 spatial Fourier coefficients of the distribution [2.107, 108]. However, if one has to allow for measuring errors in the voltage response, even fewer coefficients are accurately found. This is shown in Fig. 2.13, where the number of Fourier coefficients determined to better than 20% is given as a function of the accuracy to which $\Delta V(t)$ is measured [2.108]. The figure indicates that for a measuring accuracy of 0.1%, only 3–9 coefficients, the actual number depending on the location of the charge cloud in the dielectric, can be found. The measurement of polarization distributions is similarly affected.

2.4.6 Pressure-Pulse Method

Another recent method to determine the *charge distribution* utilizes the propagation of a pressure transient, preferably a step-function compressional wave, through the electret [2.109]. Because of the constant amplitude of such a wave, the second integral in (2.69) with ΔT replaced by Δp can be written as

$$\int_0^x \Delta p(x', t) dx' = \begin{cases} \Delta p x & \text{for } x < \bar{s}(t), \\ \Delta p \bar{s}(t) & \text{for } x \geqq \bar{s}(t), \end{cases} \tag{2.70}$$

where $\bar{s}(t)$ is the location of the pressure step. One obtains, by inserting (2.70) into (2.69) and differentiating twice with respect to time,

$$d^2 V/dt^2 \propto -\Delta p v^2 \varrho_{\rm r}(\bar{s}), \tag{2.71}$$

where $v = d\bar{s}/dt$ has been substituted. Thus, the charge distribution follows from the time derivative of the voltage in the external circuit during propagation of the step through the electret.

The advantage of this method as compared to the heat-pulse technique is that $\varrho(x)$ can be obtained uniquely since a deconvolution procedure is not needed. The method is also nondestructive but demands a relatively sophisticated experimental setup. Due to the high propagation velocity of the compressional pulse ($\sim 2 \times 10^3$ m·s^{-1} in polymers) a short rise time and good electronic resolution are required for measurements on thin-film electrets. In first experiments, self-steepening ultrasonic shock waves generated by rupture

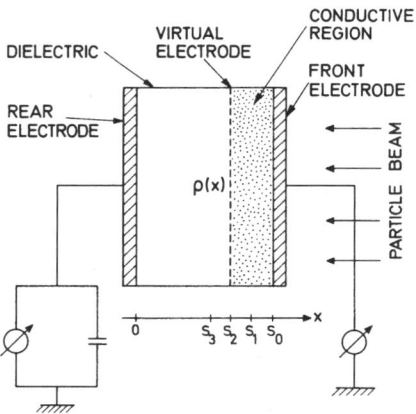

Fig. 2.14. Experimental setup for measuring charge distribution in x direction by charge-compensation method [2.110]

of a membrane in a shock tube have been used to determine potential distributions in 1 mm thick electrets with a resolution of about $10\,\mu m$ [2.109].

2.4.7 Charge-Compensation Method

Another very recent method to determine details of the *spatial distribution* of real and dipolar charges in dielectrics is based on the generation of a conductive region within a two-sided metallized and short-circuited sample by means of a particle beam, e.g., an electron beam [2.110]. The conductive region extends from the electrode of particle incidence (front electrode) to a certain depth in the material (given by the particle range), as shown in Fig. 2.14. If the front of the conductive region, which forms a virtual electrode, is swept through the sample by increasing the beam energy, all the charges originally stored within the material are progressively compensated, resulting in a release of the induction charges residing on the rear electrode. The charge distribution may then be found from

$$\varrho(s) = -\frac{d^2(qs)}{ds^2} \tag{2.72}$$

where q is the induction charge per unit area on the rear electrode and s the thickness of the nonconductive part of the sample.

The charging effects of the interrogating beam have to be analytically eliminated by means of a calibration run with a noncharged sample. It is often more convenient from an experimental point of view to increase the beam energy in steps rather than continuously. Then the second derivative in (2.72) is replaced by a finite difference coefficient.

The method is only applicable to materials in which the schubweg of holes in the nonirradiated region is smaller then the desired resolution. This is the case in some materials; in others it may be achieved by cooling.

In a modification of this method, applicable to materials and fields where the Schubweg of the holes is larger than the sample thickness, the currents $I(t)$ caused by a burst of holes moving with constant velocity v under an external field is evaluated. In this case the charge distribution is given by $\varrho(x) = -(s/av^2)dI/dt$ [2.111].

2.4.8 Indirect Methods

A number of indirect procedures for measuring *charge depth*, based on dielectric or optical degradations caused by the penetrating charges, have been described [2.112, 113]. In one method, measurements of capacitance and loss angle of the virgin and charged materials are evaluated under the assumption that only the region penetrated by charges contributes to the loss [2.112]. Such evaluations yield the thickness of the degraded layer but do not necessarily account for charge drifts or charge compensation after injection. Thus, results of this kind have to be interpreted with care.

2.5 Methods for Discriminating between Polarization and Real Charges

As discussed in Chap. 1, electret properties originate from dipole polarization, from charge displacement within molecular or domain structures, from space-charge effects or from surface charging. Knowledge of the properties of a specific material or the physics of the charging process often allows one to distinguish between these possibilities. Examples are the absence of a polarization in nonpolar materials or in materials charged at low temperatures by electron-beam injection. The charging process can also be modified by the use of blocking electrodes, e.g., a highly insulating polymer layer between electrode and dielectric. This will prevent charge injection, thus leading only to internal polarization effects [2.114]. For disk electrets, discrimination between dipole polarization and real charges is possible with sectioning procedures, as discussed in Sect. 2.4.1.

Additional information about the nature of the polarization for all kinds of electrets can be obtained from discharge experiments. Simultaneous measurement of the induction charge and the external current yields with (2.42) the interfacial current i_c [2.7]. If i_c is zero during the discharge experiment, decay of a polarization charge or internal decay of a space charge has to be assumed, while external decay of real charges manifests itself by an interfacial current.

Another method of discriminating between dipolar and space-charge effects is based on a comparison of polarization currents obtained upon repeated heating of a dielectric. If the dielectric is first polarized at linearly rising temperatures by application of a field, a polarization current I_1 is observed.

Upon cooling and renewed heating under a field of opposite polarity, a current I_2 occurs. For dipole polarization, one obtains during the second heating a superposition of a charging and a discharge current of equal magnitude, so that $|I_2| = |2I_1|$. Such a relation does not hold for charge separation and charge injection, and one has generally $|I_2| < |2I_1|$ in these cases [2.115].

In certain materials, such as alkali halides, space-charge effects can be generated by irradiation with light of band-gap energy. Thermal depolarization of such a sample will uncover the TSC spectrum due to these charges and thus allow one to distinguish it from the dipolar depolarization spectrum [2.116].

Identification of polarization effects is also possible by comparing thermally stimulated current (TSC) measurements of electrets with dielectric data. The relationship between dielectric constant and dielectric loss factor on one hand and the charges and currents released due to dipole reorientation during TSC on the other hand have been derived by *van Turnhout* [2.29] (see also Sect. 3.9). Similar relations also exist between dilatometric and TSC measurements. For example, the correspondence of TSC and dielectric or dilatometric data allows one to identify the polarization phenomena responsible for the α and β peaks in polar polymers, where they are related to dipole orientation.

Information about the nature of electret polarizations can also be gained from piezoelectric and pyroelectric measurements, as discussed in Chap. 5. Finally, other experiments such as electron-spin resonance [2.117], birefringence and singleharmonic generation [2.85], and infrared and X-ray studies [2.118] can, in principle, shed light on the electret effect. For example, far-infrared spectroscopy and X-ray studies have shown that poling of PVDF induces structural rearrangements in the crystalline parts of the polymer [2.118]. We refer, however, to the literature for a detailed discussion of these methods. Some additional information about methods for distinguishing between polarization and real charges is given in Sect. 3.4.

2.6 Permanent Dipole Polarization and Real-Charge Storage

Permanent dipole polarization and charge storage have been discussed under a variety of aspects and a host of experimental data is documented in the literature. Much of this information is, however, of purely academic interest today. To keep the material to be reviewed within reasonable limits, only some recent and representative results are presented. Much of the older information on thermoelectrets and photoelectrets is anyhow well covered in the literature [2.25].

The permanent dipole polarization and charge retention achievable in various electret-forming materials depends to a large degree on material properties and environmental conditions. Of the large number of insulators investigated with respect to electret properties, those of particular importance are certain inorganic crystals, some polymer materials and a number of (more historically interesting) waxes. To be useful for electret studies these substances have to show either suitable polar properties or extremely low conductivity due to a large number of deep trapping centers (see Sect. 1.2). Generally, both of these features are not found in the same material. It appears that most dielectrics with polar properties have an undesirably high conductivity. This can be partly attributed to the hygroscopic behavior of many polar substances which causes conductivity increases due to absorption of water. On the other hand, the lowest-conductivity materials are relatively nonpolar (see Table 1.1).

2.6.1 Retention and Decay of Dipole Polarization

The dipolar polarization of inorganic crystals may be caused by *structural properties* of the crystal lattice, as found in many of the piezoelectric substances, or it may be due to *lattice imperfection or doping*, for example in impurity–vacancy dipole systems of alkali halides. In polymers and waxes, dipole properties can similarly originate from *polar groups* in the crystalline parts of the polymer or from *imperfections or impurities*. An example for the former is the monomer group in PVDF and for the latter the carbonyl groups in polyethylene. A considerable amount of information about the nature of dipolar effects has been obtained from TSC measurements, in the case of dipole relaxation often referred to as "ionic thermocurrent" measurements (see Chap. 3).

The *saturation polarization* P_p expected after a sufficiently long polarizing time depends, according to (2.47), on the polarizing field E as

$$P_p = \varepsilon_0(\varepsilon_s - \varepsilon_\infty)E. \tag{2.73}$$

Experimental data of $P_p/\varepsilon_0 E$ for a few polymers as determined from current integrals of TSC curves are shown in Table 2.1. The results are evaluated separately for two different dipolar phenomena, referred to as the β and α relaxations, which manifest themselfes as two distinct TSC peaks. While the β relaxation is due to motions of the side groups of the molecular chains, the α relaxation is caused by joint motion of side groups and main chains (see also Sect. 3.2). A comparison with the corresponding values of the dipolar strength $\Delta\varepsilon = \varepsilon_s - \varepsilon_\infty$ indicates good agreement between $P_p/\varepsilon_0 E$ and $\Delta\varepsilon$ for the β peaks; however, for the α peaks $P_p/\varepsilon_0 E$ is higher than $\Delta\varepsilon$, which may be partially due to superposition of space-charge motions.

The proportionality between P_p and E breaks down if the dipole alignment is approaching saturation. This occurs generally for fields of about $1\,\mathrm{MV \cdot cm^{-1}}$ and values of P_p around $1\,\mathrm{\mu C \cdot cm^{-2}}$, if poling temperature and time are sufficient. For example, PVDF assumes a maximum polarization of $12\,\mathrm{\mu C \cdot cm^{-2}}$ [2.8, 119, 120]. This represents a significant alignment of the molecules available in the dielectric since the calculated maximum polarization for β-phase PVDF is about $22\,\mathrm{\mu C\,cm^{-2}}$ (see Sect. 5.7).

Table 2.1. Released polarization $P_p/\varepsilon_0 E$ and dipolar strength $\Delta\varepsilon$ for some metalized polymer foils [2.29,119]

Polymer	β relaxation		α relaxation	
	$P_p/\varepsilon_0 E$	$\Delta\varepsilon$	$P_p/\varepsilon_0 E$	$\Delta\varepsilon$
Polymethyl methacrylate (PMMA)	1.9	1.9	1.7	0.8
Polyethylene terephthalate (PET)	0.33	0.37	0.5	0.36
Polyvinyl chloride (PVC)	0.5	0.45		7
Polyfluorethylene propylene (FEP)	0.011		0.004	
Poly(vinylidene fluoride) (PVDF)			~ 100	~ 10

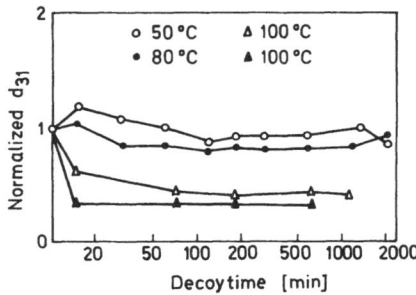

Fig. 2.15. Decay of normalized piezoelectric d_{31}-constant of PVDF (polarized at 130 °C and 800 kV cm^{-2}) at various temperatures [2.121]

A detailed account of the effect of structure and poling conditions on the polarization of PVDF and other polymers is given in Sects. 5.4, 6. The influence of space charges on the polarization has also been extensively discussed in the literature. We refer to Sect. 5.5.4 for a review of this matter.

Results on the *spatial distribution* of dipole polarizations in dielectrics, together with similar data for real charges, will be discussed in Sect. 2.6.3.

The *decay* of a permanent dipole polarization is, just as the polarizing process, controlled by the dipole relaxation frequency $\alpha(T)$ by means of (2.46), valid for the case of a single $\alpha(T)$. In the absence of an applied voltage ($E = 0$) and for isothermal conditions, this equation yields

$$P_p(t) = P_p(0)\exp(-\alpha t). \tag{2.74}$$

More general formulae for distributed relaxation frequencies and nonisothermal conditions are given in Sect. 3.6.3.

Observed decays are often nonexponential with increasing "time constants", indicating a distribution of relaxation frequencies (see, for example, [2.120]). These distributions may, in general, be due to a distribution of activation energies or a distribution of pre-exponential factors. The reasons for the existence of such distributions are discussed in Sect. 3.6.3 and Sect. 3.11. The isothermal decay of the polarization of PVDF at elevated temperatures, as evidenced by the decay of the piezoelectric constant d_{31}, is shown in Fig. 2.15

[2.121]. The nonexponential character of the decay is clearly evident. The retention and decay of dipole polarizations in dielectrics is also discussed in Chaps. 3, 5. We refer to these chapters for additional details on these topics.

2.6.2 Retention of Real Charges

Surface and space charges in electrets are stored in *trapping levels* located in the band gap between conduction and valence bands. In general, electron and hole traps are present. After a sufficiently long time, the former are in equilibrium with the conduction band and the latter with the valence band. While electron traps are neutral when unoccupied and negatively charged when occupied, hole traps are neutral when occupied and positively charged upon release of an electron. The conduction and valence bands are continuous if the material has a periodic lattice structure and if there is sufficient orbital overlap between neighboring molecules. In such materials, discrete trap levels are expected (see also Sect. 4.1.1).

These conditions are not met in amorphous, polycrystalline, or partially crystalline substances. In these materials the local energy levels are affected by their molecular environment, resulting in band structures shaped by potential barriers with every atom or group of atoms having its own energy levels (see Fig. 2.16a). Even in the presence of energetically discrete trapping levels, this randomness will cause a corresponding randomization of the trap depths, if referred to the highest conduction band level. This has often led to the assumption of continuous (in particular exponential) distributions with traps extending to a depth of about 1–2 eV [2.122, 123].

The experimental evidence from TSC and similar measurements, however, points either to *discrete trap levels* or to *bands of trapping levels* in such materials. For example, TSC measurements on Teflon FEP and other polymers show two to six discrete trapping levels emptying in the temperature range from 20 to 200 °C [2.124, 125] while measurements on Teflon and semicrystalline SiO and MgF_2 below room temperature show evidence of a distribution of trap energies about certain levels [2.126]. Finally photon-induced current spectroscopy in tantalum oxide and polycarbonate indicates a broad band of traps peaked about certain energies [2.127] and thermoluminescence data shows a similar distribution for polystyrene [2.128]. These and other results suggest that uniform or exponential trap distributions are generally not encountered in amorphous or partially crystalline materials.

A possible distribution of the density of states is shown in Fig. 2.16b (see, e.g., [2.129]). E_c and E_v are mobility edges, which replace the bottom of the conduction and valence bands; here, dramatic changes in carrier mobility occur. The implications of such distributions with respect to charge transport are reviewed in Sect. 2.6.6.

Most of the discussion on trap sites to follow will refer to polymers. In such materials, *volume traps* can in principle be due to a number of structural anomalies, such as impurities, defects of the monomeric units, chain irregula-

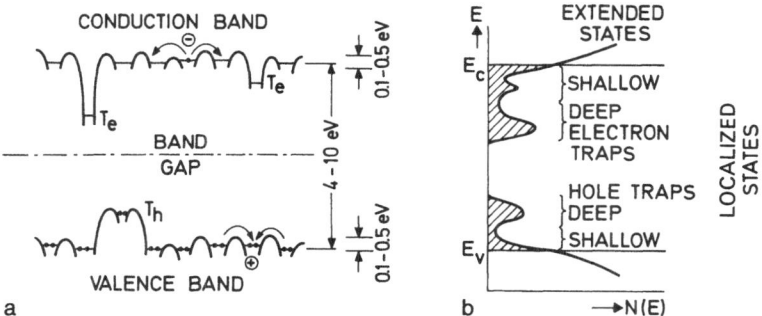

Fig. 2.16. (a) Energy diagram for a polymer. T_e electron traps, T_h hole traps [2.130]. (b) Density of states $N(E)$ for a polymer. Localized states (traps) are shaded. E_c and E_v mobility edges

rities, and imperfections of the crystallites [2.29]. The evidence for or against any of these trap categories is for most materials presently weak and often contradictory. In addition, *surface traps* exist in large numbers in polymers. They may be due to impurities, surface oxidation, or other causes (see below).

Volume traps have been most extensively studied in polyethylene (PE). The depth of these states was estimated by *Bauser* [2.130] as follows. The depth of a hole trap equals the difference of the ionization energy of an isolated PE molecule and that of the trapping molecule while the depth of an electron trap follows from the corresponding difference in electron affinity. Calculations of this kind show that structural defects such as carbonyl groups or double bonds yield shallow or intermediate trapping levels of depths up to about 1 eV while foreign molecules function as deeper traps.

Experimental data on trapping in PE has been obtained by a variety of methods, based on mobility, TSC and thermoluminescence (TL) experiments. The TL data uncovered a number of trap levels attributed to methylene groups [2.131] with depths of about 0.4 eV, in agreement with the above calculations. However, the mobility and TSC data gave trap levels which are also attributed to structural features but show depths in excess of 1 eV. For example, TSC measurements on low- and high-density PE show the importance of branching and of crystallinity on charge trapping [2.124]. These experiments clearly indicate that structural modifications, and not impurities, are responsible for trapping sites (see also below) for which activation energies of 1.2–1.7 eV were derived. A set of very similar activation energies has been obtained from mobility measurements on PE by *Davies* and *Lock* [2.132]. These measurements show an increase of the carrier mobility with the number of unsaturated terminal (vinyl) groups, indicating the contribution of these groups to carrier trapping. It appears that the experimental data on trap levels is not yet understood adequately.

Trapping effects in a series of substituted polyolefins have been investigated by *Creswell* et al. [2.124], and *Perlman* [2.133]. These experiments indicated the presence of three structural trapping levels (primary, secondary, and tertiary),

Table 2.2. Distribution of traps for negative charges in 25 μm Teflon FEP-A [2.135]

Peak temperature [°C]	Location relative to charged surface [μm]	Kind of trap
95	0–25	Energetically shallower trap active under TFL conditions
155	0–0.5	Surface trap
170	0.5–1.8	Near-surface trap
200	1.8–25	Bulk trap

Table 2.3. Greatest observed charge densities and full-trap densities in one-sided metallized dielectric films

Material	Thickness [μm]	Surface charge S or volume charge V	Projected charge density and sign [10^{-6} C cm^{-2}]	Full-trap density in volume and sign [10^{15} cm^{-3}]	Reference
FEP	12.5	Mostly S	0.5 (+,−)		[2.53]
FEP	25	V		0.14[a] (−)	[2.135, 136]
PET	3.8	S + V	1.2 (+,−)		[2.53]
PET	4–6	V		10 (?)	[2.55]
PC	2.0	S + V	1.0 (+,−)		[2.53]
SiO$_2$	0.06–0.1	S + V	4 (−) 2 (+)		[2.137]

[a] Attributable to deep traps with TSC relaxation temperature of 200 °C or higher.

but no trapping due to impurity centers. The primary levels are at atomic sites on the molecular chains. The stability of charges trapped in these sites is determined by the electronegativity of the ions and the symmetry along the chains. This is evident from the general decrease of charge stability and TSC peak temperature with decreasing electronegativity and symmetry of the storage units. For example, the fluorine, hydrogen, phenyl, and methyl groups are in order of decreasing electronegativity. At the secondary level, electrons are caged between groups of atoms in neighboring molecules and are held there due to charge affinity of these groups. The stability of this trapping level increases with packing density and decreases with branching. At the tertiary level, charge may be stored in the highly ordered crystalline regions of polymers or at crystallite–amorphous interfaces. The experimental evidence for charge trapping in a primary level is undoubtedly strong. However, the secondary and tertiary levels appear to be less well confirmed and some cautionary comments about the strength of the present evidence have been advanced [2.134].

Surface traps play an important role in electrets due to the frequent use of surface-charging techniques. For most polymers, little is known about the nature of surface traps, although chemical impurities, specific surface defects caused by oxidation products, broken chains, adsorbed molecules, or differences in short-range order of surface and bulk are believed to be responsible for the capture of charges in certain cases [2.16].

Recently, *von Seggern* [2.135] investigated surface traps for negative charges in Teflon FEP by interpretation of thermally stimulated current (TSC) data from samples charged with corona and electron-beam methods. First, a separation of surface and volume traps was achieved for corona-charged electrets by comparing short-circuit and open-circuit TSC decays. From these experiments, TSC release temperatures of 155 °C and 200 °C for the surface and volume traps, respectively, and 170 °C for an "intermediate" trap were found. In addition, a shallower trap level active under conditions where the deeper traps are filled (trap-filled limit or TFL conditions [1.44]) and discharging at 95 °C was detected. Then, information about the spatial distribution of the 155–200 °C traps was derived from open-circuit TSC experiments with samples charged in different regions with electron beams of different energies. These experiments indicated that the "intermediate" trap is a near-surface state. The results of this investigation are shown in Table 2.2.

Experimental data on *maximum stored-charge densities* and *volume-trap densities* in a number of dielectric films is given in Table 2.3. The maximum charge densities are due to the filling of volume *and* surface traps; in some polymers, such as Teflon, surface traps are known to predominate. Since breakdown or charging efficiency was the limiting factor in the experiments used for deriving the values in Table 2.3, the projected charge densities are probably smaller than those corresponding to complete trap filling. The full-trap density of 1.4×10^{14} cm^{-3} in Teflon FEP is, however, equal to the total density of traps which have a TSC relaxation temperature of 200 °C or higher. The density of shallower trap levels is not included in this value.

2.6.3 Spatial Distribution of Dipole Polarization and Real Charges

Extensive use of sectioning techniques has been made to gain insight into the spatial distribution of volume polarization and real charges in *relatively thick dielectrics*. The very early and often contradictory work reviewed in the older literature will not be discussed here. However, some of the more recent sectioning studies have provided important information about spatial distributions, as will be shown in the following.

The volume polarization of carnauba wax electrets formed by a thermal method and stored thereafter for periods of 2–7 weeks, was investigated by means of TSC measurements [2.138]. These experiments have shown that sections of a 2 cm thick sample produced the same currents as an unsectioned reference sample. This demonstrated [see (2.62)] that carnauba wax electrets prepared under the above conditions and stored for several weeks have a uniform volume polarization.

Quite different results were obtained from sectioning experiments on freshly prepared polymers electrets [2.29]. TSC measurements on thermally polarized 4.8 mm thick polymethyl methacrylate (PMMA) and 6.9 mm thick chlorinated polyether (ChPEth) showed the currents from the center sections of the samples to be smaller by a factor of 2–3 than the currents from the outer sections, indicating a nonuniform polarization. Furthermore, the outer sections yielded a considerable current due to the dissipation of space charges. Planing and subsequent TSC experiments on a 2.6 mm thick PMMA electret demonstrated that about 90% of the space charges are actually located in surface layers of 0.1–0.2 mm thickness. These near-surface layers, which are due to charge separation in the dielectric during polarization, are responsible for the nonuniform polarization since they lower the polarizing field in the interior of the sample and increase it in the surface regions. The variances between these and the earlier results discussed above have been attributed primarily to the age of the electrets since storage reduces space charges in these materials.

The distribution of real charges was also determined with planing experiments [2.99]: Faraday-cup measurements of the charge on shavings from wax electrets polarized by a thermal method showed a concentration of positive and negative space charges in the two surface layers, respectively. Experiments on a series of such electrets further indicated that these charges did not change their sign as the net charge of the electret changed its polarity due to the normal decay of the heterocharge.

On *thin-film electrets*, where sectioning does not yield sufficient resolution, the centroid and the actual distribution of dipolar and real charges have been determined by a number of other methods described in Sect. 2.4.

Centroid measurements on two-sided metallized FEP films charged by electron-beam injection were performed with the split-Faraday-cup method [2.139]. The charge depths, measured in short circuit at the beginning of beam injection and (on different samples) in open circuit at later times are shown in Fig. 2.17. For comparison, theoretical and experimental range data is also given. The initial charge depth is about 15% smaller than the practical (experimental [2.140]) range but larger than the average range. The deviation from the latter is presently not clear. The final depth exceeds the csda and practical ranges by an amount increasing in relative size with decreasing energy. This may be related to the induced conductivity which depends on the energy deposited per unit volume. Since this energy density decreases with increasing beam energy, the conductive region is relatively more extended at low beam energies. Space-charge-limited currents, of importance in PET samples, are expected to have little effect on charge depth in negatively charged Teflon. They would tend to increase the depth more at higher energies, contrary to the experimental findings. Measurements with the induction-depolarization method on one-sided metallized FEP charged by electron beam yield similar results [2.62] (see also recent computer simulations by *Berkley* [2.141]).

Centroid measurements on breakdown-, corona-, and Townsend-charged electret films have been performed with a variety of methods. Measurements

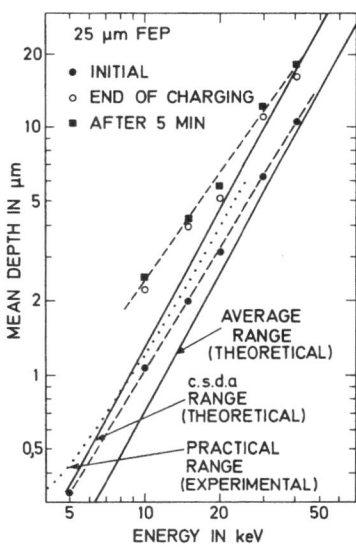

Fig. 2.17. Mean charge depth on two-sided metallized 25 μm Teflon FEP samples charged by electron-beam injection as function of the beam energy. Shown are initial depth measured in short circuit; end-of-charging depth and depth 5 min after charging measured in open circuit. *Solid lines*: theoretical ranges; *dashed lines*: best fit to experimental data; *dotted line*: experimental range from [2.140]. [2.139]

with the induction-depolarization method on negatively breakdown-charged 25 μm FEP yield charge depths of 1.8 and 5 μm for samples charged at room temperature and 180 °C respectively [2.29, 103]. Corona-charged FEP electrets appear to have smaller charge depths; On 25 μm FEP samples, capacitance measurements have shown no penetration of negative carriers while a penetration of about 1 μm was found for positive charges [2.49]. On 25 μm PET, the maximum penetration is 0.8 and 2.5 μm for positively and negatively corona-charged samples, respectively [2.29]. Finally, bridge measurements on 5 μm PET electrets formed by Townsend discharge show a mean depth of 1 μm [2.55, 102]. All these charge depths are much in excess of the initial penetration depth of the ions responsible for the charge deposition. This indicates the presence of charge drifts (probably electronic rather than ionic) through the trap-filled region at the surface, where the carrier mobility is relatively high.

Centroid measurements on positively and negatively liquid-charged FEP films were conducted with the thermal-pulse method [2.106]. The results indicate that the charges reside originally on the surface but are injected into the bulk during high-temperature aging. For positive charges this process takes place at temperatures of about 100 °C. The charges move rapidly through the bulk to the opposite side of the sample, with the centroid of the remaining charge staying close to the originally charged surface. Temperatures of about 180 °C are required to inject the negative carriers which are subsequently retrapped in the bulk. In this case, the centroid moves further away from the originally charged surface.

The *spatial distribution* of the polarization of PVDF films and PVDF–TFE copolymer films has recently been investigated by a number of authors [2.100,

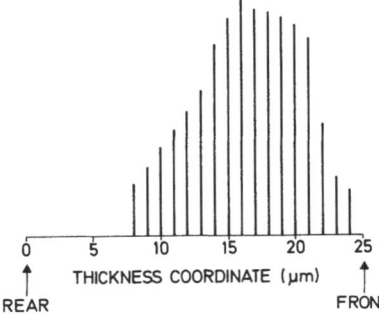

0   5   10   15   20   25
↑   THICKNESS COORDINATE (μm)   ↑
REAR   FRONT

Fig. 2.18. Charge distribution in a two-sided metallized 25 μm Teflon FEP film charged by injection of 30 keV electrons through the front surface (*Sessler* and *West*, unpublished)

107, 142]. Sectioning experiments [2.100], measurements of the pyroelectric frequency response, and evaluations of ultrasonic resonances [2.142] of thermally-charged PVDF as well as thermal-pulse experiments on corona-charged and thermally-charged PVDF–TFE [2.107] show a spatially nonuniform polarization centered near the electrode maintained at positive potential during charging. This is also evident from measurements of the piezoelectric constant of samples poled in a sandwich arrangement [2.142], which show the films poled next to the positive electrode to be the most active.

Higher poling temperatures and voltages have the effect of extending the polarization further into the material. This probably explains the almost uniform polarization determined with thermal-pulse experiments on PVDF [2.107]. More details about the distribution of the polarization in PVDF polymers are given in Chap. 5.

The spatial distribution of charges in Teflon FEP films has been investigated with the charge compensation method and the thermal-pulse technique. Both methods are subject to difficulties and give, at present, only a rough picture of the distribution (see Sects. 2.4.5, 7). As an example, the distribution in a FEP electret charged with a 30 keV electron beam and examined with the charge compensation method is shown in Fig. 2.18.

2.6.4 Analysis of the Isothermal Decay of Real Charges

Decay of real charges in an electret is either due to *internal* phenomena, such as ohmic conduction or drift and diffusion of excess charges, or to the *external* process of ion deposition. Ohmic conduction is dependent on positive or negative intrinsic carriers which are available in the valence and conduction bands of some materials, particularly at higher temperatures. The conduction process consists in the motion of these carriers in the fields persisting in an electret. Excess charges, generally due to charge injection, are subject to drifts caused by their own fields. The ohmic and excess-charge (drift) currents are represented by the terms on the right-hand side of (2.37) and usually account for most of the internal decay. According to (2.37), both phenomena are controlled by the carrier mobilities. Diffusion, which plays mostly a minor role,

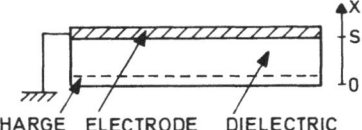

Fig. 2.19. Schematic drawing of one-sided metallized dielectric

CHARGE ELECTRODE DIELECTRIC

is caused by random motions of excess charges with the effect of diminishing concentration gradients. Finally, the external processes are due to the action of electrical fields extending outside the electret and attracting compensation charges in the form of ions from the surrounding medium. The outer electrical fields also attract polar particles, such as water molecules. Having zero net charge, these molecules do not cause an external decay but often lead to an accelerated internal decay.

Charge-decay phenomena are generally analyzed under the assumption that the external decay can be ignored. The internal decay has been discussed in the literature under a wide variety of assumptions concerning ohmic and radiation-induced conductivities, excess charges, carrier trapping and recombination, injection from surface states, external fields, etc. [2.18, 29, 143–152].

In the following, only a few *simple models of internal charge decay* are analyzed, all pertaining to one-sided metallized and grounded dielectrics with finite or zero ohmic conductivity, having an excess charge located initially (at time $t = 0$) on the nonmetallized surface. Due to the field generated by this charge, drifts of the excess carriers toward the electrode ensue. Also, ohmic currents flow which tend to compensate the excess charges. Diffusion effects are to be ignored.

The current equation (2.36) under open-circuit conditions $[i(t) = 0]$ reads, if (2.37) is considered,

$$\varepsilon_0 \varepsilon \frac{\partial E(x, t)}{\partial t} + [g + \mu \varrho_r(x, t)] E(x, t) = 0, \tag{2.75}$$

where the absence of polarization charges ($P_p = 0$) and the presence of a single type of excess charge (either positive or negative) of mobility μ and density ϱ_r has been assumed. This equation is valid if trapping is absent or if the carriers are subject to fast retrapping. In the latter case, μ is a trap-modulated mobility. Substituting ϱ_r with Poisson's equation (2.28) yields

$$\varepsilon_0 \varepsilon \frac{\partial E(x, t)}{\partial t} + \left[g + \mu \varepsilon_0 \varepsilon \frac{\partial E(x, t)}{\partial x} \right] E(x, t) = 0. \tag{2.76}$$

If the sample extends from $x = 0$ (floating surface) to $x = s$ (metallized surface), as shown in Fig. 2.19, one has for the voltage across it

$$V(t) = \int_0^s E(x, t) \, dx$$

and thus from (2.76) by integrating over the sample thickness

$$\tau\dot{V}(t)+V(t)+\tfrac{1}{2}\mu\tau E^2(s,t)=0, \tag{2.77}$$

where $\tau=\varepsilon_0\varepsilon/g$, $\dot{V}=dV/dt$, and $E(0,t)=0$ have been used.

Further progress toward the integration of (2.77) is made by introducing the concept of a *carrier-transit time*. It is defined as the time it takes the front of the excess-charge distribution, initially located at $x=0$, to reach the electrode at $x=s$. For a perfect insulator $(g=0)$, the transit time $t_{0\lambda}$ follows from the definition of the mobility $\mu=v/E$, applied to the front of the charge distribution. Setting $v=s/t_{0\lambda}$ and $E=V_0/s$, where V_0 is the initial voltage across the dielectric, one has

$$t_{0\lambda}=s^2/\mu V_0. \tag{2.78}$$

If the dielectric has a finite conductivity, the transit time is different since an increasing portion of the excess carriers are compensated, leading to a diminished driving field. The corresponding transit time t_λ can be calculated from (2.76) by introducing carrier flow lines (locations of carriers as functions of time). The analysis yields [2.150]

$$t_\lambda=\tau\ln\frac{\tau}{\tau-t_{0\lambda}}. \tag{2.79}$$

Thus, for $\tau>t_{0\lambda}$, a finite transit time exists which always exceeds $t_{0\lambda}$. For $\tau\gg t_{0\lambda}$ (small conductivity), one obtains from (2.79) $t_\lambda\approx t_{0\lambda}$, as expected. For $\tau=t_{0\lambda}$, an infinite transit time is reached, and for $\tau<t_{0\lambda}$ the carrier front never reaches the electrode. In the latter case (large conductivity), the excess carriers are neutralized before traversing the dielectric.

Solutions of (2.77) are now obtained by specifying $E(s,t)$ for $t\leq t_\lambda$ and $t\geq t_\lambda$. In the first case, the charge density $\varrho_r(s,t)$ remains zero as long as the ohmic current is supplied by the electrode; thus, (2.75) is at $x=s$

$$\tau\frac{\partial E(s,t)}{\partial t}+E(s,t)=0 \tag{2.80}$$

which yields

$$E(s,t)=E(s,0)\,\mathrm{e}^{-t/\tau}. \tag{2.81}$$

The change in $E(s,t)$ is entirely caused by the gradual compensation of the excess charges due to ohmic conduction. If (2.81) is substituted into (2.77) one obtains

$$\tau\dot{V}(t)+V(t)+\tfrac{1}{2}\mu\tau E^2(s,0)\,\mathrm{e}^{-2t/\tau}=0. \tag{2.82}$$

With the initial value $V(0)=V_0$ at $t=0$ and $E(s,0)=V_0/s$ this has the solution [2.150]

$$\frac{V(t)}{V_0}=e^{-t/\tau}[1-\tfrac{1}{2}(\tau/t_{0\lambda})(1-e^{-t/\tau})]\quad\text{for}\quad t\leq t_{\lambda}.\tag{2.83}$$

Thus, $V(t)$ is given in terms of the observable initial voltage and the material constants τ and μ. For $t=t_\lambda$ one obtains

$$\frac{V(t_{\lambda})}{V_0}=\tfrac{1}{2}\left(\frac{\tau-t_{0\lambda}}{\tau}\right).\tag{2.84}$$

For $t\geq t_\lambda$, a uniform distribution of ϱ_r exists throughout the dielectric [2.150]. Thus

$$V(t)=\tfrac{1}{2}sE(s,t).\tag{2.85}$$

Introducing this into (2.77) yields

$$\tau\dot{V}(t)+V(t)+2\mu\tau\frac{V^2(t)}{s^2}=0,\tag{2.86}$$

which, considering the initial value $V(t_\lambda)$ at $t=t_\lambda$, has the solution

$$\frac{1}{V(t)}=-\frac{2\mu\tau}{s^2}+\left[\frac{2\mu\tau}{s^2}+\frac{1}{V(t_\lambda)}\right]e^{\frac{t-t_\lambda}{\tau}},\quad\text{for}\quad t\geq t_\lambda\tag{2.87}$$

With (2.78, 79) and (2.84) this is [2.150]

$$\frac{V(t)}{V_0}=\tfrac{1}{2}\frac{t_{0\lambda}}{\tau}\frac{e^{-t/\tau}}{1-e^{-t/\tau}},\quad\text{for}\quad t\geq t_\lambda.\tag{2.88}$$

For vanishing conductivity ($g=0$, $\tau=\infty$) the solutions (2.83) and (2.88) can be written [2.144]

$$\frac{V(t)}{V_0}=1-\tfrac{1}{2}\frac{t}{t_{0\lambda}}\quad\text{for}\quad t\leq t_{0\lambda},\tag{2.89}$$

$$=\tfrac{1}{2}\frac{t_{0\lambda}}{t}\quad\text{for}\quad t\geq t_{0\lambda}.\tag{2.90}$$

In a nonconductive dielectric, the voltage decay due to excess-charge currents is thus linear with time up to the transit time at which the voltage has decayed to half the original value, and hyperbolic thereafter. According to (2.89) and (2.78),

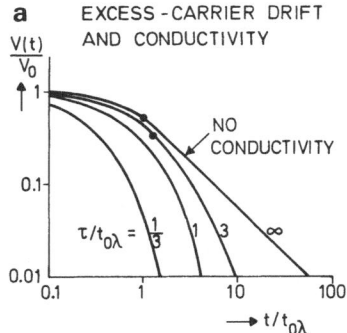
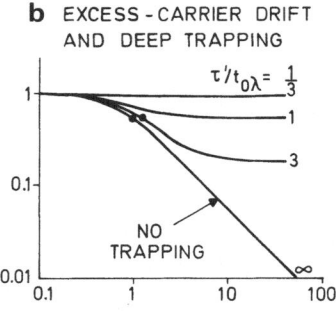

Fig. 2.20a, b. Decay of free-surface voltage of charged dielectric shown in Fig. 2.19 under the influence of excess-carrier drift and ohmic conduction in the absence of trapping (**a**) and with excess-carrier drift and deep trapping without release in the absence of ohmic conduction (**b**). The transit time t_λ, where finite, is marked by dot

the voltage change dV/dt is proportional to V_0^2 for $t \le t_{0\lambda}$. This is due to the fact that the initial decay is proportional to the field *and* the number of charges moving in the field and that both of these parameters are proportional to V_0.

So far, *trapping phenomena* have only been included to the extent that they can be expressed by a trap-modulated mobility. Full consideration of the effect of multiple trapping levels on charge decay leads to a complicated analysis which is beyond the scope of this chapter. In the following, first a discussion of the analytical results of a simple model [2.148] is presented, taking into consideration a finite number of deep traps which hold the trapped carriers indefinitely. These results will later on be compared with numerical data from a model considering trapping with a finite capture time.

The *deep-trap model* [2.148] takes into account a charge initially located at the free surface which spreads into the dielectric maintaining a rectangular distribution of trapped and free carriers. Under the assumption that the free carriers drift through the nonconductive dielectric and fill the deep traps which will hold the carriers indefinitely, the voltage across the sample is given by

$$\frac{V(t)}{V_0} = 1 - \frac{1}{2}\frac{\tau'}{t_{0\lambda}}(1 - e^{-t/\tau'}) \quad \text{for} \quad t \le t'_\lambda, \tag{2.91}$$

$$= \frac{1}{2}\frac{t_{0\lambda}}{\tau'}(1 - e^{-t/\tau'}) \qquad \text{for} \quad t \ge t'_\lambda, \tag{2.92}$$

where τ' is a relaxation time given by $\tau' = \varepsilon_0\varepsilon/\mu e N_t$ with μ representing the mobility of a free carrier and N_t the trap concentration. The transit time t'_λ follows from (2.79) by replacing τ by τ'.

Some model decays for the cases described by (2.83, 88, 91, 92) are plotted in Fig. 2.20. If excess-charge drift occurs in the absence of conductivity and deep trapping ($\tau, \tau' = \infty$), a complete discharge of the sample is observed. The

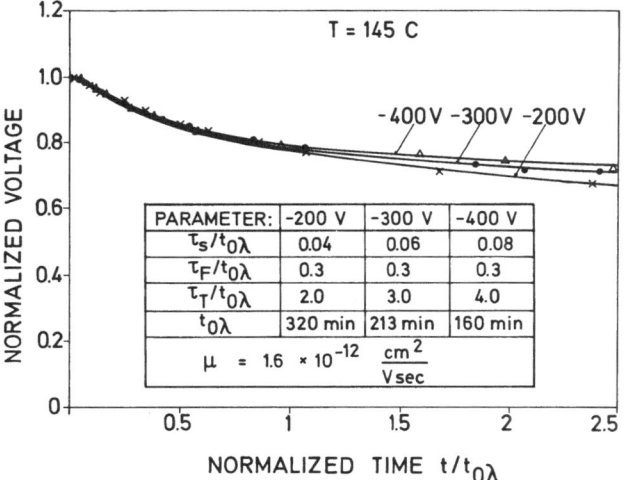

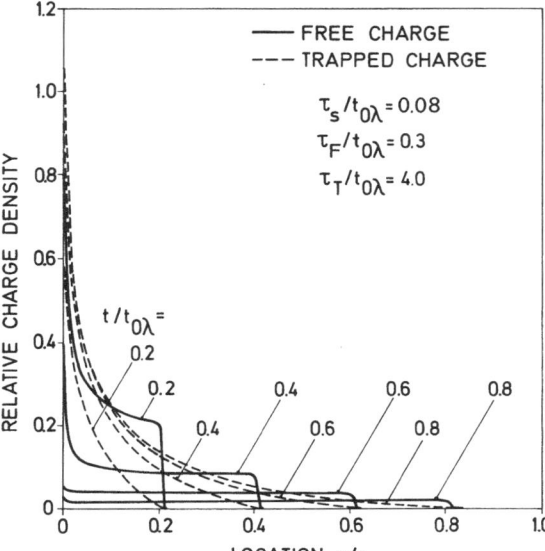

Fig. 2.21. Decay of free-surface voltage of initially surface-charged dielectric under the influence of surface release, carrier drift, and trapping with finite capture time. Experimental data for 25 µm Teflon FEP at 145 °C is also shown [H. von Seggern: J. Appl. Phys. **50**, 7039 (1979)]

Fig. 2.22. Evolution of free and trapped charge densities in an initially surface-charged dielectric under the influence of surface release, excess-carrier drift, and trapping [H. von Seggern: J. Appl. Phys. **50**, 7039 (1979)]

discharge is accelerated for finite conductivities. If the drift is accompanied by deep trapping, only a partial voltage decay is expected. The decay is more pronounced for large values of $\tau'/t_{0\lambda}$, i.e., for smaller values of the trap concentration N_t.

If the *capture time* of a carrier in a deep trap is *finite*, the dielectric will eventually discharge completely. Charge transport occurs in this case due to motion of injected carriers between traps and is characterized by three parameters: the free-carrier mobility μ between traps, the mean free time τ_F

between traps, and the mean capture time τ_T in a trap. A charge-transport model assuming such a carrier transport requires, in addition to suitably modified current and Poisson equations, a rate equation describing the change in trapped-charge density. The set of transport equations, which will not be written out, has been solved numerically under the assumption that the carriers initially located at the free surface are either immediately injected into the bulk of the nonconductive material [2.152] or that they are injected according to $\sigma_s = \sigma_0 \exp(-t/\tau_s)$, where σ_s is the carrier density at the surface and τ_s a suitable time constant.

Analytical results obtained under the assumption of gradual surface injection are depicted in Figs. 2.21, 22 [see H. von Seggern: J. Appl. Phys. **50**, 7039 (1979)]. The former figure shows the voltage across the sample as a function of normalized decay time for three initial surface potentials. Field-independent values of μ, τ_F/E, τ_S, and τ_T are assumed. As will be recognized from the figure, the voltage decay is initially slow with the slope of the curves equal to zero at $t=0$. The reason for this is the finite surface-release time. Later, the decay steepens until most of the carriers are trapped in the volume states with their long capture time. Also shown in Fig. 2.21 are experimental results for Teflon FEP which will be discussed in Sect. 2.6.5.

The spatial distributions of the free and trapped charges, calculated from this model, are shown in Fig. 2.22 for one set of values τ_s, τ_F, and τ_T. Initially, the entire charge is trapped at the surface. As it is injected and moves toward increasing x values, the free-charge cloud spreads out, loosing carriers due to trapping. Thus, the trapped-charge density increases. For $t \leq t_{0\lambda}$, the total charge in the dielectric equals the initial charge. After very long periods of time, the number ratio of free to trapped charges assumes the value τ_F/τ_T.

The mathematical models discussed above are descriptions of very special charge-decay phenomena. In particular, these models can only be applied to initially surface-charged dielectrics shielded from external decay mechanisms and external fields. Even if these conditions are met, actual decay phenomena are usually more complex because of involved trapping effects, field-dependent mobilities and other phenomena. If such complications are avoided by proper choice of materials, temperatures, surface potentials, etc., a successful description of observed voltage decays is possible with these models (see below).

2.6.5 Experimental Results of Real-Charge Decay

Under unshielded conditions, the *external charge decay* due to ion deposition can be very severe [2.153]. This is illustrated in Fig. 2.23, where the charge stability of several shielded Teflon FEP electrets is compared with that of an unshielded sample. The rapid decay on the unshielded sample is attributed to compensation of the electret charges by atomspheric ions attracted by the electret. In a large room, convection processes continuously replenish the reservoir of ions available in the vicinity of the electret. Other experiments on

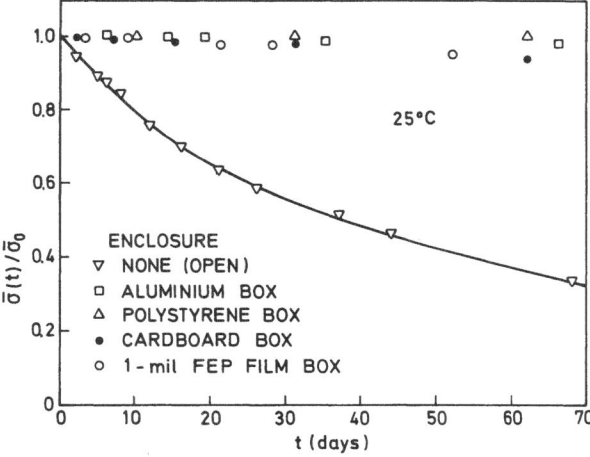

Fig. 2.23. Charge decay at 25 °C for corona-charged 25 µm Teflon FEP electrets differently stored [2.153]

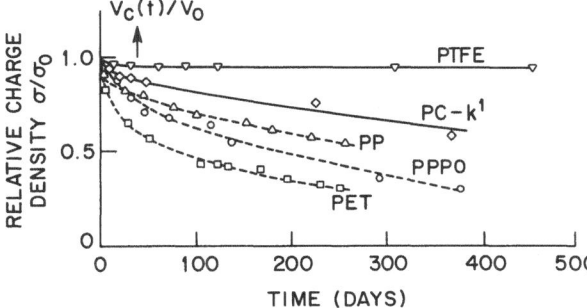

Fig. 2.24. Charge decay in dry atmosphere at room temperature for 50 µm Teflon (PTFE); 25 µm polycarbonate (PC-kl); 20 µm polypropylene (PP); 25 µm poly-2,6-diphenyl-1,4-phenyleneoxide (PPPO) 25 µm Mylar (PET). All electrets thermally charged [2.29]

polypropylene show, however, only a minor effect of air flow on charge decay [1.154]. Since corona-charged electrets were used in both experiments, the better stability in the latter case either indicates that polypropylene is less sensitive to atmospheric ions than FEP, which seems unlikely, or that air ionization or streaming and exposure conditions were different.

If the electret is stored in a small volume, the number of available ions is determined only by the atmospheric generation rate of about 10 ion pairs per $cm^3 \cdot s$. Thus, if all the ions generated in a $1\,cm^3$ volume are deposited on the surface of an electret of $1\,cm^2$ surface having an initial charge density of 10^{-8} C $\cdot cm^{-2}$, total compensation of the electret charge will take a time of 200 years. By making the storage volume small, the external decay can be diminished to the point where internal phenomena determine the loss of charge.

Internal decays at room temperature are shown in Fig. 2.24 for a number of one-sided metallized, thermally charged polymer films. The figure demonstrates the superior charge stability on Teflon PTFE, which was negatively charged in this case. Most of the discussion to follow concerns therefore this material and the similarly stable Teflon FEP.

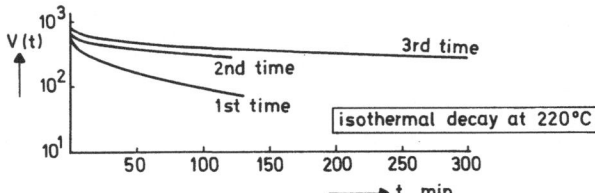

Fig. 2.25. Increase in charge stability of a laminate Teflon FEP-polyimide electret by a combination of thermal depolarization and successive charging. The number of charging processes is indicated [2.158]

The excellent charge stability of Teflon FEP and PTFE electrets is only found on negatively charged samples, while considerably faster decays are observed for positive charges [2.29, 155] (see also Fig. 2.28, below). This is due to the fact that in Teflon, because of its electronegative character, the hole mobility greatly exceeds the electron mobility [2.156] (see also below).

The charge stability of most electrets can be improved by *application of heat during charging*. This has been demonstrated for electron-beam charged [2.65] and for thermally charged [2.29] Teflon electrets. For this material, the charge stability improves with temperature during charging up to temperatures of at least 220 °C [2.29, 157].

The charge stability can also be increased by *annealing after charging*. This process results in an electret with lower charge which is, however, more stable than a comparable nonannealed sample [2.29]. By repeatedly charging and annealing such an electret, the original charge density may be exceeded and at the same time the charge stability improved. Figure 2.25 shows the results of an experiment in which a Teflon FEP-polyimide sample was charged three times and annealed in between as indicated [2.158]. Annealing of FEP electrets after charging yields time constants of the charge decay of about 200 years at room temperature [2.155].

The effect of heating during and after charging is probably due to retrapping of carriers in energetically deeper levels (see also Sect. 3.13.2). The existence of a series of trapping levels of different energetic depth in Teflon and other polymers has been demonstrated by a number of experiments [2.29, 70, 136, 156, 159].

Recent experimental results indicate that even *heating prior to charging* improves electret stability [2.160]. By annealing Teflon FEP films for one hour at 150 °C before corona charging, samples with superior TSC performance (higher-temperature peaks) were obtained. Improved stability also resulted from quenching from 250 °C prior to corona charging (see Sect. 3.13.2). The greater stability is attributed to the presence of deeper traps in the pre-annealed electrets. These are possibly generated by a change in the physical properties of the dielectric (crystalline grain size or degree of crystallinity) due to the temperature application. It is possible that such trap-generation effects are also partly responsible for the better charge stability of electrets heated during or after charging.

The *charging method* seems to have also an effect on the stability of the charge. Generally, because of the above-mentioned effects of heat on charge trapping, forming methods utilizing heat application lead to electrets with improved charge stability. However, even the different methods operating at room temperature will result in differently stable electrets. Electron-beam charged Teflon FEP electrets, for example, have a better charge stability than corona-charged electrets of the same material [2.135]. These differences can be explained in terms of the different degree of *surface- and volume-trap filling* in such samples. While in corona electrets only surface traps are filled, one obtains primarily volume trapping in electron-beam electrets. The different charge stability is thus directly related to the different TSC release temperatures which are 155 °C for surface traps and 200 °C for volume traps (see Table 2.2). The relatively strong low-temperature charge decay on electron-beam electrets, observed in other TSC work (see Fig. 3.68), must be attributed to a particularly large radiation-induced conductivity.

The charge decay in Teflon FEP under various *environmental conditions* is depicted in Fig. 2.26 for a number of thermoelectrets and electron-beam charged electrets [2.29]. Apart from demonstrating again the excellent charge-storage capabilities of this material under normal environmental conditions, the figure also shows the relatively small charge decay under high-temperature–high-humidity exposure. This can be attributed to the hydrophobic behavior of Teflon. The better stability of electron-beam electrets under high humidities is probably due to the protection of deep-seated charges from the atmosphere.

The voltage decay of electron-beam charged Teflon FEP at 150 °C is shown in Fig. 2.27. At this temperature, two exponential decay sections appear (note logarithmic time scale). The two decay phenomena with very different decay rates seem to correspond to the decay regions at about 180 and 240 °C, respectively, found in open-circuit TSC experiments on this material [2.29]. It appears that these phenomena are caused by the existence of two volume-trap levels in this material, as discussed above.

Corona-charged Teflon FEP displays a different voltage decay due to the filling of surface traps (see above). Experimental studies at 145 °C are compared in Fig. 2.21 with analytical results, discussed in Sect. 2.6.4, taking surface injection into consideration. Over the limited range of voltage decays measured, the agreement between experiment and theory is excellent.

Time constants of the voltage decay obtained at elevated temperatures can be shown on *Arrhenius-type plots*, as illustrated in Fig. 2.28. For an activated process, the time constants are on straight lines whose slopes yield the activation energy. The figure indicates the existence of such an activated process for negative and positive carriers in Teflon FEP in the temperature range 80–140 °C. An activation energy of 1.9 eV is derived for both kinds of carriers. Extrapolation of the straight-line section for negative carriers to 25 °C yields time constants in excess of 10^4 years. Since the time constants actually observed at room temperature are significantly shorter, different rate-limiting processes, such as induced conductivity due to ambient radiation, must be effective under these conditions.

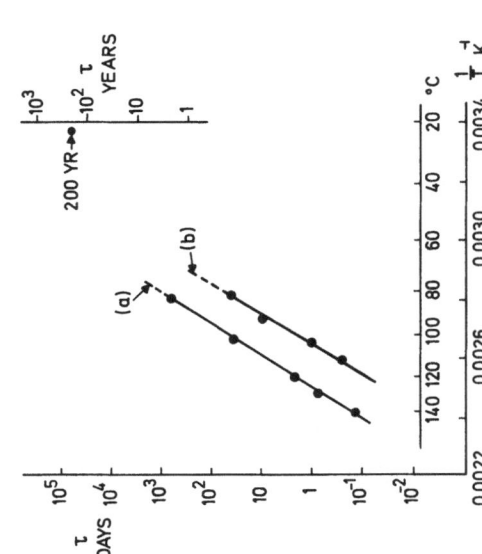

Fig. 2.28. Effective time constant of charge decay as function of temperature for 25 μm Teflon FEP electrets (a) negatively and (b) positively charged and preaged to voltage of about 250 V [2.155]

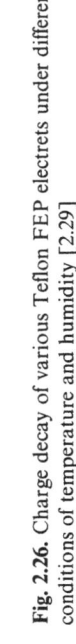

Fig. 2.26. Charge decay of various Teflon FEP electrets under different conditions of temperature and humidity [2.29]

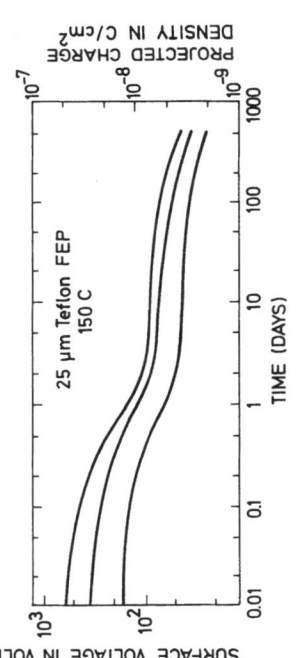

Fig. 2.27. Decay of surface potential of negatively-charged 25 μm Teflon FEP at 150 °C. Samples are aluminized on one side and charged by injection of 20 keV electrons [2.62]

The charge decay in one-sided and two-sided metallized samples is also affected by the *electrodes* if these are neutral or injecting (see also Sects. 3.2 and 4.2.1). In negatively charged Teflon electrets, for example, the injection of positive charges through an electrode will cause relatively rapid charge compensation due to the large mobility of positive carriers. The customary vacuum-deposited gold or aluminum electrodes are blocking contacts to Teflon and other materials [2.161–163]. However, sputtered gold electrodes form neutral or injecting contacts to Teflon [2.162]. Charge injection has also been observed from vacuum-deposited electrodes of gold, silver, copper, and aluminum on polyvinylacetate [2.164].

The *charge decay in polyethylene* is qualitatively different from that in Teflon. In particular, the decay in corona-charged PE at higher initial charge levels is so much faster than at lower levels that the decay curves cross each other [2.165, 166]. Several hypotheses have been advanced to explain this observation. Based on theories assuming partial injection of the carriers initially trapped at the surface [2.148, 167], it was surmised that complete injection occurs at high initial charge densities and partial injection at low densities [2.151], so that eventually more charge is left on the surface in the latter than in the former case. Since bulk charges are subject to faster decay than surface charges, the potential on the highly charged samples is expected to fall ultimately below that on the samples with lower initial charge. However, the physical processes leading to a field-dependent injection of this kind are not yet understood.

More recent investigations [2.168] pointed to the light generated by the corona process as the agent responsible for the crossovers in PE. However, later experiments indicated that on samples subjected to very rapid charging (25 ms) the crossover effect is totally absent [2.168], while other experiments show that on samples exhibiting pronounced crossover effects irradiation with light orders of magnitude more intense than the corona light does not cause an accelerated decay [2.169]. Recently, crossovers were also found on liquid-charged PE and on thin (6 µm) PTFE [2.169]. While these new findings do not offer an explanation of the effect, they show it to be more common than previously believed.

The discussion of charge decay cannot be closed without a few remarks on *lateral spreading* of surface and volume charges. Generally, the field in thin-film electrets points in a direction perpendicular to the surfaces with only minor components along the surfaces. Thus the field does not support charge spreading. Correspondingly, on FEP and PE electrets no significant spreading of charge has been observed [2.168, 170–172]. Actually, on FEP samples charged with a high-resolution pattern by electron-beam irradiation the resolution of about 10 µm originally achieved did not noticeably deteriorate after 3 months of storage [2.171]. Even in cases where the field was along the surface, no spreading was observed in this material (see Table 3.11). This has to be attributed to the high surface resistivity of FEP. In contrast, charge spreading has been seen on materials with lower surface resistance, such as *n*-

octadecane [2.163]. Also, diffusion in the bulk of PET samples appears to cause considerable charge spreading [2.173].

2.6.6 Conduction Phenomena

As discussed above, internal charge-decay processes in dielectrics are governed by *conduction phenomena* which, according to (2.37), depend on carrier mobility, carrier concentration, injection conditions at the electrodes, etc. In substances capable of quasi-permanent charge storage, trapping centers are present so that the mobilities are trap-modulated with conduction processes similarly affected. Intrinsic charge carriers may be supplemented by carriers injected across the electrode–dielectric interface.

In the following, data on the parameters affecting conduction in some pertinent dielectrics will be reviewed. However, experimental studies of conduction phenomena are considerably more complicated than investigations of charge decay since the former require additionally information about injection conditions at the electrodes and are more sensitive to the time frame in the sense that one has to distinguish between transient and steady-state effects. Also, most conduction or mobility experiments do not permit a simultaneous probing of the dielectric with respect to charge distributions. Problems of this kind together with material shortcomings (lack of structural uniformity and purity of many dielectrics) are responsible for the presently very unsatisfactory state of our knowledge about conduction phenomena. We shall therefore only review a few aspects of this field and refer to the literature for a more detailed discussion (see the review in [2.161]).

The *carrier mobility* is affected by extended (delocalized) states, by shallow traps and by deep traps (the traps are localized states, see Fig. 2.16) in different ways. An electron near the bottom of the conduction band moves between extended states, which are above the mobility edge, by quantum-mechanical hopping. This process requires no thermal activation and leads to relatively high mobilities of about $10 \, cm^2 \cdot V^{-1} \, s^{-1}$ [2.174]. If the electron is captured by a shallow trap located below the mobility edge it needs thermal energy to perform a thermally activated hopping process. The corresponding mobilities are now orders of magnitude smaller (about $10^{-3} \, cm^2 \cdot V^{-1} \, s^{-1}$). Finally, a carrier located in a deep trap experiences a very long capture time. If most available electrons are deeply trapped, the mobility is trap-modulated and very small (10^{-10}–$10^{-17} \, cm^2 \cdot V^{-1} \, s^{-1}$).

A number of methods depending on the evaluation of conductivity, transit time, charge decay, and thermally stimulated currents have been used to determine carrier mobilities [2.18, 61, 136, 156, 175–187]. For materials with shallow and deep traps, any combination of the above-mentioned mobilities can be measured, depending on the time scale of the experiment. This is obvious from the mobility data of PET. Within time periods of 20–30 ns after X-ray excitation, electron mobilities of about $10^{-3} \, cm^2 \cdot V^{-1} \, s^{-1}$ were found from conductivity measurements and from an estimate of the number of carriers [2.176]. Transit-time measurements over periods from μs to ms yield considerably smaller mobilities of about 10^{-6}–10^{-7} cm$^2 \cdot$V^{-1}s^{-1} [2.177, 178]. The decrease is believed to be due to hindrance by shallow traps. Finally, charge decay results on this material over periods of minutes or longer yield "steady-state" mobilities of the order of 10^{-14} cm$^2 \cdot$V^{-1}s^{-1} (unpubl.). These values are determined by carrier interaction with deep traps.

In Teflon, transit time measurements over periods on the order of 1 μs yield electron mobilities of 5×10^{-5} cm$^2 \cdot$V^{-1}s^{-1} and hole mobilities of 5×10^{-4} cm$^2 \cdot$V^{-1}s^{-1} [2.179]. Judging from the PET data, these values probably do not refer to free carriers but are already determined by some hindrance from shallow trapping centers. This is also suggested by the observed exponential temperature dependence (see below). The steady-state trap-modulated mobility in this material is of the order of 10^{-17} cm$^2 \cdot$V^{-1}s^{-1} for electrons [2.61], 10^{-9} cm$^2 \cdot$V^{-1}s^{-1} for holes in the bulk of the material [2.180] and 10^{-11} cm$^2 \cdot$V^{-1}s^{-1} for holes activated within the surface region as defined in Table 2.2 [2.156]. The relatively large hole mobility in the bulk of Teflon can be detected by a transit-time technique. This is operated by generating an electron–hole plasma in the surface region of the material by means of an electron beam and extracting the holes by a proper bias [2.180].

While the temperature dependence of the free mobility in insulators is weak (typically given by T^{-1} to T^{-2}) that of the trap-modulated mobility is exponential since it is controlled by carrier activation [2.175]. Arrhenius plots have demonstrated this behavior in the electret-forming materials FEP and PET [2.156, 179] and some other polymers [2.18, 181]. A field-dependence of the trap-modulated mobility has also been observed in some polymers [2.183–185]. However, experiments in FEP with electrons up to fields of 4×10^5 V$\cdot$cm^{-1} [2.152] and in PET with both types of carriers up to fields in excess of 10^6 V$\cdot$cm^{-1} [2.179] have failed to show such an effect.

The *Schubweg* of carriers in PET and FEP has been estimated from results of charge-penetration measurements. In PET, an electron Schubweg of 6 μm at 8×10^5 V$\cdot$cm^{-1} was found, and the values are proportional to the field, as expected [2.176]. In FEP, measurements of the charge centroid of electrons injected at low energies indicate a Schubweg of 0.1 μm at a few 10^5 V$\cdot$cm^{-1} [2.136]. Thus, the available evidence favors the assumption of fast retrapping of electrons in the customarily used 12–50 μm thick FEP films used in electret research. However, holes in FEP have Schubwegs of about 100 μm at fields of 10^5 V$\cdot$cm^{-1} [2.180].

The *ohmic conductivity* g of a dielectric, as introduced in (2.37), is given by $g = e(n_+\mu_+ + n_-\mu_-)$. If μ_+ and μ_- represent trap-modulated mobilities, n_+ and n_- are the total concentrations of free and trapped holes and electrons, respectively. As shown above, the mobilities μ_+ and μ_- are very small in electret-forming polymers. Similarly, because of the large band gap in such materials (see Fig. 2.16), the concentrations n_+ and n_- of intrinsic carriers (trapped or free) are also small as long as radiation effects are absent (for a discussion of the effects of radiation and light see Chap. 4 and [2.188]). Thus, ohmic (intrinsic) conduction must be excluded as a source of conduction currents in these polymers. For example, the Teflon materials PTFE and FEP show steady-state conductivities unmeasurable by conventional techniques. Charge-decay measurements giving time constants of $\tau \approx 200$ years at room temperature [2.155] imply $g = \varepsilon/\tau = 3 \times 10^{-23}\ \Omega^{-1}\,cm^{-1}$, assuming that the decay is due to ohmic conduction. Since the decay is mostly due to excess-carrier drifts, this number has to be considered an upper limit. Conductivities of this order can be attributed to ionization due to ambient radiation [2.189].

In many insulators devoid of intrinsic carriers, however, carriers may be supplied by *electrode injection* (see also above). If more carriers are supplied than can be transported through the dielectric, the currents are space-charge limited; in the reverse case they are electrode limited. Both types of limitation appear to be of importance in electroded polymers [2.161].

Space-charge limitation occurs only when a sufficiently large carrier supply is available from an electrode. Conduction studies on two-sided metallized samples with an applied field have shown that on polyolefins the currents are space-charge limited [2.190, 191] but subject to a hopping process with a distribution of relaxation times [2.149]. Space-charge limited currents are modified by the presence of high fields. In this case the field gradient causes a lowering of the trap barriers even if the traps are neutral when unoccupied [2.192] (Poole–Frenkel effect [2.193]). This yields an additional dependence similar to that for Schottky emission given by (2.57), but with β replaced by 2β. Evidence of Poole–Frenkel conduction has been found in some dielectrics [2.194]. Another kind of limitation occurs in the case of very small mobilities (due to charge trapping, for example), encountered in some of the electret-forming materials. In this case, the injected charge may form a barrier close to the electrode which causes a t^{-n} dependence of the current and eventually prohibits further injection, as seen from (2.58). This mechanism could be of importance in PET [2.195, 196]. Similarly, trapping effects in the bulk may cause a decrease of the current in polymers such as PE [2.197].

Indications are that in some metallized dielectrics currents are *electrode-limited*. In many cases, the electrodes appear to be blocking and injection is completely absent (see p. 71). If injection occurs, it is in the simplest case controlled by Schottky emission according to (2.57). Its field dependence is thus difficult to distinguish from Poole–Frenkel type conduction. Furthermore, the effect is frequently modified by trapping at the interface which yields a time dependence similar to that for bulk trapping [2.31]. Due to the similarity of the observable field and time dependencies of bulk and electrode effects a direct distinction is often not possible.

Other experiments with partially penetrating electron beams on PET and mica show currents orders of magnitude greater than normal electrode currents [2.161, 198–200] and thus indicate electrode limitation in this material. A barrier-type limitation, as discussed above, is not observed.

Although this indicates that answers have been obtained to some problems related to conduction phenomena in dielectrics, the state of our knowledge about this subject is, on the whole, presently still very unsatisfactory.

Acknowledgements. The author is grateful to Professor Bernhard Gross, to whom the book is dedicated, for many stimulating discussions on the topics of this chapter. He is also indebted to his colleagues at Bell Laboratories, Dr. D. A. Berkley and Mr. J. E. West, for numerous helpful comments on the manuscript, and to Mrs. E. Kubli deceased for her help with the literature search.

References

2.1   A. N. Gubkin: Sov. Phys.-Tech. Phys. **2**, 1813 (1958)
2.2   G. M. Sessler: J. Appl. Phys. **43**, 405 (1972)
2.3   G. Morgenstern: Appl. Phys. **9**, 209 (1976)
2.4   G. Dreyfus, J. Lewiner: J. Appl. Phys. **46**, 4357 (1975); Phys. Rev. B **14**, 5451 (1976)
2.5   G. Morgenstern: Appl. Phys. **11**, 371 (1976)
2.6   O. D. Jefimenko: Proc. W. Va. Acad. Sci. **40**, 345 (1968); IEEE Trans. ASSP-**23**, 497 (1975)
       T. B. Jones: IEEE Trans. ASSP-**22**, 141 (1974); ASSP-**23**, 498 (1975)
       J. F. Hoburg, J. R. Melcher: IEEE Trans. ASSP-**23**, 500 (1975)
2.7   B. Gross: J. Chem. Phys. **17**, 866 (1949)
2.8   P. D. Southgate: Appl. Phys. Lett. **28**, 250 (1976)
       D. K. Das Gupta, K. Doughty: J. Phys. D **11**, 2415 (1978)
2.9   B. Gross: Br. J. Appl. Phys. **1**, 259 (1950)
2.10   J. Roos: J. Appl. Phys. **40**, 3135 (1969)
2.11   P. S. H. Henry: Br. J. Appl. Phys. **4** (Suppl. 2), 31 (1953)
2.12   H. Bauser: In *Elektrostatische Aufladung* (Verlag Chemie, Weinheim 1974) pp. 11–28
2.13   D. Seanor: In *Electrical Properties of Polymers*, ed. by K. C. Frisch, A. V. Patsis (Technomic, Westport 1972) pp. 37–58
2.14   I. I. Inculet: In *Electrostatics and Its Applications*, ed. by A. D. Moore (Wiley, New York 1973) pp. 86–114
2.15   H. Bauser: Het Ingenieursblad **44**, 321 (1975)
2.16   H. Fuhrmann: J. Electrostat. **4**, 109 (1978)
2.17   W. R. Harper: *Contact and Frictional Electrification* (Oxford University Press, London 1967)
2.18   D. K. Davies: In *Static Electrification* (Institute of Physics, London 1967) pp. 29–36; Br. J. Appl. Phys. (J. Phys. D) **2**, 1533 (1969)
2.19   I. I. Inculet, E. P. Wituschek: In *Static Electrification* (Institute of Physics, London 1967) pp. 37–43
2.20   A. Chowdry, C. R. Westgate: J. Phys. D. **7**, 713 (1974)
2.21   H. Bauser, W. Klöpffer, H. Rabenhorst: *Adv. Stat. Electric.*, Vol. I (Auxilia, Brussels 1970) pp. 2–9
2.22   D. K. Davies: In *1972 Annual Report, Confer. Electric. Insul. and Diel. Phenom.* (NAS, Washington 1973) pp. 1–8
2.23   H. J. Wintle: J. Phys. D **7**, L 128 (1974)
2.24   A. Chowdry, C. R. Westgate: J. Phys. D **7**, L 149 (1974)
2.25   V. M. Fridkin, I. S. Zheludev: *Photoelectrets and the Electrophotographic Process* (Consultants Bureau, New York 1961)
       V. M. Fridkin: *Photoferroelectrics*, Springer Series in Solid State Science, Vol. 9 (Springer, Berlin, Heidelberg, New York 1979)
2.26   B. Gross: *Charge Storage in Solid Dielectrics* (Elsevier, Amsterdam 1964)
2.27   M. Latour: Ann. Phys. (Paris) **7**, 115 (1972)
2.28   P. K. C. Pillai, K. Jain, V. K. Jain: Phys. Status Solidi **13**, 341 (1972)
2.29   J. van Turnhout: *Thermally Stimulated Discharge of Polymer Electrets* (Elsevier, Amsterdam 1975)

2.30   R.G.Vyverberg: "Charging Photoconduction Surfaces", in *Xerography and Related Processes*, ed. by J.H.Dessauer, H.E.Clark (Focal Press, London 1965) pp. 201–216

2.31   P.Röhl, P.Fischer: Kolloid Z. Z. Polym. **251**, 947 (1973)
       H.J.Wintle: Solid State Electron. **18**, 1039 (1975); IEEE Trans. EI-**12**, 424 (1977)

2.32   J.Euler: Elektrotech. Z. A **90**, 600 (1969)

2.33   A.C.Lilly, L.L.Stewart, R.H.Henderson: J. Appl. Phys. **41**, 2007 (1970); also
       E.Sacher: J. Macromol. Sci., Phys. B**6**, 365 (1972)

2.34   Y.Asano, T.Suzuki: Jpn. J. Appl. Phys. **11**, 1139 (1972)

2.35   G.M.Sessler, J.E.West: J. Electrochem. Soc. **115**, 836 (1968)

2.36   M.M.Perlman, C.W.Reedky: J. Electrochem. Soc. **115**, 45 (1968)

2.37   A.N.Gubkin, T.S.Yegorova, L.M.Kokorin, N.Y.Zitser: Vysokomol soyed A**12**, 602 (1970)

2.38   E.Fukada, T.Sukarai: Polym. J. **2**, 656 (1971)

2.39   M.Campos, S.Mascarenhas, G.Leal Ferreira: Phys. Rev. Lett. **27**, 1432 (1971)

2.40   E.B.Podgorsak, P.R.Moran: Science **179**, 380 (1973)

2.41   B.Gross, L.F.Denard: Phys. Rev. **67**, 253 (1945)

2.42   D.K.Das Gupta, K.Joyner: J. Phys. D **9**, 2041 (1976)

2.43   C.S.Bhatnagar: Indian J. Pure Appl. Phys. **2**, 331 (1964); **4**, 355 (1966)
       K.L.Khare, C.S.Bhatnagar: Indian J. Pure Appl. Phys. **8**, 700 (1970)

2.44   R.W.Tyler, J.H.Webb, W.C.York: J. Appl. Phys. **26**, 61 (1955)

2.45   C.F.Carlson: US Patent 2,588,699 (1952)

2.46   M.M.Shahin: J. Appl. Opt., Suppl. 3 (Electrophotography, 1969) pp. 106–110

2.47   D.W.Vance: In *1970 Annu. Rep. Conf. Electr. Insul. Dielectr. Phenom.* (NAS, Washington D.C. 1971) pp. 1–7

2.48   M.Ieda, G.Sawa, U.Shinahara: Electr. Eng. Jpn. **88**, 67 (1968)
       G.Sawa, D.C.Lee, M.Ieda: Jpn. J. Appl. Phys. **14**, 643 (1975)

2.49   R.A.Moreno, B.Gross: J. Appl. Phys. **47**, 3397 (1976)

2.50   R.A.Creswell, M.M.Perlman: J. Appl. Phys. **41**, 2365 (1970)

2.51   B.Gross, G.M.Sessler, J.E.West: J. Appl. Phys. **46**, 4674 (1975)

2.52   J.H.Dessauer, H.E.Clark (eds.): *Xerography and Related Processes* (Focal Press, London 1965)

2.53   G.M.Sessler, J.E.West: J. Appl. Phys. **43**, 922 (1972)

2.54   K.Ikezaki, I.Fujita, K.Wada, J.Nakamura: J. Electrochem. Soc. **121**, 591 (1974)

2.55   H.Seiwatz, J.J.Brophy: In *1965 Annu. Rep. Conf. Electr. Insul. Dielectr. Phenom.* (NAS, Washington D.C. 1966) pp. 1–3

2.56   R.E.Collins: AWA Tech. Rev. **15**, 53 (1973)

2.57   P.W.Chudleigh: Appl. Phys. Lett. **21**, 547 (1972); J. Appl. Phys. **47**, 4475 (1976)

2.58   S.Engelbrecht: J. Appl. Phys. **45**, 3421 (1974)

2.59   G.M.Sessler, J.E.West: Appl. Phys. Lett. **17**, 507 (1970)

2.60   R.F.Wills, D.K.Skinner: Solid State Commun. **13**, 685 (1973)
       S.Gair: Proc. IEEE Annu. Conf. Nucl. Space Radiat. Eff. (1974) p. 177

2.61   B.Gross, G.M.Sessler, J.E.West: J. Appl. Phys. **45**, 2841 (1974)

2.62   G.M.Sessler, J.E.West: J. Electrostatics **1**, 111 (1975)

2.63   G.M.Sessler, J.E.West: J. Acoust. Soc. Am. **53**, 1589 (1973)

2.64   R.E.Collins: Proc. IREE **34**, 381 (1973)

2.65   M.M.Perlman, S.Unger: Appl. Phys. Lett. **24**, 579 (1974)

2.66   B.Gross: J. Electrostatics **1**, 125 (1975)

2.67   P.V.Murphy, S.C.Ribeira, F.Milanez, R.J.de Moraes: J. Chem. Phys. **38**, 2400 (1963)

2.68   D.E.Fields, P.R.Moran: Phys. Rev. Lett. **29**, 721 (1972)

2.69   E.B.Podgorsak, P.R.Moran: Appl. Phys. Lett. **24**, 580 (1974)

2.70   B.Gross, G.M.Sessler, J.E.West: J. Appl. Phys. **47**, 968 (1976)

2.71   B.Gross: Z. Phys. **155**, 479 (1959); J. Appl. Phys. **36**, 1635 (1965); IEEE Trans. NS-**25**, 1048 (1978)

2.72   G.M.Sessler, J.E.West: Polym. Lett. **7**, 367 (1969)

2.73   J.R.Freeman, H.P.Kallmann, M.Silver: Rev. Mod. Phys. **33**, 553 (1961)

2.74   P.K.C.Pillai, S.K.Arya: Solid State Electron. **15**, 1245 (1972)
2.75   R.Andreichin: J. Electrostatics **1**, 217 (1975)
2.76   H.Kallmann, B.Rosenberg: Phys. Rev. **97**, 1596 (1955)
2.77   J.van Turnhout: In *Advances in Static Electricity*, Vol. 1, ed. by W.de Geest (Auxilia, Brussels 1971) pp. 56–81
2.78   P.K.C.Pillai, V.K.Jain: J. Sci. Ind. Res. **29**, 270 (1970)
2.79   M.Oshiki, E.Fukada: Jpn. J. Appl. Phys. **15**, 43 (1976)
       M.Tamura, S.Hagiwara, S.Matsumata, N.Ono: J. Appl. Phys. **48**, 513 (1977)
2.80   A.M.Glass: J. Appl. Phys. **40**, 4699 (1969)
2.81   M.G.Broadhurst, C.G.Malmberg, F.I.Mopsik, W.P.Harris: In *Electrets, Charge Storage and Transport in Dielectrics*, ed. by M.M.Perlman (Electrochemical Society, Princeton 1973) pp. 492–504
2.82   H.Saito: Jpn. J. Appl. Phys. **4**, 886 (1965)
2.83   J.Dow, S.V.Nablo: IEEE Trans. NS-**14**, 231 (1967)
2.84   J.D.Cross, C.Smalley: J. Phys. E **2**, 633 (1969)
2.85   J.H.McFee, J.G.Bergman, G.R.Grane: IEEE Trans. SU-**19**, 305 (1972)
2.86   J.G.Bergman, J.H.McFee, G.R.Crane: Appl. Phys. Lett. **18**, 203 (1971)
2.87   B.Gross: Am. J. Phys. **12**, 324 (1944)
2.88   H.Krämer, D.Messner: Kunststoffe **54**, 696 (1964)
2.89   D.K.Davies: J. Sci. Instrum. **44**, 521 (1967)
       D.K.Davies: In *Elektrostatische Aufladung* (Verlag Chemie, Weinheim 1974) pp. 45–52
2.90   T.R.Foord: J. Phys. E **2**, 411 (1969)
       H.J.Wintle: J. Phys. E **3**, 334 (1970)
2.91   G.M.Sessler, J.E.West: Rev. Sci. Instrum. **42**, 15 (1971)
2.92   W.A.Zisman: Rev. Sci. Instrum. **3**, 367 (1932)
       L.A.Freedman, L.A.Rosenthal: Rev. Sci. Instrum. **21**, 896 (1950)
2.93   G.M.Sessler, J.E.West: Proc. 4th Int. Cong. Acoust., Copenhagen (1962) paper N 55
2.94   C.W.Reedyk, M.M.Perlman: J. Electrochem. Soc. **115**, 49 (1968)
2.95   G.Dreyfus, J.Lewiner: J. Appl. Phys. **45**, 721 (1974)
       D.Legros, J.Lewiner: J. Acoust. Soc. Am. **53**, 1663 (1973)
2.96   P.A.Thiessen, A.Winkel, K.Hermann: Phys. Z. **37**, 511 (1936)
2.97   K.Antenen: Z. Angew. Math. Phys. **6**, 478 (1955)
2.98   B.Gross: In *Static Electrification* 1971 (Institute of Physics, London) pp. 33–43
2.99   D.K.Walker, O.Jefimenko: In *Electrets, Charge Storage and Transport in Dielectrics*, ed. by M.M.Perlman (Electrochemical Society, Princeton 1973) pp. 455–461
2.100 M.Latour: J. Phys. Lett. **41**, L35 (1980)
2.101 B.Gross, J.Dow, S.V.Nablo: J. Appl. Phys. **44**, 2459 (1973)
2.102 H.Seiwatz: Bull. Am. Phys. Soc. **8**, 539 (1963)
2.103 G.M.Sessler, J.E.West: In *1970 Annu. Rep. Conf. Electr. Insul. Dielectr. Phenom.* (NAS, Washington, D. C. 1971) pp. 8–16
       G.M.Sessler: J. Appl. Phys. **43**, 408 (1972)
2.104 W.Moore, M.Silver: J. Chem. Phys. **33**, 1671 (1960)
2.105 A.D.Tavares: J. Chem. Phys. **59**, 2154 (1973)
2.106 R.E.Collins: Appl. Phys. Lett. **26**, 675 (1975); J. Appl. Phys. **47**, 4804 (1976); Rev. Sci. Instrum. **48**, 83 (1977)
       B.Andreß, P.Fischer, P.Röhl: Prog. Colloid Polym. Sci. **62**, 141 (1977)
2.107 A.S.De Reggi, C.M.Guttman, F.I.Mopsik, G.T.Davis, M.G.Broadhurst: Phys. Rev. Lett. **40**, 413 (1978);
       G.T.Davis, A.S.De Reggi, M.G.Broadhurst: In *1977 Annu. Rep. Conf. Electr. Insul. Dielectr. Phenom.* (NAS, Washington, D. C. 1978)
2.108 H. von Seggern: Appl. Phys. Lett. **33**, 134 (1978)
2.109 P.Laurenceau, G.Dreyfus, J. Lewiner: Phys. Rev. Lett. **38**, 46 (1977)
       P.Laurenceau, J.Ball, G.Dreyfus, J.Lewiner: C. R. Acad. Sci. Ser. B**283**, 135 (1976)
2.110 G.M.Sessler, J.E.West, D.A.Berkley, G.Morgenstern: Phys. Rev. Lett. **38**, 368 (1977)

2.111 G. M. Sessler: In *1978 Annu. Rep. Conf. Electr. Insul. Dielectr. Phenom.* (NAS, Washington, D. C. 1978) pp. 3–10

2.112 L. Badian, B. Ai, R. Lacoste, C. Mayoux: C. R. Acad. Sc. Paris **261**, 2181 (1965)

2.113 L. A. Harrah: Appl. Phys. Lett. **17**, 421 (1970)

2.114 G. Pfister, M. A. Abkovitz: J. Appl. Phys. **45**, 1001 (1974)

2.115 J. van Turnhout: Personal communication (see also Sect. 3.4)

2.116 S. W. S. McKeever, D. M. Hughes: J. Phys. D**8**, 1520 (1975)

2.117 S. Bhoraskar, R. Abe: Jpn. J. Appl. Phys. **15**, 1471 (1976); M. Legrand, G. Dreyfus, J. Lewiner: J. Phys. Paris **38**, L439 (1977)

2.118 J. P. Luongo: J. Polym. Sci. A-2 **10**, 1119 (1972) M. Latour: Polymer **18**, 278 (1977)

2.119 R. G. Kepler: In *Electrophotogr.; 2nd Int. Conf.*, ed. by D. R. White (Society of Photographic Scientists and Engineers, 1974) pp. 167–170

2.120 M. Tamura, K. Ogasawara, N. Ono, S. Hagiwara: J. Appl. Phys. **45**, 3768 (1974)

2.121 N. Murayama, K. Nakamura, H. Obara, M. Segawa: Ultrasonics **14**, 15 (1976)

2.122 A. Rose: RCA Rev. **12**, 362 (1951)

2.123 J. F. Fowler: Proc. R. Soc. London A**236**, 464 (1956)

2.124 R. A. Creswell, M. M. Perlman, M. A. Kabayama: In *Dielectric Properties of Solids*, ed. by F. E. Karasz (Plenum, New York 1972) pp. 295–312

2.125 J. van Turnhout: In *Electrets, Charge Storage and Transport in Dielectrics*, ed. by M. M. Perlman (Electrochemical Society, Princeton 1973) pp. 230–251

2.126 T. L. Rokoske, P. P. Budenstein: In *Electrets, Charge Storage and Transport in Dielectrics*, ed. by M. M. Perlman (Electrochemical Society, Princeton 1973) pp. 84–95

2.127 J. D. Brodribb, D. M. Hughes, T. J. Lewis: In *Electrets, Charge Storage and Transport in Dielectrics*, ed. by M. M. Perlman (Electrochemical Society, Princeton 1973) pp. 177–187

2.128 R. J. Fleming, L. F. Pender: J. Electrostat. **3**, 139 (1977)

2.129 N. F. Mott, E. A. Davis: *Electronic Processes in Non-Crystalline Materials* (Clarendon Press, Oxford 1971)

2.130 H. Bauser: Kunststoffe **62**, 192 (1972)

2.131 R. H. Partridge: J. Polym. Sci. A**3**, 2817 (1965)

2.132 D. K. Davies, P. J. Lock: J. Electrochem. Soc. **120**, 266 (1973)

2.133 M. M. Perlman: J. Electrochem. Soc. **119**, 892 (1972)

2.134 H. J. Wintle: J. Acoust. Soc. Am. **53**, 1578 (1973)

2.135 H. von Seggern: J. Appl. Phys. **50**, 2817 (1979)

2.136 G. M. Sessler: In *International Symposium on Electrets and Dielectrics*, ed. by M. S. de Campos (Academia Brasil Ciencas, Rio de Janeiro 1977) pp. 321–335

2.137 R. Williams, M. H. Woods: J. Appl. Phys. **44**, 1026 (1973)

2.138 B. Gross, R. J. de Moraes: J. Chem. Phys. **37**, 710 (1962)

2.139 B. Gross, G. M. Sessler, J. E. West: J. Appl. Phys. **48**, 4303 (1977)

2.140 J. R. Young: J. Appl. Phys. **27**, 1 (1956) T. Matsukawa, R. Shimizu, K. Harada, T. Kato: J. Appl. Phys. **45**, 733 (1974)

2.141 D. A. Berkley: J. Appl. Phys. **50**, 3447 (1979)

2.142 G. W. Day, C. A. Hamilton, R. L. Peterson, R. J. Phelan, Jr., L. O. Mullen: Appl. Phys. Lett. **24**, 456 (1974) H. Sussner, K. Dransfeld: J. Polym. Sci. **16**, 529 (1978)

2.143 A. Many, G. Rakavy: Phys. Rev. **126**, 1980 (1962)

2.144 A. Reiser, M. W. B. Lock, J. Knight: Trans. Faraday Soc. **65**, 2168 (1969)

2.145 J. H. Calderwood, B. K. P. Scaife: Philos. Trans. R. Soc. London **269**, 217 (1971)

2.146 G. F. L. Ferreira, B. Gross: J. Nonmetals **1**, 129 (1973)

2.147 H. Seki, I. P. Batra: J. Appl. Phys. **42**, 4207 (1971)

2.148 H. J. Wintle: J. Appl. Phys. **43**, 2927 (1972); Thin Solid Films **21**, 83 (1974)

2.149 H. Scher, E. W. Montroll: Phys. Rev. B**12**, 2455 (1975) G. Pfister, H. Scher: Phys. Rev. B**15**, 2062 (1977)

2.150 K. K. Kanazawa, I. P. Batra, H. J. Wintle: J. Appl. Phys. **43**, 719 (1972)

2.151 T.J.Sonnonstine, M.M.Perlman: J. Appl. Phys. **46**, 3975 (1975)
2.152 P.W.Chudleigh: J. Appl. Phys. **48**, 4591 (1977)
2.153 E.W.Anderson, L.L.Blyler, G.E.Johnson, G.L.Link: In *Electrets, Charge Storage and Transport in Dielectrics*, ed. by M.M.Perlman (Electrochemical Society, Princeton 1973) pp. 424–435
2.154 J. van Turnhout, C. van Bochove, G.J.van Veldhuizen: Staub-Reinhalt. Luft **36**, 36 (1976)
2.155 P.W.Chudleigh, R.E.Collins, G.D.Hancock: Appl. Phys. Lett. **23**, 211 (1973)
2.156 G.M.Sessler, J.E.West: J. Appl. Phys. **47**, 3480 (1976)
2.157 K.Ikezaki, K.Wada, J.Fujita: J. Electrochem. Soc. **122**, 1356 (1975)
2.158 J. van Turnhout: J. Electrostat. **1**, 147 (1975)
2.159 M.M.Perlman (ed.): *Electrets, Charge Storage, and Transport in Dielectrics* (Electrochemical Society, Princeton, N. J. 1973)
2.160 K.Ikezaki, M.Hattori, Y.Arimoto: Jpn. J. Appl. Phys. **16**, 863 (1977)
2.161 T.J.Lewis: In 1976 Ann. Rpt. Conf. Elec. Insul. Diel. Phen. (NAS, Washington 1978) pp. 533–561
2.162 G.M.Sessler, J.E.West, R.W.Ryan, H.Schonhorn: J. Appl. Polym. Sci. **17**, 3199 (1973)
2.163 E.A.Baum, T.J.Lewis, R.Toomer: J. Phys. D**11**, 703 (1978)
2.164 P.C.Mehendru, K.Jain, P.Mehendru: J. Phys. D**11**, 1431 (1978)
2.165 M.Ieda: Jpn. J. Appl. Phys. **9**, 727 (1970)
2.166 I.B.Jordan: J. Electrochem. Soc. **122**, 290 (1975)
2.167 I.P.Batra, K.K.Kanazawa, B.H.Schechtman, H.Seki: J. Appl. Phys. **42**, 1124 (1971)
2.168 E.A.Baum, T.J.Lewis, R.Toomer: J. Phys. D**10**, 487/2525 (1977)
2.169 G.M.Sessler: In *1978 Annu. Rpt. Conf. Electr. Insul. Dielectr. Phenom.* (NAS, Washington, D. C. 1978) pp. 3–10
2.170 G.M.Sessler, J.E.West: In *Electrophotogr. 2nd Int. Conf.*, ed. by D.R.White (Society of Phot. Scientists and Engineers, 1974) pp. 147–151
2.171 J.Feder: J. Appl. Phys. **47**, 1741 (1976)
2.172 R.J.Atkinson, R.J.Fleming: J. Phys. D**9**, 2027 (1976)
2.173 E.A.Baum, T.J.Lewis, R.Toomer: J. Phys. D**11**, 963 (1978)
2.174 N.Mott: Electron. Power 321 (1973)
2.175 W.E.Spear: J. Noncryst. Sol. **1**, 197 (1969)
2.176 R.C.Hughes: In *Electrophotography, Second Internat. Conf.*, ed. by D.R.White (Soc. of Phot. Scient. and Engineers, Washington, D. C. 1974) pp. 147–151
2.177 E.H.Martin, J.Hirsch: J. Appl. Phys. **43**, 1001 (1972)
2.178 J.Hirsch, E.H.Martin: J. Appl. Phys. **43**, 1008 (1972)
2.179 K.Hayashi, Y.Yoshino, Y.Inuishi: Jpn. J. Appl. Phys. **14**, 39 (1975)
2.180 B.Gross, G.M.Sessler, H. van Seggern, J.E.West: Appl. Phys. Lett. **34**, 555 (1979)
2.181 T.Tanaka, J.H.Calderwood: J. Phys. D**7**, 1295 (1974)
2.182 D.K.Das Gupta, T.Noon: J. Phys. D**8**, 1333 (1975)
2.183 W.D.Gill: J. Appl. Phys. **43**, 5033 (1972)
2.184 J.Mort, A.I.Lakatos: J. Noncryst. Sol. **4**, 117 (1970)
2.185 P.Fischer, P.Röhl: Prog. Colloid Polym. Sci. **62**, 149 (1977)
2.186 J.A.Wall, E.A.Burke, A.R.Frederickson: In *Proc. Spacecraft Charging Technol. Conf.*, ed. by C.P.Pike, R.R.Lovell (NASA, Cleveland, Ohio 1977) pp. 569–581
2.187 J.Mort, D.M.Pai: *Photoconductivity and Related Phenomena* (Elsevier, Amsterdam 1976)
2.188 H.J.Wintle: IEEE Trans. EI**12**, 97 (1977)
2.189 G.Jaffe: Ann. Phys. (Leipzig) **36**, 25 (1911); **28**, 326 (1909)
2.190 P.Fischer: J. Electrostat. **4**, 149 (1978)
2.191 G.Weber: *Untersuchungen zum Leitungsmechanismus in Polyäthylen*; Dissertation, Darmstadt (1976)
2.192 A.K.Jonscher: Thin Solid Films **1**, 213 (1967)
2.193 J.Frenkel: Phys. Rev. **54**, 647 (1938)
2.194 R.M.Hill: Philos. Mag. **23**, 59 (1971)

2.195 D.M.Taylor, T.J.Lewis: J. Phys. D**4**, 1346 (1971)

2.196 H.J.Wintle: J. Non-cryst. Solids **15**, 471 (1974)

2.197 V.Adamec, J.H.Calderwood: J. Phys. D**11**, 781 (1978)

2.198 L.M.Beckley, T.J.Lewis, D.M.Taylor: J. Phys. D**9**, 1355 (1976)

2.199 W.E.Spear: Proc. Phys. Soc. London B**68**, 991 (1955)

2.200 B.Gross, L.N.de Oliveira: J. Appl. Phys. **45**, 4724 (1974); **46**, 3132 (1975)

3. Thermally Stimulated Discharge of Electrets*

J. van Turnhout

With 85 Figures

A general review is presented of thermally stimulated discharge (TSD), a technique that has contributed significantly to the current understanding of the charge-storage and charge-decay processes in electrets. As such, TSD has become indispensable not only for the quality control of existing electrets, but also for the development of better electrets and charging methods.

The principal concern of TSD research is to study the charge decay by heating the electret at a constant rate. The decay processes are thus investigated as a function of temperature instead of time. This has the obvious advantage of unravelling the nature of the various decay processes very quickly, which makes TSD an attractive proposition. Other features that have contributed to its popularity are its convenience and its high sensitivity and resolving power.

Although TSD has a relatively short history it has already evolved into a basic tool for the identification and evaluation of dipole reorientation processes and of trapping and recombination levels. Its rapid growth has been spurred on by the fact that charge-trapping and charge-transport phenomena are not only of vital importance for electrets, but also for materials used in thin films, photoconductors, electro-optical devices, etc.

The many items of interest to TSD are compiled in the block diagram below, which outlines the contents of this review. It was impractical to deal with all items in depth. Instead we have tried to give a concise and coherent exposition of the basic aspects of TSD along with a critical account of its applicability and its merits. The main topics are: the mechanisms underlying a thermo-accelerated decay; the equipment devised to perform the various TSD experiments; the physical models and theories advanced for describing the decay processes; the manipulation and mathematical analysis of TSD data for optimal information retrieval; and the various applications of TSD. In addition, experimental results of a wide range of materials are given to highlight the versatility of the technique. An attempt has been made to present an up-to-date text with emphasis on the latest ideas and developments. Early activities, which date from 1936, are mentioned only briefly; they have been adequately described in previous reviews.

It was also not possible to fully cover the enormous body of literature on the subject. Perforce, a selection has been made of articles in which important results or novel developments are reported. Nonetheless, this has resulted in an

---

* To Bernhard Gross in friendship and gratitude.

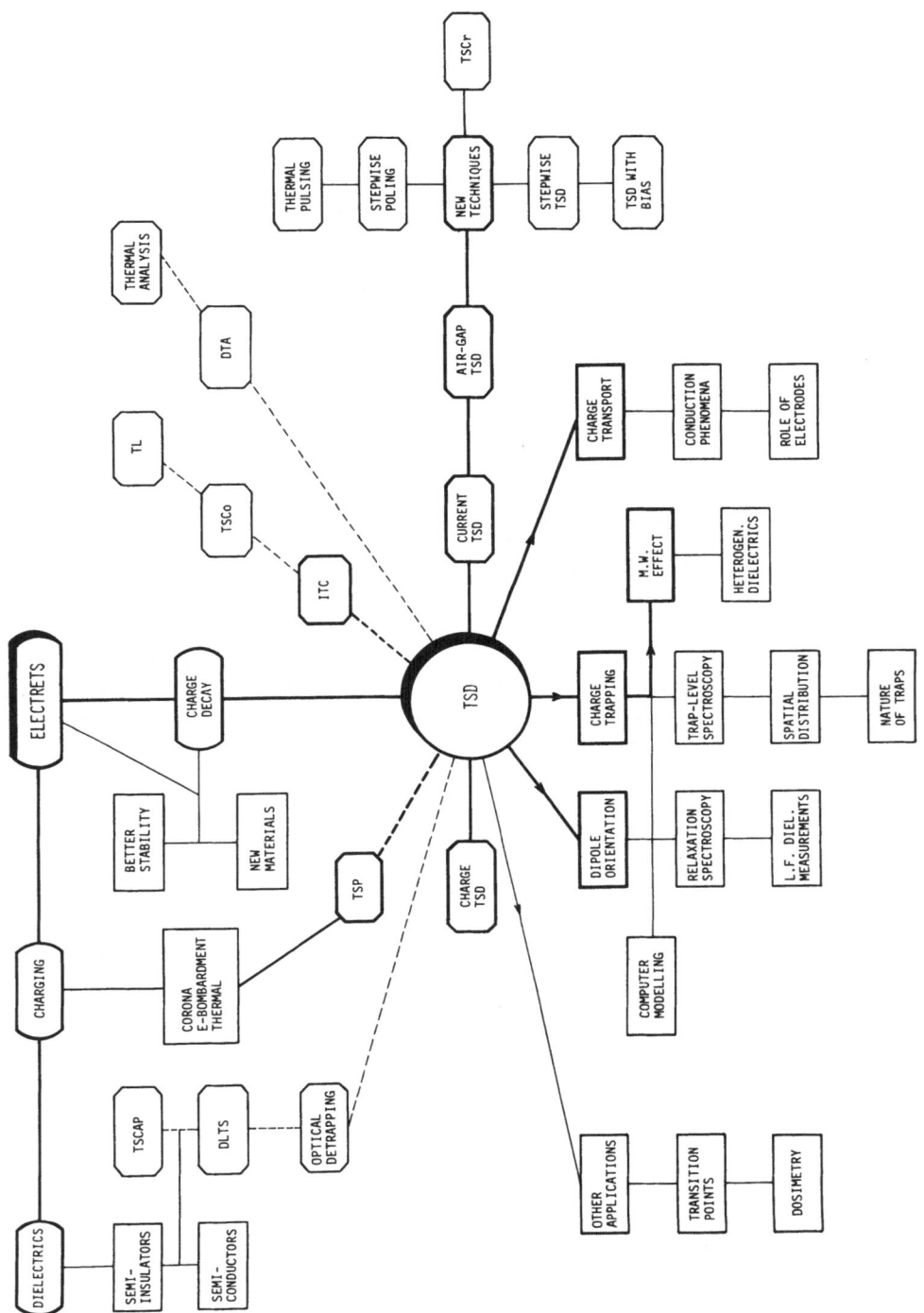

extensive list of references, which should be consulted for additional information.

One of the aims of this chapter is to assess the usefulness and merits of the various models and concepts available for analyzing and interpreting the experimental results. Often arriving at a sound interpretation is a problem, particularly when several decay processes compete. One method which partially obviates these difficulties is to refine the experimental procedure (e.g., by the introduction of stepwise charging or discharging). Another approach worth considering is not to use TSD in isolation but to compare TSD results with those of complementary techniques. Hence, ample attention is given to techniques that are either closely related to TSD or lie at its root (TL and TSCo).

It appears that the principal features of TSD by dipole disorientation are well established and that its mathematical formulation is essentially complete. The situation of TSD originating from detrapping and transport of charges is less satisfactory. However, encouraging progress has been made in recent years. In particular, the direction which further experimental and theoretical developments should take is fairly clear. In this area the simulation of the complicated processes on a computer is of great help, because it allows the validity of a particular model to be checked prior to experimental investigations. The proliferation of minicomputers in research laboratories will certainly help advance this work.

Another important challenge is understanding the storage and transport phenomena in molecular terms. In this regard, we are only beginning to appreciate the full range of virtuosity displayed by insulators. More insight may well be won from changing the morphology of the electret materials systematically. Current research centred on this topic is briefly considered in Sect. 3.13.2.

3.1 Introduction

As discussed in Sect. 2.2, the charge of electrets may be generated by various mechanisms: orientation of permanent dipoles (in polar materials), trapping of charges by structural defects and impurity centers, and build-up of charges near heterogeneities such as the amorphous–crystalline interfaces in semicrystalline polymers, and the grain boundaries in polycrystalline materials. In this chapter

---

◄ Family tree of TSD, a technique developed for studying charge-decay processes in solid materials. The upper part shows the relation of TSD with electrets, the materials to which the technique applies, and some common methods for charging them. It also shows the intimate links between TSD and other electrical and nonelectrical measurements that use thermal stimulation. The entries in the central part depict the various ways TSD can be carried out; they also include some recently proposed refinements and modifications. The lower part shows the multitude of information gained from TSD measurements, and also indicates some other applications of TSD. (The abbreviations are explained in Table 3.1)

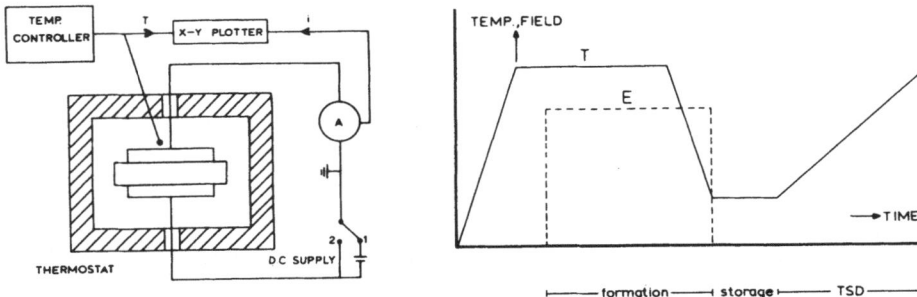

Fig. 3.1. *Left:* Electret charging and discharging unit (position *1*: charging and position *2*: discharging). *Right:* Field and temperature programme during charging, storage (optional) and TSD

we shall discuss experiments aimed at gaining insight, in a relatively *short* time, into the various mechanisms of charge storage and charge decay in electrets.

At room temperature, charge decay measurements are rather time consuming, because at such low temperatures the dipoles and charges remain virtually immobile (cf. Figs. 2.24–27). However, when an electret is heated, the dipoles and charges quickly regain their freedom of motion. *Thermal stimulation* of the discharge therefore shortens the measurement considerably. During such a heat-stimulated discharge, an electret connected to two electrodes generates a weak *current* that shows a number of peaks when recorded as a function of temperature. The shape and location of these peaks are characteristic of the mechanisms by which electrets store their charges. Analysis of the peaks yields detailed information on the permanent dipoles (density, relaxation time, activation energy) and trapping parameters (energies, concentration, and capture cross section of traps). In spite of the fact that the charge of the electret is destroyed, the method has come into extensive use for the analysis of electrets, as well as the development of electrets having a longer lifetime.

The method has its analogues in various branches of solid-state physics, namely, thermally stimulated conductivity and thermally stimulated phosphorescence (thermoluminescence) [3.1–6]. It was first applied systematically by *Fieschi* and his group [3.7], although work that was in principle similar was done as early as 1936 [3.8; see also 3.9, 10]. Using the new technique, *Fieschi* et al. investigated the dielectric relaxation of impurity–vacancy dipole complexes in ionic crystals. They called their method ITC (ionic thermoconductivity or ionic thermal currents), but in our opinion TSD (*thermally stimulated discharge* or *depolarization*) is a more descriptive name [3.11]. Other investigators use still other names [3.12–14], such as electret thermal analysis [3.15], thermally stimulated currents (Chaps. 2, 4, and [3.16]), thermally stimulated dielectric relaxation [3.17], thermally assisted discharging, thermally activated depolarization [3.18], and dielectric depolarization spectroscopy [3.19]. This variety of names illustrates the widening interest that TSD has aroused in science, but it also shows the need for a more consistent nomenclature [3.14].

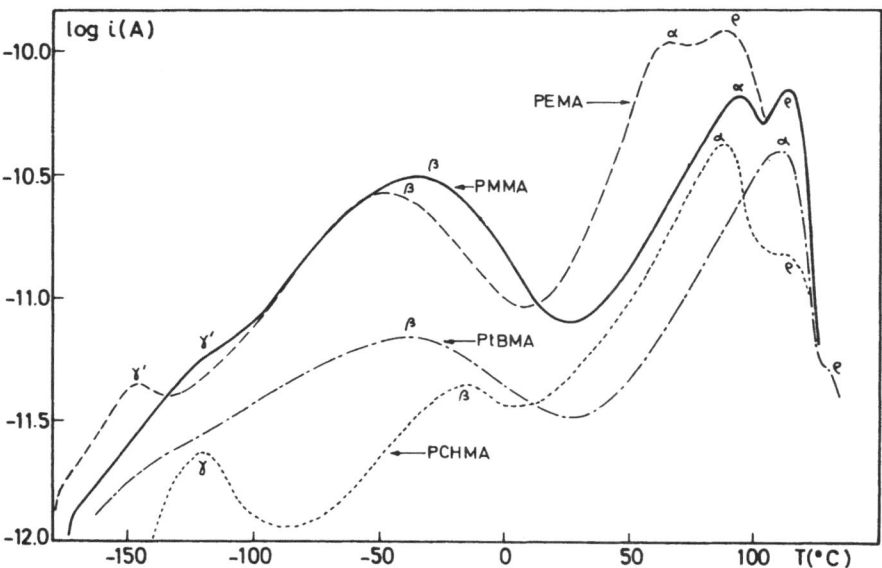

Fig. 3.2. Current spectra of poly(methyl, ethyl, tert. butyl, and cyclohexyl) methacrylate, which have the common structure

$$-CH_2-\underset{\underset{COOR}{|}}{C(CH_3)}- \text{ with } R = -CH_3, -C_2H_5, -C(CH_3)_3 \text{ and } -\langle H \rangle-.$$

The electrets had a diameter of 5 cm, and were poled during 30 min with $10\,kV\,cm^{-1}$, at $127°$ (PMMA), $90°$ (PEMA), $140°$ (PtBMA), and $120\,°C$ (PCHMA). The spectra were recorded at $5\,°C\,min^{-1}$. [3.20]

The principle of the method is outlined in Fig. 3.1. The sample (in disc form) is metallized on both sides, and charged by application of a dc field at a high temperature. This field is maintained, and the sample cooled to room temperature or below. Next, it is short-circuited and reheated at a linear rate of, say, $1\,°C\,min^{-1}$, the discharge current generated being measured with an electrometer, and recorded as a function of temperature by an x–y plotter. As an illustration of the results obtained, Fig. 3.2 shows the current–temperature plots of a series of methacrylic polymers. All curves exhibit three or four peaks. As will be explained later, the peaks labelled α, β, and γ are mainly due to dipoles, whereas the ϱ peaks are due to space charges.

The method has lately been extended in that the *charging* is also performed while the sample is *heated linearly* [3.18, 21–24]. In addition to current peaks due to dipole orientation or filling of traps, one now observes a steadily increasing conduction current at high temperatures. It is important to know how strong the conduction in an electret is, because if it is too strong it prevents the subsequent TSD from being properly performed, the charges stored being

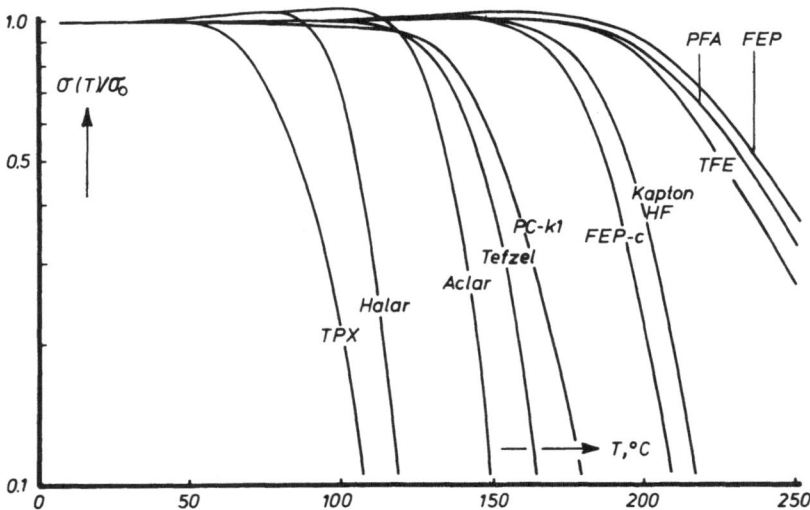

Fig. 3.3. Charge thermograms of several commercial foils, 25 μm thick and charged homopolarly with −1.7 kV in 1 min at the following temperatures: TPX (ICI) 80°, Halar and Aclar-33C (Allied Chemical Corp.) at 130°, PC-K1 (Eastman-Kodak) 200°, and Tefzel, Kapton-HF (12.5 μm laminate of FEP and pure Kapton), Teflon-TFE, Teflon-PFA, and Teflon-FEP (all DuPont) also 200 °C. The recordings were made at 1 °C min⁻¹

too quickly neutralized[1]. Apart from revealing the charge build-up and the conduction process, *thermally stimulated polarization* (TSP) has the advantage that it avoids overheating the sample.

In a variant of the method, the *decay* of the *charge* of electrets is measured during the heating [3.11, 16, 25]. In this measurement the electret is metallized on one side only. As shown in Fig. 3.3, the charge-temperature recordings have less structure than the current plots, but they directly reveal the thermal stability of the charge. This information is important for assessing the practical usefulness of electrets. The two methods of investigating the thermal discharge of electrets will be referred to as *current TSD* and *charge TSD*.

Before TSD, the sample need not be charged as illustrated in Fig. 3.1. Any of the charging methods reviewed in Sect. 2.2, will do. Current and charge TSD have therefore been applied to hetero- and homoelectrets. Homoelectrets, which are obtained by charge injection, are at most metallized on one side. For such electrets we have proposed a special current TSD, in which an air gap is introduced between the measuring electrode and the nonmetallized side of the electret [3.11, 26–28]. *Current TSD* with an *air gap* or, alternatively, with an insulating insert has also been used for TSD and TSP investigations on semiconductors [3.29, 30].

---

1 For this reason the TSD of semiconductors can only be measured at temperatures below room temperature [3.1–3].

It is seen in Fig. 3.2 that the location and shape of the current peaks are quite different for the various polymers. The peaks of polymethyl methacrylate, for example, appear at much higher temperatures than those of polyethyl methacrylate. It is the aim of TSD to deduce from the thermograms those kinetic parameters which determine a specific charge release, and to pinpoint the molecular origin of the dipoles and the trapping sites. In pursuing this goal it is advisable to compare the results of TSD with a variety of other measuring techniques, such as the well-known dielectric and mechanical relaxation measurements, about which a large body of literature is available [3.19, 31–42]. In making comparisons, one should realize, however, that TSD is a non-isothermal technique yielding data in the *ultralow* frequency range, so that one has to be careful in comparing it with dynamic measurements at medium or high frequencies [3.11, 43, 44].

If the carriers in the sample can be excited *optically*, it is often illuminating to compare TSD data with data obtained from thermally stimulated con-ductivity (TSCo) or from thermoluminescence (TL) measurements [3.2, 12–14, 45, 46]. The latter two methods are in concept very similar to TSD, and have been in use for many years.

Thermally stimulated conductivity studies the trapping and detrapping of electrons or holes in photoconductors and semiconductors [3.1–3]. The traps are filled at low temperatures by excitation of the photoconductor or semicon-ductor with *light*, which liberates nonequilibrium carriers that are subsequently trapped. Next, the photoconductor is heated linearly in the dark, and its conductivity is measured. The escape of electrons or holes from their traps increases the dark conductivity, and so this shows specific peaks characteristic of the trapping sites (or impurity centres) in the photoconductor. TSCo is also applied frequently to dielectrics in which a store of nonequilibrium carriers has been formed by nuclear radiation (cf. Sect. 4.6.4).

Thermoluminescence is suitable for those materials which show natural phosphorescence, or which have been made phosphorescent by doping with activators (acceptors) and coactivators (donors) [3.47, 48]. The specimen is again excited at low temperatures by light [3.4–6] or atomic radiation [3.49, 50], whereby its traps are filled. During the subsequent heating, electrons or holes escape from their traps and recombine radiatively (or at least some of them do) with carriers of opposite charge in recombination centres, thus giving rise to luminescence, the intensity of which is measured with a photomultiplier. The glow peaks appearing in the emitted light vs temperature curves are specific for the activator and coactivator centres in the phosphor.

In those materials in which *light* efficiently *releases* electrons and holes from their traps, optical stimulation can also be used directly as a diagnostic tool for trap-level spectroscopy [3.51–53]. The discharge is then stimulated by photons rather than by phonons. In the *photon-stimulated discharge*, the sample (which is biased with a charge collecting voltage) is kept at a constant temperature (e.g., 20 °C) and illuminated with monochromatic light, its wavelength being varied continuously, by means of an adjustable monochromator, from the far

Table 3.1. Rapid methods for studying electric and dielectric processes in solids

Method	Excitation	Sweep parameter	Phenomenon observed	Materials
TS Depolarization, TSD	Electrical	T	Discharge current or charge decay induced by dipole disorientation or charge migration	Dielectrics semi-insulators
Special version of current TSD	Electrical or optical	T	Current originating from release of carriers in a depletion layer	Photoelectrets, semiconductors
Ionic Thermal Currents, ITC	Electrical	T	Discharge current accompanying the disorientation of impurity-vacancy dipoles	Ionic crystals
Photon-Stimulated Discharge, PSD	Electrical or optical	λ	Photocurrent due to optically released carriers	Dielectrics, semiconductors
TS Polarization TSP	Electrical	T	Charging current induced by orientation or charge migration	Dielectrics, semi-insulators
TS Conductivity, TSCo	Light or radiation	T	Increase in dark conductivity due to carriers lifted into the conduction band	Photo- and semiconductors
Thermoluminescence, TL	Light or radiation	T	Light emitted by recombination of detrapped charge carriers	Phosphors
TS Exoelectron Emission, TSEE	Atomic or electron radiation	T	Electron emission from the surface of the sample into a vacuum	Insulating or semi-insulating layers

infrared to the band-gap wavelength. At certain wavelengths, the photocurrent develops peaks whose shape and position are determined by the trapping levels in the sample. Optical depopulation of traps has, of course, been commonly practised for discharging photoconductors and photoelectrets [3.1, 54, 55], but it was not until recently that it was used systematically for trap-level spectroscopy in dielectrics.

The various methods for studying electric and dielectric processes in solids with thermal or optical stimulation are summarized in Table 3.1, and in a more conceptual way in Fig. 3.4. These summaries include yet another technique, viz., *thermally stimulated exoelectron emission*, TSEE [3.45, 49, 56, 57]. In TSEE the number of thermally detrapped electrons that have gained sufficient energy to escape from the surface of a sample in a vacuum is counted.

TSD can be seen in an even wider perspective, because mathematically it is related to other methods of thermal analysis not based on charge decay [3.13, 58]. These physicochemical methods are listed in Table 3.2 [3.59, 60].

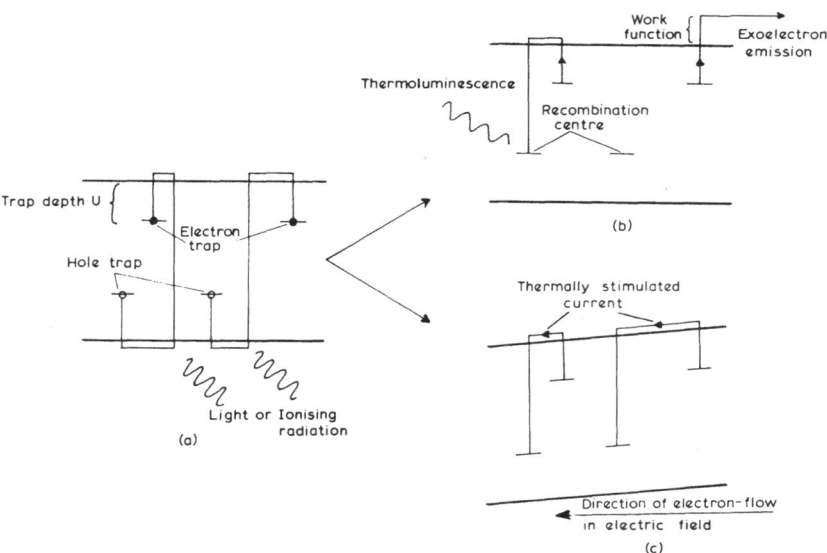

Fig. 3.4a–c. Simple energy-band diagram of an insulator, illustrating the processes of: (a) excitation and trapping; (b) thermal release of electrons followed by luminescence (TL) and exoelectron emission (TSEE); (c) thermal release of electrons followed by a transient current due to a drift in an electric field of the freed electrons (TSCo and TSD). [3.56]

Table 3.2. Other thermal analysis techniques

Technique	Response measured
Differential thermal analysis	Temperature difference between sample and standard
Differential scanning calorimetry	Energy difference between sample and standard
Thermogravimetry	Loss of weight
Thermomechanical analysis	Dimensional changes (dilatometry), modulus changes, damping
Electrothermal analysis	Resistivity, dielectric loss
Thermomagnetometry	Magnetic susceptibility
Thermal desorption	Changes in gas pressure
Evolved gas detection	Total volatiles evolved by decomposition
Evolved gas analysis	Individual volatiles evolved

In this review various aspects of TSD will be discussed. After an outline of the underlying charge decay mechanisms and the experimental set-ups required, a description will be given of the basic theories advanced for analyzing the TSD data and deducing from them the required information about the trapping levels and the dipole relaxations. Several results for the various types of materials investigated will be given. Further, a comparison will be made between TSD and other techniques for studying charge transport processes and charge storage in solid dielectrics.

Some of the topics discussed have been reviewed earlier. For instance, the work on ionic crystals has been reviewed by *Fieschi* et al. [3.61–63], that on bioelectrets by *Mascarenhas* (Chap. 6), and that on polymers by the author [3.11]. Instructive reviews of general aspects of the technique are also available [3.12, 16, 45, 64–68]. Much information on TSD has appeared in the proceedings of several conferences [3.69–77], in particular in that of a recent workshop devoted entirely to thermally stimulated processes [3.72]. For reviews of related topics, viz., TSCo, TL, and TSEE, we refer to [3.1–6, 48–50, 78, 79]. On these topics, too, conferences are regularly held [3.80, 81].

3.2 Mechanisms Responsible for TSD

Several processes contribute to the discharge of electrets, but the driving force of them all is the restoration of charge neutrality. In electrets made from polar materials the *disorientation* of *dipoles* plays a prominent role. This disorientation tends to destroy the persistent dipole polarization by redistributing all dipoles at random.

The disorientation of dipoles involves the rotation of a coupled pair of positive and negative charges, and requires a certain energy, which in solids may amount to a few eV per dipole. It is for this reason that discharge by dipole disorientation is thermally activated, and so can be speeded up by heating.

Often the disorientation energy (or *activation energy*) is not the same for all dipoles. The current-temperature plot will then consist of several peaks (cf. Fig. 3.2), because the dipoles with a low activation energy will disorient at low temperatures, while those with a high activation energy will only respond at higher temperatures. If the differences in the various activation energies are not large, it is more appropriate to assume a continuous distribution of activation energies, for which all individual peaks overlap and merge into a broad peak.

Such broad peaks are often seen as a result of disorientation of polar side groups in polymers at low temperatures (cf. the β peaks in Fig. 3.2). Another possible cause for the appearance of broad peaks is a difference in the *rotational mass* of the dipoles. These differences occur in, e.g., a polymer when heated to its softening temperature, where the dipoles are disoriented by the motion of main chain segments. This disorientation is responsible for the α peaks in Fig. 3.2, which are located at the glass–rubber transition temperature T_g.[2]

In addition to dipoles, the electret usually contains immobilized *space charges*. These are nonuniformly stored, often residing near the electrodes. During heating, they will be mobilized, and neutralized either at the electrodes,

---

2 More recent theories ascribe the broad α peak of polymeric glasses to a distribution in structural ordering parameters, which determine the free-volume relaxations required by the main-chain rearrangements [3.82, 83]. However, because the relaxation times determining the structural reordering are all assumed to have the same temperature dependence, such a distribution formally resembles one in pre-exponential factors.

or in the sample by recombination with charges of opposite sign. The forces driving the charges are their *drift* in the local electric *field*, and *diffusion*, which tends to remove concentration gradients. In general, the field-controlled self-drift prevails [3.84]. Recalling Fig. 3.2, we see that the space-charge or ϱ peaks appear at higher temperatures than the α and β peaks due to dipole disorientation. This is so because the disorientation of dipoles merely requires local rotation, whereas the neutralization of space charges requires them to move over many atomic distances.

At high temperatures, the self-motion of space charges becomes accompanied by a second neutralization mechanism, namely, recombination with thermally generated carriers. These carriers are generated uniformly in the entire specimen by dissociation of neutral entities. They are responsible for the *conductivity* of the material. The conductivity can be either electronic or ionic; in polymers it seems to be (impurity) ions which contribute most to the ohmic conduction, because polymers show an appreciable conduction only above T_g where enough free volume is available for ions to move [3.85–89]. ([3.86–89] also review electronic processes.) The loss of space charges owing to recombination with thermal carriers is not always noticeable. In the TSD of shorted electrets it passes completely *unnoticed*, the net conduction current being zero, because there is no voltage across the sample.

A significant increase in the number of free carriers, and therefore in conductivity, can also be induced by light or by nuclear radiation. Materials whose conductivity changes appreciably when they are illuminated are called photoconductors [3.1, 55]. Consequently, the only kinds of electrets sensitive to light are photoelectrets, which are made from photoconductors ([3.54, 90]; see also Sect. 2.2.7). However, nearly all electrets are sensitive to nuclear radiation, which is strongly ionizing ([3.91, 92]; see also Sect. 2.2.6, and Chap. 4).

Experiments with electrets made by electron bombardment have shown that the *radiation-induced conductivity* persists for a long time (see Fig. 4.15) and should properly be accounted for in TSD ([3.93, 94]; and Sect. 4.6). Obviously, the production of electron–hole pairs by radiation need *not* be uniform; they are formed only in that part of the sample which absorbs the radiation.

Returning now to the *self-motion* of the space charges, we can describe this process in two ways. If the charges are ions, they are generally considered to be free to move with a thermally activated mobility. One can then visualize them as hopping from one vacancy to another across a potential barrier equal to the activation energy. If, however, the charges are electrons or holes, it is more appropriate to visualize them as being immobilized in local traps, from which the heating releases them into a band of energies in which they can freely diffuse to the electrodes.

If the electronic carriers are trapped at a single energy level, the TSD current will show one peak. Normally, however, there is more than one peak, indicating that the carriers occupy different energy levels. This is illustrated in Fig. 3.5 for electrets obtained by electron bombardment, which exhibit two current peaks above room temperature.

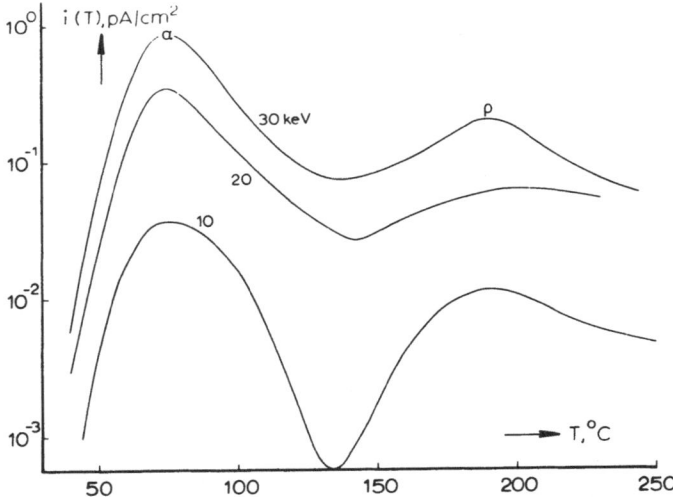

Fig. 3.5. Current thermograms of metallized Teflon-FEP bombarded with electrons of three different energies. The deposited charge was 20 nC cm^{-2}, and the heating rate 1 °C min^{-1}. [3.11]

Usually, the detrapping of electronic carriers is analyzed with the band-gap model developed for crystalline solids. However, most electret materials are amorphous or at most partially crystalline. The high degree of positional (and compositional) disorder in these materials results in many localized states in the forbidden energy gap [3.95–97]. These trapping states differ in number and depth; the higher the disorder, the fewer there are and the deeper they tail off into the energy gap. The energy gap is thus replaced by a pseudogap, in which the density of states remains finite. As a result the band edges are blurred, so that it is difficult to mark the energy at which a carrier is completely nonlocalized.

Mott et al. [3.95–97] have therefore proposed to base the definition of the transport bands in amorphous materials on the *mobility* of the carriers. This is illustrated in Fig. 3.6; the critical energy E_c at the mobility edge (where the mobility drops by three orders of magnitude) separates localized from de-localized states. Clearly, it is only above E_c that electrons can move through the solid without the help of thermal energy. In addition to the *disorder-induced* band-tail states (which stretch out monotonically from the band edges), clearly recognizable gap states may be formed by *intrinsic structural* defects. In polymers, intrinsic defects are, e.g., branching, chain entanglements, chain ends, pendant groups, etc. [cf. Ref. 3.11, p. 271; 3.98][3].

However, the actual situation being unknown for most amorphous materials, the conventional band-gap model is still used as a first approximation

---

3 The reader's attention should also be drawn to *Fabish* and *Duke*'s model [3.99] of the electronic structure of polymers, in which molecular ion states are considered to be the intrinsic localized states in polymers.

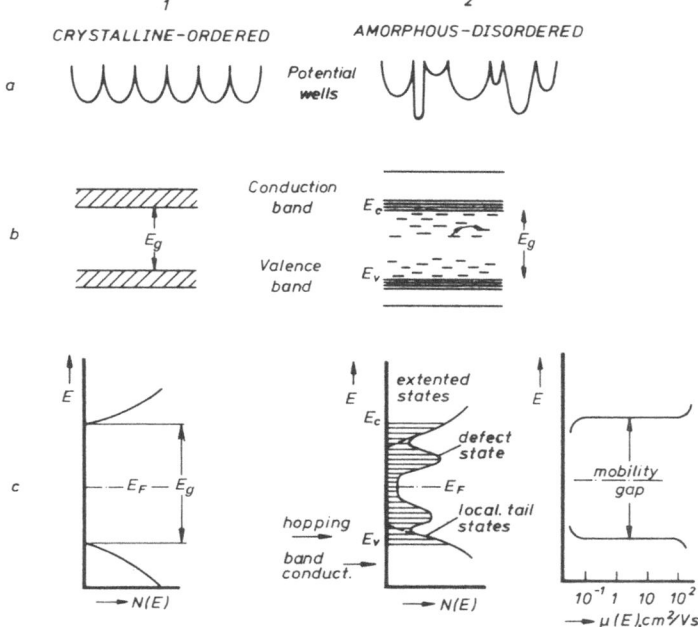

Fig. 3.6. Band formation for (*1*) crystalline and (*2*) amorphous insulators (**a**) potential wells, (**b**) band structure, (**c**) density of states $N(E)$ and mobility gap. E_F is the Fermi level, E_c the critical energy for band motion, and E_g the forbidden energy gap. The areas hatched horizontally in *2b* and *2c* represent localized states in the band tails, which are created by fluctuations in the potential wells and interatomic distances owing to lack of long-range order. [3.97]

(cf. Fig. 4.1). With this model it is easy to visualize the other two processes a carrier may undergo after its thermal release from a trap. These are *retrapping*, when on its way to the electrodes it meets another trapping centre, and *recombination* with an opposite charge, when it is caught by a recombination centre.

At low temperatures, the multitude of trapping sites in disordered materials allows the electronic carriers to move in the direction of the field by *hopping* across the localized gap states instead of moving freely along extended band states. An interesting analysis on this basis of the transport of charges, which can only proceed via thermal activation, has been presented by *Scher* et al. [3.100], who treat the hopping of carriers between the randomly distributed localized sites as a stochastic process. In order to account for easy hops and more difficult ones, for which a carrier wavers, they assume a distribution of waiting times between the sequential hopping events (sequential hopping is also discussed in [3.101]); in [3.102] they extend their random-walk theory to cover trapping. Recently, a successful alternative (the "trap-limited band-transport model") was proposed, in which the transport is assumed to proceed by multiple trapping [3.103].

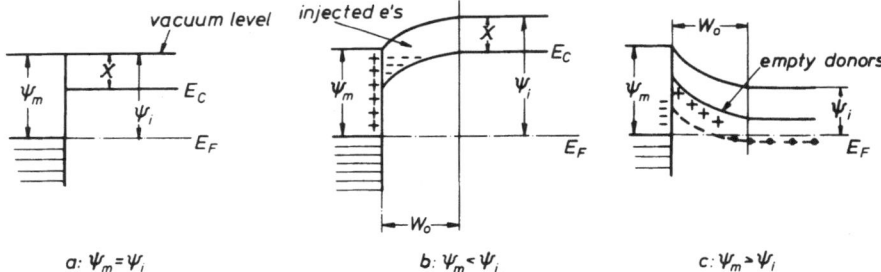

Fig. 3.7a–c. Energy diagrams showing the various types of metal-insulator contacts (**a**) neutral, (**b**) ohmic or injecting, and (**c**) blocking. W_0 represents an accumulation layer in case (**b**) and a depletion layer in case (**c**); further, ψ_m is the work function of the metal, and χ the electron affinity and ψ_i the work function of the insulator. [3.105b]

The above description of charge trapping applies to homogeneous materials. In *heterogeneous materials*, consisting of, e.g., amorphous and crystalline phases, another kind of trapping occurs, viz., interfacial trapping. The interfacial charges are due to the difference in conductivity of the phases (Maxwell–Wagner effect). When such a material is charged, carriers will either accumulate or be depleted near a particular interface, depending on whether the incoming local conduction current is greater or smaller than the outgoing one. The differences in the local conduction currents are also responsible for the dissipation of the charges in the subsequent TSD, because the currents then flow in the opposite direction.

The nature of the *electrodes* plays an important part in the neutralization process ([3.104]; and Sect. 4.2.1). As Fig. 3.7 depicts, a distinction is made between neutral, ohmic, and blocking electrodes [3.1, 105, 106]. Neutral electrodes do not hinder the neutralization of charges arriving from the sample during TSD. The same holds for ohmic electrodes which, in addition, may inject carriers of opposite sign into the charged sample. Blocking electrodes, on the other hand, act in just the opposite way, preventing any injection and neutralization of charges. With blocking electrodes, the neutralization must therefore proceed entirely within the sample. Whether an electrode is an electron-injection or an electron-blocking one depends on whether its work function has a smaller or greater value than that of the sample. If it is greater, a Schottky barrier is formed. Such a blocking barrier enables one to study the TSD of semi-insulators and semiconductors, in which the high conduction current would otherwise obscure the emptying of any traps [3.67, 107]. A sample can be deliberately blocked by insertion of a highly insulating layer (air, Teflon) between the sample and the metal electrodes [3.29, 30, 68].

The magnitude of the TSD current of course depends on the charge retained by the electret. At first sight one would expect the time integral of the current to be equal to the charge originally present. This need *not* be so, however, because not all decay processes contribute fully to the external current ([Ref. 3.11, p. 13]; and Sect. 4.2.3). We have already mentioned that charge neutralization by ohmic conduction in shorted electrets does not contribute to the external current at all. In addition, the self-drift of the charges is a rather inefficient current-generating process, particularly when neutralization of the charges requires them to move over short distances only. Diffusion is likewise an inefficient current-generating process. With ohmic electrodes the charges diffuse symmetrically outwards, and no current is generated at all. The only process whose efficiency is 100 % is the disorientation of dipoles.

In air-gap current TSD, by contrast, all the decay processes contribute to the external current, including that of charge neutralization by conduction, because the mean electric field in the electret is now *nonzero*. The same applies to charge TSD, in which one also works with an air gap. These methods further allow the detrapping of charges from *surface states* to be studied. The various methods for investigating TSD are thus supplementary to one another.

3.3 Experimental Techniques

Diagrams of the current and charge TSD measuring set-ups are shown in Fig. 3.8. In current TSD, the current released by a shorted bilaterally metallized electret is measured, whereas in charge TSD the charge remaining on the nonmetallized side of an electret is measured during heating in open circuit (see also Table 3.3).

To measure the current, one simply connects the electrodes with a sensitive ammeter (electrometer). Measurement of the charge is less simple, particularly if it has to be recorded as a function of temperature. It is least complicated for foils, whose charge (or rather surface potential) can be measured by the vibrating capacitor method (cf. Sect. 2.3.4), in which the external field of the vibrating electret foil is cancelled by a backing voltage. This is generated by an electronic feedback system and the measurement of the surface potential of the electret thus reduces to a recording of the backing voltage [3.11, 25, 108]. TSD charge measurements on thick samples are less convenient; to these, the lifted electrode method or a rotating vane system (with backing voltage) are applicable [3.109].

Figure 3.8 also shows the principle of air-gap current TSD. In the presence of a finite air gap, the charge on the nonmetallized side of the electret induces part of its image charge on the lower electrode[4], the result being a non-zero internal field, which at high temperatures causes a net conduction current to flow in the electret. Air-gap current TSD (or as it is sometimes called "TSD with a contactless electrode" [3.27, 28]) thus enables one to observe the neutralization of space charges by ohmic conduction. The air-gap technique is

---

4 As the loss of induction to the lower electrode increases with the width of the air gap, this should not be too large, because it then weakens the discharge current too much. A width of 0.2 times the thickness of the sample is about correct [Ref. 3.11, p. 165, 3.110]. If the air gap becomes too narrow, e.g., with foil electrets, Teflon or some other highly insulating foil can be inserted into it.

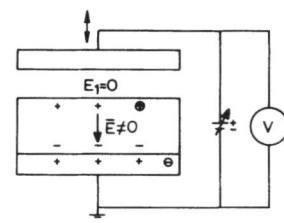

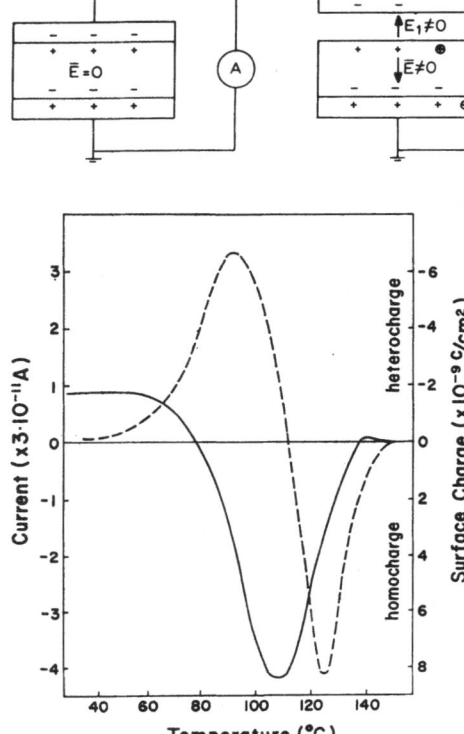

Fig. 3.8. Left to right: normal current TSD with shorted evaporated electrodes, current TSD with an air gap, and charge TSD in open circuit

Fig. 3.9. Change in the surface charge (*solid line*) of a polyethylene terephthalate electret during air-gap TSD. The accompanying discharge current is also shown (*broken line*). The electret was formed at 130 °C in a field of $16\,\mathrm{kV\,mm^{-1}}$, and heated at a rate of $3\,^\circ\mathrm{C\,min^{-1}}$. [3.28]

Table 3.3. Characteristics of the three TSD methods[a]

Current TSD with shorted electrets	$\int_0^s E\,dx = 0$	$i^* \neq 0$
Current TSD with an air gap, s_1	$\int_0^{s+s_1} E\,dx = 0, \quad \int_0^s E\,dx \neq 0$	$i^* \neq 0$
Charge TSD in open circuit	$V^* = \int_0^{s+s_1} E\,dx = \int_0^s E\,dx$	$i = 0$

[a] The quantities measured are labelled with an asterisk.

also valuable in that it reveals whether the electret contains a heterocharge or a homocharge or both. A homocharge is usually seen to persist to higher temperatures than a heterocharge. As a result, the discharge current of a bipolarly charged sample (Fig. 3.9) changes sign during the heating.

In current TSD, the samples usually are about 0.5 mm thick, but this can be reduced to 25 µm if they consist of polymers, from which foils can easily be made. The samples are provided with two identical electrodes, except for charge TSD and air-gap current TSD, which require only *one* contacting electrode. Asymmetric contacts are employed particularly for semi-insulators and semi-conductors, one of whose contacts is made *blocking* so as to form a highly resistive region in the sample [3.67, 107, 111]. The usual method of metallization is to evaporate the electrode material onto the sample in a vacuum[5]. If the sample is insensitive to organic solvents, it can be painted with colloidal solutions of metals. Electron-beam contacts have been useful in TSD studies of the phosphors in vidicon tubes [3.112]. The metals applied are resistant to oxidation (gold or aluminium). The active electrode area is 10–20 cm^2. Often, one of the electrodes is provided with a guard ring, which reduces field fringing and surface leakage. This arrangement is also valuable for comparing the charge transport along the surface and through the bulk of the sample [3.113].

Figure 3.10 shows a block diagram of the author's equipment for current TSD[6]. The sample is heated in a thermostat whose temperature can be controlled linearly by electronic circuits [3.11, 114–116][7]. Some investigators prefer inverse linear heating, because this simplifies evaluation of the data [3.118]. A heating rate of $1\,°C\,min^{-1}$ is the most common, this being low enough to prevent temperature lags and to guarantee a good resolution of peaks. As several materials show current peaks at very low temperatures (cf. Fig. 3.2), it is advisable to start at the temperature of liquid nitrogen, which can be attained by the use of nitrogen vapor as the flow gas.

The design of the thermostat is not critical, except that the feed-throughs should be well insulated and that thermal gradients across the sample should be avoided, because these may generate spurious currents ([3.119]; and Sect. 4.6.6). For supplementary measurements with thermally stimulated conductivity and thermoluminescence, the sample should be easily accessible to radiation and radiation detectors [3.120].

The current measurement requires a sensitive ammeter such as a Cary, Vibron, or Keithley electrometer, which has a lower detection limit of 10^{-15} A. Keithley also manufactures a logarithmic picoammeter, which is very convenient because the current often changes by orders of magnitude during a TSD. The current and temperature are plotted on a strip chart recorder or on an *x-y* plotter. With more sophisticated data acquisition equipment, the data can be punched on tape for storage and further evaluation with a computer [3.121].

---

5 The combination of an unmetallized sample and pressure contacts is not recommended, because it leads to results that are less reproducible. However, when a sample is equipped with evaporated electrodes it is always slightly heated. Therefore, when precharged samples (with a free surface) are to be studied, pressure contacts can still be useful [3.93].

6 Commercial equipment for current TSD such as the Electret Thermal Analyzer (Toyo Seiki, Seisaku-sho, Tokyo) has recently become available.

7 It is easiest, for strictly linear heating, to have a linear output of the temperature sensor; linearization circuits for a thermocouple are given in [3.116] and for a platinum thermometer in [3.11, 117].

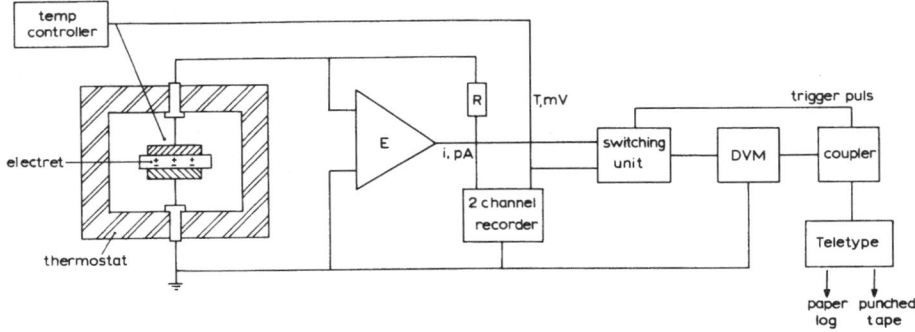

Fig. 3.10. Main components of fully automated equipment for current TSD

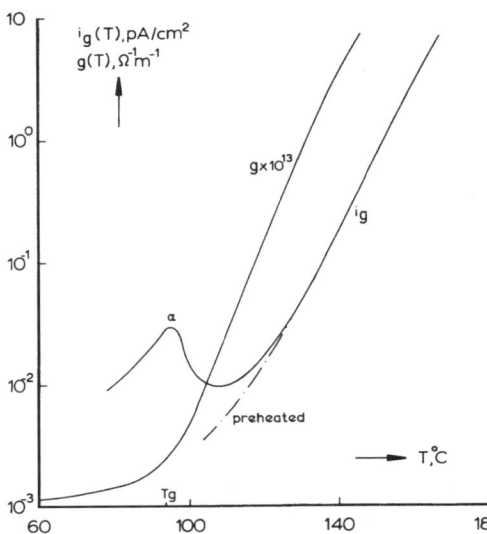

Fig. 3.11. Parasitic galvanic current, i_g, in a noncharged polyethylene terephthalate-c foil heated at 0.5 °C min^{-1}. This current reappears in a second heating and is only observed well above T_g, when the ohmic conduction g becomes high. [3.11]

In the current TSD arrangement, the sample can be charged if a voltage source is connected to the electrodes (cf. Fig. 3.1). Before being charged, the sample should be freed from any spurious charges generated during its manufacture and handling [3.122]. This is done by running a blank TSD on the sample just before it is charged. A typical result is shown in Fig. 3.11. Apart from a transient peak at T_g, the thermogram shows a current which, above T_g, steadily increases with temperature. This high-temperature current is found in most materials. It probably is spontaneously generated by a small *difference* in *contact potentials* of the two electrodes, which induces a weak conduction current at high temperature (Ref. 3.11, p. 199; [3.123]). Chemical degradation may also generate currents, as *Carr* et al. [3.124] have shown for unpoled polyacrylonitrile. Typical for the spurious galvanic current, which is not observed in air-gap current TSD, is that it *reappears* when the heating is repeated (dot–dash line).

3.4 Methods for Unravelling the Discharge Processes

The deceptive similarity between dipole and space charge peaks makes it difficult to distinguish between the two, and one often has to resort to other methods or techniques for additional information (cf. Sect. 2.5).

A first clue may be the nature (composition, structure, etc.) of the material. Both dipole and space charge relaxations may occur in polar materials, but in nonpolar ones, which do not contain permanent dipoles, only space charges can be active. The polarizability of a material can be readily measured dielectrically. A comparison of TSD with dielectric data, in particular those in the *low-frequency* range may therefore be helpful to identify the origin of a peak. Because dielectric losses generally do not manifest themselves in ϱ peaks, these are easy to recognize (cf. Fig. 3.33). Similarly, TSD can be related to data of mechanical measurements.

This comparison can be extended to certain novel techniques, such as *thermally stimulated creep*, in which a mechanical stress is frozen in instead of a polarization [3.125]. In this technique, the relief of the mechanical stress is monitored during the heating. The peaks observed in the rate of creep vs temperature curves very much resemble the corresponding TSD current peaks. TSD is also closely related to the low-frequency measurements of *Oshiki* and *Fukada* [3.126] on the real and imaginary parts of the *piezocoefficients* of piezoactive polymers and biomaterials. The relaxation peaks of these piezocoefficients are associated with the electrical as well as the mechanical properties. Another new technique is the *piezostimulated current* method [3.127, 128], in which the relaxing units, after being oriented by an electric field, are locked in by a very high *pressure* rather than by the cooling of the sample. Upon removal of the poling field and a gradual lowering of the pressure, current peaks are observed at pressures at which the relaxing units have regained enough mobility for disorientation.

We have already mentioned that it is also worthwhile comparing TSD with TSCo, TL, and TSEE, because such a comparison may enable a TSD space-charge peak to be identified as arising from either the motion of ions or that of electrons or holes, the latter two being the only entities that contribute to the transition processes involved in these measurements.

Recently, *McKeever* et al. [3.22] showed that the combined use of TSCo and TSP can be used to isolate dipole and charge release peaks. Figure 3.12 depicts their measurement of the charging current in LiF during a series of thermal polarizations. Thermal polarization preceded by photoexcitation with uv light produces electron and hole pairs, which settle down in traps, upon release from which they give rise to a transient ϱ peak. Without previous photoexcitation, only the dipoles are aligned, but they will do so only once, failing to produce dipole peaks in runs C and D.

The dipole peaks which normally occur in polar materials can be suppressed by the use of special charging techniques for filling the traps in these

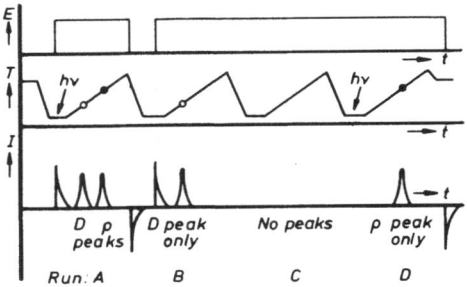 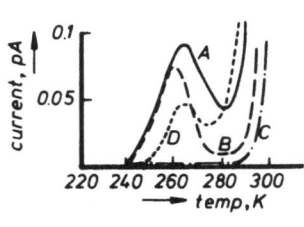

Fig. 3.12. *Left*: Procedure for isolating dipole (*D*) and space-charge (ϱ) peaks in dielectrics. *Right*: Current spectra produced by the special poling scheme in LiF. *Run A* ——— impurity-vacancy dipoles and space charges; *B* – – – impurity-vacancy dipoles; *C* —·—· neither; and *D* – – – space charges. [3.22]

Table 3.4. Effect of poling conditions on dipole and space charge peaks of amorphous polymers

Poling parameters	Effect on dipole peak	Effect on space charge peak
T_p	Peak shifts when $T_p < T_m$; no shift when $T_p > T_g$	Peak develops fully when $T_p > T_g$; its position is independent of T_p, but its height passes through a maximum
t_p	Independent of t_p when $T_p > T_g$	Peak passes through a maximum
E_p	Linear dependence on E_p	Nonlinear
Electrodes	No effect	Electrodes affect position and height of peak
Sample thickness	No effect	Peak is higher and appears at lower temperatures for thinner samples
Preparation of sample	No effect	Hydration, swelling and doping have pronounced effects
Nuclear radiation	No effect	Peak is lowered
Reproducibility	Good	Moderate

materials. Such techniques are, e.g., electron bombardment and corona charging at low temperatures. These charging methods, however, are limited to foils, because the penetration depth of the injected carriers is small. Another expedient that may help to distinguish between dipole and space-charge peaks is their dependence on the poling field, which for the space charge peak is expected to be nonlinear. Table 3.4 lists some more differences between the two types of peaks ([3.11, 20]; and Sect. 3.11.9).

We are currently investigating yet another method for distinguishing between dipole and space-charge peaks. It is based on measuring the charging currents as a function of temperature for positive and negative voltages one after the other (see Fig. 3.13). For dipoles, the second charging current should be twice the first, but for space charges this need not be so ([3.129]; and Sect. 2.5).

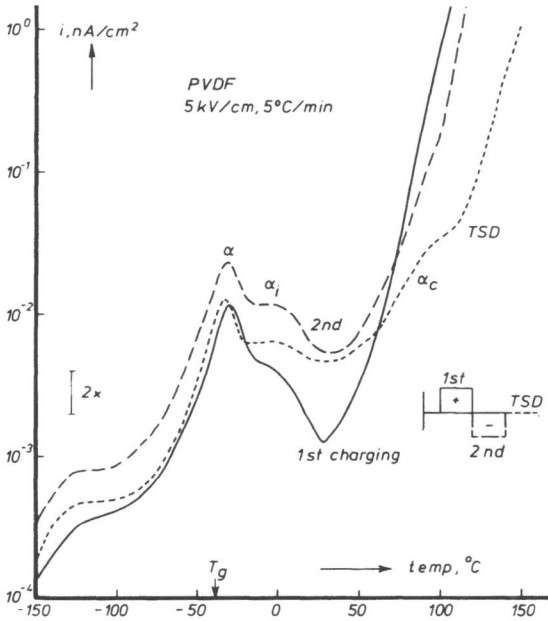

Fig. 3.13. Charging current in unstretched polyvinylidene fluoride – Kureha (25 μm) during a positive and a negative voltage sequence. For the low-temperature α and $α_i$ peaks, the positive current is half the negative one, whereas for the high-temperature $α_c$ peak, which only shows up in TSD, it surprisingly exceeds the negative current. This clearly shows that dipoles dominate at low temperatures and space charges at high temperatures

If the electret is thick enough, the nature of its peaks can also be established by cutting it into thin slices ([3.9, 11, 43, 66] and Sects. 2.4.1, 2.6.3). If the TSD of the slices is not the same, the peaks are definitely due to space charges. Methods for measuring the charge profile within foils have only been proposed recently ([3.130, 131]; and Sects. 2.4.5–7). They are based on determination of the changes in surface potential attending exposure of a foil to *heat* or *pressure pulses*. A destructive, but more precise, method of determining the charge distribution in foils has been developed by *Sessler* et al. [3.132]. They use an electron beam of increasing energy to stepwise destroy the space charge by radiation-induced conduction. Their method resembles the measurement of the charge distribution in photoelectrets with an optical probe ([3.133], Sect. 2.4.4).

In Sect. 3.2 we noted that a dipolar TSD peak may arise either from a single dipole relaxation or from a distributed relaxation, i.e., one with many relaxation times. To find out which is the case, we study the dependence of the discharge current on the *forming* and *storage conditions*. If there is a single relaxation, the shape and position of the current peak do not depend on those conditions, but if there is a distribution of relaxations, they do [3.11, 134]. This is so, because the fast and slowly reacting dipoles of a distributed relaxation will be affected differently, e.g., by a too short formation or a too long storage. This is illustrated in Fig. 3.14 for a PMMA electret.

Another way to establish that a relaxation is distributed is to perform the *TSD stepwise*, or to *charge* the sample *stepwise*. A simple method of doing the latter has been described by *Vanderschueren* [3.135], who charges the sample by

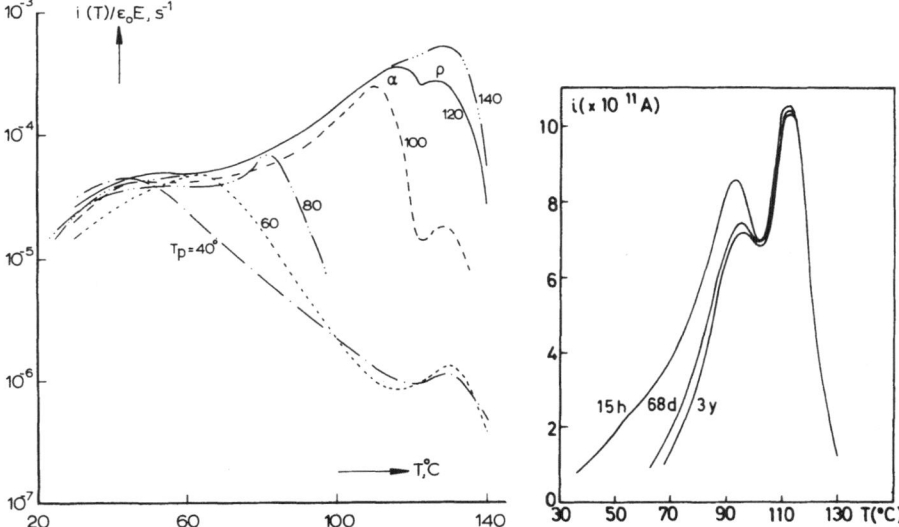

Fig. 3.14. The curves in the left-half of the figure show that for low poling temperatures, the thermograms of polymethyl methacrylate are cut off at temperatures above T_p, so that the dipolar α peak does not emerge at the glass-rubber transition proper. ($E_p = 100 \text{ kV cm}^{-1}$, $t_p = 1.5 \text{ h}$, $\beta = 0.5 \,^{\circ}\text{C min}^{-1}$. [3.11]). The curves on the right show that the α peak is also affected by prolonged storage. Remarkably, the ϱ peak has not changed after storage, though it usually does [Ref. 3.11, p. 243] ($E_p = 20 \text{ kV cm}^{-1}$, $t_p = 0.5 \text{ h}$, $T_p = 138 \,^{\circ}\text{C}$, $\beta = 3 \,^{\circ}\text{C min}^{-1}$. [3.20]

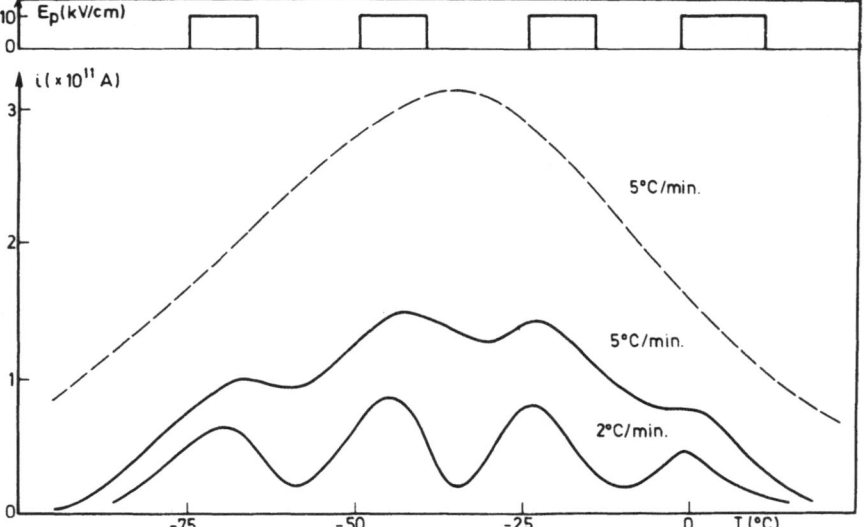

Fig. 3.15. Fractional polarisation of the β peak of polymethyl methacrylate by application of voltage pulses during linear cooling. *Below:* The resulting partial TSD peaks are compared with the fully polarized β peak (dashed curve, obtained after poling at 40 °C during 0.5 h). [3.135]

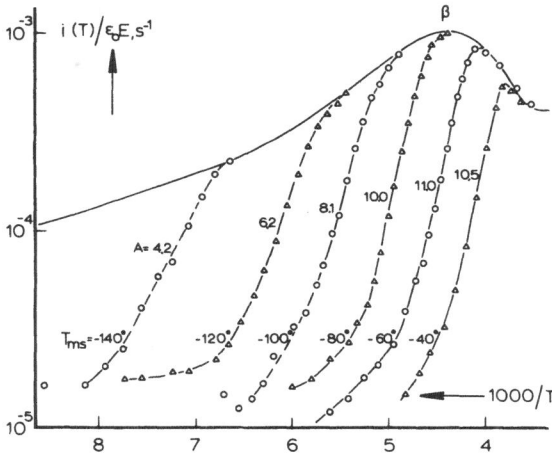

Fig. 3.16. Multistage TSD of the β peak of polymethyl methacrylate. T_{ms} is the temperature at which the electret is preheated for 1 h, and A is the activation energy in kcal/mol calculated from the initial slope. The electret was formed at $20°$–$8\,kV\,cm^{-1}$–1 h, and $\beta = 3\,°C\,min^{-1}$

exposing it to short voltage pulses in narrow temperature intervals, when it is cooled down from a high temperature (Fig. 3.15). When the dipoles are distributed, only groups of them with specific relaxation times are activated in this way. Another way to activate parts of a distribution has been advanced by *Chatain* et al. [3.136a]. They polarize the sample at a certain temperature and depolarize it at a slightly lower temperature. Poling in a narrow temperature window can also be achieved by simple removal of the field in the cooling phase [3.136b]. An alternative method of stepwise poling has been proposed by *Ponevski* et al. [3.137]; they keep the temperature constant and programme the field.

TSD performed in steps also enables one to establish whether a dipole distribution arises from a distribution in activation energies or from one in natural frequencies. The method is based on the fact that the *initial slope* of the discharge current depends on the activation energy; the steeper the current rises, the higher the activation energy. The method, which we have called *multistage TSD*, (Ref. 3.11, p. 61; [3.134]), but which is also known as decayed TSD [3.1], stepwise TSD, thermal cleaning or partial heatings [3.138], is illustrated in Fig. 3.16. We see that in multistage TSD, the heating is interrupted at increasingly higher temperatures. Since the current rise is found to steepen during the series of heatings, it is likely that the dipoles of the β peak of PMMA are distributed in activation energy, the dipoles with a low activation energy disorienting first, and those with a higher activation energy later, as the temperature rises. *Kryszewski* et al. [3.139] have shown by model calculations that fractional polarization can similarly be invoked to determine the type of distribution.

As discussed in Sect. 3.3, most materials at high temperatures generate a spontaneous current, probably a conduction current that must be ascribed to the difference in contact potentials between the electrodes and the sample. Such a difference always exists, because electrodes can never be made quite the same.

However, the increase in current may also signal the onset of a real TSD peak. It is therefore important to know at what temperature the conduction begins to become significant. This temperature can be derived from the TSP charging current. The best thing to do is to charge the sample in *two* heating runs, while the field is maintained. This procedure ensures that the only current measured in the second run is the conduction current (cf. Figs. 3.11, 21). Another, less clear-cut method to distinguish a spurious TSD conduction current from the onset of a TSD relaxation current is to heat the sample at different rates. Faster heating increases the height of a TSD peak, but not that of the conduction current, because this arises from equilibrium carriers.

3.5 Applications of TSD

TSD is a powerful tool for finding and developing better electret materials, because it reveals in a short time the mechanisms responsible for the decay of the charge of an electret [3.11]. It is also useful in optimizing the charging conditions. Air-gap current TSD and charge TSD provide additional information; moreover, they are suitable for investigation of electrostatically charged samples, such as triboelectrically charged polymers and gun-sprayed powder coatings [3.110, 140].

TSD is, in fact, developing into an essential source of information on the storage and transport of charge carriers in insulators and semiconductors generally [3.1–3, 11, 16, 67–76]. It helps to elucidate the nature of the defect and impurity centres that function as traps. It also provides information on the nature of the electrode contacts, the interfacial barrier heights and the majority carrier [3.67]. TSD has also been extensively applied in the study of the formation and aggregation of impurity–vacancy dipoles in ionic crystals, such as the alkali halides, because it detects relaxation peaks at the lowest possible temperature and thus avoids aggregation of the dipoles during the measurement [3.61–63]. Part of its attraction lies, of course, in its high sensitivity and the simplicity of the equipment needed.

Current TSD can further be used to study the low-frequency dielectric relaxations in solids [3.11, 43] and supercooled liquids [3.64a, 141]. This is an interesting development, because the dielectric behaviour of most solids is complex and needs to be investigated in an as broad a frequency range as possible. The techniques available for the ultralow frequency range require much more time than TSD. In view of its low-frequency range, TSD is capable of separating relaxation peaks with a difference in activation energies of only 10 % [Ref. 3.11, p. 59]. Moreover, overlapping peaks can easily be resolved by fractional polarization or depolarization.

TSD extends dielectric research in other directions as well. In the previous section we have seen that it enables differentiation between dipole relaxations and space charge relaxations, and between dipole relaxations distributed in activation energies and natural frequencies, respectively.

Table 3.5. Glass-rubber transitions of some polymers[a]

Polymer	TSD value	Literature value
Poly-2-chloroethyl methacrylate	69	103
Polycyclohexyl methacrylate	85	90
Polyethylmethacrylate	66	65
Poly-3-ethyl-3-pentyl methacrylate	81	83
Polymethyl methacrylate	105	105
Polymethyl methacrylate (isot.)	50	50
Polyphenyl methacrylate	106	105
Poly-t-butyl methacrylate	110	113
Chlorinated polyether (Penton)	− 3	
Polycarbonate-n	152	149
Polyethylene terephthalate	88	81
Tetrafluoroethylene co hexafluoropropylene (Teflon-FEP)	75	77
Polyvinylchloride (rigid)	75	74

[a] From [3.11, 20].

The very fact that TSD measures relaxation peaks at very low frequencies allows it to be used for determining the glass-rubber transition temperature of polymers [3.26]. Some results of this application are collected in Table 3.5.

Sectioning of the electret before TSD reveals the distribution of the charge within it. This technique has been used for determining the charge distribution in polyethylene high-voltage cables [3.66]. Novel techniques have been devised for probing the charge profiles in foils [3.130].

The purity of a material can also be checked with TSD [3.142], because impurities generally increase the number of ionizable carriers and therefore promote temporary storage of charges considerably. In view of its high sensitivity TSD can likewise be used to detect changes in the dielectric properties induced by chemical degradation, atomic radiation, swelling, moisture and other agents [3.11, 19, 20, 63, 143–146].

Recently, TSD has been applied to study the effect of *physical aging* in amorphous polymers (Ref. 3.19, Chap. 7; [3.146, 147]). This aging occurs when a polymer is cooled to below its glass–rubber transition, and is due to continued vitrification, which reduces the free space available for molecular motions, and which therefore, according to the free-volume theory [3.148], increases the relaxation times. Figure 3.17 shows the change in TSD response due to this *isothermal increase* in relaxation times. When a PVC sample has aged for 3 h at 60 °C, the number of dipoles that can be aligned in 1 h is much smaller than immediately after the quench. As a result, the TSD current after aging is much weaker. These time effects might open a way to the use of TSD in archaeological and geological dating (see also [3.149]).

TL has long been used for this purpose [3.150, 151]. Here the dating is based on the progressive filling up of deep traps in phosphorescent inclusions in

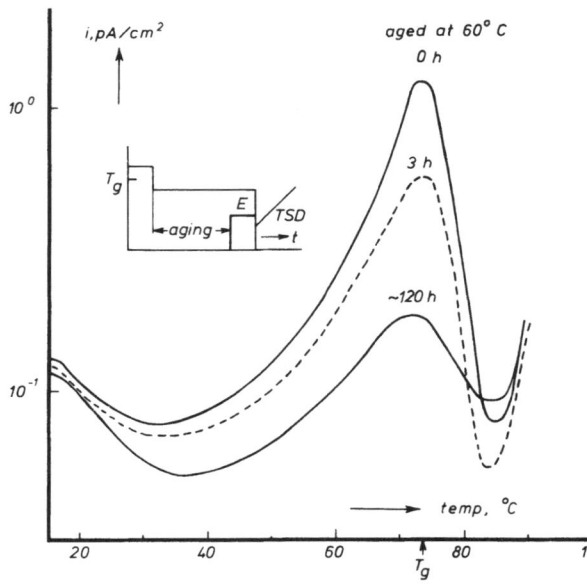

Fig. 3.17. Effect of physical aging on the current spectra of unplasticized PVC. The sample was quenched from 100 to 60 °C, aged at 60 °C for 0, 3, and 120 h, and next polarized at 60 °C with 3.5 kV cm^{-1} in 1 h. [3.147]

the sample owing to its exposure to atomic radiation from its surroundings. The extent of the filling is revealed by comparison of the light emitted by the naturally acquired TL with that of TL created artificially by means of a known dose of, e.g., X-rays. The age of the sample is given by

$$\text{age} = \frac{\text{natural TL} \times \text{artificial dose}}{\text{artificial TL} \times \text{natural dose rate}},$$

the natural dose rate being calculated independently.

As discussed in Chap. 4 and [3.56, 152–154], TSD finds further application in radiation dosimetry, a field in which TSEE and particularly TL are already well established ([3.155] see also [3.156] for the evaluation of the data).

3.6 Theory of Current TSD by Dipole Disorientation

3.6.1 TSD of Dipoles with One Relaxation Time

We shall first consider the discharge of a shorted electret in which the dipoles have a *single* relaxation time. Typical for such a discharge is that the TSD peak appears in a narrow temperature interval and that its shape and position are unaffected by changes in the poling conditions. Such a behavior is found for

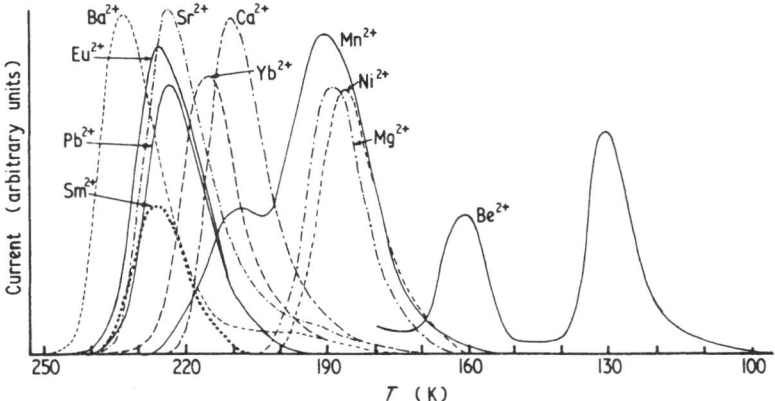

Fig. 3.18. I–V dipole peaks in X^{2+}-doped KCl crystals. [3.157]

most of the I–V dipoles formed by divalent *impurity* ions and their accompany-ing cation *vacancies* in alkali halides.

I–V peaks observed in KCl crystals doped with different divalent impurity ions are shown in Fig. 3.18. The peaks are attributed to a redistribution of the possible positions of the cation vacancy that jumps around the impurity ion. The change in position takes place mainly from next-nearest-neighbor to nearest-neighbor sites, and the redistribution process is therefore dominated by a single relaxation frequency [3.157, 158]. Accordingly, most of the curves show only one peak.

If a polar material contains N dipoles per m^3 with a dipole moment p, these will be oriented during the charging to produce a final polarization $P_0 = Np\overline{\cos\theta}$, where θ is the angle the dipoles subtend with the direction of the field. The alignment that is forced on the dipoles by the charging field E is counteracted by their thermal motion. For noninteracting dipoles (i.e., when the dipole concentration is sufficiently low), we have for the average orientation $\overline{\cos\theta} = pE/3kT$. During the subsequent TSD, the aligned dipoles will ran-domly disorient at a rate proportional to the number of dipoles still aligned. The polarization will therefore decay according to the Debye rate equation

$$dP(t)/dt + \alpha(T)P(t) = 0, \tag{3.1}$$

where $\alpha(T)$ is the reciprocal relaxation time, or relaxation frequency, which is assumed to be the *same* for all dipoles[8].

The density of the current generated by the decay in polarization equals

$$i(t) = -dP(t)/dt = -\alpha(T)P(t), \tag{3.2}$$

---

8 Note that α depends not only on T but, by virtue of the T–t heating programme, also implicitly on t. However, for the sake of briefness we have simply written $\alpha(T)$ instead of $\alpha\{T(t)\}$.

where $P(t)$ follows from (3.1) after integration[9],

$$P(t) = P_0 \exp\left[-\int_0^t \alpha(T)\,dt\right].$$ (3.3)

Since the temperature is raised linearly with time, the current density can also be written as a function of temperature,

$$i(T) = -\alpha(T)P_0 \exp\left[-h\int_{T_0}^T \alpha(T)\,dT\right],$$ (3.4)

where h is the *inverse* heating rate dt/dT.

As at low temperatures $\alpha(T)$ is small, electrets will retain their charge for a long time when stored at room temperature. During heating $\alpha(T)$ increases strongly. This increase often obeys an Arrhenius equation

$$\alpha(T) = \alpha_0 \exp(-A/kT)$$ (3.5)

where α_0 is the natural relaxation frequency and A is the activation energy needed to disorient a dipole. The energy A can be seen as a potential barrier, which the dipole has to surmount before it can readjust its direction [3.34–36]. Equation (3.5) broadly applies to relaxations involving the rotation of small molecular groups. It does not apply to the major relaxation in polymers, which occurs when they pass from the glassy to the rubbery state. This glass–rubber transition involves the configurational rearrangement of parts of the long main chains, which in view of their bulkiness require some space to move. Their relaxation rate thus depends on the unoccupied or *free volume*. The relaxations then obey the WLF equation proposed by *Ferry* and co-workers [3.19, 34, 40–42, 82, 83, 163],

$$\alpha(T) = \alpha_g \exp[c_1(T - T_g)(c_2 + T - T_g)^{-1}] \quad (T > T_g).$$ (3.5′)

For most amorphous polymers, $\alpha_g = 7 \times 10^{-3}\,\text{s}^{-1}$, $c_1 = 40$, and $c_2 = 52\,\text{K}$.

Our discussion will be based mainly on (3.5); for a description of TSD based on (3.5′), we refer to [3.11, 26, 164]. After substituting (3.5) for $\alpha(T)$, we can

---

9 Since P is a memory function, it is not uniquely defined by t or T alone if these variables vary independently. The interchange of t with T is therefore possible only by virtue of the interdependence of T and t through the heating programme. Because T is a function of t, P is in fact a *functional* of t during TSD, and so it would be more appropriate to write, e.g., $P_T(t)$ instead of simply $P(t)$ in (3.1). It would be incorrect, however, to write $P(t, T)$, which implies that P depends explicitly on t, *and on T*. This is the view of *Scaife*, who uses a partial derivative in (3.1). These points, which are of fundamental interest have led to some controversy [3.159–162]. It seems that *Scaife*'s theory [3.159] is applicable only to *pyroelectric* materials, in which P does depend explicitly on T, because the change in P with T now does *not* result from a *relaxation* process, but from the volume expansion or contraction of the material. In pyroelectric materials P therefore changes even when T decreases (cf. Chap. 5, and Fig. 3.56).

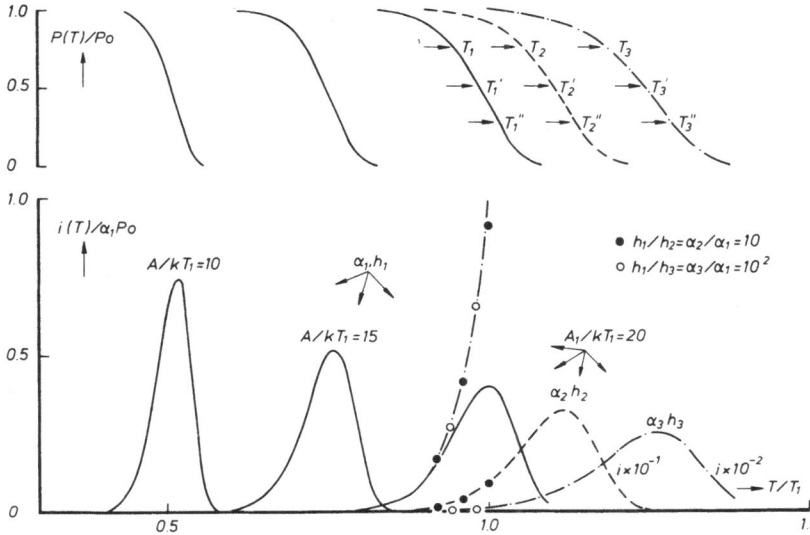

Fig. 3.19. Retained polarization and released current for a single Debye relaxation. The trend is shown for three activation energies, three natural frequencies and two heating rates [Notice that $P(T)$ cannot be observed in metallized electrets; it can, however, be calculated by integration of the current]. The upper curves also serve to illustrate a method for calculating the activation energy from experimental data. It is based on collecting a set of temperatures $T_{n1,2}$ at which the polarization is the same for the two heating rates [cf. Sect. 3.6.4].

approximate the integral of (3.4) in the following way:

$$h \int_0^T \alpha(T)\,dT \simeq \alpha(T)\,hkT^2(1+0.682\,kT/A)(A/kT+2.663)^{-1}, \qquad (3.6)$$

which has a relative error of 0.17‰ for $8 \leqq A/kT \leqq 362$. More sophisticated approximations are given in [3.11, 26, 165].

Model curves of the polarization and current density during TSD are plotted in Fig. 3.19. The polarization is seen to decrease gradually to zero. The current shows more structure: it first increases when the dipoles regain their mobility, passes through a maximum, and then drops sharply when the disorientation rate becomes high and the number of oriented dipoles becomes exhausted. The bell-shaped current peak is thus asymmetric. We further see that dipoles bound with a lower activation energy disorient at a lower temperature. Moreover, such dipoles produce a stronger current, because they disorient more or less en bloc in a smaller temperature interval. When the dipoles react more quickly, i.e., when α_0 is higher, the current maximum likewise appears at a lower temperature.

By differentiating (3.4), we find that the current maximum occurs when

$$\frac{d}{dT}\left(\frac{1}{\alpha(T)}\right) = -h, \qquad (3.7)$$

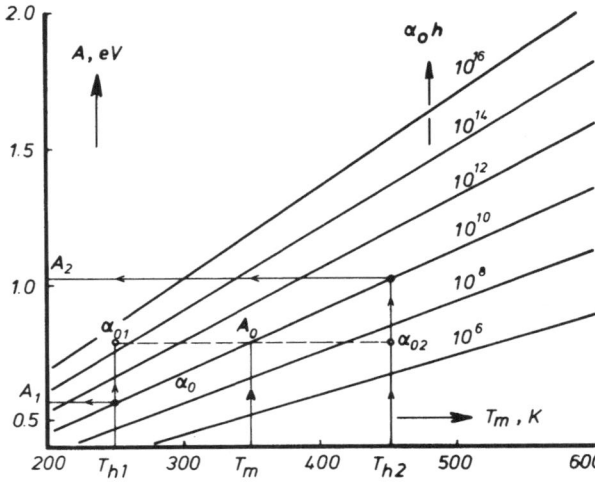

Fig. 3.20. A vs T_m plots, with $\alpha_0 h$ as the parameter [upon insertion of the half-value temperatures, $T_{h1,2}$ of a peak, such plots readily yield the limits of a distribution in activation energies ($A_{1,2}$) or in natural frequencies ($\alpha_{01,2}$); cf. Sect. 3.6.3]

which, for an Arrhenius type of α–T relation, transforms into

$$\alpha(T_m)\,hkT_m^2/A = 1 \,. \tag{3.8}$$

In Fig. 3.19 we showed that the temperature of the current maximum depends on A. This dependence is almost linear, as is evident from Fig. 3.20. *Simmons et al.* [3.166] have shown that, for practical values of the parameters,

$$A \simeq cT_m - 0.016\,\text{eV} \tag{3.9}$$

where $c = 1.92 \times 10^{-4}\,[\log(\alpha_0 h) + 1.67]$.

Figure 3.19 also shows that the current maximum increases for faster heating rates. This is logical, because the charge is released in a shorter time when dT/dt is higher; the curves coincide only in the beginning, when $h\int_{T_0}^{T} \alpha(T)/dT \simeq 0$. The sensitivity of TSD can thus be improved by an increase in the heating rate. However, too high rates are impracticable, because the temperature of the sample lags behind excessively. Moreover, nearby peaks become less well resolved. We see that the heating rate also affects the position of the maximum. The slower the heating, the more time the dipoles have to get disoriented, and the lower T_m.

3.6.2 Thermally Stimulated Charging of a Polar Sample

The common practice is to charge the sample isothermally at a sufficiently high temperature before it is subjected to TSD (Fig. 3.1). It can also be charged, however, while being heated linearly [3.21–24, 67, 68]. Such a thermally stimulated polarization (TSP), has several advantages. One of these is that the measured TSP current reveals how the orientation of the dipoles is proceeding. A second advantage is that the search for the optimum poling temperature is eliminated, and that unnecessary overheating of the specimen is avoided. In addition, TSP current measurements reveal the temperature at which the ohmic conduction becomes significant.

Since the charging is carried out in the presence of an electric field E, (3.1) has to be replaced with

$$dP(t)/dt + \alpha(T) P(t) = \alpha(T) P_e(T). \tag{3.10}$$

In (3.10), $P_e(T)$ is the equilibrium value of P, i.e., $P_e(T) = \varepsilon_0 (\varepsilon_s - \varepsilon_\infty)_T E$. The temperature dependence of the relaxation strength, $\varepsilon_s - \varepsilon_\infty$, follows from

$$\varepsilon_s - \varepsilon_\infty = Np^2/3kT, \tag{3.11}$$

where N and p have been defined before.

Assuming that the sample is initially uncharged, we shall use the following approximation for a solution of (3.10):

$$P(t) \simeq P_e(T) \left\{ 1 - \exp\left[- \int_0^t \alpha(T) dt \right] \right\}, \tag{3.12}$$

the change in $\varepsilon_s - \varepsilon_\infty$ being small compared with that of $\alpha(T)$. The exact solution of (3.10) is discussed in [3.167].

Equation (3.2) for the current density should be modified to include the conduction current,

$$i(t) = dP(t)/dt + g(T)E, \tag{3.13}$$

Like $\alpha(T)$, the ohmic conductivity $g(T)$ usually assumes an Arrhenius type of temperature dependence, i.e., $g(T) = g_0 \exp(U/kT)$. Combining (3.10), (3.12), and (3.13), we have for the charging current density,

$$i(t) = \alpha(T) P_e(T) \exp\left[- \int_0^t \alpha(T) dt \right] + g(T)E. \tag{3.14}$$

The two contributions to the charging current behave differently as functions of T. The orientation of the dipoles is a transient process giving rise to a peak, whereas the conduction current, which derives from the motion of equilibrium carriers, steadily increases with temperature. This is illustrated in Fig. 3.21. The conduction current appears at a higher temperature than the dipole orientation current, the former being due to a motion of charges over relatively large distances. In the temperature range of low conduction, the TSP and TSD currents are equal. But when the temperature eventually becomes high, the conduction current dominates in TSP.

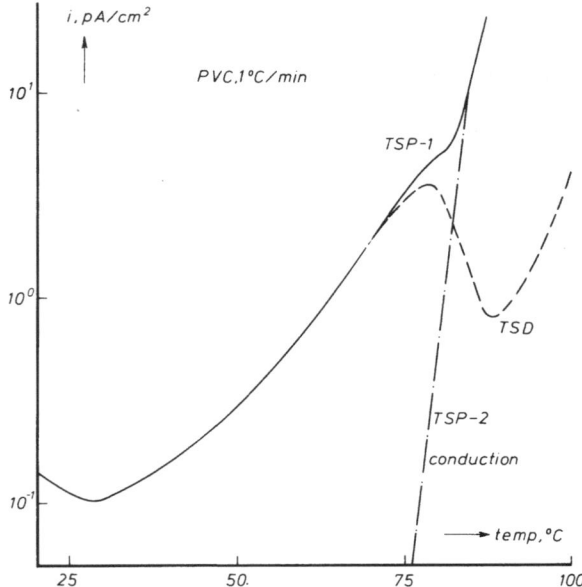

Fig. 3.21. TSP and TSD currents of the α peak of rigid PVC. Note that in a second charging only the conduction current contributes to the TSP current (dot-dash line). $E_p = 3.5\,\mathrm{kV}$ cm^{-1}, poling conditions for TSD: $T_p = 100\,°\mathrm{C}$, and $t_p = 1\,\mathrm{h}$

Figure 3.21 also illustrates how the contributions of the dipole orientation and the conduction can be separated. When the sample is reheated for a second time in the same field, no more dipoles will orient, and so only a conduction current will be observed. Another way to separate the two contributions is to change the heating rate. This will alter the shape and position of the dipole peak (cf. Fig. 3.19), but the conduction current remains the same.

3.6.3 TSD of Dipoles with a Distribution of Relaxation Times

As shown in Fig. 3.2, the low-temperature β peaks of methacrylic polymers are much broader than the I–V dipole peaks shown in Fig. 3.18. These β peaks therefore cannot be ascribed to a single Debye relaxation. They are due to the disorientation of the polar *side groups* in the polymer, which proceeds by a local rotation of the side groups around their C–C bond with the main chains. Considering the different conformations macromolecules may have, it is likely that the rotation of the side groups does not proceed in the same environment and therefore does not require the same *activation energy*. As a result the relaxation frequencies of all dipoles will differ:

$$\alpha_i(T) = \alpha_0 \exp(A_i/kT).$$

The differences might also arise from different values of the *pre-exponential factor* α_0, as is the case when the rotational masses of the dipoles are not equal. We then have

$$\alpha_i(T) = \alpha_{0i} \exp(A/kT).$$

Differences in α_0 are more likely for the α peaks in Fig. 3.2, because these peaks are associated with the glass–rubber transition, where the polar side groups move in unison with parts of the main chains differing in mass [3.34, 161b]. More generally, however, differences will occur in both α_0 and A.

The TSD of a dielectric with a number of relaxation frequencies shows several peculiarities. In Fig. 3.14 is illustrated what happens to the α peak of PMMA when this is charged at various temperatures. If the charging temperature is too low, only the fast dipoles are oriented. The slow ones then will not contribute to the peak and so this is cut off at high temperatures. Only if the poling temperature is chosen *above* the TSD maximum will all dipoles be aligned properly and the peak develop fully. Similarly, if the electret is stored too long before TSD, or stored at too high a temperature, the contribution of the fast dipoles will be lost, and the peak will be truncated on the low-temperature side (see also Fig. 3.14).

In these respects the TSD of a distributed polarization differs markedly from that of a single Debye polarization. When an electret with a single relaxation time is charged at too low a temperature or for too short a time, or when it is stored for too long, all $i(T)$ values are lowered in the *same* proportion. The Debye peak is thereby lowered, but its shape and location remain the same.

The marked difference in dependence on the *conditions* of *charging* and *storage* can serve as a criterion for differentiating between a distributed and a nondistributed peak. The shift of a peak with *heating rate* may also be helpful in deciding whether a broad peak is due to a single dipole relaxation with a low activation energy, or rather to a distributed dipole relaxation. When the shift is slight, the peak is undoubtedly due to a distributed polarization.

A distribution in relaxation frequencies can be either discrete or continuous. If it is the former, we can divide the dipoles into groups with the same relaxation frequency. Each group i will then contribute with a polarization and current given by (3.3) and (3.4) and the total response will be the algebraic sum of all responses of the individual groups.

One might ask for what differences in α_{0i} and A_i the constituent peaks appear just separated. Calculations have shown that the resolution limits are $\alpha_{0i}/\alpha_{0i-1} \geqq 5.8$ and $A_i/A_{i-1} \geqq 1.09$ [Ref. 3.11, p. 59]. Apparently, TSD enables resolution of relaxations whose kinetic parameters are quite close.

The distribution becomes continuous when all dipoles have different relaxation frequencies. The contributions of the various subpolarizations then have to be summed by an integral. For a distribution in α_0, we have

$$P(t) = P_0 \int_0^\infty f(\alpha_0) \exp\left[-\alpha_0 \int_0^t \exp(-A/kT)\,dt\right] d\alpha_0 ; \qquad (3.15)$$

and for one in A,

$$P(t) = P_0 \int_0^\infty g(A) \exp\left[-\alpha_0 \int_0^t \exp(-A/kT)\,dt\right] dA , \qquad (3.16)$$

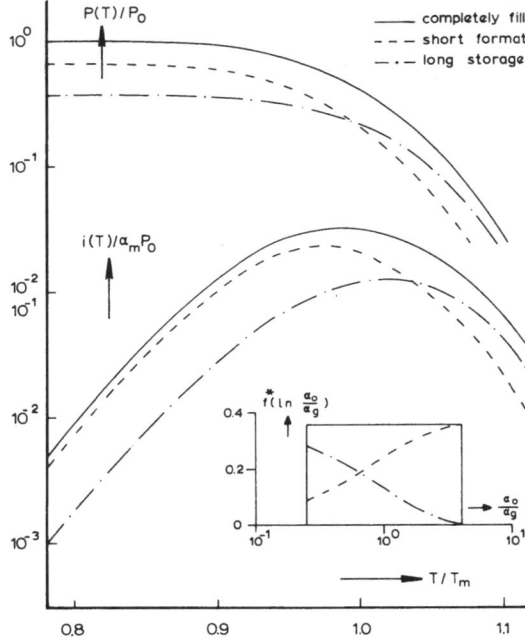

Fig. 3.22. Changes caused by a too short formation and a too long storage in the thermograms of an electret with a Gevers distribution in α_0. The insert gives the effective distribution function at the beginning of each of the TSD runs. [3.11]

where $f(\alpha_0)$ and $g(A)$ are the distribution functions in natural frequencies and in activation energies. These functions represent the contributions of the subpolarizations between $\alpha_0 \to \alpha_0 + d\alpha_0$ and $A \to A + dA$ to the total dipolar relaxation strength $\varepsilon_s - \varepsilon_\infty$. Actually, they are the relative contributions to $\varepsilon_s - \varepsilon_\infty$, because they are normalized, so that

$$\int_0^\infty f(\alpha_0)\,d\alpha_0 = \int_0^\infty g(A)\,dA = 1\,. \tag{3.17}$$

The corresponding expressions for the TSD current are obtained by differentiation of (3.15) and (3.16):

$$i(t) = P_0 \exp(-A/kT) \int_0^\infty \alpha_0 f(\alpha_0) \exp\left[-\alpha_0 \int_0^t \exp(-A/kT)\,dt\right] d\alpha_0\,, \tag{3.18}$$

and

$$i(t) = P_0 \alpha_0 \int_0^\infty g(A) \exp\left[-A/kT - \alpha_0 \int_0^t \exp(-A/kT)\,dt\right] dA\,. \tag{3.19}$$

At each temperature T, the TSD current can be thought of as being due to a Debye relaxation of which the current has its maximum at T. This argument enables one to readily find the limits of the range of A or α_0 for a specific i–T

curve. The method is demonstrated in Fig. 3.20. The range of A so found is slightly too large, and should on either side be lowered by 1.2 $kT_{h1,2}$, where $T_{h1,2}$ are the half-value temperatures. In a similar way we can find the range of α_0 when A is constant.

According to (3.18) and (3.19), the TSD current does not depend on the poling conditions. This independence follows from our assumption that the sample is *fully* charged, with all dipoles contributing to the fullest possible extent. If not, they do not contribute to the polarization according to $f(\alpha_0)$ or $g(A)$, but according to an *effective* distribution $f^*(\alpha_0)=f(\alpha_0)\,FS(\alpha_0)$ or $g^*(A)=g(A)\,FS(A)$ where FS represents the filling state of $f(\alpha_0)$ and $g(A)$ attained during the formation. FS also accounts for any disorientation of the dipoles during storage. For the completely filled state, FS = 1. Expressions for FS covering partial charging are given in (Ref. 3.11, p. 32; [3.26, 134]).

The integral transforms appearing in (3.18) and (3.19) usually have to be solved by numerical integration. A distribution for which (3.18) can be solved analytically is the box or *Gevers–Fröhlich* distribution [3.35, 168], for which[10]

$$f(u) = \frac{1}{u_2 - u_1} \quad \text{for} \quad u_1 \leqq u \leqq u_2 \,,$$
$$= 0 \qquad \text{for} \quad u_1 \geqq u \geqq u_2 \,; \tag{3.20}$$

where $u = \ln(\alpha_0/\alpha_g)$, and $\alpha_g = \sqrt{\alpha_1 \alpha_2}$.

An example of the TSD thermograms for such a distribution in α_0 is given in Fig. 3.22, which also shows one of the consequences of the incomplete filling resulting from a too short formation or a too long storage. The insert shows that in either case the effective "box" distribution is strongly deformed. This clearly has its repercussions for the current peaks, which are also deformed, and again stresses the importance of *complete charging*.

Various other empirical distribution functions have been proposed, e.g., by *Cole* and *Cole*, *Fuoss* and *Kirkwood*, and *Wagner* [3.11, 34–36, 169–171]. These distribution functions are symmetric, and suitable for describing the low-temperature β and γ peaks of polymers, whereas the high-temperature α peaks can be fitted best with asymmetric distribution functions [3.34–36], e.g., those of *Cole–Davidson* [3.172], and *Havriliak–Negami* [3.173], the latter of which has the disadvantage that it requires adjustment of two parameters, one for the width and one for the skewness of the distribution. To analyze TSD data with these empirical distributions, the integrals (3.18) and (3.19) are evaluated with a computer for the appropriate $f(\alpha_0)$ and $g(A)$ functions. In these calculations the width and skewness parameters are changed until a good fit with the experimental results is obtained; examples of such calculations are given in (Ref. 3.11, p. 49; [3.134, 174]).

---

10 Note that $f(u)$ represents the logarithmic distribution function, which is related to $f(\alpha_0)$ by $f\{\ln(\alpha_0/\alpha_g)\} = (\alpha_0/\alpha_g)f(\alpha_0/\alpha_g)$.

Williams and co-workers [3.141] have suggested another way to describe the asymmetric α peaks of polymers. They assume the TSD of a polarization to proceed according to

$$P(t) = P_0 \exp\left\{-\left[\int_0^t \alpha(T)dt\right]^n\right\}, \qquad 0 \leq n \leq 1, \tag{3.21}$$

where n is a parameter that measures the broadness of the peak. The smaller the value of n, the broader the peak; for a Debye peak, $n = 1$. Differentiating this equation with respect to time, we have for the current density,

$$i(t) = -n\alpha(T)P(t)\left[\int_0^t \alpha(T)dt\right]^{n-1}. \tag{3.22}$$

The evaluation of asymmetric peaks with (3.22) is less complicated than the use of the integral transforms (3.18) and (3.19).

Figure 3.23 shows that (3.22) can readily be made to fit the TSD peaks of a supercooled liquid. It should be borne in mind that a polarization according to (3.21) only holds after a *complete* charging. The same holds for *Jonscher's* TSD theory [3.175], which is based on the screened hopping model. The polarization after an *incomplete* charging can be calculated with Boltzmann's superposition principle.

3.6.4 Methods for Distinguishing Between a Distribution in A and in α_0

It would be interesting to see which of the two distributions is active in a material. In Fig. 3.16 we showed that a distinction can be made by means of *multistage TSD*, in which the heating is interrupted at increasingly higher temperatures. After each interruption the electret is cooled and reheated for another TSD. The distributed polarization is thus investigated in parts. In the first TSD runs the fast dipoles are mobilized, while in the later runs, when the temperature becomes higher, the slow dipoles get a chance to disorient. Now we can easily show, by differentiating (3.4), that the initial slope of the current is proportional to the activation energy. Thus, when the activation energy is constant, as is the case for a distribution in α_0, the initial slope in the successive heating runs does not change, but when A is distributed, the initial slope increases steadily. The method of *thermal sampling* or polarization in parts can likewise be used to see which of the two quantities is distributed [3.139].

Another method of proving whether a distribution in α_0 is present or not has recently been proposed by *Solunov* et al. [3.176]. It is based on measurement of the TSD currents at *two* different *heating rates*, $h_{1,2}^{-1}$, and integration of the currents to calculate the course of the corresponding polarizations. The

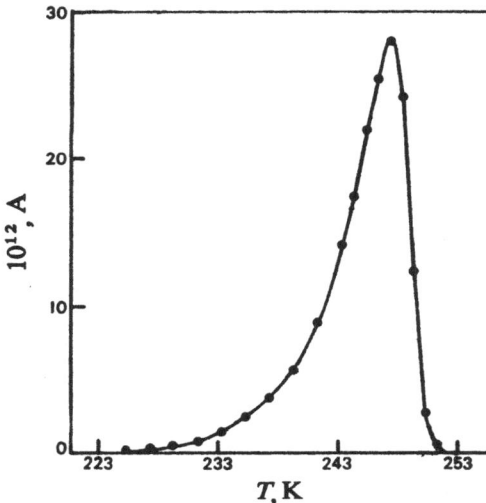

Fig. 3.23. TSD current in a supercooled liquid solution of 1.3 % of tri-n-butylammonium picrate in o-terphenyl ($T_p = 260$ K, $V_p = 1$ kV/1.5 mm, and $\beta = 1.2$ K min^{-1}). The theoretical fit through the experimental points is based on $n = 0.64$ and $A = 1.7$ eV. [3.141]

authors then collect a set of temperatures T_{1n} and T_{2n}, in such a way that the two polarizations at these temperatures are equal (see Fig. 3.19). For this to be true we must have, according to (3.3),

$$h_1 \int_0^{T_{1n}} \exp(-A/kT)dT = h_2 \int_0^{T_{2n}} \exp(-A/kT)dT , \qquad (3.23)$$

or, approximately

$$h_1 T_{1n}^2 \exp(-A/kT_{1n}) \simeq h_2 T_{2n}^2 \exp(-A/kT_{2n}) . \qquad (3.24)$$

For the ratio of the currents at $T_{1,2n}$, we have

$$i(T_{1n})/i(T_{2n}) = \exp(-A/kT_{1n} + A/kT_{2n}) . \qquad (3.25)$$

This simple relation is due to the integral transforms in (3.18) being equal for the two currents in view of (3.23). The activation energy of an α_0 distribution can be calculated from (3.24) and (3.25), provided that $T_{1,2n}$ can be determined accurately. The latter requirement is the main limitation of the method.

Whether the assumption of a distribution in α_0 is correct can most easily be checked with the equality

$$h_1 T_{1n}^2 i(T_{1n}) = h_2 T_{2n}^2 i(T_{2n}) , \qquad (3.26)$$

which results from substitution of (3.24) in (3.25). Thus, when $hT^2 i(T)$ vs $P(T)$ is plotted for the two heating rates, the resulting curves should coincide if the distribution is one in α_0.

3.7 Theory of Current TSD by the Self-Motion of Charges

A dielectric never insulates perfectly, particularly not at high temperatures, when an increasing number of thermal carriers are generated. During poling these carriers will move in the applied field and, when the dielectric is cooled, some of them will be trapped. Owing to this process, thermally formed electrets usually contain a space-charge polarization. Space charges can also be injected from the electrodes, especially at high field strengths [3.106, 177].

At low temperatures the space charges remain frozen in, but when the temperature is raised they are remobilized. During TSD they will move in the internal field of the electret, and restore its charge neutrality. This neutralization can proceed in two ways: (I) the charges move to the electrodes and recombine with their image charges; (II) they recombine with their counterparts within the electret. The latter will of course not occur if the electret has accumulated charges of *one* sign only. Such *unipolarly* charged electrets are obtained by, e.g., injection of electrons into the sample by electron bombardment or corona charging. Since these electrets are now produced by the million, we shall here deal with the self-motion of charges in this type of electret. (For an analysis of bipolarly charged dielectrics, see [Ref. 3.11, p. 127, [3.178–180].)

Neglecting diffusion, we can write for the decrease in, say, a positive space-charge density ϱ during TSD,

$$\partial\varrho(x,t)/\partial t = -\mu(T)\partial\varrho(x,t)E(x,t)/\partial x - g(T)\partial E(x,t)/\partial x. \qquad (3.27)$$

In this continuity equation, $\mu(T)$ is the mobility of the space charges and $g(T)$ the ohmic conductivity. Both quantities are thermally activated, and usually change with temperature according to an Arrhenius equation. We thus imagine the space charge to move virtually freely, like ions, by hopping across a potential barrier. Such a motion is similar to that of electrons in a band-gap model with strong retrapping. A more general description of the vicissitudes of detrapped electrons is given in Sect. 3.10. We further stress that we assume a single potential barrier, so that all space charges have the *same* $\mu(T)$. In this respect, however, the model can be easily generalized [3.178].

E and ϱ are related by Poisson's equation

$$\partial E(x,t)/\partial x = \varrho(x,t)/\varepsilon_0\varepsilon. \qquad (3.28)$$

The density of the current released by the space-charge motion is given by

$$i(t) = \varepsilon_0\varepsilon\partial E(x,t)/\partial t + [\mu(T)\varrho(x,t) + g(T)]E(x,t). \qquad (3.29)$$

This expression can be integrated with respect to x, and simplified by application of the short-circuit condition,

$$\int_0^s E(x,t)dx = 0. \qquad (3.30)$$

We thus obtain

$$i(t) = [\varepsilon_0 \varepsilon \mu(T)/2s][E^2(s, t) - E^2(0, t)], \tag{3.31}$$

where $E(0, t)$ and $E(s, t)$ are the boundary values of E obtained by integration of (3.28) and use of (3.30). Evidently, the external current results only from the *self-motion* of the space charges, and neither the displacement current nor the conduction current contributes to it.

The mathematical description of the space-charge motion requires two variables, namely space and time, and so is more complex than that of dipoles, whose position is irrelevant because they are not displaced by a field.

The *partial* differential equations describing the motion can only be solved analytically for very simple charge distributions, such as a space charge cloud whose initial charge density $\varrho(x, 0)$ is constant up to a depth r_0. This distribution has the interesting property that it keeps the form of a box during TSD, but its height reduces as it expands [3.11, 181–184].

To calculate the decrease in height we particularize (3.27) at the zero-field point $x_0(t)$ where $E(x_0, t) = 0$. This simplifies the equation because it reduces the term $\partial \varrho(x, t)E(x, t)/\partial x$ to $\varrho^2(x_0, t)/\varepsilon_0 \varepsilon$. Moreover, since for a uniform distribution $\varrho(x, t)$ is the same up to the leading front, we also have $\partial \varrho(x_0, t)/\partial t = d\varrho(x_0, t)/dt$. The continuity equation can thus be recast into

$$d\varrho(x_0, t)/dt = -[\mu(T)\varrho(x_0, t) + g(T)]\varrho(x_0, t)/\varepsilon_0 \varepsilon. \tag{3.32}$$

This *ordinary* differential equation can readily be integrated, either analytically for $g(T) = 0$, or numerically if $g(T) \neq 0$.

The expansion of the space charge cloud can be found from the argument that a charge in an electric field acquires a velocity of μE. The leading front will therefore expand to the back electrode at a velocity of

$$dr(t)/dt = \mu E(r, t). \tag{3.33}$$

In this equation,

$$E(r, t) = E(s, t) = \varrho(x_0, t)r^2(t)/2\varepsilon_0 \varepsilon s, \tag{3.34}$$

as follows from (3.28) and (3.30). The set of equations is now sufficiently complete for calculation of the discharge current density from

$$i(t) = -\mu(T)\varrho^2(x_0, t)r^2(t)[1 - r(t)/s]/2\varepsilon_0 \varepsilon s. \tag{3.35}$$

By integrating this current we can find how much *charge* is *released* during the heating. We shall here restrict ourselves to calculating the charge ultimately released. This will obviously be a maximum when no space charges are lost by ohmic conduction, i.e., when $g(T)$ is zero. It will also be clear that the charge release will stop as soon as the charge is uniformly spread throughout the sample. For then the space charges will leave the sample at the same rate, at both electrodes, and $i(t) = 0$.

For times less than the crossing time t_λ, (3.29) simplifies to

$$i(t) = \varepsilon_0 \varepsilon d E(s, t)/dt , \tag{3.36}$$

because $\varrho(s, t) = 0$ and $g(T) = 0$. By integrating this equation we find for the final value of the charge density released,

$$q(t_\lambda) = \varepsilon_0 \varepsilon [E(s, t_\lambda) - E(s, 0)] , \tag{3.37}$$

in which the two fields can be specified with the aid of (3.32–34). We thus obtain

$$q(t_\lambda)/p_0 r_0 = -(r_0/2s)\{1 - (s^2/r_0^2)\exp[2(1 - s/r_0)]\} , \tag{3.38}$$

which reduces to

$$q(t_\lambda)/p_0 r_0 \simeq -r_0/2s \tag{3.39}$$

for $r_0 < s$. On this equation *Sessler* ([3.185]; Sect. 2.4.3) has based his method of finding the mean *spatial depth*.

Equations (3.38) and (3.39) show that the charge release is determined by the *thickness* of the *charge layer* relative to that of the sample. Thin samples will therefore release more charge than thick ones, because the low-energy electrons used for electrets do not penetrate to a depth of more than a few μm. From (3.38), we can calculate that the maximum charge recoverable in the TSD of space charges is 14% of the initial charge. It goes without saying that the TSD current also depends on the thickness of the charge layer. From Fig. 3.24 we see that the current released is strongest when $r_0/s = 0.5$. The effect of the ohmic conductivity is also shown; this lowers the ensuing current and charge because it neutralizes part of the space charges.

In practice the shape of the initial charge distribution will generally differ from that of a box. For electron-bombarded foils, a distribution residing within the material is more likely for the following reason. The outer layers of the sample will receive no electrons, because these need a certain distance to slow down. But on their way through the sample they gradually lose energy and therefore penetrate only up to a certain distance.

For such a floating space charge distribution, i.e., not touching the electrodes, *Leal Ferreira* et al. [3.186] have shown that the set of equations can be combined in the following way (see also Sect. 4.2.6). From (3.28), we have

$$\hat{\sigma}(t) = \varepsilon_0 \varepsilon [E(s, t) - E(0, t)] , \tag{3.40}$$

where $\hat{\sigma}(t) = \int_0^s \varrho(x, t)dx$ is the total space charge stored. This stored charge will remain constant if $g(T) = 0$ and as long as the charge fronts have not yet reached the electrodes. And so, we have $\sigma(t) = \hat{\sigma}_0$, for $t < t_\lambda$.

During this time (3.31), (3.36), and (3.40) give

$$dE(s, t)/dt = [\mu(T)\hat{\sigma}_0/2\varepsilon_0\varepsilon s][2E(s, t) - \hat{\sigma}_0/\varepsilon_0\varepsilon], \tag{3.41}$$

from which $E(s, t)$ can be solved. Substitution of $E(s, t)$ into (3.36) gives for the TSD current during $0 \leq t \leq t_\lambda$;

$$i(t) = [\mu(T)\hat{\sigma}_0/s][E(s, 0) - \hat{\sigma}_0/2\varepsilon_0\varepsilon] \exp\left[(\hat{\sigma}_0/\varepsilon_0\varepsilon s) \int_0^t \mu(T)dt\right]. \tag{3.42}$$

This current has a temperature dependence similar to that of a single dipole relaxation, which implies that the initial part of the TSD current due to the self-motion of floating space charges can be analyzed in the same way as that of a Debye peak. In contrast with that of dipoles, the current does *not*, however, depend *linearly* on the initial charge. Instead, if $\hat{\sigma}_0 > \varepsilon_0\varepsilon E(s, 0)$, it increases quadratically with $\hat{\sigma}_0$. Such a quadratic dependence is to be expected, because the space charges supply the carriers for the current as well as the driving field (in other words the self-motion is *space-charge limited*, and therefore is also called SCL drift [3.11]).

The change in TSD current due to a threefold increase in the initial charge of a box distribution is illustrated in Fig. 3.25. In agreement with (3.42) the initial current increases by a factor of 9. The current maximum, by contrast, is seen to increase by no more than about three times. This means that it varies almost *linearly* with the initial charge. As a consequence it is hard to decide whether a peak arises from space charges or not solely by comparison of the height of the current maxima for different poling fields. More decisive is the *shift* of the peak to lower temperatures, which is due to the higher charge speeding up the drift of the charges to the electrodes.

The current for space-charge clouds of any other shape must be calculated by numerical solution of the two hyperbolic partial differential equations (3.27) and (3.28). The calculation can be carried out by either of two methods, viz. the *method of characteristics* [3.180, 181] and the method of replacing differential equations with *difference equations* (Ref. 3.11, p. 296; [3.25]). Suitable difference schemes were found to be those of *Lax–Wendroff–Richtmyer* and *Wendroff–Thomée* [3.187–189]. Using these schemes we made model calculations of the temperature dependence of the electron distribution one would expect for electron-bombarded foils [3.190]. The results are shown in Fig. 3.26. We see that the bell-shaped distribution eventually tends to become uniform. More results of such calculations are given in [Ref. 3.11, p. 123]. Another numerical solution has been described by *Chen* [3.178], who uses first-order difference equations, which he solves by successive iterations.

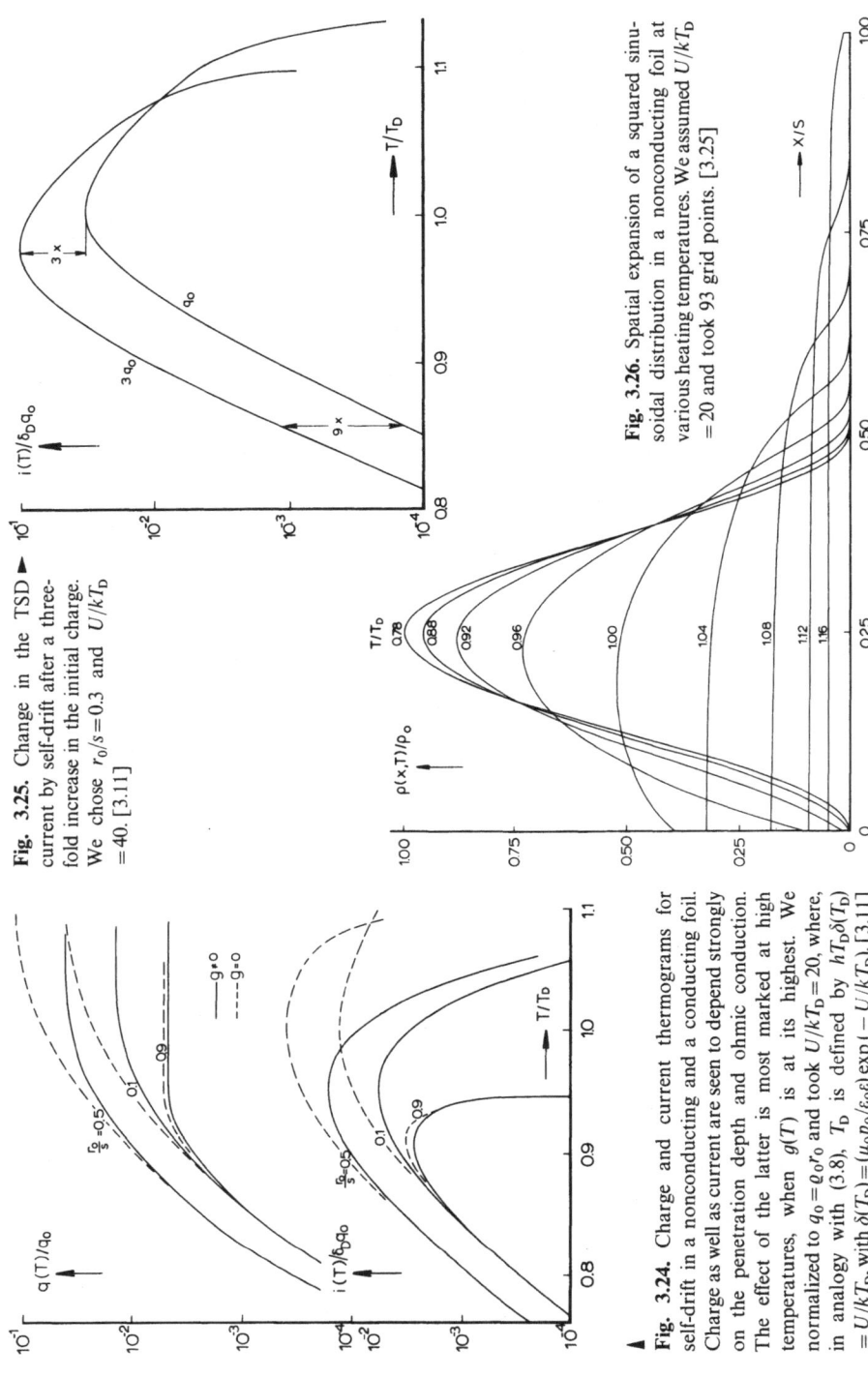

Fig. 3.25. Change in the TSD current by self-drift after a three-fold increase in the initial charge. We chose $r_0/s = 0.3$ and $U/kT_D = 40$. [3.11]

Fig. 3.26. Spatial expansion of a squared sinusoidal distribution in a nonconducting foil at various heating temperatures. We assumed $U/kT_D = 20$ and took 93 grid points. [3.25]

Fig. 3.24. Charge and current thermograms for self-drift in a nonconducting and a conducting foil. Charge as well as current are seen to depend strongly on the penetration depth and ohmic conduction. The effect of the latter is most marked at high temperatures, when $g(T)$ is at its highest. We normalized to $q_0 = \varrho_0 r_0$ and took $U/kT_D = 20$, where, in analogy with (3.8), T_D is defined by $hT_D\delta(T_D) = U/kT_D$, with $\delta(T_D) = (\mu_0 p_0/\varepsilon_0 \varepsilon)\exp(-U/kT_D)$. [3.11]

In the method of characteristics, the partial differential equations are replaced with ordinary differential equations. These are solved in x–t space along the characteristics, which coincide with the particle trajectories or flow lines, and which are defined by $x_t = x(t)$. Here, (3.27) reduces to the following set of differential equations:

$$\frac{dt}{1} = \frac{dx_t}{\mu(T)E(x_t, t)} = -\frac{\varepsilon_0 \varepsilon d\varrho(x_t, t)}{\varrho(x_t, t)[\mu(T)\varrho(x_t, t) + g(T)]}. \qquad (3.43)$$

These ordinary differential equations can readily be solved either analytically or numerically. For more details about this method for calculating the discharge current of electrets, we refer to the work by *Leal Ferreira* et al. [3.191–193]. A similar approach is used by *Zahn* [3.180], who focusses on problems in which rather than the initial charge distribution the current in or the voltage across the sample is prescribed. He starts from (3.29) and uses the set

$$\frac{dt}{1} = \frac{dx_t}{\mu(T)E(x_t, t)} = \frac{\varepsilon_0 \varepsilon dE(x_t, t)}{i(t) - g(T)E(x_t, t)}. \qquad (3.44)$$

Mention should also be made of a recent congress on "Space Charges in Dielectrics" where some Polish workers came up with a clever manipulation of the method of characteristics for deriving the internal charge distribution from TSD data [3.194]. An alternative technique for solving the set of partial differential equations has been described by *Lonngren* [3.195]. Most of the authors just cited analyse the current transients due to the collapse of a space-charge-limited type of charge distribution [3.192, 193, 195].

We have so far taken the conductivity $g(T)$ to be *independent* of x, implying that it arises solely from ohmic conduction. This assumption is, however, unrealistic for electron-bombarded foils, the irradiated part of which is much more conductive than the nonirradiated part owing to the formation of electron–hole pairs [3.91, 92][11]. As a result of this radiation-induced conductivity (RIC) and of the self-motion of the deposited electrons, the initial distribution of electrons may be quite different from that shown in Fig. 3.26. As explained in Chap. 4 and [3.196], the entire charge accumulates as a planar charge near r_0, provided that the RIC is assumed to be uniform up to r_0. However, if we make the more realistic assumption that the x dependence of the RIC resembles that of the primary electrons [3.197], we find a deposition profile like that in Fig. 3.26, though much more confined (cf. Fig. 3.72).

Although the RIC gradually declines, it does persist for quite some time, and so is likely to be effective during storage and TSD. According to *Sessler* et al. [3.93, 94], its dependence on *time* and temperature causes the TSD peaks to

---

11 A nonuniform conductivity can also be induced by asymmetric heating. It has been analyzed by *Coelho* et al. ([2.198], see also Sect. 4.6.6).

shift with storage time. We shall briefly consider this matter (which is discussed in full in Chap. 4) in Sect. 3.10. In this section, and in Sect. 3.11.9, we shall present some experimental results originating from the detrapping and self-motion of charges.

3.8 Evaluation of TSD Current Data

The aim of the evaluation is to derive the magnitude of the molecular parameters responsible for the TSD. For dipoles, these are the activation energy, the natural orientation frequency, the relaxation strength and the distribution function. We shall first discuss several methods for calculating the activation energy of dipoles with a *single* relaxation frequency. The methods reviewed here [cf. Ref. 3.11, Chap. 3] are very similar to those used for obtaining the trap depth from TSCo and TL curves [3.2, 12, 13, 199–204] and for finding the activation energy from thermoanalytical data [3.59, 60, 205].

By differentiating (3.4) with respect to $1/T$, we obtain for the initial current rise $\left[\text{when } h \int_0^T \alpha(T)dT \simeq 0\right]$

$$\frac{d}{d(1/T)} \ln i(T) = -A/k. \tag{3.45}$$

Thus, by plotting $\ln i$ vs $1/T$, we can find the activation energy A (cf. Fig. 3.27). This can also be calculated from

$$\alpha(T) = i(T)/P(T) = i(T)/h \int_T^\infty i(T)dT. \tag{3.46}$$

This equation was suggested as early as 1930 by Urbach (the originator of TL), but is now often credited to *Fieschi* and co-workers [3.7]. An example of a *Bucci–Fieschi* plot is given in Fig. 3.27.

A third method for calculating A is the use of the halfwidth of the current peak. The higher A, the narrower the peak at a particular T_m, because the electret charge has to be released in a narrower temperature interval. For the halfwidth, we can derive

$$\Delta T/T_m \simeq 2.47 \, kT_m/A. \tag{3.47}$$

More elaborate formulas are given in (Ref. 3.11, p. 67; [3.13, 200]).

If, finally, the heating rate is changed from h_1^{-1} to h_2^{-1}, we can deduce A in yet another way. The deduction is based on the shift of the temperature maxima $T_{1,2}$ which, according to (3.8), obey

$$h_1 T_1^2 \alpha(T_1) = h_2 T_2^2 \alpha(T_2) = \text{const}. \tag{3.48}$$

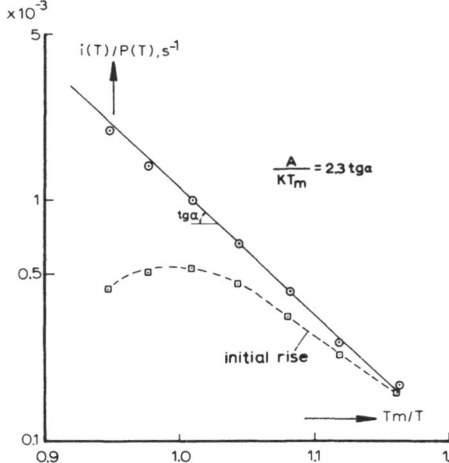

Fig. 3.27. BF plot (*solid line*) and initial rise (*broken line*) of the γ peak of a polycyclohexyl methacrylate electret

If the heating rate is changed more than twice, it is advantageous to find A from a $h_m T_m^2$ vs $1/T_m$ plot [3.78]. Such a plot need not be based solely on the TSD maxima. The other points of the TSD peaks may also be invoked. This is done in the method of *Solunov* et al. [3.137, 176], for which two heating rates suffice to find A. Another way to use all points of a peak is to fit (3.2) to the experimental data with a computer (Ref. 3.11, p. 75; [3.206, 207]).

Once the activation energy is known, α_0 can be calculated by means of (3.8). The kinetic quantities for describing the dipole reorientation are then known. We have deliberately given several methods for calculating A. It is always advisable to check whether the calculated values of A are consistent. If not, one might expect the dipoles to react with more than one relaxation frequency, as is the case for most TSD peaks of polymer electrets.

From the data we can finally calculate the polarization originally stored by integrating the current:

$$P_0 = h \int_{T_0}^{\infty} i(T)dT. \tag{3.49}$$

If the sample has been charged by a field E to its equilibrium value, we also have

$$P_0 = \varepsilon_0(\varepsilon_s - \varepsilon_\infty)E, \tag{3.50}$$

from which the relaxation strength $\varepsilon_s - \varepsilon_\infty$ can be found. Equation (3.11) shows how this quantity is related to the molecular parameters N and p.

For distributed polarizations formulas, (3.45–47) have to be modified. These modifications can be derived by integration of (3.18) for the distributions best known (Ref. 3.11, p. 54; [3.134]). In this way we have shown that for the α_0

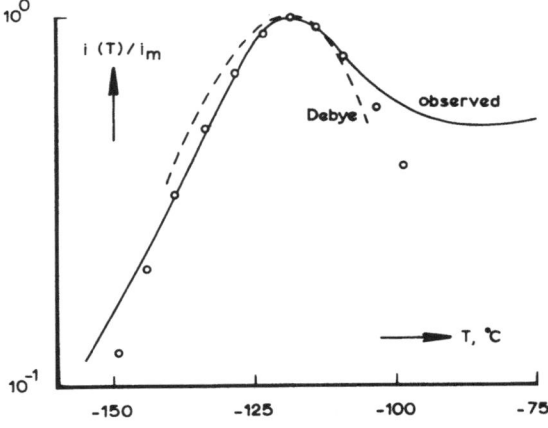

Fig. 3.28. A *Fuoss-Kirkwood* fit (circles) to the γ peak of poly-cyclohexyl methacrylate. Note that a simple Debye fit (*dashed line*) deviates rather substantially from the data

distribution proposed by *Cole* and *Cole*, and *Fuoss* and *Kirkwood* [3.169–171], (3.45) has to be changed into

$$\frac{d}{d(1/T)}\ln i = -nA/k, \qquad 0 \leqq n \leqq 1 \tag{3.51}$$

where n is a distribution parameter that determines the width of the distribution. This parameter can often be obtained from the literature or from dielectric loss measurements [3.34–36, 169–173]. For broad distributions, with a low n, the initial rise of the current is therefore considerably *less* than that of a Debye relaxation. This effect is obvious, because a broad distribution forces the TSD current to be spread over an extended temperature range. Consequently, the current maximum is also lowered by a factor of about n, in comparison with a Debye peak. The same holds for the ratio between the halfwidth temperatures, so that n must also be introduced into (3.47).

The initial currents for an α_0 distribution of *Gevers* (box) and *Wagner* (Gaussian) are less affected by the broadness of the distribution; for these (3.45) still holds. This is probably so, because these distributions have no extended tails like the *Cole–Cole* and *Fuoss–Kirkwood* distributions.

The experimental results of most materials with a distributed polarization, however, show a low current rise and are therefore more compatible with a *Cole–Cole* or a *Fuoss–Kirkwood* distribution.

An illustration of this is given in Fig. 3.28, which shows that the TSD peak due to the γ relaxation of polycyclohexyl methacrylate can be nicely fitted to a *Fuoss–Kirkwood* distribution with $n = 0.4$ and $A = 8.4$ kcal/mol.

The expressions for the initial slope for the *Cole–Davidson* and *Havriliak–Negami* distributions [3.172, 173] which have been advanced to describe asymmetric relaxation peaks are the same as given by (3.51), except that for the *Havriliak–Negami* distribution, which has two parameters, n equals the product of the two. Differentiating the expression of *Williams* for asymmetric peaks (3.22) shows that here the initial slope also satisfies (3.51).

3.8.1 Calculation of Distribution Functions

All dielectric distributions suggested in the literature are empirical, and so are not derived from physical principles. One can therefore question the correctness of assuming a specific distribution function. It is more appropriate not to prescribe this function but to derive it from the TSD data (Ref. 3.11, p. 77; [3.26]).

For a distribution in α_0, this can be done by modification of the approximations of *Schwarzl* and *Staverman* for isothermal mechanical measurements [3.208]. We can also adopt the method due to *Alfrey* et al. [3.209]; this is applicable to both α_0 and A distributions. It approximates

$$\exp\left[-\alpha_0 \int_0^t \exp(-A/kT)dt\right] \equiv \exp(-X)$$

as follows:

$$\exp(-X)=1 \quad \text{for} \quad 0 \leq X \leq 1, \tag{3.52}$$
$$=0 \quad \text{for} \quad X \geq 1;$$

and reduces (3.15) and (3.16) to

$$P(t) \simeq P_0 \int_{\alpha_1}^{\infty} f(\alpha_0)d\alpha_0, \tag{3.53}$$

$$P(t) \simeq P_0 \int_{A_1}^{\infty} g(A)dA, \tag{3.54}$$

respectively, where α_1 and A_1 form the borderline for which X equals one. By differentiating (3.53) and (3.54) we find

$$dP/d\alpha_1 \simeq P_0 f(\alpha_1), \tag{3.55}$$

$$dP/dA_1 \simeq P_0 g(A_1). \tag{3.56}$$

Putting $X = 1$, which conforms with (3.8) for the maximum of a Debye peak, we can replace the differentiation with respect to α_1 and A_1 by one with respect to T:

$$P_0 f(\alpha_1) \simeq i(T)/\exp(-A/kT), \tag{3.57}$$

$$P_0 g(A_1) \simeq hTi(T)/A_1. \tag{3.58}$$

These first-order approximations for the two distributions are simple and convenient. However, instead of $f(\alpha_1)$ one frequently uses the logarithmic

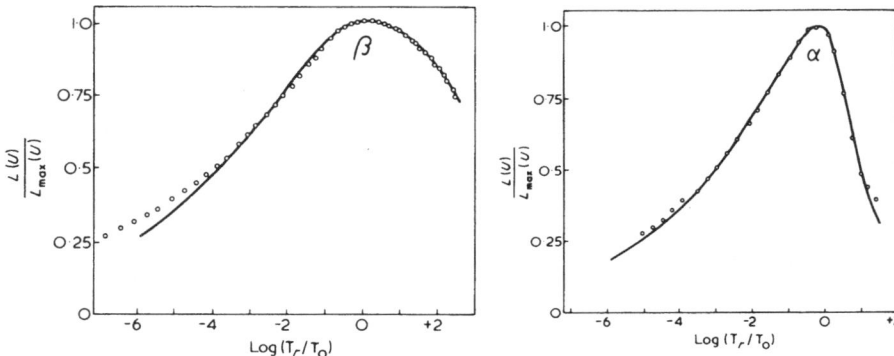

Fig. 3.29. Logarithmic distribution functions calculated from the β and α peaks in PMMA and fitted with a *Fuoss-Kirkwood* and a *Havriliak-Negami* distribution. [3.211]

distribution $f(\ln \alpha_1)$, in view of the vast range of α_0 values involved. This distribution can be calculated from

$$P_0 f(\ln \alpha_1) \simeq hkT^2 i(T)/A . \tag{3.59}$$

By combining (3.58) and (3.59) we find that

$$f(\ln \alpha_1) \simeq kTg(A_1) . \tag{3.60}$$

We can afterwards show that this expression is exact by equating (3.15) and (3.16). After calculation of one of the two distributions from the TSD data, it is therefore an easy matter to find the other distribution, by using the interrelation (3.60).

Equations (3.58) and (3.59) have been used by *Vanderschueren* and by *Kryszewski* et al. [3.210, 211] to calculate the distributions of the β and α peaks of some methacrylic polymers. *Kryszewski's* results are shown in Fig. 3.29. It appears that the distribution of the β peaks can be fitted well with a *Fuoss–Kirkwood* distribution and that of the α peaks with a *Havriliak–Negami* distribution.

Equation (3.58) has also been derived by *Fischer, Pfister*, and *Ikeda* et al. [3.137, 212, 213], and by *Simmons* et al. [3.214]. The last-mentioned actually give their derivation for the detrapping of electrons out of distributed traps, and calculated the trap distribution function with (3.58). A different approach for calculating a distribution in A has been attempted by *Hino* [3.215]. He approximates a continuous distribution by a discrete one of, say, 30 Debye relaxations and calculates the different subpolarizations and energies by a matrix inversion.

A very different attack is as follows: Instead of calculating the distribution function from a single TSD run, it is also possible to use fractional polarization or depolarization and to decompose the distributed response experimentally into its elementary components. This procedure has the obvious disadvantage

Fig. 3.30. Fractional glow curves of a ZnS phosphor activated with Cu and Ga (*left*). From these curves the trap spectrum on the right has been calculated (the group of traps at 0.95 eV has not been drawn). [3.216]

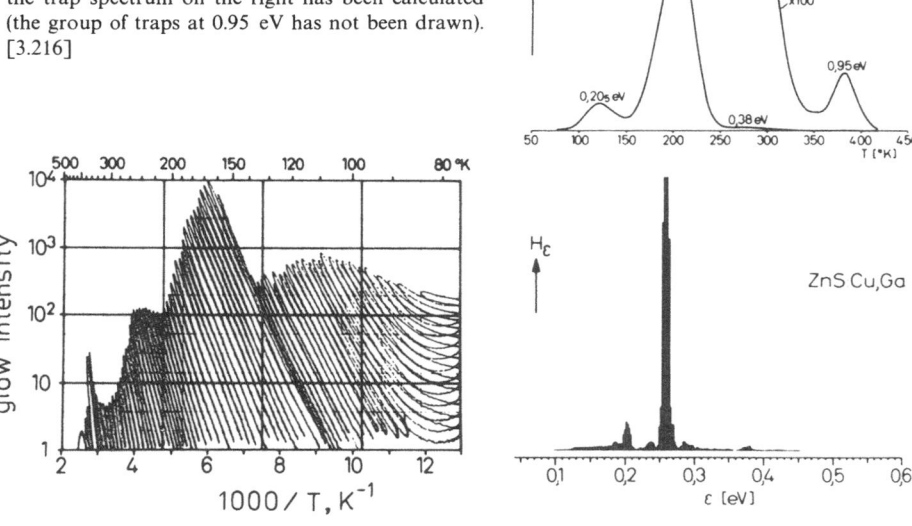

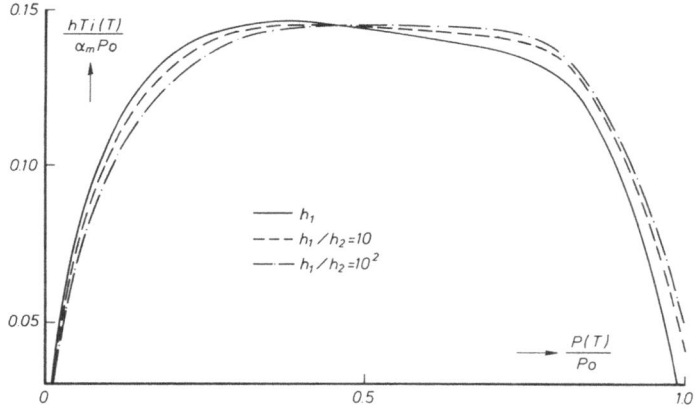

Fig. 3.31. $hTi(T)$ vs $P(T)$ plots illustrating how the presence of a distribution in A can be checked using (3.61). We took $A_1/A_m = A_m/A_2 = 0.82$ and $A_m/kT_m = 20$

that the TSD must be repeated several times to obtain sufficient resolution. *Gobrecht* et al. [3.216] have, however, managed to avoid this in an ingenious way by superimposing a large number of *small temperature oscillations* upon the linear TSD heating. They so obtain a large number of initial rise curves, from which the distribution in A is readily found (cf. Fig. 3.30).

In Sect. 3.6.4, we discussed *Solunov*'s method [3.176] for checking the presence of a distribution in α_0 by measuring the change in current with heating rate. We are now in a position to indicate how $i(T)$ will change for a distribution in A. Again collecting a set of currents and temperatures at which the polarizations for the two heating rates are the same, we apparently have from (3.58)

$$h_1 T_{1n} i(T_{1n}) \simeq h_2 T_{2n} i(T_{2n}). \tag{3.61}$$

This formula only holds approximately, whereas the corresponding expression for a distribution in α_0, (3.26), holds exactly. There is a slight difference in the two expressions, and this might enable one to distinguish between a distribution in α_0 and one in A.

An impression of the accuracy of (3.61) for a *Gevers*'s distribution in activation energies is given in Fig. 3.31. We see that the $hTi(T)$ vs $P(T)$ plots almost coincide for a difference in heating rate of a factor of 100.

3.8.2 Evaluation of Data Pertaining to the Self-Motion of Space Charges

When the TSD is due to the self-motion of space charges, we are interested in parameters such as carrier mobility, spatial distribution, and penetration depth. If all carriers have the same mobility, the activation energy (or trapping depth) can be calculated according to Sect. 3.8 (Ref. 3.11, p. 120; [3.178]). But if the carriers have different mobilities owing to different activation energies, recourse must be had to (3.58) for calculating the distribution in activation energies. The mean penetration depth can be estimated with (3.39) as is discussed in Sect. 2.4.3.

Experimental methods for finding the spatial distribution will be outlined in Sect. 3.13.3. Very recently, it has been shown that the spatial distribution can be obtained directly from the TSD data if the current measurement is combined with surface charge measurements, so that $E(0,t)$ and $E(s,t)$ are known in addition to $i(t)$ [3.194].

3.9 Current TSD and Dielectric Measurements

In Sect. 3.4 we emphasized the importance of comparing TSD with dielectric measurements. Such a comparison might be helpful in elucidating the origin of TSD peaks, because the science of dielectric loss phenomena is more advanced. This is exemplified in several books and reviews which have appeared on dielectric measurements [3.19, 31–36, 173b, 217–224]. However, in making such comparisons, it should be borne in mind that the dielectric relaxations are measured in different ways. TSD typically is a nonisothermal measurement in which time and temperature are varied simultaneously, while dielectric measurements are traditionally done either at a constant frequency or at a constant temperature. The height and position of the relaxation peaks may therefore be *quite* different.

In most dielectric measurements one basically measures the current response of a sample to which a *sinusoidal* voltage is applied. As a rule, the current is found to lag behind the voltage, because the permanent dipoles cannot respond to the voltage instantaneously. The slower the dipoles are, the larger the phase shift. The magnitude of the current depends on the number and strength of the dipoles. The response of a dielectric can conveniently be described by a complex dielectric constant

$$\varepsilon = \varepsilon' - i\varepsilon'', \tag{3.62}$$

where the real and imaginary parts are functions of temperature and frequency. For dipoles with one relaxation time $\tau(=1/\alpha)$, these dependences are given by the celebrated Debye equations

$$\varepsilon' = \varepsilon_\infty + (\varepsilon_s - \varepsilon_\infty)(1 + \omega^2\tau^2)^{-1}, \tag{3.63}$$

$$\varepsilon'' = (\varepsilon_s - \varepsilon_\infty)\omega\tau(1 + \omega^2\tau^2)^{-1}, \tag{3.64}$$

where ω is the angular frequency. Differentiating (3.64) with respect to ω or τ, we find that the loss peak occurs at

$$\omega\tau = 1. \tag{3.65}$$

To investigate the evolution of the loss peak at a constant temperature we generally have to change the frequency over a considerable range. This range cannot be spanned by a single technique. For medium frequencies we use bridges; for high frequencies, resonant circuits; and for very high frequencies, microwaves [3.19, 34, 36, 173b]. The use of different techniques is time-consuming and costly. If, on the other hand, the frequency is kept constant and the temperature changed, the required change is much less. This is so, because the relaxation time varies by orders of magnitude with the temperature for thermally activated processes.

As in TSD such dielectric measurements can at present be done during linear heating, now that *automatic balancing* bridge circuits are available (e.g., General Radio type 1680-A, Hewlett-Packard 4265A). A refined method for automatic dielectric measurements at several frequencies simultaneously has been developed by *Wada* et al. [3.225]. They apply a multifrequency excitation to the sample and analyze its response by digital signal processing with a minicomputer.

Obviously, TSD is most closely linked to dielectric measurements as a function of temperature. To relate them we compare the conditions for the ε'' maximum and the current maximum, i.e., (3.8) and (3.65). Both maxima will appear at the same temperature if the ε'' measurements are made at the angular frequency

$$\omega = A/hkT_m^2 \,. \tag{3.66}$$

This frequency definitely lies in the *ultralow* frequency range. For example, for the β peak of polymethyl methacrylate for which $A = 0.8\,\text{eV}$, $h = 1\,\text{min}/°\text{C}$, and $T_m = 225\,\text{K}$, we find a frequency of $10^{-4}\,\text{Hz}$. By comparing the magnitude of the two maxima we find the following relation

$$\varepsilon_0 \varepsilon_m'' E \simeq 1.36\, i_m hkT_m^2/A \,. \tag{3.67}$$

Interestingly, this simple relation holds for the entire TSD peak, i.e., for every set of i and T values, if one takes a factor of 1.47 instead of 1.36 in the equation [Ref. 3.11, p. 88]. With this modification we can convert all TSD data to ε'' data and vice versa. Such a conversion is not restricted to a single Debye peak; it can also be applied to distributed polarizations in α_0 or A. However, for the latter, A is not a constant but a function of the conversion temperature. This temperature dependence can be found from (3.9) and Fig. 3.20 if α_0 is known. In [Ref. 3.11, p. 88] we have shown that it is also possible to convert TSD data into ε' data.

In Fig. 3.32 we compare ε'' data calculated with (3.67) from TSD results and direct ε'' measurements at 0.1 Hz. The agreement is reasonable, but the ε'' values calculated appear at a lower temperature, because they pertain to a frequency substantially lower than 0.1 Hz. This again demonstrates that for a good comparison the dielectric data should also be taken at *low* frequencies. Otherwise the comparison might be confusing, in particular if the dielectric relaxation spectrum contains several peaks with different activation energies, such peaks generally shifting differently and tending to unite at high measuring frequencies.

Relaxation peaks are in fact best resolved when the measuring frequency is low. This circumstance makes TSD so attractive as a dielectric research tool. The more so, because the ultralow frequency range, where transformer bridges cannot be used, is not easily accessible to conventional ε'' measurements. A common practice is to study this range with *dc-step response* measurements, in which the current response is measured as a function of time after a dc voltage is applied to or removed from the sample [3.19, 34–36]. These measurements are quite similar to TSD, except for the temperature being kept constant. However, if the measurements are made at various temperatures, it is possible to collect the dc-step data at a specific time and to plot them as a function of temperature. Figure 3.33 clearly shows that such data closely resemble the TSD results, which can be shown to fit in at an equivalent time of hkT_m^2/A [3.11, 26, 134, 226].

By conversion of the dc-step transients into the frequency domain with a Fourier transform, it is possible to compare TSD also with *low-frequency ac*

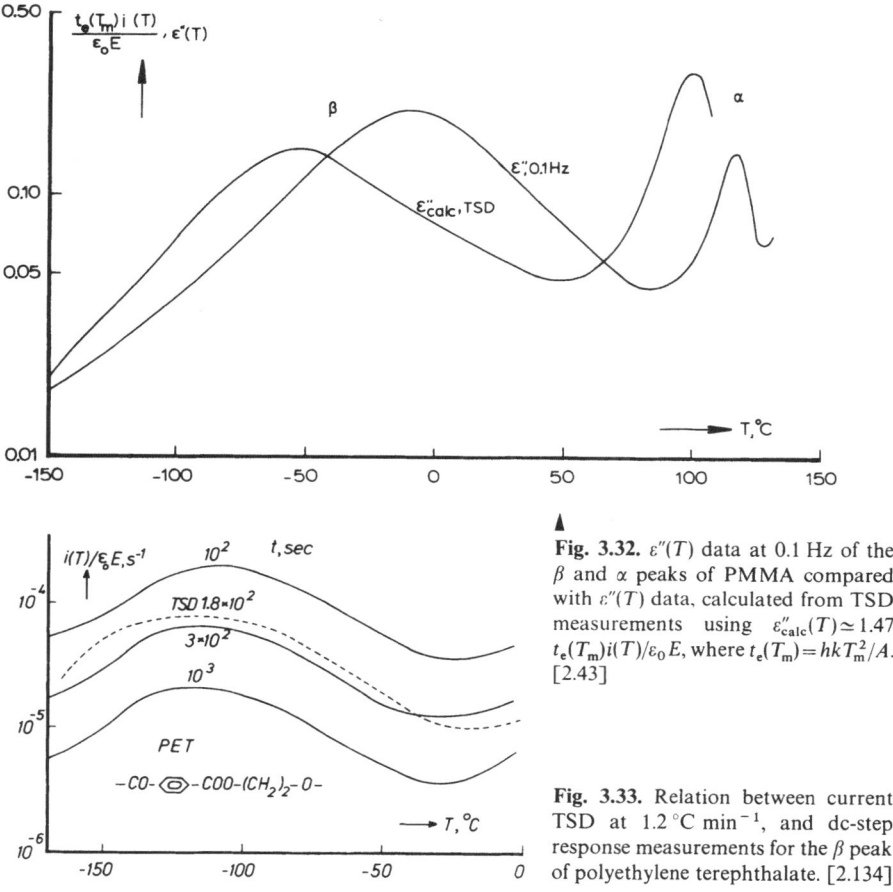

Fig. 3.32. $\varepsilon''(T)$ data at 0.1 Hz of the β and α peaks of PMMA compared with $\varepsilon''(T)$ data, calculated from TSD measurements using $\varepsilon''_{\mathrm{calc}}(T) \simeq 1.47$ $t_e(T_m)i(T)/\varepsilon_0 E$, where $t_e(T_m) = hkT_m^2/A$. [2.43]

Fig. 3.33. Relation between current TSD at $1.2\,°C\ \mathrm{min}^{-1}$, and dc-step response measurements for the β peak of polyethylene terephthalate. [2.134]

data. The conversion can be done numerically (Ref. 3.11, p. 88; [3.227]), but more straightforwardly by application of sophisticated electronic circuits performing the Fourier transform automatically [3.228].

Owing to recent developments, however, it is no longer necessary to conduct dc-step measurements as a means to obtain low-frequency ac measurements; interesting progress has also been made in direct low-frequency measurements [3.19]. One method, introduced by *Harris* [3.229] uses an operational amplifier with a four-quadrant sine-wave generator. A different type of operational amplifier bridge has been proposed by *Fukada* et al. [3.230]. Also worth mentioning is the simple set-up devised by *Van Roggen* [3.231], in which the sample is excited by a triangular voltage. Yet another approach is to return to first principles and to measure the amplitude and phase angle of the ac current, using a low-frequency generator, an electrometer, and a time-interval counter [3.43]. With the latter method, the 0.1 Hz results plotted in Fig. 3.32 were obtained.

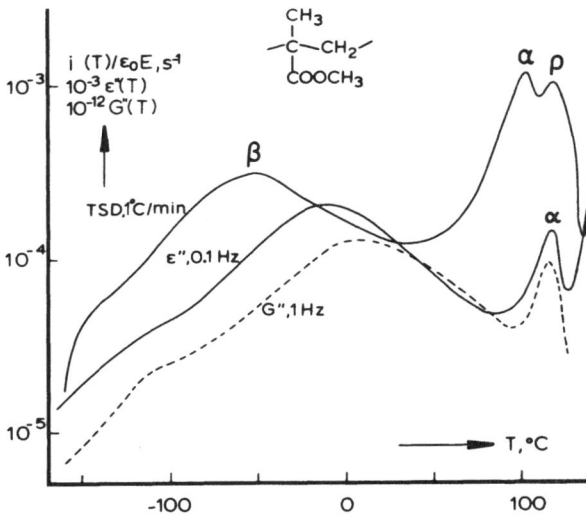

Fig. 3.34. Comparison between current TSD, and ε'' and G'' data for PMMA. The electret was formed at $140\,°C - 75\,kV\,cm^{-1} - 1.5\,h$. Note that the ϱ peak is missing in the ε'' and G'' results. [3.11]

Automatic low-frequency measurements can also be performed with a correlator technique for finding the in-phase and out-phase components of the current [3.232]; commercial equipment with a frequency range down to $10^{-4}\,Hz$ based on this principle is available (Solartron, 1170 Frequency response analyser). Quite a different, but less sensitive, low-frequency method has been advocated by *Chatain* et al. [3.233]. They determine the rotation of a sample between two pairs of perpendicular noncontacting electrodes to which two out-of-phase voltages are supplied.

We limited ourselves to the conversion of dipole peaks, because peaks entirely due to space charges are rarely found in dielectric measurements, because here they are normally obscured by losses due to ohmic conduction. This situation is illustrated for PMMA in Fig. 3.34, where the TSD ϱ peak is seen to replace the exponentially increasing conduction losses in the ε'' curve. We note that the TSD data are now compared without conversion. The result is that the heights of the β and α maxima are quite different in TSD and in ε'' because of the large difference in activation energy of the two peaks. The TSD current at the α peak is, so to speak, sharpened by the high activation energy of the peak.

Hino, and *Perlman* et al. [3.215, 234] have also described a conversion of TSD into ε'' data. They used the TSD data to calculate both $\varepsilon_s - \varepsilon_\infty$ and τ, and then calculated ε'' for various ω from (3.64). A disadvantage of their method is that it allows the conversion of overlapping TSD peaks only if the overlapping peaks can be decomposed into their elementary components [cf. 3.234–236]; otherwise the sum of all the individual ε'' responses cannot be calculated correctly [3.139].

Results of mechanical measurements at 1 Hz are also plotted in Fig. 3.34. They too exhibit only a β and α peak, the latter deriving from the glass–rubber transition, because it is at this temperature that the real part of the modulus (not shown) drops most sharply. We further note, from the positions of the peaks, that the β peak shifts the most, clearly because it has a lower activation energy. It goes without saying that there is also a wealth of information available on mechanical measurements with which TSD data can be compared [3.34, 37–42, 237–239].

3.10 Current TSD Arising from the Detrapping of Charges

In this section we deal with the theory of corona or electron bombarded foils and with TSD peaks evolving from electron or hole detrapping.

In Sect. 3.7 we assumed the accumulated charges to be liable to move through the sample with a certain thermally activated mobility. Such a concept can be used for describing the motion of ions, which in view of their dimensions require the presence of vacancy sites to which they can hop by surmounting the intermolecular barriers.

Electrons, however, can in principle move in the sample as in a continuum because of their small size and their wave properties, provided that their energy is high enough to release them from their parent atoms. The energy range in which electrons can propagate freely without thermal activation can be grouped together in a conduction band.

The free electron motion is interrupted by disturbances in the molecular regularity of the sample by structural defects or impurities (cf. Figs. 2.16, 3.6). The electron is then trapped or localized at an energy level below the conduction band. The sites at which trapping takes place can be neutral or charged; in the latter case the capture cross section is about 10^{-12} cm^2 which far exceeds interatomic dimensions. Several electron volts below the conduction band there is another band of energies in which free propagation of holes (missing valence electrons) is possible. The localized states at which the holes are trapped lie above the valence band. In between the two transport bands, any delocalized carrier motion is forbidden.

Nowadays, most electrets are made of amorphous or semicrystalline polymers, with a high degree of disorder showing many structural irregularities and a high number of traps in particular of deep traps. They are therefore very suitable for the production of electrets, because deep traps can store injected electrons for prolonged periods[12]. From Sect. 3.2 we recall that the detrapping in such materials should be analyzed with the energy-band model on the right-hand side of Fig. 3.6. Regrettably, we do not know the structural details of most of these materials, so that we have to make do with a much simpler model.

In the model of Fig. 3.35 we have assumed that the (unipolarly charged) electret contains an excess of electrons all trapped at the same energy. In this

12 An electron pinned down at a depth of, e.g., 2 eV (at room temperature) will remain trapped for longer than the age of the universe.

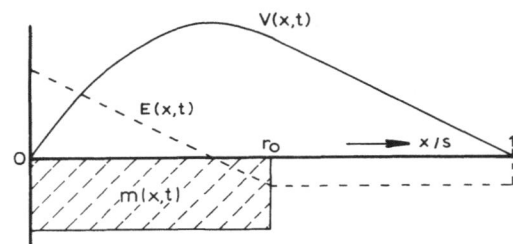

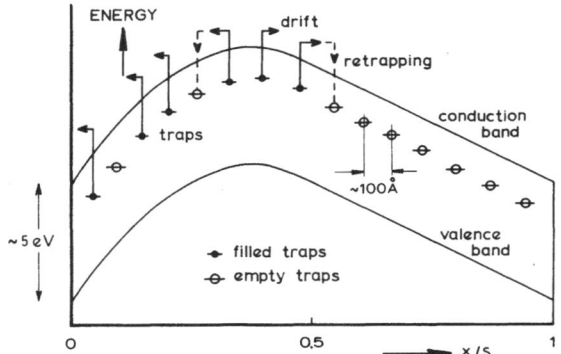

Fig. 3.35. Energy band diagram *(below)* for excess electrons trapped uniformly up to a depth r_0 at a single trap level in a shorted electret. Above, the profiles of the corresponding field and voltage are shown. The conduction and valence band are referred to the internal voltage, and therefore they are also concave. Accordingly, the freed electrons fall from a potential hill in their drift to the electrodes. [3.11]

model the TSD can be considered to proceed as follows. During the heating the electrons trapped at the localized states gain enough energy to jump into the conduction band. At this energy level they are transported by the internal field to the shorted electrodes. On their way there they can fall back into traps (retrapping) and ultimately they recombine with their image charges at the electrodes[13].

To describe these processes we have the following equations:

$$\varepsilon_0 \varepsilon \partial E(x,t)/\partial x = -e[n(x,t)+m(x,t)], \tag{3.68}$$

$$\partial n(x,t)/\partial t = \mu_0 \partial n(x,t) E(x,t)/\partial x - \partial m(x,t)/\partial t, \tag{3.69}$$

$$\partial m(x,t)/\partial t = C_m n(x,t)[M-m(x,t)] - v(T)m(x,t), \tag{3.70}$$

where e is the electronic charge, n the concentration of free electrons, μ_0 their mobility, m the concentration of trapped electrons, M the total number of available traps with a capture cross section C_m, and $v(T)$ the escape rate for detrapping. Equation (3.68) is Poisson's equation; (3.69), the continuity equation; while (3.70) describes the trapping kinetics – its first term represents the retrapping, and the second the thermal detrapping. Compared with $v(T)$, the quantities μ_0 and C_m barely change with temperature and so they are assumed

---

13 In a nonconducting unipolarly electron-charged electret, no neutralization with holes in recombination centres can occur [3.240].

to be constant. For a trap depth U the escape rate obeys, according to Boltzmann statistics,

$$v(T) = v_0 \exp(-U/kT). \tag{3.71}$$

From the partial differential equations (3.68–70), n has to be solved and substituted into (3.29) for the discharge current; this equation can be integrated with respect to x to give

$$i(t) = (e\mu/s) \int_0^s n(x, t) E(x, t) dx. \tag{3.72}$$

In their general form the partial differential equations cannot be solved analytically, and they must be solved by approximations. The first approximation is that

$$m > n, \tag{3.73}$$

because in a good electret the density of trapped electrons is usually much higher than that of the free electrons [3.2, 6, 11–13].

In line with (3.73) we further take

$$\partial m(x, t)/\partial t \gg \partial n(x, t)/\partial t. \tag{3.74}$$

With these assumptions, (3.68) and (3.69) reduce to

$$\varepsilon_0 \varepsilon \partial E(x, t)/\partial x = -em(x, t) \tag{3.75}$$

$$\mu_0 \partial n(x, t) E(x, t)/\partial x = \partial m(x, t)/\partial t. \tag{3.76}$$

These simplified forms have been solved for two extreme cases: *slow* retrapping and *fast* retrapping [Ref. 3.11, p. 277]. In the latter case the equations can be shown to reduce to those of Sect. 3.7. This is to be expected because electron motion with fast retrapping can be thought of as carrier hopping across a potential barrier U. Hence, we can abandon here any further discussion of fast retrapping. The analytical solution for slow retrapping can only be given for simple trapped charge distributions as has been discussed by *Perlman* et al. [3.241], and in [3.11].

If no constraints are made on the shape of the charge distribution, the equations have to be solved numerically. Such numerical calculations have been done for isothermal measurements by *Monteith* et al., and for TSD by the present author [3.11, 242]. The reader is referred to these references for further details. The equations solved were the simplified versions (3.75) and (3.76), in which recombination is neglected. The calculations were done for various retrapping rates.

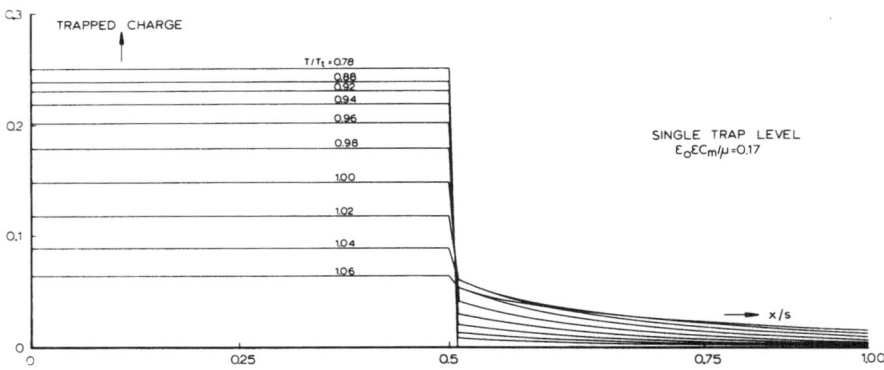

Fig. 3.36. Spatial distribution of trapped and free charges at various temperatures during TSD of a unipolar electron-charged electret having a single trap. The retrapping, which is moderate, is characterized by the ratio $\varepsilon_0\varepsilon C_m/\mu$; there is no recombination. We took $U/kT_t = 20$, where in analogy with (3.8) T_t is defined by $hT_t\nu(T_t) = U/kT_t$. [3.11]

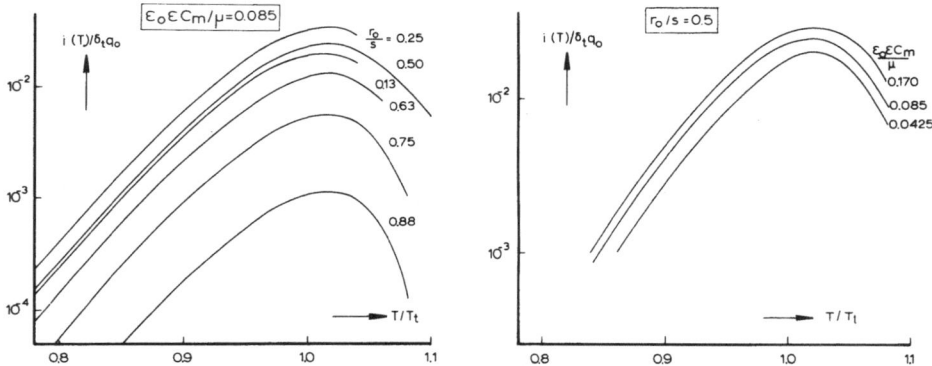

Fig. 3.37. Current thermograms of a unipolar electron-charged electret with a single trap level. *Left:* for various penetration depths. *Right:* for different degrees of retrapping. We took $m_0/M = 0.25$ and $U/kT_t = 20$. [The factor $\delta_t q_0$ with which the current is normalized equals $\nu(T_t)m_0r_0$]. [3.11]

The calculated spatial distributions of n and m for moderate retrapping are shown in Fig. 3.36. We see how m, which originally was a box, develops a tail to the right. At the same time n builds up, its shape becoming similar to that of m, from which it derives, so it is also nonuniform beyond the penetration depth. Figure 3.37 illustrates that a higher current is released when retrapping becomes stronger. The plots also show the effect of the penetration depth on the current release. In contrast with the results for strong retrapping in Sect. 3.7, the position of the current maximum is almost *independent* of the penetration depth. Using (3.45) we can calculate the trapping depth from the initial slope of the current.

We have already stressed that polymers, which contain a variety of structural defects and impurities, cannot be assumed to have a single trapping level. *Sessler* [3.243] has recently indicated how the kinetic parameters can be derived when a polymer contains two or more discrete trapping levels. For a continuous distribution of traps one has to resort to numerical calculations with a computer [Ref. 3.11, p. 278]. Several analyses have further been given for current transients in insulators in which the released carriers are caught in *deep traps*, from which they are *not* re-emitted [3.244–246].

3.10.1 Experimental Results of Corona- and Electron-Beam-Charged Electrets

Figure 3.38 shows the current TSD of Mylar-c corona-charged positively and negatively. We see that the current after a negative charging is substantially higher than after a positive one. This is probably due to electrons being more mobile than holes in polyethylene terephthalate [3.247], so that they penetrate more deeply into the bulk of the electret when this is charged. The first of the peaks coincides with the glass–rubber transition T_g. This indicates that the traps are closely linked to or form part of the *native* polymer structure; the electrons are, so to speak, shaken out of their traps by the strong internal molecular motion at T_g. In other words, the detrapping is a *dual* process arising from the thermal excitation of the electrons in the traps as well as from the erosion of the traps by the molecular motion.

Results for electron-bombarded Teflon were shown in Fig. 3.5. We recall that the height of the currents changes with injection energy, because this changes the penetration depth. Again two peaks appear, the first occurring at T_g. According to the latest views this peak very probably arises from a neutralization of part of the injected electrons by the nonuniform radiation-induced conductivity, RIC ([3.94], and Sect. 4.6). The appearance of a peak at T_g then implies that the detrapping of the secondary electrons and holes responsible for RIC is also intimately related to the onset of the segmental motions in the polymer. We further note that both peaks are rather broad and appear to originate from carriers released from a distribution of traps. Clear evidence for this is given in Fig. 3.39, which shows that when the TSD is

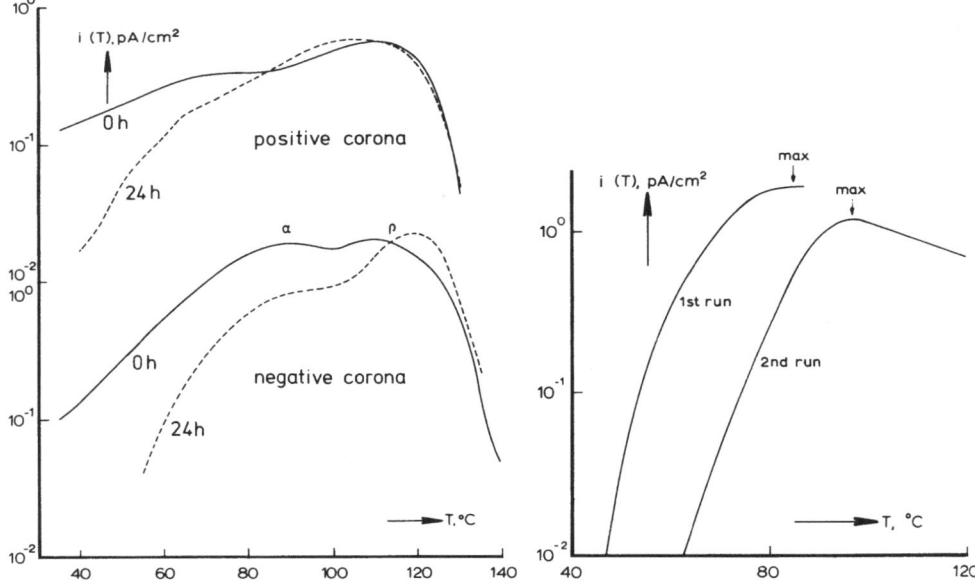

Fig. 3.38. Current thermograms of PET-c films corona-charged positively and negatively. The recordings were made (on metallized samples) just after the corona charging, and 24 h later. From [3.11]

Fig. 3.39. Results of a two-stage current TSD of electron-bombarded Teflon-FEP foil. The injection energy was 30 keV and the deposited charge was $20\,\mathrm{nC\,cm^{-2}}$. [3.25]

interrupted and restarted, the α peak shifts appreciably. This shift might also in part be due to the lowering of the driving field by the partial discharge.

Recently, *Sessler* et al. [3.93, 94] have shown that the α peak in electron-bombarded Teflon-FEP also shifts to a higher temperature with storage time (cf. Sect. 4.6.2 and Fig. 4.22). They attribute the shift to the time dependence of the RIC, which, being large shortly after the irradiation, neutralizes the injected electrons, but gradually declines with time; see (4.2) and (4.30). This *time-dependent* RIC, $g_i(T,t)$, is only present in the irradiated part; its effect can readily be revealed by modification of the equations of Sect. 3.7.

Let us consider the deposited electrons to be trapped uniformly up to r_0, and assume, as *Sessler* et al. [3.93] have done, that during storage and TSD, the decay of these primary electrons is dominated by RIC, so that their self-motion can be neglected, and that the conductivity of the nonirradiated part can be taken zero. Integrating (3.29) up to r_0 we get

$$i(t)r_0 = \varepsilon_0\varepsilon\frac{d}{dt}\int_0^{r_o}E(x,t)dx + g_i(T,t)\int_0^{r_o}E(x,t)dx, \qquad (3.77)$$

where $\int_0^{r_o}E(x,t)dx$ can be replaced with $-E(s,t)(s-r_0)$ and $i(t)$ with $\varepsilon_0\varepsilon dE(s,t)/dt$, using (3.30) and (3.36). Substitution of the resulting expression for

$E(s, t)$ into (3.36) yields

$$\frac{i(t)}{\varrho_0 r_0} = -\frac{r_0}{2s}\left(1 - \frac{r_0}{s}\right)\beta_i(T, t)\exp\left[-\left(1 - \frac{r_0}{s}\right)\int_0^t \beta_i(T, t)\,dt\right] \qquad (3.78)$$

where $\beta_i(T, t) = g_i(T, t)/\varepsilon_0\varepsilon$. This equation differs fundamentally from the earlier expression (3.35) without RIC.

Integrating (3.78) we find for the released charge density

$$\frac{q(t)}{\varrho_0 r_0} = -\frac{r_0}{2s}\left\{1 - \exp\left[-\left(1 - \frac{r_0}{s}\right)\int_0^t \beta_i(T, t)\,dt\right]\right\}, \qquad (3.79)$$

and so ultimately a charge density of

$$q(\infty)/\varrho_0 r_0 = -r_0/2s \qquad (3.39a)$$

is released. Interestingly, this expression is the same as (3.39), which means that for small penetrations *Sessler*'s method for finding the mean spatial depth [3.185] also applies to electrets with a substantial nonzero RIC.

Sessler et al. [3.93] performed their TSD with electrodes pressed on electron-beam charged Teflon-FEP electrets that had been stored in *open circuit*. Repeating their experiments in a different way with evaporated electrodes on electrets stored in *short-circuit*[14], the present author observed not only a shift in the position of the α peak, but also a lowering in its height; both findings are in agreement with [3.78] and [3.79]. Interestingly, when the decay by RIC is the most active, i.e., for short storage times, the α peak is found to appear below the actual glass–rubber transition. From the author's data, which extend up to a storage of 12 weeks, it is not clear whether for extremely long storage times, T_g is exceeded.

Peak temperatures higher than T_g certainly will not arise if the detrapping of holes responsible for the RIC decay of the primary electrons is induced by the onset of the segmental motions at T_g[15]. Such a close interrelation of RIC with the conformational relaxation peak of the main chain segments is apparent in, e.g., Figs. 4.26 and 4.27, and in view of this it would pay to extend *Sessler*'s analysis to a RIC involving multiple traps [3.248]. However, a clear-cut interrelation will only exist when the RIC decay proceeds chiefly in the amorphous part of the Teflon-FEP samples. By contrast, any decay in the

---

14 It should be noted that we had applied the evaporated electrodes already before the bombardment. Lately, *Sessler* et al. [J. Appl. Phys. **50**, 3328 (1979)] have also used evaporated electrodes which they applied to their electrets (stored in open circuit) just before TSD. They took extreme care not to discharge the samples during the evaporation and found that the α peak shifted with storage time in the same way as did that of pressed-on electrodes.

15 Concurrently, the detrapping of primary electrons in the amorphous part of the samples becomes stronger, so that their self-motion gains in importance and is also likely to produce a peak at T_g. However, *Sessler* et al. conjecture that in Teflon-FEP this peak is a small one.

crystalline part is generally unrelated with T_g. Yet, it may well be that this part becomes more important after a certain storage time, namely, when the percolations made in the crystallites during the bombardment have disappeared and when the conductivity has virtually been restored to its intrinsic low value, because most of the holes have been detrapped and removed. This might explain the surprising result of *Sessler* et al. [3.93], who found that the α peak appears above T_g after very long storage times (cf. Fig. 4.22).

Sessler et al. [3.249] have also shown that the RIC can be removed by thermal annealing. They have further studied the effect of a bias field during electron bombardment and during TSD; the 80 and 190 °C peaks in Teflon-FEP are affected in different ways by the bias (for more details, see [3.94] and Sect. 4.5). *Ieda* et al. [3.250] have also applied a bias during TSD. Their results for electron-beam bombarded polyethylene terephthalate, polyethylene naphthalate, and polystyrene are different for a positive and a negative charge-collecting bias. Electron-bombarded polyethylene terephthalate has also been investigated by *Taylor* [3.251].

Perlman et al. [3.252] performed the bombardment in a cryostat and so were able to scan the low-temperature part of the spectrum as well. They showed that, for Teflon-FEP, the peak temperature increases with the injection temperature. Recently, *Hampe* [3.253] has compared the TSD of electron-bombarded polyethylene, PE, with that of electrically poled PE. The TSD of corona-charged PE was studied by *Ieda* et al. [3.254], who again applied a bias field to strengthen the current release. Very recently, they determined the RIC and TSP current in X-ray irradiated oxidized PE and in X-ray irradiated ethylene-vinylacetate copolymers [3.255].

In Sect. 3.4 we have already mentioned that *Sessler* et al. [3.132] have made an interesting use of RIC for revealing the internal charge distribution in foil electrets. Their method, based on a piecewise destruction of the trapped space charges by increasing the width of the RIC region in steps, seems to be more precise than any of the other existing methods for scanning the internal charge profile (cf. Fig. 2.13 and Sect. 2.4.7).

3.11 Some Illustrative Results of Current TSD of Heterocharged Electrets

In this section we shall focus on the current TSD of materials charged by heating in a field with two evaporated adhering electrodes, so that they are likely to contain a *heterocharge* only. Such a charging can be expected to be most closely related to the intrinsic properties of the material because it originates from charges and dipoles present *in* the sample, unless charge injection from the electrodes would have occured.

3.11.1 Inorganic Solids

Of the inorganic materials, the ionic crystals, and in particular the alkali halides, have been investigated extensively with TSD. This work was initiated by *Fieschi* and his goup [3.61–63]. They concentrated their TSD or "Ionic Thermal Currents" (ITC) research on the impurity–vacancy (I–V) dipoles formed when divalent impurities are added to monovalent crystals. A typical example is KCl into which Ca^{2+} can be introduced by addition of $CaCl_2$ to a KCl melt. The Ca^{2+} ions are incorporated into the KCl lattice by substitution at a K^+ site. Charge neutrality is maintained by formation of a K^+ vacancy near the Ca^{2+} ion. This arrangement constitutes an I–V dipole, which can be oriented in a field because the vacancy can occupy different positions – 12 nearest-neighbor (nn), and 6 next-nearest-neighbor (nnn) positions. The orientation thus proceeds mainly by jumps of the vacancy around the impurity ion. A direct exchange of position between the impurity ion and a vacancy site is less likely, because the impurity ion is more strongly bound and less mobile [3.158, 256]. An illustration of the various jumping modes possible is given in Fig. 3.40.

An impurity content of, say, 50 ppm already gives marked I–V peaks. These appear at temperatures below room temperature. A typical example of a series of X^{2+} doped KCl crystals is given in Fig. 3.18. The peaks were obtained from samples quenched from the melt to avoid aggregation of the I–V dipoles. The peaks are rather sharp and can usually be described with one relaxation time, because in general one of the possible relaxation modes shown in Fig. 3.40 dominates; in this particular series the dominating mode is mode 3 [3.157, 158].

Some calculated values for the relaxation parameters A and α_0 are collected in Table 3.6. These values compare well with those given in parentheses derived from dielectric measurements by *Dryden* and *Meakins* [3.257]. The A values

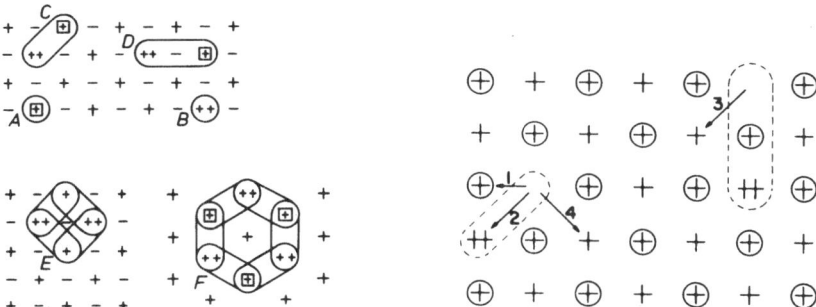

Fig. 3.40. *Left:* Point defects in alkali halides with divalent impurities. Cation vacancy, A; free impurity, B; *I–V* dipole (C in nn and D in nnn coordination); quadrupole dimer, E and hexagonal trimer F. *Right:* Possible vacancy jumps. Arrow 1 corresponds to an nn↔nn jump, arrow 2 to an impurity–vacancy exchange, arrow 3 to an nnn→nn and arrow 4 to an nn→nnn jump (note that circled cations lie above and below the plane of the page). [3.158, 289]

Table 3.6. Dipole relaxation parameters for $KCl:X^{2+}$

Ion	Radius [Å]	T_m [K]	A [eV]	α_0 [s^{-1}]
K^+	1.33	233	0.71 (0.70)	3.1×10^{13} (5×10^{13})
Ba^{2+}	1.34	224	0.65 (0.69)	4.8×10^{12} (5×10^{12})
Pb^{2+}	1.20	223	0.63 (0.65)	2.7×10^{12} (2×10^{12})
Sr^{2+}	1.12	227	0.70	3.7×10^{13}
Sm^{2+}	1.11	224	0.68	2.7×10^{13}
Eu^{2+}	1.09	211	0.61 (0.64)	3.3×10^{12} (5×10^{12})
Ca^{2+}	0.99	215	0.67	9.1×10^{13}
Yb^{2+}	0.93	192	0.49	1.6×10^{11}
Mm^{2+}	0.80	186	0.46	8.6×10^{10}
Ni^{2+}	0.69	189	0.49	2.6×10^{11}
Mg^{2+}	0.66	161	0.45	2.0×10^{12}
Be^{2+}	0.35	133	0.25	1.0×10^{8}

Table 3.7. Summary of recent work on the ITC of ionic crystals

Host crystal	Impurity	Reference	Host crystal	Impurity	Reference
AgCl	Cd^{2+}, aggregation	[3.259]	LiD	Mg^{2+}	[3.278]
CaF_2	$Gd^{3+}, Lu^{3+}, Tb^{3+}$	[3.260]	LiH	Ca^{2+}	[3.278]
CaF_2	Ce^{3+}, Gd^{3+}	[3.261]	LiF	Be^{2+}	[3.271]
CaF_2	$Ce^{3+}, Dy^{3+}, Ho^{3+}, Nd^{3+}, Yb^{3+}$	[3.262]	LaF_3		[3.279]
CaF_2	Gd^{3+}, Tm^{3+}	[3.263]	NaCl	$Cd^{2+}(OH^-)$	[3.274]
CdF_2	Eu^{3+}	[3.264]	NaCl	Mn^{2+}	[3.280]
CsBr	$Ba^{2+}, Ca^{2+}, Pb^{2+}, MnO_4^{2-}$	[3.265, 266]	NaCl	Sr^{2+}, aggregation	[3.281]
KCl	CrO_4^{2-}	[3.266]	NH_4Cl		[3.282, 283]
KCl	Cu^{2+}, Mn^{2+}	[3.267]	NH_4Br		[3.283]
KCl	Pb^{2+}, optical stimulation	[3.268–270]	K_2SnCl_6		[3.283]
KCl	Pb^{2+}, aggregation	[3.271]	RbBr	Cu^+	[3.284]
KCl	Sr^{2+}, aggregation	[3.272]	RbCl	Cu^+	[3.284]
KCl	Sr^{2+}, irradiation	[3.273]	RbCl	Sr^{2+}	[3.285]
KCl	$Sr^{2+}, (OH^-)$	[3.274]	SrF_2	Gd^{3+}	[3.263, 286]
KCl	$Ca^{2+}, Pb^{2+}, Sr^{2+}$, pressure	[3.275]	SrF_2	Tb^{3+}	[2.287]
K I	S^{2-}, optical stimulation	[3.276]	SrF_2	all RE^{3+}	[3.288]
K I	$(Cl^-, Br^-), Ba^{2+}, Ca^{2+}, Sr^{2+}$	[3.277]			

decrease almost linearly with the radius of the impurity ion; only Be deviates from this rule. *Bucci* [3.258] attributes this to the fact that the small Be ion probably resides in an off-centre position.

Several alkali and alkaline earth halides have been studied by now. A summary of the results up to 1971 has been given by *Nowick* and by *Crawford* et al. [3.158, 256]. The ITC research on ionic crystals has also been amply reviewed by *Fieschi* and co-workers [3.61–63]. Table 3.7 enumerates the investigations up to 1977; only references to the most recent work are given. For references up to 1972 and 1975, [3.61–63, 158, 256] should be consulted.

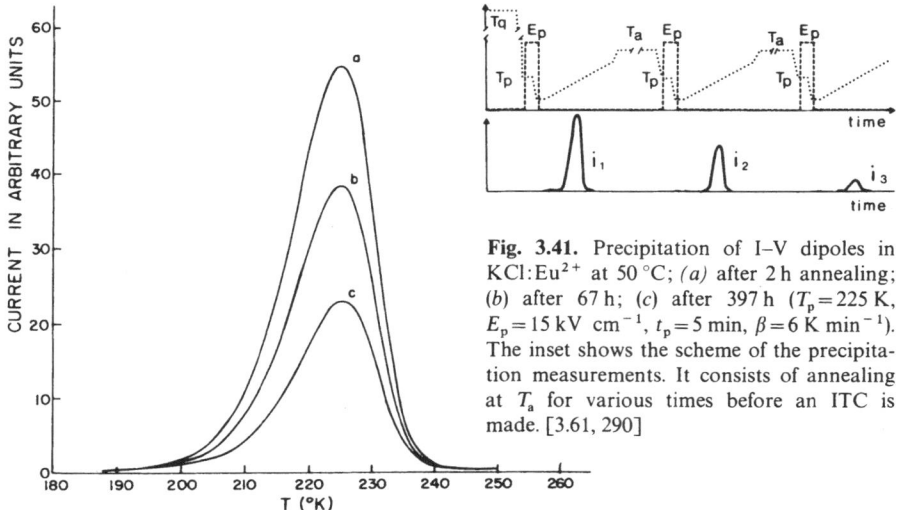

Fig. 3.41. Precipitation of I–V dipoles in KCl:Eu^{2+} at 50 °C; *(a)* after 2 h annealing; *(b)* after 67 h; *(c)* after 397 h ($T_p = 225$ K, $E_p = 15$ kV cm^{-1}, $t_p = 5$ min, $\beta = 6$ K min^{-1}). The inset shows the scheme of the precipitation measurements. It consists of annealing at T_a for various times before an ITC is made. [3.61, 290]

We noted that the ITC peaks are most pronounced just after quenching, because the I–V dipoles eventually tend to *cluster* together to form dimers and trimers (cf. Fig. 3.40) and *aggregates* of higher order. The last mentioned eventually precipitate and disappear from the solid solution [3.269].

This aggregation or precipitation can be revealed by TSD, because the dipole complexes are dielectrically rather inactive, and so the ITC current will be lowered when the crystals are annealed or aged [3.61–63, 259, 264, 271, 272, 281, 289, 290]. There has been some controversy about the order of the aggregation kinetics. The latest view is that the dipoles first aggregate to form dimers and then trimers [3.289, 291]. According to *Perlman* et al. [3.292] it is also essential to consider the dissociation of the aggregates. Figure 3.41 illustrates to what extent the number of dipoles in KCl:Eu^{2+} decreases by precipitation during isothermal annealing.

The *solution* kinetics of impurities in ionic crystals can also be studied with ITC. Much of this work has again been done by the Parma group [3.61–63, 264, 274, 289, 293]. Some results are given in Fig. 3.42, which shows that the solubility increases when the temperature of the unsaturated solid solution is raised from T_{a1} to T_{a3}.

Radiation damage due to, e.g., γ, n, or X-rays, can likewise be studied by ITC, because the radiation produces electrons and holes and changes the number of dipole complexes. Generally, nuclear radiation induces aggregation and causes a decrease in polarizability, but sometimes new dipolar defects are formed [3.61–63, 273, 290, 294, 295].

Recently, the Parma group has established the *photon-induced* reorientation of I–V dipoles [3.268–270, 276]. This is possible when the I–V dipoles form color centres and exhibit optical absorption bands due to electronic transitions between the ground and excited states. This behaviour is shown by I–V centres

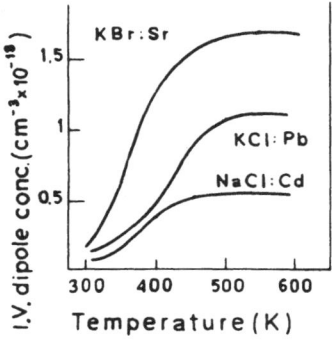

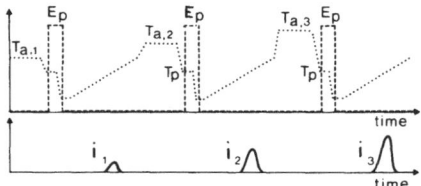

Fig. 3.42. Solubility of I–V dipoles vs temperature in various alkali halides. The inset outlines the scheme of the solubility measurements, and shows that the sample is kept at different temperatures before an ITC is run. [3.61]

with Pb, Co, or rare-earth impurity ions. When such samples are kept at low temperatures, the dipoles can be excited optically, and be oriented in an electric field. Similarly, the orientation of dipoles can be destroyed by illumination of a sample that was polarized in the dark.

Much attention has also been paid to the TSD of alkaline earth fluoride crystals like LiF and CaF_2, because of their importance in *radiation dosimetry* (Sects. 4.6.8). Most of this work has been carried out by *Moran* and co-workers [3.18, 152, 296]. The procedure for measuring dose rates is the following. The sample is exposed to e.g., X-rays at room temperature. The X-rays create electrons and holes which are randomly stored in traps. When the sample is next heated in an electric field, the trapped carriers are remobilized and drift in the collecting bias, thus generating an RIC current whose height increases with the radiation dose. *Moran*'s group have investigated two additional methods, in one of which the field is applied during irradiation. The electrons and holes are separated and then driven to the electrodes and trapped there. They are next released from their traps by heating in short-circuit. The third method is to make a thermo-electret, which is cooled and irradiated with X-rays at room temperature, and next reheated in short circuit. The various methods are outlined in Fig. 3.43.

Typical thermograms obtained by the last method (in which the radiation-induced carriers are separated and trapped under the action of the *internal* electret field) are shown in Fig. 4.28 from which we see that irradiation with 100 R clearly generates two additional peaks in the TSD spectrum of CaF_2 [3.296]. The method is quite sensitive. The group has also investigated the TSD and TSP of Mg-doped LiF [3.18], and of alumina Al_2O_3, the latter showing some promise for TSD dosimetry [3.151, 152][16]. Other recent work on CaF_2 has been done by *Kiessling, Lacabanne*, and *Popov* et al. [3.260–262].

Two other groups investigating the TSD of ionic crystals are those of *de Souza* and *Crawford*. *De Souza* and co-workers have investigated, e.g., the

---

16 Note that CaF_2 and LiF are also employed in TL dosimetry [3.49, 151a, 155]. Another interesting material for dosimetry, in particular for TSEE dosimetry, is BeO, the TSD of which has been studied by *Muccillo* et al. [3.297].

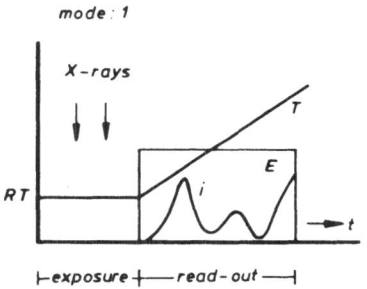

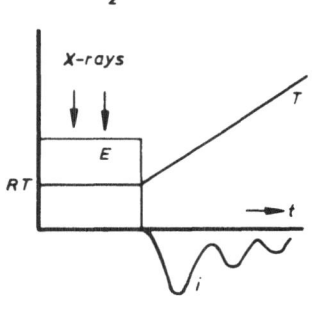

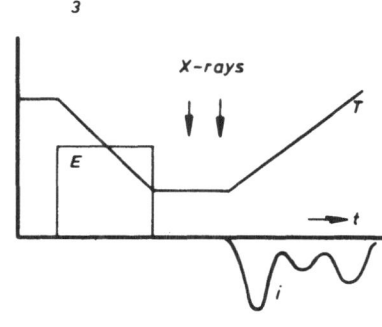

Fig. 3.43. Three different modes of thermocurrent dosimetry (note that in the second mode a radioelectret is made)

Cu^+ off-center defects in potassium halides [3.284, 298], divalent doped mixed crystals of $KI + KCl$ [3.277] and $KCl + KBr$ [3.299], and the effect of the addition of H^- and OH^- on the aggregation of divalent impurities [3.300]. *Crawford* and co-workers [3.263, 286, 301, 302] have paid attention to alkaline earth fluorides doped with trivalent rare earths (CaF_2, etc.). In these systems it is not a vacancy, but rather an interstitial F^- ion which forms part of the dipolar defect pair. In fact, two different *interstitial–impurity* dipoles are created depending on whether F_i^- is in an nn or in an nnn position with respect to the impurity ion. *Crawford* et al. have investigated both types of complexes (see also [3.260–264, 287, 288]).

Silver halides (AgBr and AgCl) have also been studied [3.259, 303–305]; these are interesting materials because of their application in photography.

Roth [3.306] has shown that the ITC response of ionic crystals need *not* be *isotropic*. He studied $ZnF_2 : LiF$ in which the dipole is formed by a Li^+–F^- ion pair. After poling in the [001] direction he observed a much stronger discharge current than after poling in the [100] or [110] directions. His latest view is that this anisotropy is produced by an interstitial motion of the Li^+ ion [3.307].

Another interesting inorganic solid is ice, on which a great deal of well-documented dielectric work has been done [3.308–310]. This work has revealed, for example, the importance of two orientational or *Bjerrum defects* in the hydrogen bonding of the H_2O molecules, viz. the doubly occupied and the vacant hydrogen bond (–OH ... HO– and –O ... O–). In addition, two ionic defects arise by self-ionization, viz. H_3O^+ and OH^-, which may lead to a space-charge polarization.

The TSD of ice has been studied by various workers. *Mascarenhas* and *Dansas* et al. [3.311, 312] have measured the TSD of pure and doped ice, the dopant being HF; with HF the predominant electret discharge takes place by proton transfer at H_3O^+ ions. *Dansas* et al. used blocking electrodes by inserting mica between sample and metal electrode. *Johari* et al. [3.313] have done TSD measurements on H_2O and D_2O under a high pressure and found three peaks. Their view is that the low-temperature peak at 110 K is not due to a ferroelectric ordening of protons, but to the relaxation of water dipoles.

Several other inorganic solids have been studied with TSD. Particular attention is being paid to amorphous inorganics with semiconducting properties, such as Si [3.3] and Se [3.314], the chalcogenide glasses As_2S_3, As_2Se_3, As_2Te_3 [3.30, 65, 68, 315–319], II–VI compounds CdS, CdTe [3.1, 3, 320], and III–V compounds GaAs, GaP [3.1, 3, 321–323]. The compounds studied often have photoconductive properties and are investigated by TSCo [3.1–3] or by invoking the photoelectret state [3.65, 315]. Alternatively, bias excitation is used in conjunction with a Schottky barrier (cf. [3.320–323], and Sect. 3.14). It is well known that the conductivity of chalcogenide glasses can be changed dramatically with a bias voltage, which makes them suitable for switching devices. Recently, the TSD in the off and on state of a switching layer of $Si_{12}Ge_{10}As_{30}Te_{48}$ has been reported [3.324]. Films of carbides, oxides, and nitrides such as Al_2O_3, Ta_2O_5, SiO_2, and Si_3N_4 have also been investigated [3.325], because of their importance as insulating layers in capacitors and semiconductor devices (see further Sect. 3.14).

3.11.2 Organic Solids

A typical representative of organic materials is carnauba wax, which has been the work horse in electret research for a long time. It is the first material of which the TSD was recorded, in 1936 [3.8a]. More recent TSD studies on this material were made by *Gross, Caserta, Perlman*, and *Takamatsu* et al. [3.9, 326–328].

Above room temperature, its thermogram shows three overlapping peaks. *Perlman* has shown that the shape of the thermograms strongly depends on the forming conditions. This indicates that the polarization is distributed kinetically in line with the fact that carnauba wax consists of many different components (esters, alcohols, and acids [3.329]).

There is quite some controversy about the origin of the different peaks. *Caserta* et al. [3.326] attribute their peaks to the detrapping of nonuniformly stored space charges, this view being at variance with the findings of *Gross* et al. [3.9], who found that the polarization is uniformly stored, indicating that it is due to dipoles. To reveal the internal charge distribution, *Gross* et al. removed parts of their sample by planing, and compared the charge released from the

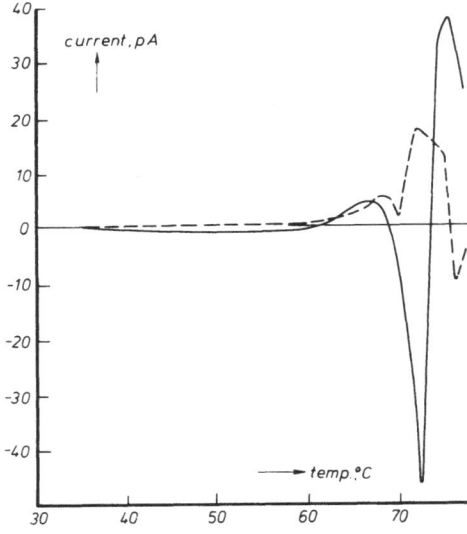

Fig. 3.44. TSD of a carnauba wax magneto-electret poled at 70 °C with 0.47 Wb m^{-2} and recorded with and without interchanging the electrometer connections. Solid line, face adjacent to N pole, and broken line face adjacent to S pole connected to electrometer terminal. [3.331]

planed and unplaned sample. They found no difference. However, using a Faraday cup as the charge measuring device, *Walker* et al. [3.330] found with the planing technique a nonuniform charge storage. In our view as well, the peaks, at least the higher temperature ones, are more likely to be due to space charges than to dipoles, and for that reason the stored charge is relatively high ($\varepsilon \simeq 230$, if $T_p = 70$ °C).

Currents released by magnetoelectrets of carnauba wax have been measured by *Bhatnager* and co-workers [3.331]. In contrast with thermoelectrets, the TSD current of magnetoelectrets becomes appreciable only when the heating temperature exceeds 60 °C (cf. Fig. 3.44).

Lately, the TSD of corona-charged rosin has been published [3.332]. Other organic materials which have been subjected to TSD are molecular crystals such as anthracene [3.333], naphthalene [3.334], and organic dyes [3.335]. These substances have mostly been investigated with TSCo, because they are photoconductors.

3.11.3 Organic Solids Incorporated in Clathrates

Clathrates, or inclusion compounds, have a special *cagelike* crystal lattice in which other molecules can be caught. Typical examples are the complexes of urea, NH_2CONH_2, with long-chain ketones and bromides studied by *Dansas* et al. [3.336]. In the presence of a long-chain compound, the urea crystallizes in a helical hexagonal lattice [3.35, 36], in which the polar groups of the trapped compound can assume six equilibrium positions. *Dansas* et al. have studied the TSD of four types of entrapped long-chain molecules below room temperature.

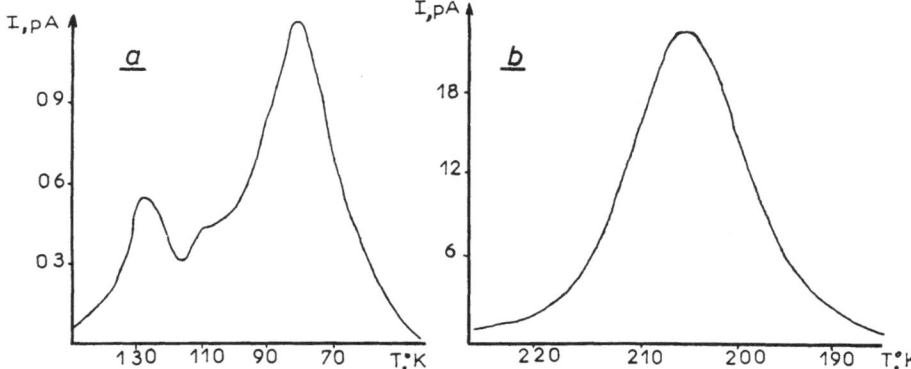

Fig. 3.45. Depolarization thermocurrents for 3-hexadecanone in urea. (*a*) $T_p = 130$ K, and (*b*) $T_p = 229$ K ($E_p = 5$ kV cm^{-1}, $\beta = 6$ °C min^{-1}). [3.336]

Following *Lauritzen* [3.337] they have tried to explain their results with a six-potential-well model. Some typical thermograms are shown in Fig. 3.45.

3.11.4 Supercooled Organic Liquids

Dielectric measurements on supercooled liquids have recently been reviewed by *Williams* [3.338]. He shows that the dielectric behavior of small-molecule organic glasses obtained by supercooling closely resembles that of solid amorphous polymers (see also [3.339]).

The TSD of these substances was studied by *Dansas, Williams, Johari*, and *Heitz* et al., *Dansas* et al. [3.340] investigated 2-ethylhexanol-1 and tricresyl phosphate, both of which are strongly polar liquids. The hexanol shows a sharp TSD peak at the glass transition, whereas the tricresyl phosphate shows a broad asymmetric peak typical of a distributed polarization, which has been fitted by a *Cole–Davidson* distribution.

The supercooled liquids studied by *Williams* et al. [3.141] were solutions of alkylammonium salts and mixtures of alkylammonium salts and anthrone in *o*-terphenyl. Their TSD results for a 1.3 % solution of tri-*n*-butylammonium picrate in *o*-terphenyl are reproduced in Fig. 3.23. The thermogram again shows a sharp peak at T_g, which was described by a *Williams–Watt* distribution. The materials studied by *Johari*, and *Heitz* et al. [3.341, 342] were molecular liquid mixtures of, e.g., chlorobenzene and pyridine, and some alkyl halides such as 1-chloro-2-ethylhexane.

3.11.5 Liquid Crystals

Interest in these crystals is strongly growing now that they have found widespread use in display units [3.343]. A review of dielectric measurements on liquid crystals has lately appeared [3.344]. Liquid crystals consist of long or

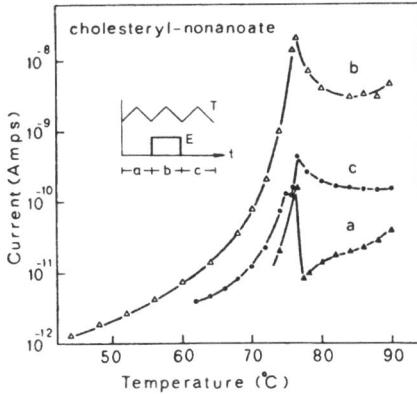

Fig. 3.46. TSD current of a liquid crystal of cholesteryl nonanoate. Curves a, b, and c correspond to the stages shown in the insert (E_p = 2.5 kV cm^{-1} and β = 7 °C min^{-1}). [3.345]

rodlike organic molecules which, when heated, do not pass directly into the liquid state. The molecules remain lined up because they are constrained to rotate around their long axis. There are three liquid crystalline phases, viz. the smectic, nematic, and the cholesteric phase.

Some TSD results obtained by *Capelletti* et al. [3.345] are shown in Fig. 3.46. Curve a represents the current generated spontaneously without a field, curve b is the polarization current, and curve c the depolarization current. In all curves a sharp current burst is seen at the solid–liquid crystal transition at 76 °C. The spontaneous current (a) is probably due to the *Costa–Ribeiro* effect [3.346].

Some smectic liquid crystals were investigated by *Takamatsu* et al. [3.347]. The current was found to appear again at the phase transition, as was unequivocally shown by comparison with DTA results.

3.11.6 Biological Materials

Currently, a great deal of research is being performed on biological materials. These very important organic materials, of which all living organisms are composed, are difficult to study, because they are easily denatured or dehydrated by the slightest rise in temperature or by loss of water. They can therefore be advantageously analyzed by TSD instead of dielectric measurements, because TSD registers the relaxation peaks at the lowest possible temperature. Moreover, TSD is less sensitive to the relatively high conductivity of biological materials.

Dielectric measurements on biomaterials have been reviewed by *Schwarz* and *Rosenberg* et al. [3.348] and conductivity measurements by *Eley* ([3.349], see also [3.348b, 350]). The latter has proposed some interesting ideas about electron tunneling in biological materials.

Much of the TSD work on biological materials, such as bone (of different species), enzymes (trypsin, urease), polynucleotides (DNA), polysaccharides

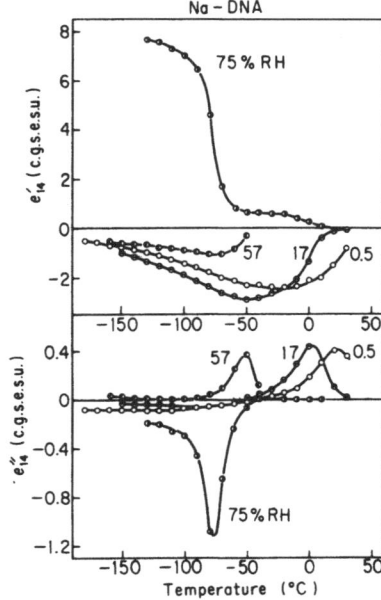

Fig. 3.47. Temperature dependence of the piezo-electric stress coefficients e'_{14} and e''_{14}, for oriented films of Na-DNA equilibrated at different r.h. (1 c.g.s. unit ≡ 3.33 µC m^{-2}). [3.357]

(chitin), and proteins (collagen) has been done by *Mascarenhas* (see Chap. 6). He has indicated several possibilities for charge storage in bioelectrets and also hinted that TSD might be useful for the diagnosis of diseases. He discussed the role played by bound water and studied the effects of denaturation and dehydration. It seems that the high-temperature peaks arise from space charges and it is interesting to note that their position often is close to the body temperature. Other TSD work on biological materials has been done by *Reichle* et al. [3.351], who studied horse hemoglobin; by *Chatain* et al. [3.352], who investigated polyamino acid derivates (polyglycine and polyproline); by *Pillai* et al. [3.353], who studied β carotene; and by *Ledwith* et al. [3.354], who worked on pharmacological materials (dibenz[*b*, *f*]azepine derivates).

Also worth mentioning are *Menefee*'s observations [3.355a] on porcupine quill keratin, which behaves as a *natural* electret and *spontaneously* generates a thermal current, probably due to the disordering ("melting") of the α-helical protein coils. The TSD of polarized keratin of human hair has been studied by *Leveque* et al. [3.355b]. So far, most of the investigators have not tried to correlate their TSD results with those of dielectric, conductivity, or TL measurements [3.356].

The piezo- and pyroelectric properties of biomaterials are also of fundamental interest to life processes. Measurements of the piezoelectric coefficients vs T were made by *Fukada* and co-workers [3.357] on several important biomaterials. Some of their data on DNA of salmon sperm are plotted in Fig. 3.47.

3.11.7 Polymers

A great deal of TSD research has been done on polymers which, by virtue of their insulating properties and high concentration of deep traps, are good charge-storing materials, therefore yielding the best electrets for practical applications.

In a polymer a large number of molecules called monomers are joined together to form a long macromolecular chain, which can assume various conformations in space. Depending on the degree of polymerization the chains consist of some 1000 or more units. A typical example of a polymer with a carbon chain is polyethylene, which consists of a linear sequence of CH_2 groups.

If the monomer is symmetric the polymer will crystallize readily; if it is not, the polymer remains amorphous (disordered). But even the structure of a crystalline polymer is far from perfect, because it is very difficult for *all* macromolecules to align themselves in ordered arrays [3.163, 358, 359]. Crystalline polymers are therefore always semicrystalline and consist of lamellae of folded (or sometimes extended) chains, interlinked by tie molecules, and alternating with regions of a disordered amorphous phase. Only from a very few polymers, such as polyethylene and polyoxymethylene, have small single crystals been grown, but even these contain many structural defects (cf. Fig. 3.48). Charge trapping on a *structural* level can therefore be quite effective.

In polymers various types of molecular relaxations are possible [3.19, 34, 37, 39–41, 217, 219, 221–223, 237, 238, 361]. The only motions possible at low temperatures are local motions of molecular groups, e.g., rotation of *side groups*, or internal motions within side groups. At high temperatures segments of the *main chain* become mobile. The main chains then become flexible, so that the polymer softens and becomes rubbery. The temperature at which these conformational rearrangements of the main chain segments set in is therefore called the glass–rubber transition. In semi-crystalline polymers relaxations involving the *crystalline* parts or the *interzonal* phase are possible [3.19, 34, 37, 39–41, 219, 362, 363].

Let us illustrate the situation for an amorphous polymer like PMMA, the TSD thermogram of which is shown in Fig. 3.34. The polymer clearly has a dipolar side group which can rotate locally or in unison with the main chain. The local motion gives rise to a β peak and the cooperative motion with the main chain to the α peak. The two dipole peaks are rather broad and arise from a distribution in relaxation times. Above the α peak, which coincides with the glass–rubber transition, a third peak appears that is due to the motion of space charges.

In Table 3.8 we list several types of polymers studied with TSD, and in Table 3.9 the effects of various treatments and structural features on their TSD. Apparently, a large number of polymers and their various aspects have been investigated over the past few years, and this trend is expected to continue, for it

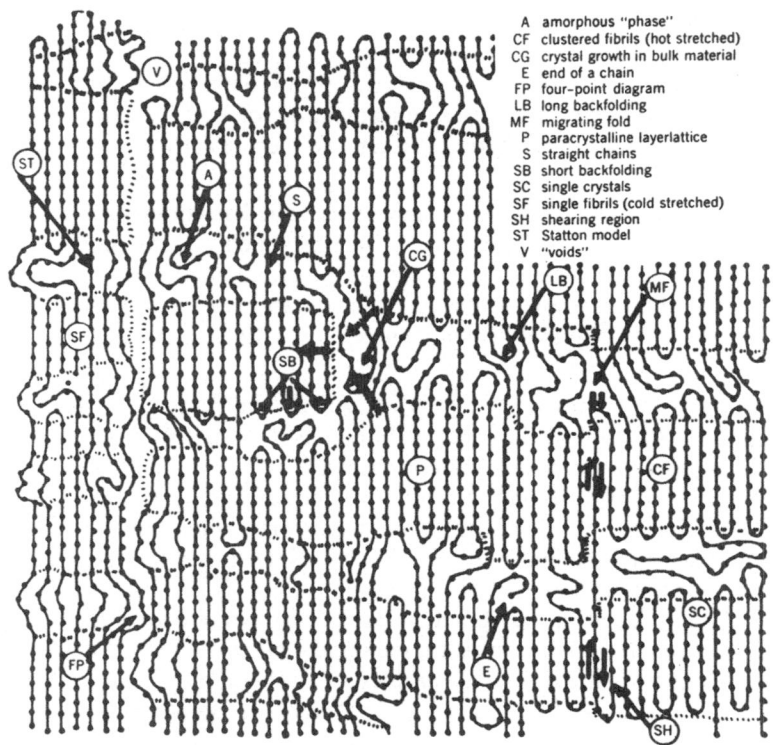

A	amorphous "phase"
CF	clustered fibrils (hot stretched)
CG	crystal growth in bulk material
E	end of a chain
FP	four-point diagram
LB	long backfolding
MF	migrating fold
P	paracrystalline layerlattice
S	straight chains
SB	short backfolding
SC	single crystals
SF	single fibrils (cold stretched)
SH	shearing region
ST	Statton model
V	"voids"

Fig. 3.48. Imperfections in a polymer crystal. [3.360]

is in the field of polymers that the use of TSD is expanding most rapidly and most successfully.

In order to obtain pure polymers whose structures can be systematically modified, they are synthesized in the laboratory. As illustrated in Fig. 3.34 the TSD thermograms obtained can often be compared with dielectric and mechanical data available in the literature.

Comprehensive studies on a large number of polymers and copolymers have been conducted by *Vanderschueren* and the author [3.11, 20]. Since most of the work of the former has not been published, we shall here mainly review his work. Following the present author, *Vanderschueren* has systematically varied the temperature and duration of poling. His results confirm those described in [3.11, 134] and show that the temperature and time of poling can be used to achieve incomplete filling and thus to decide whether a peak is distributed or not. He also explored the method of fractional polarization to prove that the β peak of, e.g., PMMA is distributed [3.135]. His method is illustrated in Fig. 3.15, from which we recall that in the partially activated TSD spectrum, appreciable currents appear only in the temperature windows where the sample has been charged.

Table 3.8. Types of polymers studied with TSD

Group	Reference
Acrylic polymers	[3.20]
Methacrylic polymers	[3.11, 20, 21, 26, 135, 211, 364–368]
Cellulose derivates	[3.369, 370]
Halogen polymers	[3.11, 19, 20, 25, 27, 110, 145, 147, 212, 371–381]
Oxide polymers	[3.11, 20]
Polyamides	[3.20, 213, 328–383]
Polycaprolactones	[3.20]
Polyesters and polycarbonates	[3.11, 20, 26, 28, 177, 215, 226, 368, 384–388]
Polyolefins	[3.11, 20, 27, 66, 136, 137, 389–393]
Vinyl polymers	[3.11, 20, 124, 394–398]
Polyvinyl esters and related polymers	[3.20, 399–402]
Ionic membranes	[3.403, 404]
Resins	[3.405, 406]
Rubbers and elastomers	[3.407, 408]

Table 3.9. Effects of structural features and treatments on the TSD of polymers

Subject	References
Additives (dopes)	[3.11, 20, 110, 401, 409]
Aging, physical	[3.19, 146, 147]
Blends	[3.397]
Composites (fillers)	[3.410]
charge-transfer complexes	[3.411]
Corona treatment	[3.143, 412]
Copolymerization	[3.11, 20, 368, 378, 402]
Cross-linking	[3.407, 413]
Crystallization (single crystals)	[3.20, 28, 385, 386] [3.414]
Curing	[3.405, 407, 415]
Electrodes	[3.11, 20]
Hydration (moisture)	[3.11, 20, 415, 416]
Ionizing radiation	[3.11, 20, 145, 255, 391, 392, 417]
Oligomers	[3.418]
Orientation (stretching)	[3.147, 396, 419]
Oxidation	[3.420]
Plastification	[3.369, 408]
Polymerization	[3.11, 421]
Swelling	[3.20, 407]
Stereoregularity	[3.20, 422]
Thermal treatment	[3.19, 20, 146, 147, 388b, 419]
Ultrathin films	[3.398, 409, 423]

Using the approximations (3.58) and (3.59), he calculated the dipole distributions of a number of polymers, and compared the results with those obtained by dielectric measurements [3.210]. He also systematically studied the effect of absorbed water on the thermograms and showed that it mainly lowers the ϱ peak. This lowering was also demonstrated in [3.11, 415, 416]. *Vanderschueren* further investigated the influence of swelling agents. His findings for methanol vapour are shown in Fig. 3.49. Again it is the ϱ peak which is affected most. The same happens when a different electrode material is chosen, or when the electret is irradiated with nuclear radiation (cf. Table 3.4). He confirmed that TSD is a convenient way to measure the *glass–rubber transition* (cf. Table 3.5). The method is more sensitive than a dilatometric or DTA analysis, because it gives a *peak* rather than a change in slope.

He clearly demonstrated the effect of the three spatial arrangements of the polar vinyl groups attached to the main chain of PMMA: they can all lie on the same side of the main chain (isotactically), alternately on each side (syndiotacti-

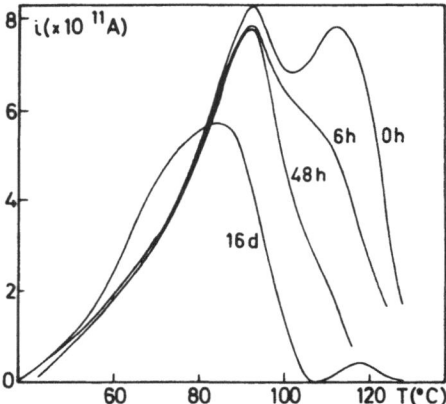

Fig. 3.49. Effect of exposing a PMMA electret for various times to methanol vapour ($T_p = 123\,°C$, $E_p = 20\,kV\,cm^{-1}$, $t_p = 30$ min). [3.20]

cally) or be arranged at random along the main chain (atactically). Figure 3.50 shows that the syndiotactic type generates a stronger β current than the isotactic type; the reverse is true for the α peaks. The *steric* arrangement of the dipolar groups thus markedly affects the polarizability of the polymer.

Figure 3.51 shows the effect of *crystallinity* on the β peak of polycarbonate [3.20]. The higher the crystallinity, the lower the β peak. The TSD of polycarbonate has also been described in [3.11, 226, 388]. Very recently, *Vanderschueren* et al. have compared the TSD and TL of polymers [3.44].

Of the polymer group in Toulouse, *Chatain* [3.424] has focussed on polyamides (aliphatic, semi-aromatic and aromatic) and polypeptides (polyglycine). The TSD of these hydrophylic polymers was found to depend strongly on their water content. *Lacabanne* [3.425] studied the TSD of two polysulphoamides and critically reviewed the use of the WLF equation for peaks at and above the glass–rubber transition. Several papers about this research have now been published [3.164, 352, 366, 382, 383]. The group also investigated ultrathin layers of polystyrene formed in a vacuum by a high-frequency glow discharge [3.423a]. To separate nearby peaks they advocate the use of fractional polarization in a narrow temperature window [3.136]. By this method they believe that they have shown that the broad low-temperature peaks in polymers can be decomposed into a number of single Debye peaks, but this view is questioned by others [3.135a]. One of their latest activities has been the development of the mechanical parallel of TSD, TS Creep [3.125].

There is considerable interest in the TSD of polyethylene, because of its use for high-voltage cables in power transmission. This work has been reported by *Fischer* et al. [3.137, 390], *Lacabanne* et al. [3.136], and *Marchal* et al. [3.392]. *Perret* et al. [3.66, 389] performed TSD measurements on sectioned polyethylene cables to reveal the internal charge distribution. Attention has been paid even to *single* crystals of polyethylene [3.414].

Measurements on polyethylene terephthalate were done by the author [3.11, 26], by *Asano* et al. [3.28], and by *Hino* et al. [3.215, 387]. The last-

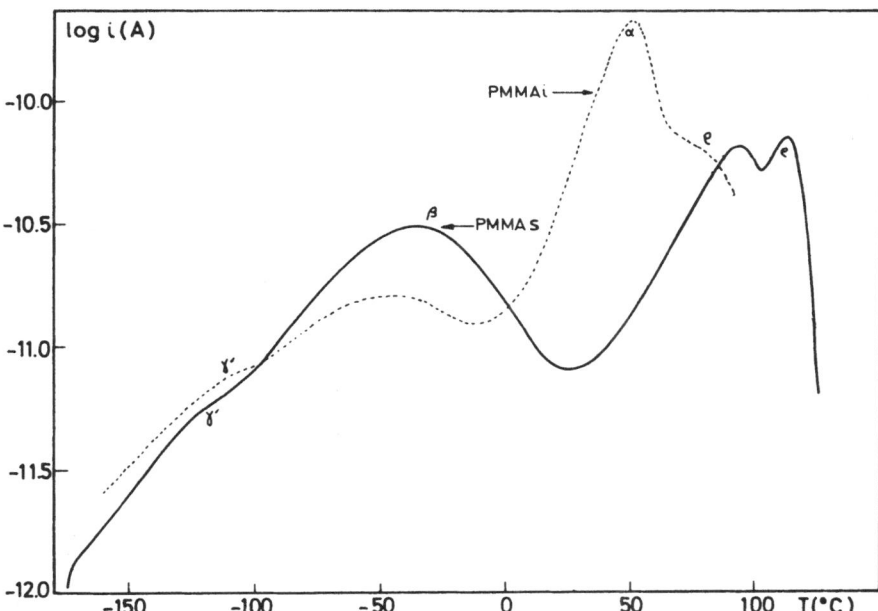

Fig. 3.50. Effect of the stereoregularity on the TSD of PMMA. The poling conditions were: $T_p = 127\,°C$ (PMMA-s), $T_p = 65\,°C$ (PMMA-i), $E_p = 10\,kV\,cm^{-1}$, $t_p = 30\,min$ and the TSD was recorded at $5\,°C\,min^{-1}$. The γ peak in these spectra also appears in dielectric and mechanical measurements (cf. Fig. 3.34), and is often attributed to impurities (e.g., absorbed water; cf. [3.34]). [3.20]

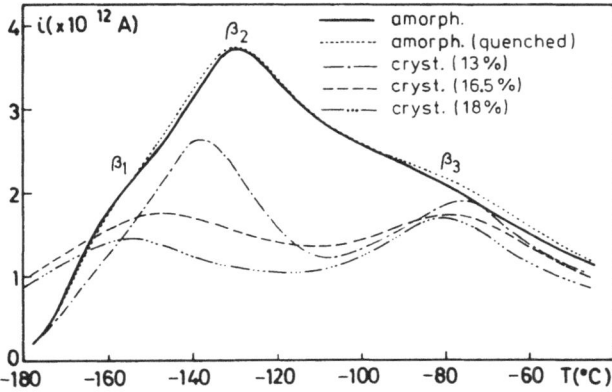

Fig. 3.51. Influence of crystallinity on the β peak of polycarbonate ($T_p = 22\,°C$, $E_p = 10\,kV\,cm^{-1}$, $t_p = 30\,min$, $\beta = 5\,°C\,min$). [3.20]

mentioned authors described the β and α peak with a distribution in activation energies. The polymer has also been studied by *Sacher* [3.122], who recorded the spontaneous current release due to charges introduced in the manufacturing process. *Pillai*'s team has mainly contributed papers on doped and undoped polyvinyl esters and on the effect of hydration [3.375, 399–401, 409, 413, 416].

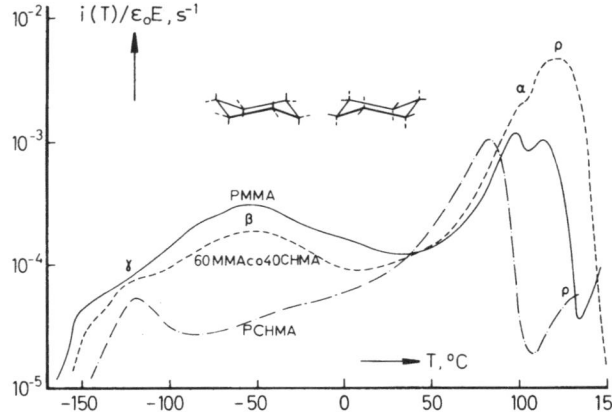

Fig. 3.52. Current spectra of methacrylic homo- and copolymers with the cyclohexyl side group. For comparison the thermogram of PMMA is also shown. [3.11]

Murphy and *Latour* [3.373] have reported the TSD of Aclar (polychlorotrifluoroethylene), a polymer which has been used beside Teflon-FEP in electret microphones [3.426]. Since 1974 the TSD of polymers has regularly been reported on at the annual conferences on "Electrical Insulation and Dielectric Phenomena" [3.74].

Figure 3.52 illustrates that TSD is also capable of revealing the motion *within* a *nonrigid* side group [3.11, 26]. The figure concerns the chair–chair flipping motion of the cyclohexyl ring in polycyclohexyl methacrylate, the mechanical relaxation of which has been studied extensively by *Heijboer* [3.238]. The two chair conformations of the six-ring membered are shown in the insert of the thermogram. We see that the hydrogens attached to the ring are oriented either equatorially or axially. On inversion of the chair the axial hydrogens become equatorial and vice versa. That this inversion manifests itself in TSD arises from the fact that the flipping motion of the nonpolar ring exerts a certain drag on the neighboring carboxyl group and so causes it to move slightly.

We note in passing that the copolymer of MMA and CHMA releases a much higher current above room temperature than each of the homopolymers. This trend is a general one, the higher charge storage of copolymers probably being linked with their ability to trap charges with a higher efficiency.

In Sect. 3.5 we indicated that we are currently engaged in studying the effect of *thermal history* on the charge storage of amorphous polymers. This effect is likely to be important, because amorphous polymers are not in thermal equilibrium at temperatures below T_g, as is evident from, for instance, the *spontaneous* volume relaxation (densification) they undergo. This is owing to the gradual continuation of the glass formation (*vitrification*) that sets in around T_g [3.148, 427].

To be more specific, we recall that the volume expansion of an amorphous polymer changes abruptly at T_g. Above T_g the polymer is in the rubbery state and has a large expansion coefficient; below T_g it behaves like a glass and has a low expansion coefficient. When a polymer is cooled rapidly from a high temperature to below T_g (e.g., during the manufacturing process), it acquires a specific volume that is too high because the macromolecules cannot adjust their configurations fast enough, so that their structural arrangement remains largely the same as in the rubbery state. Accordingly, the polymer is in a thermodynamic nonequilibrium state, but evidently it tends to assume its equilibrium glassy volume.

During the continued structural relaxation (which proceeds the more slowly the lower the temperature) the polymer changes nearly all its physical properties (hence the name "physical aging"), particularly its relaxation times, which *increase* with *time*. This *isothermal* increase is due to the inability of the main chains to rearrange properly, unless there is sufficient free volume. This, however, decreases continuously during aging. Between the free volume and the relaxation time there exists the following mathematical relation [3.82, 83, 148, 427]:

$$df/dt = -f/\tau(T,f), \tag{3.80}$$

$$\tau(T,f) = \tau_0 \exp[f^{-1}(T) - f_0^{-1}]. \tag{3.5''}$$

Equation (3.5'') corresponds to the WLF equation (3.5'), and $f(T)$ denotes the fractional free volume: $f(T) = f_0 + \alpha_f(T - T_0)$, where α_f is the thermal expansion coefficient of f.

The effect of the structural relaxation at various constant temperatures on the mechanical creep of amorphous polymers has been studied in detail by *Struik* [3.148]. Dielectrically the effects were first described by *Kästner* et al. [3.428]. Physical aging has now also been studied by TSD [3.147]. Some results are given in Fig. 3.17, the curves showing how the TSD current decreases as the storage or aging time increases, the cause of the decrease being that fewer and fewer dipoles are oriented within the chosen charging time.

Figure 3.53 illustrates the effect in a different way. It compares the TSD after application of the two charging schemes shown in the inset. In scheme *a*, the dipoles are first oriented at 100 °C, and are next allowed to age at 60 °C. In scheme *b*, the dipoles are mainly oriented just before the quench so that the oriented dipoles get no chance to age. Consequently, they are disoriented more easily, giving rise to a higher TSD current at low temperatures. These results lead to the expectation that the cooling rate during charging affects the subsequent TSD as well. This has been proved by *Hedvig* [3.19, 146]. He also studied the effect of various aging times, but his results do not wholly agree with those of the author and co-workers, the probable reason being that he quenched from a much higher temperature, so that his thermograms may have been disturbed by the ϱ peak.

The effect of quenching on the TSD of the β peak of a crystalline polymer, polycarbonate, has been discussed by *Vanderschueren* [3.20] and by *Aoki* et al. [3.388b]. The TSD of chemically aged polymers has been pursued by *Motyl*

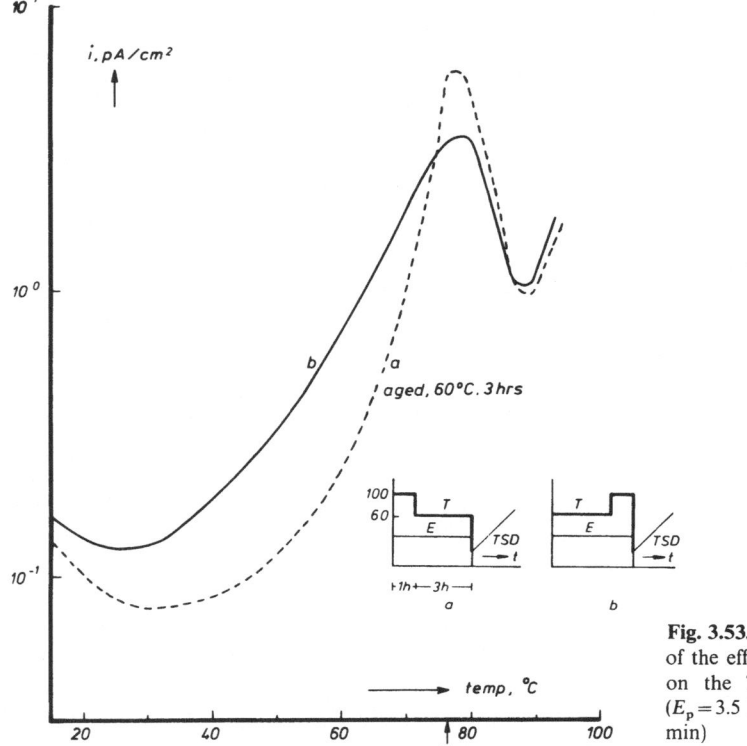

Fig. 3.53. Another illustration of the effect of physical aging on the TSD of rigid PVC ($E_p = 3.5$ kV cm^{-1}, $\beta = 1$ °C min)

[3.143], who investigated the effect of thermal aging and the degradation by electrical spark discharges and uv light.

Finally, we should point out that (physical) aging has a favorable effect on the service life of electrets, for in slowing down the charge decay it improves their long-term stability. The study of thermal treatments and annealing therefore seems to be a rewarding topic of TSD research, the more so because the thermal history also has its repercussions on the dielectric properties in general.

On the other hand, when we wish to prevent our TSD investigations from being affected by physical aging, we should precondition the sample before poling by simply heating it to above T_g. The sample is then also cleaned from spurious charges arising from the manufacturing process (cf. Fig. 3.11).

To sum up some other recent activities, we note from Table 3.9 that TSD has been applied to investigations into cross-linking, curing oxidation, and the polymerization process. Special types of polymers, e.g., oligomers, elastomers, ionic membranes of polyelectrolytes, resins, and ultrathin films have also been the subject of TSD research.

Interesting work has been devoted to the effect of crystallization. A recent study has concentrated on the compatibility of polymers in blends [3.397].

Carr et al. [3.124, 396] have worked on the spontaneous current release by polyacrylonitrile. They suggest that this current is due to chemical degradation. The effect of doping polymers with, e.g., iodide, charge-transfer complexes and dyes has been studied by *Jain* et al. [3.401, 409], *Kryszewski* et al. [3.411], and *Vanderschueren* et al. [3.44].

3.11.8 Polyvinylidene Fluoride

When foils of PVDF, which is a semicrystalline polymer, are uniaxially stretched before poling, it turns out that the electrets made from them become piezoelectric (cf. Chap. 5 and [3.429]). Such electrets have found widespread use in, for example, headphones and push buttons and for this reason there is much interest in the polymer. The stretching is necessary to create crystallites of the β form; without stretching the crystallization takes the nonpiezoactive α form. In the β form the main chains have a planar zig-zag structure, in which the strongly polar CF_2 groups point in the same direction, and they also do so in the β crystal. In the α form the main chains have a helix structure, in which the dipoles cancel when they are packed together in the α crystal (cf. Fig. 5.12).

The TSD of PVDF has been studied extensively [3.212, 376–381]. The current spectrum of unstretched PVDF is shown in Fig. 3.13. The α peak at $-42\,°C$ corresponds to the glass–rubber transition, and is due to a dipolar relaxation in the amorphous phase. The α_c peak at about $80\,°C$, which also appears in dielectric and mechanical measurements [3.34, 37c, 430], and in the piezoelectric relaxation measurements by *Fukada* et al. [3.27], is probably due to a dipole relaxation in the crystalline phase or in the region intermediate between the crystalline and amorphous phases. However, because this peak appears well above the glass–rubber transition, it may also arise (at least partially) from the destruction of charges accumulated near the crystal boundaries (Maxwell–Wagner charging).

Evidence for the relation of the α_c peak with the crystalline phase is given in Fig. 3.54, which shows that the peak shifts to higher temperatures when the poling temperatures increases, the probable reason being that high poling temperatures promote formation of more perfect and thermally more stable crystallites.

Moreover, the high-temperature α_c peak is broad and therefore involves a distributed polarization. The foot of the peak is therefore truncated with increasing storage time, in particular when the PVDF is charged at a low temperature [3.380]. According to *Pfister* et al. [3.377] a strong intermediate peak at about $12\,°C$ is found for the other piezoactive crystalline form of PVDF, the γ phase (note that an intermediate α_i peak is also visible in the spectrum of the α form; see Fig. 3.13).

Stretched and unstretched samples give different thermograms. Notably, stretching increases the TSD current by an order of magnitude. It likewise increases the piezoelectric strain coefficient; see Fig. 3.55, [3.27, 126]. These increases clearly demonstrate the much higher polarizability of the β phase.

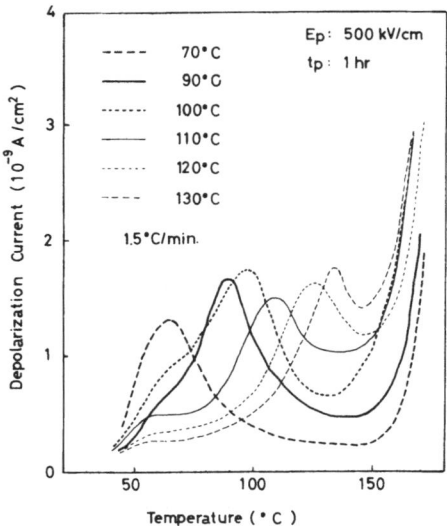

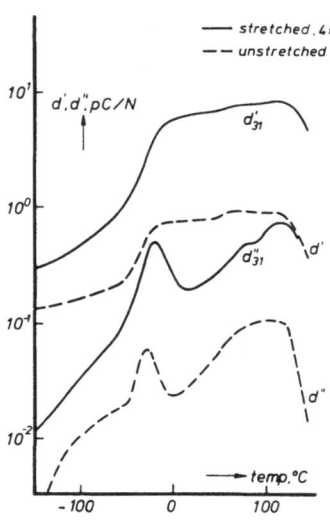

Fig. 3.54. Depolarization current of stretched β-PVDF for various poling temperatures. The corresponding piezoelectric strain coefficients, d_{31}, range from 16 to 27 pC/N when T_p changes from 70 to 130 °C. [3.379]

Fig. 3.55. Temperature dependence of the real and imaginary part of d_{31}, for stretched and unstretched PVDF electrets. $T_p = 83$ °C, $E_p = 400$ kV cm^{-1}. [3.27]

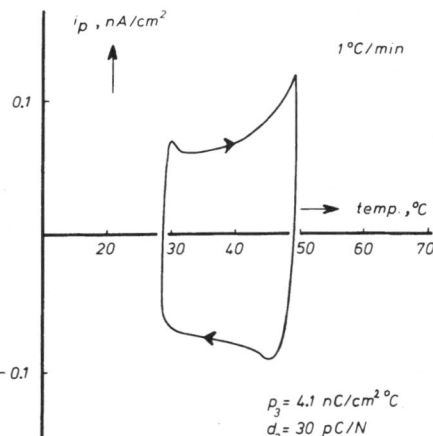

$p_3 = 4.1$ nC/cm^2 °C
$d_{31} = 30$ pC/N

Fig. 3.56. Pyroelectric current of a stretched PVDF electret

The dependence of the piezoelectric properties of PVDF electrets on the poling conditions as well as the *mechanical history* of the sample renders the investigation of PVDF electrets fascinating. In order to improve their piezoelectric properties, one has to optimize both the poling and the stretching conditions. TSD might help to achieve this goal and to elucidate the underlying processes.

Another interesting property of PVDF electrets is that they are *pyroelectric*. This is illustrated in Fig. 3.56; the PVDF electret not only generates a current when it is heated, but also when it is *cooled*. The polarity of the current is then reversed, but its magnitude is virtually the same as during heating. This behavior differs markedly from that of normal electrets, in which the current does not change sign and rapidly falls to zero, when the sample is cooled, owing to the sharp increase in relaxation times. As discussed in Chap. 5 the pyroelectricity of PVDF is probably due to a volume expansion and contraction during the heating and cooling. The effect is utilized for the detection of infrared radiation.

Polymers with a structure similar to that of PVDF are: polyvinyl fluoride, polyvinylidene chloride, polyvinyl chloride, the TSD of which is also reasonably well documented [3.11, 20, 110, 371, 372]. However, the piezocoefficients of electrets of these polymers are much smaller than those of PVDF (Chap. 5).

3.11.9 TSD Peaks Arising from the Self-Motion of Space Charges

Above T_g most polymers show a ϱ peak that is attributed to the motion of space charges. Recalling Fig. 3.34, we see that in dielectric measurements the ϱ peak replaces the conduction losses, which derive from the oscillatory motion of ohmic carriers in the applied ac field. Naturally, these conduction losses do not show a peak, but increase exponentially with temperature.

In TSD, however, the sample is short-circuited. The charges then migrate in their own internal field. This self-motion of charges does give rise to a peak, because the number of charges stored becomes eventually exhausted. Yet, the self-motion basically is a conductive motion. This is confirmed by the fact that the activation energy of the ϱ peak neatly agrees with that for dc conduction, as is illustrated for several polymers in Table 3.10.

Another fact supporting the view that the ϱ peak arises from space-charge motion is its dependence on the electrode materials [3.11, 20]; see Fig. 3.57.

Table 3.10. Activation energies of the ϱ peak and the ohmic conductivity [3.11, 20]

Polymer	U_ϱ [eV]	U_c [eV]
Chlorinated polyether (Penton)	1.2	1.5
Polycarbonate	3.5	2.2–3.4
Polyethylene	1.5	1.5
Polyethylene terephthalate	1.8–2.2	1,7–2.2
Polyimide (Kapton)	1.2	1.3
Polyethyl methacrylate	1.7	2.0
Poly-3-ethyl-3-pentyl methacrylate	1.6	1.8
Polymethyl methacrylate	1.9–2.3	1.9–2.4
80 Methyl methacrylate co 20 dimethyl itaconate	1.8	2.0
80 Styrene and 20 acrylonitrile	3.3	3.2

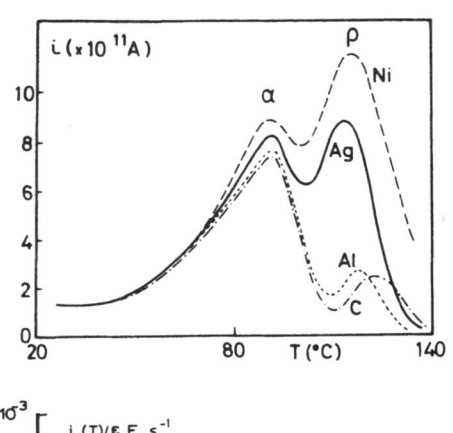

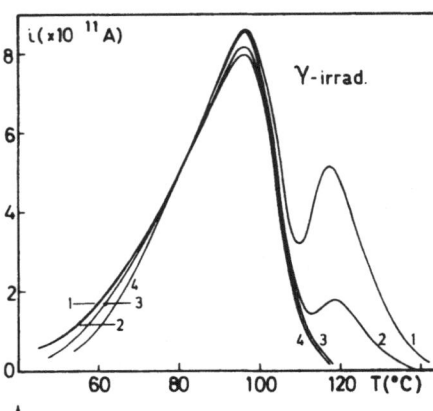

Fig. 3.57. Effect of various electrode materials (graphite, Al, Ag, and Ni) and of γ-irradiation on the current TSD of PMMA. Gamma doses used: curve *1* none, $2: 4.5 \times 10^4$, $3: 4.3 \times 10^6$, and *4*: 2×10^7 rad. Poling conditions: 125 °C, $20 \, \mathrm{kV \, cm^{-1}}$, 30 min. [3.20]

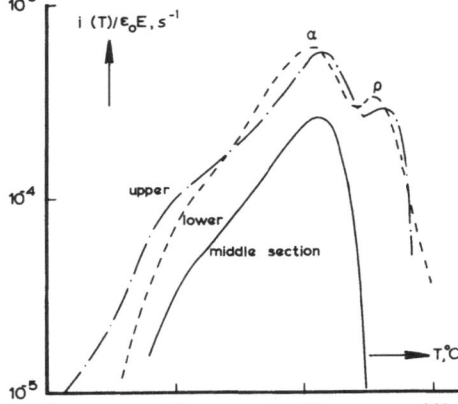

Fig. 3.58. Current thermograms of the three sections of a PMMA electret. The electret was formed at 140 °C – $20 \, \mathrm{kV \, cm^{-1}}$ – 1.5 h; its original thickness was 4.8 mm, and $\beta = 0.5$ °C $\mathrm{min^{-1}}$ [3.11]

The different electrodes probably cause different degrees of blocking owing to differences in their work functions (Schottky barriers). The figure also shows that the ϱ peak is lowered by γ-irradiation [3.11, 20], obviously because of the neutralization of the space charges by RIC. The ϱ peak is also altered drastically by impurities, dopants, absorbed water, and swelling agents, as is illustrated in Fig. 3.49. The additives clearly increase the number of charges stored, probably because they dissociate or promote dissociation.

Incomplete polarization or excessively long storage affects the ϱ peak in a different way than the dipole peaks. When the poling temperature is too low, the ϱ maximum remains in the same position, only its height being affected. A similar change is observed after a too long storage (cf. Fig. 3.14).

Measurements on sectioned polymer electrets have revealed that space-charge motions generate not only the ϱ peak, but might also contribute to the α peak (cf. Fig. 3.58).

Space-charge peaks have also been observed in other materials, e.g., ionic crystals [3.7, 431]. *Podgorsak* et al. [3.296] postulate that the ϱ peak in CaF$_2$

obeys second-order kinetics. Later *Bugrienko* et al. [3.432] tried to formulate a theory, but this can easily be disproved (see next section and [3.433]). We believe that space-charge peaks in ionic crystals should be analyzed either along the lines set forth in Sect. 3.7 or in terms of Schottky barrier formation (Sect. 3.14).

3.12 Current TSD of Heterogeneous Systems

3.12.1 TSD by the Maxwell–Wagner Effect

We have so far assumed the electrical properties to be the same throughout the sample. This does not hold, however, for a heterogeneous sample consisting of different *components* or *phases*. If the components have different dielectric constants and conductivities, charges accumulate near the interfaces when the sample is heated and subjected to a field (Maxwell–Wagner effect). These charges are frozen-in if the field is maintained during cooling.

In a subsequent TSD the interfacial charges are destroyed because the local fields in the dissimilar parts of the sample are reversed so that neutralizing charges of opposite sign are conveyed to the interfaces by ohmic conduction. This neutralization process clearly is a transient one, giving rise to s specific ϱ peak. Maxwell–Wagner peaks can be expected in, e.g., *semicrystalline* polymers, the amorphous parts of which have higher conductivities than the crystalline parts. Such peaks also appear in the TSD of *electron-bombarded* foils if the electrons are dissipated by the nonuniform radiation-induced conductivity ([3.93, 94] Sect. 3.10.1, Chap.4).

Moreover, heterogeneous systems are important because they allow TSD analysis of dielectrics with a high conductivity, such as liquid crystals [3.345] and some types of ionic crystals [3.303, 304, 434], and because they also allow investigation of semi-insulators and semiconductors [3.29, 30, 68].

The simplest system we have to consider is a structure consisting of two layers having dielectric constants ε_1, ε, and conductivities $g_1(T)$, $g(T)$, and thicknesses s_1 and s [Ref. 3.11, Chap. 6]. Assuming that the layers are *nonpolar*, we exclude dipole relaxations and take ε_1 and ε to be *independent* of time. We further suppose that during charging a planar charge σ is formed at the interface. During TSD the planar charge will create in both layers a field. These fields obey

$$\sigma(t) = \varepsilon_0 \varepsilon E(t) - \varepsilon_0 \varepsilon_1 E_1(t) \tag{3.81}$$

according to Gauss's law. The voltage across the sample being zero, we also have

$$E_1(t)s_1 + E(t)s = 0 . \tag{3.82}$$

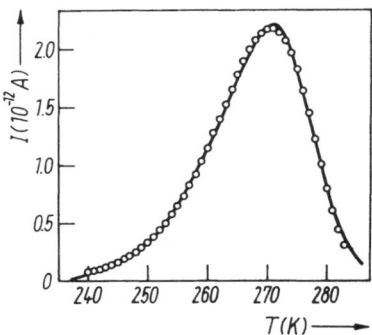

Fig. 3.59. TSD of (the chalcogenide) As_2Se_3 recorded with an insulating insert of Teflon. The measured peak is plotted as a solid line. The circles represent values calculated using (3.86). [3.436]

In the layers, the fields generate ohmic conduction currents which neutralize the interfacial charge

$$d\sigma(t)/dt = g_1(T)E_1(t) - g(T)E. \tag{3.83}$$

Eliminating E_1 and E with (3.81) and (3.82), we can rewrite (3.83) as

$$d\sigma(t)/dt + \beta_g(T)\sigma(t) = 0, \tag{3.84}$$

where

$$\beta_g(T) = \frac{g(T)/s + g_1(T)/s_1}{\varepsilon_0(\varepsilon/s + \varepsilon_1/s_1)}.$$

Evidently, the charge at the interface decays like a Debye relaxation, cf. (3.1). The TSD current generated by this decay can be found from

$$i(t) = \varepsilon_0\varepsilon_1 dE_1(t)/dt + g_1(T)E_1(t) = \varepsilon_0\varepsilon dE(t)/dt + g(T)E(t). \tag{3.85}$$

Expressing $E_1(t)$ in $\sigma(t)$, we finally obtain

$$i(t) = \frac{d\sigma}{dt} \frac{[\varepsilon/\varepsilon_1 - g(T)/g_1(T)]}{[s/s_1 + g(T)/g_1(T)](1 + \varepsilon s_1/\varepsilon_1 s)}. \tag{3.86}$$

Depending on the ratios $\varepsilon/\varepsilon_1$ and g/g_1, the TSD current may be *positive* or *negative*. Furthermore, since $i \neq d\sigma/dt$, the charge ultimately recovered during TSD will be *less* than the initial charge stored. This is so because the conduction currents neutralizing σ oppose each other in the upper and lower layer[17]. Accordingly, heterogeneous systems do *not* obey *Gross's* charge invariance principle [3.11, 435]. They will do so only when one of the layers is

---

17 However, the internal conduction currents do not exactly cancel each other, as in homogeneous electrets.

perfectly insulating; for if $g_1 = 0$, we have

$$\int_0^\infty i \cdot dt = \sigma_0 (1 + \varepsilon s_1 / \varepsilon_1 s)^{-1}. \tag{3.87}$$

All the charge induced on the upper electrode by the original charge σ_0 will then be recovered without loss. The induction is strongest, and the charge released highest, when the insulating layer is thin, i.e., when $s_1/\varepsilon_1 \ll s/\varepsilon$.

Figure 3.59 shows as an example, the TSD performed by *Müller* [3.436] on a sandwich of the amorphous semiconductor As_2Se_3 and a Teflon foil.

3.12.2 Air-Gap Current TSD by Ohmic Conduction and Dipole Disorientation

We have several times emphasized that in current TSD of *homogeneous* electrets with shorted electrodes, the neutralization of space charges by ohmic conduction is not observed. It can, however, be observed in a two-layer system in which an electret metallized on *one side* is shorted together with an air gap (Ref. 3.11, Chap. 7; [3.26–28, 391]). In the electret there then exists a *net* field, which causes the space charges to be neutralized by a net unidirectional ohmic conduction current that is externally observable. This neutralization gives rise to a current peak that can be expected to resemble the ϱ peak of shorted electrets. This similarity is exemplified in Fig. 3.60 for two polyethylene terephthalate electrets.

Air-gap TSD is also relevant to the practical use of electrets, for in most applications they are used in combination with an air gap, so that their charges generate an external field. However, the electrets then usually remain at a *constant* temperature. Nevertheless, we can still apply the equations derived below, except for the fact that all rate parameters must be taken constant [3.109].

To analyze air-gap TSD, we can use the equations of the last Sect. 3.12.1. However, for the purpose of including the disorientation of *dipoles*, we shall generalize the equations by assuming that the sample is *polar* and contains a persistent dipole polarization. On the other hand, the equations can be simplified because the conductivity of air is zero (i.e., $g_1 = 0$) and, in addition, we can take $\varepsilon_1 = 1$.

We shall assume that the charges in the sample are neutralized by ohmic conduction only, i.e. that they get no chance of moving during the discharge[18]. Their position thus remaining the same, we can avoid the use of space-dependent variables, and simply use the *average* values of the internal variables.

---

18 This assumption seems to be a fair one because we have found experimentally [Ref. 3.11, p. 316] that for most materials the neutralization by ohmic conduction dominates that by self-drift in an air-gap system. For a discussion of self-drift in air-gap TSD the reader should consult (Ref. 3.11, p. 312; [3.179, 437]). In the first reference suggestions are made on how to discriminate between the two decay modes.

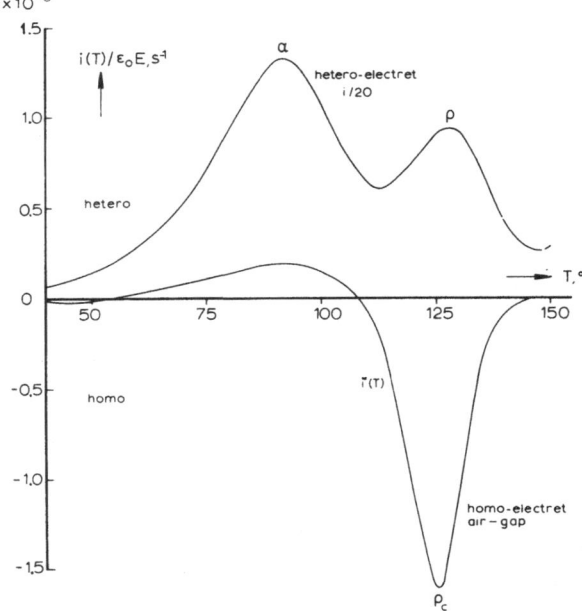

Fig. 3.60. Current TSD of a PET-c electret with a net homocharge, discharged with an adjacent air gap. It shows current peaks of opposite sign that correspond to the α and ϱ peaks of a heteroelectret recorded with normal TSD without an air gap. [3.11]

To account for the contribution of the space charges to the external current, we use the projected surface charge density

$$\hat{\sigma}_1(t) = \int_0^s \varrho(x,t)(1-x/s)dx \tag{3.88}$$

introduced in Sect. 2.1.2. Apart from the space charges, the sample may also support a real surface charge σ_r on its charged side. The sum of $\hat{\sigma}_1$ and σ_r will be denoted by σ.

In view of the presence of a persistent dipole polarization, $P(t)$, (3.81) and (3.85) change into

$$\sigma(t) = \varepsilon_0 \varepsilon_\infty E(t) + P(t) - \varepsilon_0 E_1(t), \tag{3.89}$$

and

$$i(t) = \varepsilon_0 \varepsilon dE_1(t)/dt = \varepsilon_0 \varepsilon_\infty dE(t)/dt + dP(t)/dt + g(t)E(t). \tag{3.90}$$

Using (3.82) to eliminate E, we find from the first identity of (3.90)

$$i(t) = \frac{d[P(t) - \sigma(t)]/dt}{1 + \varepsilon s_1/s}. \tag{3.91}$$

Clearly, the TSD current generated is proportional to the drop in the net charge $P - \sigma$, of which P and σ are assumed to have opposite signs, so that P

corresponds to a dipolar heterocharge and σ to a homocharge. When the reorientation rate of the dipoles is greater than the neutralization rate of the homocharges by ohmic conduction, then we first see a heteropolar peak, and next a homopolar peak of opposite sign. This means that air-gap TSD can provide unambiguous proof for the presence of two opposite charges. The method is therefore extremely useful for checking *Gross's* two-charge theory [3.438]. An example of the current reversal of bipolarly charged electrets is given in Figs. 3.9 and 3.60.

For calculating the actual course of $i(t)$, it is necessary to know $dP(t)/dt$ and $d\sigma(t)/dt$. Assuming the permanent dipoles to react with a *single* relaxation frequency, we can use (3.10) to describe the decay of P. The right-hand side of this equation accounts for the fact that not all dipoles are subject to disorientation, part of them remaining aligned as long as there is a field in the sample[19]. Eliminating this field from (3.10) with (3.82) and (3.89), we obtain

$$(s + \varepsilon_\infty s_1)dP(t)/dt + (s + \varepsilon_s s_1)\alpha(T)P(t) = (\varepsilon_s - \varepsilon_\infty)\alpha(T)s_1\sigma(t), \tag{3.92}$$

which shows that P and σ do not decay independently. The change in σ can be found from (3.83) when g_1 is taken zero in view of the perfectly insulating air gap.

Eliminating E as before, we get

$$d\sigma(t)/dt + \beta_g(T)\sigma(t) = \beta_g(T)P(t), \tag{3.93}$$

where β_g now equals $\beta_g = \{g(T)/\varepsilon_0\varepsilon\}(1 + s/\varepsilon_\infty s_1)^{-1}$. Equation (3.93) again reveals the link between the decay of P and σ. The decay rate of the homocharge σ is apparently *slowed down* by the heterocharge P, because this weakens the field created in the electret by σ.

If $i(t)$ is to be calculated by (3.91), Eqs. (3.92) and (3.93) have to be integrated simultaneously. The integration usually has to be done numerically, except for special cases, such as nonpolar samples, which contain no dipoles [$P(t) = 0$], so that the equations are considerably simplified. Such samples show a single current peak that is due to the neutralization of the homocharges by ohmic conduction.

The air-gap system can, of course, also be used to *charge* a sample [3.11, 29, 110]. When a sample is subjected to such a thermally stimulated charging (TSP), the charging current will display peaks that are identical to those of TSD. If, for instance, the sample is polar, we first observe a TSP current peak due to the orientation of the dipoles, and next a peak due to the build-up of an interfacial charge by the conduction of the sample. As in discharging, dipole orientation and conductive charge flow are transient processes, this time because, during charging, the field in the sample is bound to vanish if for another reason. It now vanishes not because the sample tends to become

---

19 The theories of *Swann* and *Gubkin* [3.439] about the isothermal decay of electrets neglect this effect, which may lead to serious errors, because the internal field can be quite strong.

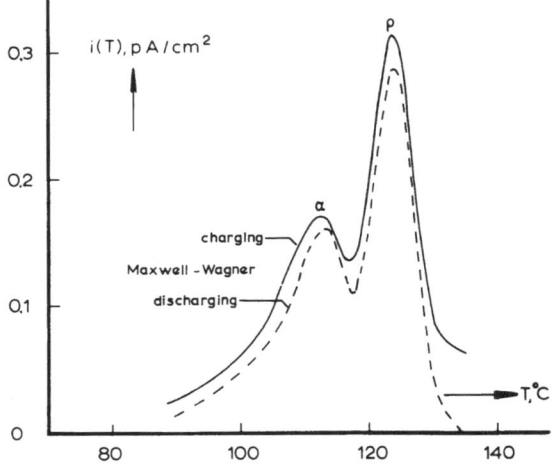

Fig. 3.61. Thermally stimulated Maxwell–Wagner charging and discharging currents of 80 styrene co 20 acrylonitrile layered together with an air gap. [3.11]

neutral, but because the perfectly insulating air gap tends to consume the *entire* charging voltage.

In Fig. 3.61 we compare the charging and discharging currents obtained with an air-gap system. The two currents almost coincide, as they should, because the same charges that develop during charging are destroyed during discharging.

3.13 Theory and Practice of Charge TSD

In charge TSD, the charge retained by the electret is measured. Here we shall focus on a convenient and well-defined method for measuring charges, namely, that described in Sects. 2.3.4 and 3.3. This method is the converse of current TSD, because it does not operate in short circuit but in *open* circuit, i.e., with a *zero* external current. This condition is fulfilled by cancellation of the external field of the (unilaterally metallized) electret with a backing voltage. This is continuously adjusted to equal the surface potential of the electret [3.11, 18, 25].

The measurement is especially suited to foil electrets. For thick electrets the backing voltage becomes too high and cannot be generated by a simple feedback system. A block diagram of the necessary electronics is given by Fig. 3.62. A similar arrangement has been designed by *Dreyfus* et al. [3.108]; *Vossteen* [3.440] has described the commercial equipment marketed by Monroe Electronics (Lyndonville, USA).

In open-circuit charge TSD, the field conditions are virtually the same as those which electret foils generate in electret transducers. For analyzing the decay of such foils at room temperature, we can therefore simply take the isothermal versions of the equations derived below (cf. Sect. 2.6.4). Some of the

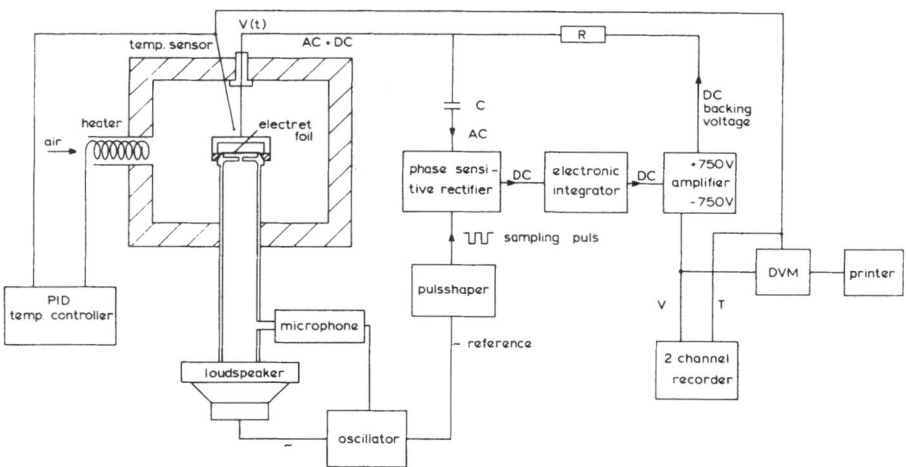

Fig. 3.62. Set-up for fully automated charge TSD measurements on foil electrets. [3.25]

equations are also useful for describing the space-charge limited charge decay in photoconductors [3.441–443][20].

3.13.1 Some Theoretical Notes

We first imagine the electret foil to contain permanent *dipoles* and *space charges*, the latter of which we assume to be *immobile*, so that they are only neutralized by conduction. This assumption allows us to consider them formally as a projected surface charge. The measurable quantity is the surface potential $V(t)$, which is determined by the net charge density of the electret, $\sigma(t) - P(t)$. From (3.89), in which we put $E_1(t) = 0$, and from the line integral of the electric field [which yields $V(t) = E(t)s$ for the two-layer system in hand] we obtain the relation

$$V(t) = [\sigma(t) - P(t)]s/\varepsilon_0\varepsilon_\infty , \qquad (3.94)$$

where $\sigma(t)$ is the sum of the projected surface charge density of the space charges and any real surface charge density. The actual course of the surface potential during the heating can be calculated by solution of P and σ from (3.92) and (3.93), in which s_1 should be put equal to infinity to account for $E_1(t) = 0$.

---

20 The equations can also be used to analyze the build-up of a return voltage on two-sided metallized electrets briefly shorted after being poled [3.435, 444]. This type of charge measurement has been abandoned, because it is much less practical than charge TSD. It is, however, occasionally used to determine the thermally stimulated photovoltaic emf of photoconductors [3.445].

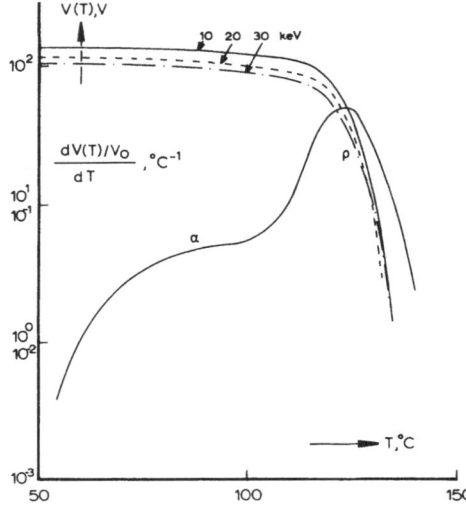

Fig. 3.63. Charge TSD of electron-bombarded Mylar-c foil. The figure also shows the derivative of the electret potential. This curve exhibits the α and ϱ peak known from current TSD measurements (cf. Fig. 3.60). The deposited charge density was 20 nC cm^{-2}. [3.11]

Figure 3.63 shows a thermogram of V vs T for Mylar. The figure also shows the derivative of V, which is much more structured than V itself, because it displays the changes in σ and P. In fact, by differentiating (3.94), we have

$$\varepsilon_0\varepsilon_\infty dV(t)/dt = d[\sigma(t)-P(t)]/dt. \tag{3.95}$$

This expression is identical with that for the current observed in air-gap TSD, except for the factor $1+\varepsilon s_1/s$. This explains why the derivative exhibits peaks that correspond to those of the TSD current recordings.

For *nonpolar* samples [in which $P(t)=0$] we find, by combining (3.93–95) that during charge TSD

$$V(t) = V_0 \exp\left[-\frac{1}{\varepsilon_0\varepsilon}\int_0^t g(T)dt\right]. \tag{3.96}$$

Hence, in samples without dipoles the decay proceeds by ohmic conduction alone.

Applying (3.96) to the *isothermal decay* at room temperature, we find that a decay time of ten years[21] requires a conductivity of $g = 5.6 \times 10^{-20}\ \Omega^{-1}\ m^{-1}$, if $\varepsilon = 2$. The conductivity of a stable electret should therefore be extremely low. The recent finding [3.147] that physical aging lowers the conductivity of an electret is therefore interesting, because it implies that an electret may lengthen its own lifetime.

---

21 If the detrapping of electrons is to proceed at a similar rate, an attempt-to-escape frequency of, say, $10^{12}\ s^{-1}$ requires a trap depth of 1.22 eV.

If (3.96) holds, the activation energy U_c for conduction can readily be found from the TSD data by plotting $V^{-1}(t)dV(t)/dt$ against $1/T$, since this expression equals $g(T)/\varepsilon_0\varepsilon$. It is simpler to calculate U_c from the slope of $v = \ln \cdot \ln[V_0/V(t)]$ vs $1/T$. Putting $L = T \, dv/dT$ we have

$$U_c/kT \simeq L - 1.989 + 1.602/L$$

which has a relative error of only 0.11‰ for $A \geq 8 \, kT$. U_c can also be found by changing the heating rate from h_1^{-1} to h_2^{-1}. We can then apply *Solunov*'s method [3.176] discussed in Sect. 3.6.4 and collect the temperatures for which $V(T_{1n}, h_1)$ and $V(T_{2n}, h_2)$ are equal. For these temperatures, (3.24) will hold. It is essential that the initial charge of the electret at the start of the two heatings is the same. This is not difficult to realize when the corona charging is performed with a control grid between the corona wires and the sample (Sect. 2.2.3 and [3.446]).

We next consider the case of the space charges being *mobile*, so that during charge TSD their own field will drive them towards the back electrode, where they are neutralized. During this *self-motion*, some of the charges will be neutralized by ohmic conduction. This should not be too strong, because otherwise the charges will be neutralized before moving out of the sample. For a low conductivity the sample must be nonpolar, so we take $P(t) = 0$. This assumption simplifies the analysis to some extent but, on the other hand, the presence of mobile space charges entails the complication that the variables in the electret become space dependent.

As in Sect. 3.7 we suppose that the space charges are not trapped, and are *free* to move with a *thermally activated mobility*. The theory set forth below can, however, be readily extended to include trapping, as we have done in Sect. 3.10 for current TSD [3.447–450]. We further assume the electret to be charged unipolarly. This is true for most of the electret foils which, when used in practice, are either charged with a corona, with electron bombardment, or with a liquid electrode on its nonmetallized side.

If we assume the injected charges to be positive, we can calculate from Poisson's equation (3.28) how the external variable $V(t)$ depends on the stored space charge density. Since, in view of the zero external field $E(0, t) = 0$, we find after two integrations

$$\varepsilon_0\varepsilon V(t) = \int_0^s dx \int_0^x \varrho(x, t)dx = \int_0^s \varrho(x, t)(s - x)dx, \tag{3.97}$$

the latter identity being obtained by integration by parts. The other basic equation is (3.29). Integrating this, and noting that $i(t) = 0$, we find for the change in $V(t)$

$$\varepsilon_0\varepsilon dV(t)/dt = -gV(t) - \mu(T) \int_0^s \varrho(x, t)E(x, t)dx, \tag{3.98}$$

from which $\varrho(x, t)$ can be eliminated with (3.28). Recalling that $E(0, t) = 0$ we obtain

$$\varepsilon_0 \varepsilon dV(t)/dt = -gV(t) - \tfrac{1}{2}\mu(T)\varepsilon_0 \varepsilon E^2(s, t) . \tag{3.99}$$

This equation shows that in charge TSD, as in every other air-gap system, the decay proceeds by *conduction* as well as by the *self-motion* of charges. Comparison with (3.31) further shows that the self-motion generates a *stronger* internal displacement current than it does in current TSD with shorted electrodes, because all charges are now driven to the same electrode. The field at the back electrode, $E(s, t)$, it determined by the total amount of space charges stored; from Poisson's equation (3.28), we have

$$\varepsilon_0 \varepsilon E(s, t) = \int_0^s \varrho(x, t)dx = \hat{\sigma}(t) . \tag{3.100}$$

All space charges initially residing near $x = 0$ will gradually spread out and eventually reach the back electrode. The space-charge density $\varrho(s, t)$ will then be zero up to the transit time t_λ of the leading front of the charges. We can calculate the time dependence of $E(s, t)$ up to this time by particularizing (3.29) for $x = s$. Recalling that $i(t) = 0$, this equation yields, after integration with respect to t

$$E(s, t) = E(s, 0) \exp\left[-\frac{1}{\varepsilon_0 \varepsilon} \int_0^t g(T)dt \right] \qquad (t \leq t_\lambda) . \tag{3.101}$$

By solving (3.99) and (3.101) we can calculate the surface potential $V(t)$. The integration has to be performed numerically, except when $g(T) = 0$, i.e., when there is no charge neutralization by ohmic conduction.

For times exceeding t_λ, the time dependence of $E(s, t)$ cannot be given explicitly, unless the original charge distribution has a simple shape, such as that of a box. For $t \geq t_\lambda$, the charge cloud then fills the entire sample uniformly, and from combining of (3.97) and (3.100) it follows that

$$E(s, t) = 2V(t)/s . \tag{3.102}$$

Substituting this equation into (3.99) we can calculate $V(t)$ for times greater than the transit time. When $g(T) \neq 0$, this calculation again requires a numerical integration of the differential equations.

For the isothermal case, however, the general solution can be given explicitly [3.451], and even diffusion can be accounted for [3.452]. The isothermal solution for $t < t_\lambda$ was given in (2.83), which in our notation reads

$$\frac{V(t)}{V_0} = \exp\left(-\frac{g(T_0)t}{\varepsilon \varepsilon_0} \right) \left\{ 1 - \frac{\varepsilon_0 \varepsilon \mu(T_0)E^2(s, 0)}{2g(T_0)V_0} \left[1 - \exp\left(-\frac{g(T_0)t}{\varepsilon \varepsilon_0} \right) \right] \right\} . \tag{2.83'}$$

Equation (2.83′) shows that at a constant temperature T_0 the voltage decays according to two exponentials, the decay rates of which differ by a factor of 2.

If the isothermal decay proceeds fast enough and if $g(T_0) \simeq 0$, we can easily derive the *mobility* of the carriers from isothermal measurements, either directly from the $V(t)$ data or after differentiating them numerically. We then have the following set of equations

$$\frac{dV(t)}{dt} = -\tfrac{1}{2}\mu(T_0)E^2(s,0) \rightarrow V(t) = V_0 - \tfrac{1}{2}\mu(T_0)E^2(s,0)t \qquad t \leq t_\lambda,$$

$$\frac{d}{dt} \cdot \frac{1}{V(t)} = \frac{2\mu(T_0)}{s^2} \rightarrow 1/V(t) = 1/V(t_\lambda) + 2\mu(T_0)(t - t_\lambda)/s^2 \qquad t \geq t_\lambda,$$

(3.103)

where $E(s,0) = V_0(s - r_0/2)^{-1}$. The graph of the derivative versus time shows a marked cusp when the carrier front reaches the back electrode. The cusp thus occurs at the transit time, which obeys the equation

$$t_\lambda = (s - r_0)/\mu(T_0)E(s,0) \simeq s/\mu(T_0)E(s,0).$$

The approximation on the right-hand side holds for $r_0 < s$; using this condition we can simply calculate $\mu(T_0)$ from t_λ.

However, it is just as easy to calculate $\mu(T)$, or rather $\mu(T)$ vs T, from charge TSD data [if $g(T) \simeq 0$] by utilization of the nonisothermal version of (3.103) and approximation of $\int_0^t \mu(T)dt$ by $hkT^2\mu(T)/U$.

Some model charge TSD curves for $g(T) = 0$ are presented in Fig. 3.64. The lower curves on the left show the influence of the carrier penetration depth. We see that the electret charge decays the more rapidly the deeper the carriers have penetrated, because they then reach the back electrode sooner. This is demonstrated more clearly in the lower curves on the right, which depict the position of the carrier front, $r(T)$. For $r_0/s < 0.5$ the carrier front is seen to reach the back electrode when the derivative of the charge, dV/dt, attains its maximum. Up to the transit, the slope of the dV/dt curve is solely determined by the *activation energy* of the mobility. These simple facts form the basis for calculating U and $\mu(T)$. The upper curves on the left show that more strongly charged electrets decay more *rapidly* than weakly charged ones, because the field driving out the carriers is stronger. However, as soon as the levels of the two charges retained become equal at high temperatures, the charge decay curves join, provided that the initial penetration depths are equal. In practice (particularly after prolonged corona charging) the curve of the more strongly charged electret is sometimes found to *cross* that of the more weakly charged one [3.446b, 453]. In the present theory the crossing can occur only when the carriers of the strong electret have penetrated more deeply into its bulk.

The problem of crossing curves has recently been discussed by *Perlman* et al. [3.454], who modified the theory in two ways. First they assumed the mobility

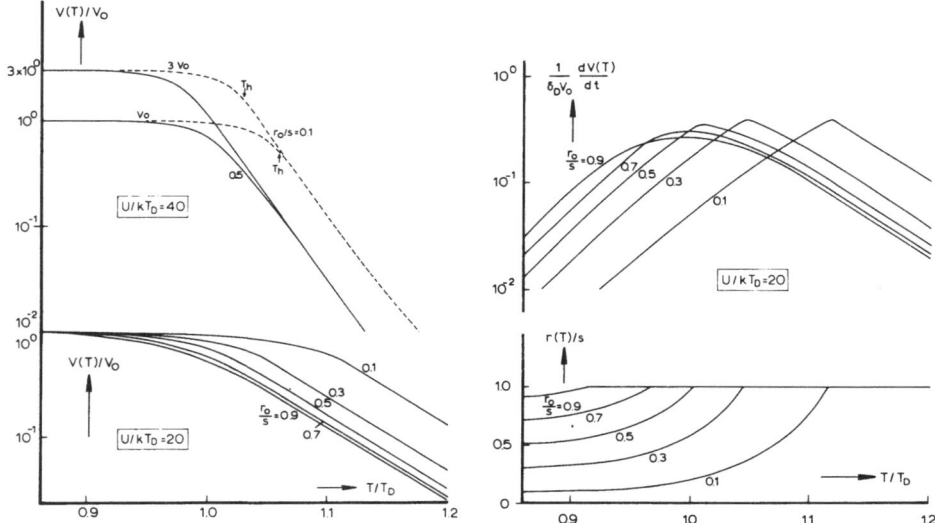

Fig. 3.64. Influence of the initial charge V_0 and the carrier penetration depth r_0 on the charge TSD by self-drift in a nonconducting electret foil. In the curves on the right the derivative of the charge retained and the position of the carrier front are also presented. We took $U/kT_D = 20$ and 40. [3.11]

of the carriers to be *field-dependent*. They further assumed that most or all of the injected carriers are trapped in deep *surface traps*. They finally assumed the *emission rate* from these traps into the bulk of the electret to be field dependent. They elaborated these modifications in a number of papers, the latest of which [3.454c, d] also concerns pre-existing trapped space charges. Recently, *Lewis* et al. [3.453a] formulated a different explanation embodying the view that the uv light attending the corona charging may be the cause of the crossover, because it induces *photoinjection* into the bulk of the polymer. Their view is not supported by recent findings of *Gerhard* [3.453b].

The general solution for a charge distribution that is initially *nonuniform* can be found along the lines discussed in Sect. 3.7, i.e., either by finite-difference methods or by the method of characteristics. Using the latter, *Nunes de Oliveira* et al. [3.449] were able to derive the general solution for a space charge of the shape $\varrho(x) \propto x$. This method is also applied in [3.180, 193, 448]. Results of calculations using a difference method have recently been published [3.450].

In the theory just given we have assumed that all carriers have the same mobility. Most experimental results, however, indicate a distribution of mobilities or its equivalent, a detrapping from several energy levels. It would therefore be worthwhile to extend the theory to multiple trapping. A first attempt in this direction was made by *Sessler* [3.243]. Other theories include deep traps of zero emission rate as a first approximation [3.447–449]. A more general analysis of trapping, making allowance for re-emission, has been given by *Chudleigh* ([3.450], Sect. 2.6.4).

Note that our theory further assumes that the back electrode is perfectly *blocking*, because we leave out of account injection of countercharges from this electrode, as is done in theories on charge decay of photoconductors [3.443]. However, our assumption seems a reasonable one for polymer–metal contacts (cf. Sect. 4.2.1 and [3.450]), unless the polymer surface has been exposed to a drastic corona pretreatment improving the adhesion of the contacting metal layer.

3.13.2 Experimental Results

In Fig. 3.3 we showed the charge TSD of a number of negatively charged polymers. We found that the fluorocarbon polymers (Teflon, Teflon-FEP, Teflon-PFA) are by far the most stable.

Figure 3.65 shows the increase in thermal stability when a Teflon-FEP electret is charged *repeatedly* to the same value. After three chargings the charge becomes extremely stable. This striking improvement suggests that the charge decay by ohmic conduction in Teflon-FEP is negligible (or that the small number of free holes present have been depleted by the electron injection). The results also suggest that the electrons injected are not trapped at a single level. Apparently, they are trapped at *deeper* and *deeper* levels as they migrate to the back electrode [3.243]. The electrons deeply trapped in the *bulk* of the electret will remain *immobile* in the next decay run and will slow down the drift of the freshly injected electrons at the surface. When the immobilized electrons are dispersed well into the bulk, the fresh electrons (which make up for the loss of charge in the previous decay) will in fact be driven to the back electrode by their own field only. As a result the charge decay rate is substantially lowered.

The analytical description of this decline in decay rate is the following. Taking $g(T)=0$, we have from (3.103) for the initial decay rate after the first charging

$$dV(0)/dt = -\tfrac{1}{2}\mu(T)E^2(s,0),\tag{3.104}$$

where

$$E(s,0) \simeq V(0)/s,$$

provided that $r_0 < s$. Since for $x \geqq r_0$, $\varrho(x,0)=0$, we can take $E(r_0,0)$ instead of $E(s,0)$. With this substitution, (3.104) also holds for the initial decay after the second charging, provided that for the deeply trapped electrons $\mu(T)=0$. Obviously, $E(r_0,0)$ is again related to the initial surface potential, but we now have

$$E(r_0,0) = [V(0)-V_s]/s,$$

where V_s represents the contribution of the immobilized deeply trapped electrons. We thus find that the initial decay rate is decreased by a factor of

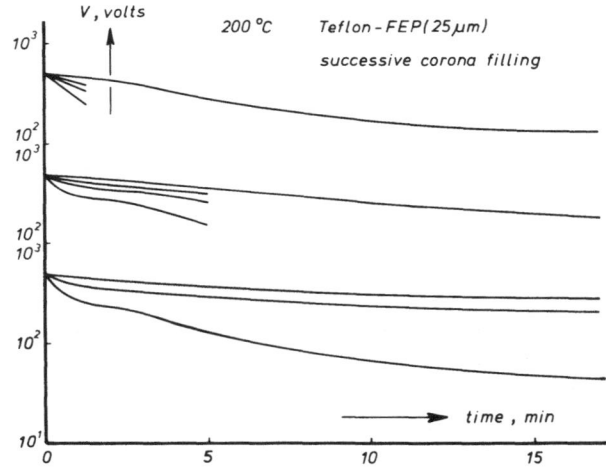

Fig. 3.65. Increase in charge stability of Teflon-FEP by repeated corona charging at room temperature after different isothermal annealings at 200 °C

$1 - V_s^2/V^2(0)$. Clearly, the decay rate is reduced to zero when V_s tends to $V(0)$[22]. An increased stability be repeated charging has also been found for other materials, e.g., photoconductors [3.443].

Recently, we have found another method for improving the stability of Teflon-FEP electrets, cf. Fig. 3.66. This method simply consists in *quenching* a foil from a high temperature to room temperature just before it is subjected to corona charging[23]. This finding indicates that the detrapping in Teflon-FEP is closely linked to its *crystalline* structure. For it is likely that the high-temperature annealing produces more perfect and thermally more stable crystallites, the electrons trapped near the crystallites not being detrapped until the latter partially melt.

In Fig. 3.67 a comparison is made between the thermal stability of Mylar-c electrets charged by corona and by a method advocated by *Baum* et al. [3.455]. In this method the corona wires are replaced with a Nernst *glow bar*, which ionizes the air owing to its high temperature. We see that the charge decay curves of electrets prepared by the two methods are almost the same, and so *Baum*'s method presents no advantages over previous methods, the less so because it is more time-consuming.

Figure 3.68 compares the charge decay of electrets obtained by electron bombardment and corona charging. We see that the former begin to decay at lower temperatures. This reduced thermal stability of electron-bombarded foils may have three causes:

---

22 The decrease in decay might be due in part to the detrapping being field assisted (Poole–Frenkel effect) [3.17, 101, 105b, 453a]. Since the number of field-generating electrons injected with each charging falls off sharply, it is particularly the emission of electrons from deep *surface* traps into the bulk that is likely to be cut short most. But also the detrapping rate in the bulk itself will be diminished when more and more electrons spread out across the sample.

23 The editor has pointed out to me that the same method has been found independently by *K. Ikezaki* et al., Jpn. J. Appl. Phys. **16**, 863–864 (1977).

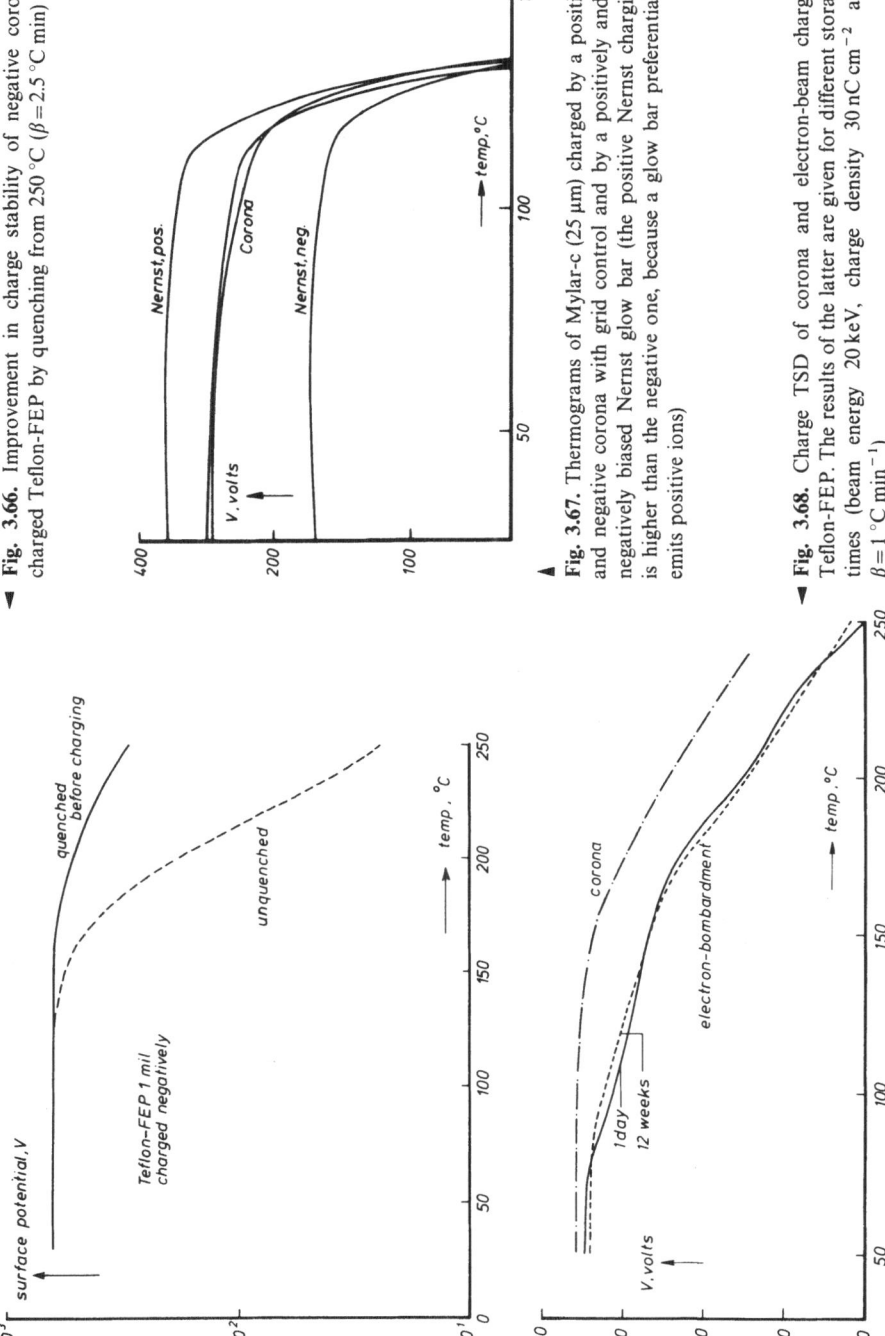

Fig. 3.66. Improvement in charge stability of negative corona charged Teflon-FEP by quenching from 250 °C ($\beta = 2.5$ °C min)

Fig. 3.67. Thermograms of Mylar-c (25 μm) charged by a positive and negative corona with grid control and by a positively and a negatively biased Nernst glow bar (the positive Nernst charging is higher than the negative one, because a glow bar preferentially emits positive ions)

Fig. 3.68. Charge TSD of corona and electron-beam charged Teflon-FEP. The results of the latter are given for different storage times (beam energy 20 keV, charge density 30 nC cm^{-2} and $\beta = 1$ °C min^{-1})

Table 3.11. Charge decay of two electrets with different surface leakage routes

Type of electret	Surface potential [V]	
	at $t=0$	after 5 min at 200 °C
Normal	499	145
With metal grid deposited	508	134

(I) The electrons are not caught in deep surface traps, because they are injected straight into the bulk;

(II) they are shot much deeper into the electret, and therefore reach the back electrode sooner;

(III) the electrons suffer from decay by RIC.

The decay of RIC can readily be calculated from the equations given[24]. Neglecting $g(T)$ the conductivity in the nonirradiated region and neglecting the self-motion of the injected charges, and assuming the RIC, $g_i(T,t)$, to be active only up to r_0 we get instead of (3.99)

$$\varepsilon_0 \varepsilon dV(t)/dt = -g_i(T,t) \int_0^{r_0} E(x,t)dx, \tag{3.105}$$

for which we can also write

$$\varepsilon_0 \varepsilon \frac{d}{dt} \int_0^{r_0} E(x,t)dx = -g_i(T,t) \int_0^{r_0} E(x,t)dx, \tag{3.106}$$

or, after integration,

$$\int_0^{r_0} E(x,t)dx = \exp\left[-\frac{1}{\varepsilon_0 \varepsilon} \int_0^t g_i(T,t)dt\right] \int_0^{r_0} E(x,0)dx. \tag{3.107}$$

For a uniform charge injection up to r_0, we have at $t=0$

$$\int_0^{r_0} E(x,0)dx = V_0 r_0 (2s-r_0)^{-1}. \tag{3.108}$$

Substituting (3.107) and (3.108) into (3.105), and integrating, we find for the RIC-induced potential decay,

$$V(t)/V_0 = 1 - \frac{r_0}{2s-r_0}\left\{1 - \exp\left[-\frac{1}{\varepsilon_0 \varepsilon} \int_0^t g_i(T,t)dt\right]\right\}. \tag{3.109}$$

---

24 Note that g_i depends not only on T, but also *explicitly on* t, because the radiation-induced conductivity fades away with time.

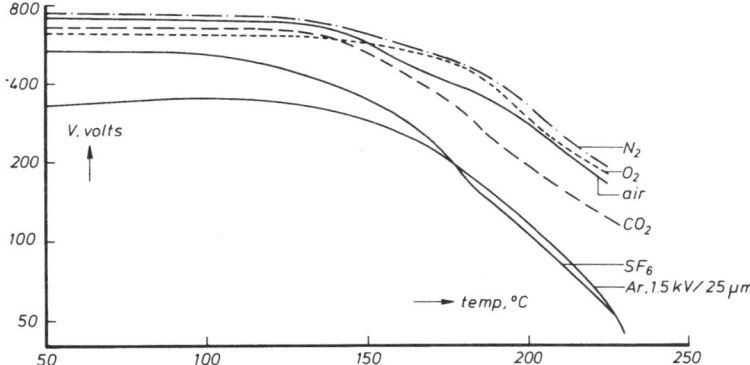

Fig. 3.69. Charge TSD of Teflon-FEP electrets formed by negative spark discharges in different gases. Formation conditions: 20 C – 4 kV/25 μm – 1 min. TSD recorded at 1 C min^{-1}

Physically, the decrease in surface potential is due to the conversion of the uniform space charge into a planar charge at r_0.

Since the magnitude of $g_i(T, t)$ depends on the time elapsed after the bombardment, one would expect the charge TSD of the electron-beam charged foils to be different after different waiting times. However, the curves of Fig. 3.68 show only minor differences, indicating that the conversion into a planar charge has already taken place during or directly after the bombardment [3.196]. This is also apparent from the initial surface potentials, which are almost equal for the different waiting times. Consequently, during TSD the charge decay is mainly due to the self-drift of the planar charge to the back electrode and thus is the same for all curves. (Note that our results confirm the open-circuit current measurements in Fig. 4.22.)

Table 3.11 compares the charge decay of two Teflon-FEP foils, one of which was provided with earthed *metal strips* on its charged side. The distance between the strips was 1 mm. We see that the decays are almost identical. From this we can conclude that main self-drift of the charges is *not along* the *surface*, and that the ohmic *surface conductivity* is *not* significantly higher than the bulk conductivity. Our results are corroborated by scanning of the lateral charge motion with probes [3.453, 455] and by the high resolution obtained in electrostatic recording [3.456]. This result is an important one, because it allows electrets to be made in *small* dimensions, a typical example being the very stable electret fibres used in air filters [3.457] on which positive and negative charges are separated by a mere 6 μm. Lateral spreading of charges has only been observed in configurations where the electric field along the surface vastly exceeds that in the bulk [3.455, 458].

Figure 3.69 shows the thermograms of Teflon-FEP foils charged in different gases. There is no significant difference in decay, which indicates that the deposited negative ions pass their electrons to the polymer, so that the decay is determined by the *bulk* properties of the Teflon-FEP rather than the type of

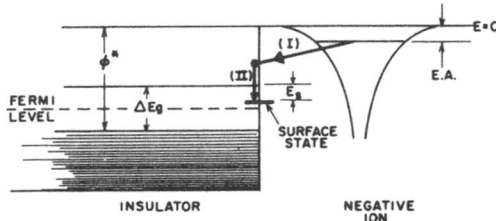

Fig. 3.70. Energy-level diagram depicting location of electronic energy levels at an insulator surface when a negative ion approaches. The solid arrows indicate electronic transitions corresponding to ion neutralization (*I*) and electron capture by a deep surface state (*II*). [3.459a]

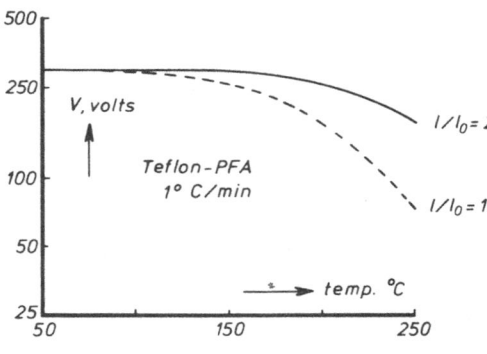

Fig. 3.71. Increase in charge stability of Teflon-PFA foil by stretching before corona charging

ions deposited. The charge exchange may proceed as presented in Fig. 3.70 [3.459].

The effect of mechanical *stretching* on Teflon-PFA films is shown in Fig. 3.71. We see that the stretching improves the thermal stability of the corona-charged film, probably because the molecular orientation induced by the uniaxial stretching changes and deforms the crystalline structure by compressing the polymer chains, and so hindering charge transport and deepening the traps. In addition, due to the alignment, the physical bonds between the molecules become stronger so that the crystallites, at which the charges are preferentially trapped, melt at a higher temperature.

3.13.3 Charge Decay Induced by Heat Pulses

Recently, *Collins* [3.130] has described a new method for probing the distribution of charges inside foils. It measures the small drop in surface potential versus time when the electret foil is heated by a *short* intense light pulse (cf. Fig. 2.12, Sect. 2.4.5).

We can explain his method by considering (3.97), in which x and s are functions of time, because the heat pulse causes the electret to expand. The expansion is *nonuniform*, starting on the charged side of the foil, where the heat pulse enters, and eventually involving the whole foil. During the expansion the total charge $\hat{\sigma} = \int_0^{s(t)} \varrho[x(t), t]\,dx(t)$ is, however, conserved, and so we can rewrite (3.97) as

$$V(t) = \hat{\sigma}[s(t) - \bar{r}(t)]/\varepsilon_0 \varepsilon, \tag{3.110}$$

where $\bar{r}(t)$ is the mean position of the charge cloud. From the change in $V(t)$ at $t=0$, and at $t=\theta$, when the heat pulse has just passed and the temperature of the foil again becomes uniform, we can find the mean depth $\bar{r}_0$ at which the charges resided, by using

$$\Delta V_{t=\theta}/\Delta V_{t=0}=(s_0-\bar{r}_0)/s_0 .\tag{3.111}$$

This equation can be derived as follows. At $t=0$, $(s_0-\bar{r}_0)$ changes by $\Delta(s_0-\bar{r}_0)$ due to the local expansion of a narrow zone z of the foil near $x=0$,

$$\Delta(s_0-\bar{r}_0)=\alpha z\Delta T' .\tag{3.112}$$

This limited expansion displaces the entire charge cloud away from the back electrode. In (3.112), α is the thermal expansion coefficient and $\Delta T'$ the local increase in temperature. At $t=\theta$ the foil has expanded uniformly, so that $\Delta(s-\bar{r})$ equals

$$\Delta(s-\bar{r})=\alpha(s-\bar{r}_0)\Delta T,\tag{3.113}$$

where ΔT is the uniform increase in temperature of the foil. Because the foil is assumed not to loose any heat during the pulsing, we further have from the heat equation as a relation between $\Delta T'$ and ΔT,

$$z\Delta T'=s_0\Delta T\tag{3.114}$$

Combination of (3.112), (3.113), and (3.114) leads to (3.111).

By the heat-pulse method, *Collins* was able not only to find the mean position of the charges, but also, by deconvolution of the $V(t)$ vs t curves, the main features of their spatial distribution. His results for a number of electrets are shown in Fig. 3.72. Unfortunately, the deconvolution is not accurate enough for the fine details of a distribution to be resolved [3.130, 460]. The distributions shown are therefore *not* unique. Recently, *Andresz* et al. [3.461] have applied the heat-pulse technique to current measurements.

A more direct and nondestructive method for finding the internal charge distribution has been developed by *Lewiner* and co-workers [3.131]. They use *pressure* pulses instead of heat pulses to expand the electret locally. This method, which can also be used in either the charge or the current measuring mode, does not require intricate deconvolutions to find the charge distribution, because the local compression during the propagation of the pressure wave through the electret is the same for all x values. On the other hand it requires more advanced equipment (cf. Sect. 2.4.6).

Lately, *Hino* et al. [3.462] have shown that insight into the internal charge distribution can also be gained from a modified version of charge TSD. This version is based on the idea of gradually changing the direction in which the charges of a particular distribution move through the sample. They use a two-

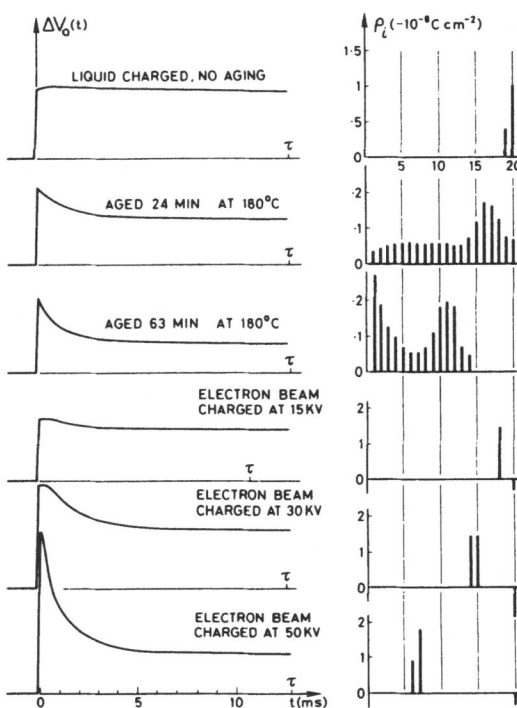

Fig. 3.72. Voltage signals obtained in the thermal pulsing experiment from four electrets charged and aged as noted. Typical charge distributions, calculated by deconvolution of these curves, are also given. (The positive surface charge on the electron-beam electrets is presumably due to secondary electron emission or to collection of ions from back discharges when the pressure is raised to normal.) [3.130b]

sided metallized sample to which a charge collecting potential is briefly applied just before a TSD is made. By changing the collecting potential from positive to negative, the *zero-field-point* is gradually shifted along the width of the sample, cf. (4.7), this shift altering the self-drift of the charges to either of the electrodes. For each of the voltages applied the electret is charged anew and its charge TSD is determined. From the set of curves so obtained we can deduce the total charge and its mean depth. It should be noted that this interesting method can only be applied to samples in which the charge decay by ohmic conduction is small. It is also essential that the charge profiles after each of the rechargings are ideally the same.

The idea of changing the collecting voltage for gaining insight into the spatial distribution has earlier been applied in optical detrapping experiments [3.463]. Another, probably less practical method, for displacing the zero-field point is a system with an adjustable air gap [3.437].

3.14 TSD of Thin Films and Semiconductor Devices

Thin-film insulators are used in various electronic components such as capacitors, passivation coatings, thin-film transistors, solid-state detectors and semiconductor devices, such as MOS-FET's (metal-oxide-semiconductor field-

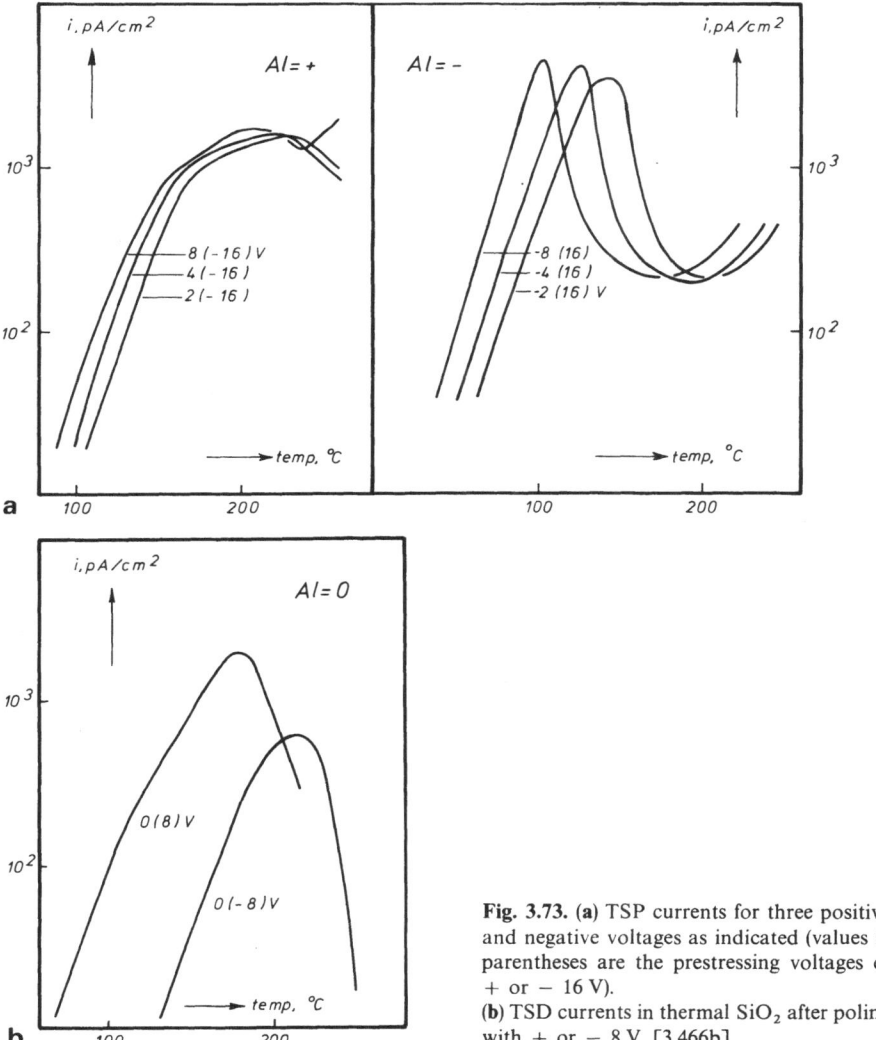

Fig. 3.73. (a) TSP currents for three positive and negative voltages as indicated (values in parentheses are the prestressing voltages of + or − 16 V).
(b) TSD currents in thermal SiO_2 after poling with + or − 8 V. [3.466b]

effect transistors), MIM (metal-insulator-metal), and MIS (metal-insulator-semiconductor) diodes, and MNOS (metal-nitride-oxide-semiconductor) memories [3.105b, 464]. Knowledge about the traps in such devices is of paramount importance, because the traps control the performance, yield, and reliability of the device. Thin films are as a rule characterized by measurement of the capacitance and current–voltage curves [3.464], but TSD often provides important new information and can easily be performed on test devices from the shelf [3.3, 107, 111, 321, 465]. The insulators used are about 0.1 μm thick and are amorphous inorganic materials such as SiO_2, Si_3N_4, Al_2O_3, Ta_2O_5, CdTe, CdS, and CdSe.

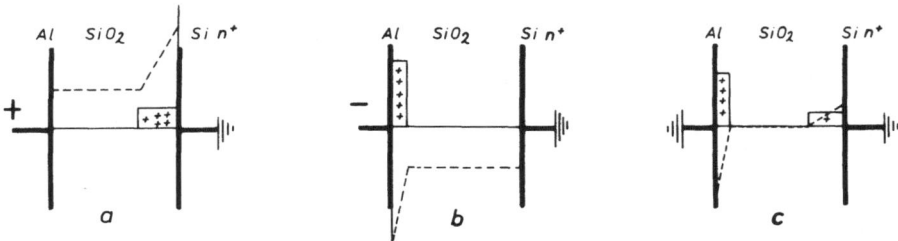

Fig. 3.74. Final ion and field distributions in thermal SiO_2 with blocking electrodes after (**a**) TSP with a positive voltage, (**b**) TSP with a negative voltage, and (**c**) TSD (zero voltage). [3.466b]

As an example we shall discuss results obtained by TSD measurements above room temperature on thermally grown 0.5 μm thick SiO_2 wafers that were incorporated in a MOS structure of Al–SiO_2–Si ([3.24]; see also [3.466]). The Si was heavily n-doped and can be regarded as a conductor. From Fig. 3.73 we see that the TSD current after a positive poling greatly exceeds that after a negative poling.

In addition, the plots show the TSP charging currents which are also dissimilar. Nonetheless the current–temperature integrals for the three charging voltages are the same. This means that the charge (1.55 μC/cm²) displaced in the successive charging runs is the same. We further note that the currents for negative charging shift appreciably with increasing voltage to lower temperatures.

Agreeing with *Hickmott* [3.23] we believe that the results must be ascribed to the motion of *sodium ions*, which are known to form a persistent impurity in thermal SiO_2 [3.467]. *Hickmott* imagines the trapping of the Na^+ ions in the SiO_2 to be *spatially* nonuniform. The ions are strongly trapped near the interfaces, whereas throughout the bulk, which only contains shallow traps, they can move almost freely. At the interfaces, however, the trap densities are of quite different magnitude. Apparently, the Al–SiO_2 interface has a much higher trap density than the SiO_2–Si interface. Hence, the layer containing the ions near the Al–SiO_2 interface is thinner than that near the other interface. No ions being lost during the series of heating runs, the electrodes must be *blocking*. We thus get a situation as depicted in Fig. 3.74, in which we indicate that during TSD (Fig. 3.74c) the ionic motion stops as soon as the field in the bulk becomes zero.

The traps near the two interfaces must also differ in nature, in view of the different voltage dependence for positive and negative chargings. Evidently, the depth of the traps near the Al–SiO_2 interface is *lowered* by the applied field E and so, when the charging voltage is raised, the traps are filled at a lower temperature. This lowering of trap depth is known as the *Poole–Frenkel effect* [3.101, 105b]. For a Coulombic trap, we have

$$U - U_0 \propto E^{1/2} .$$

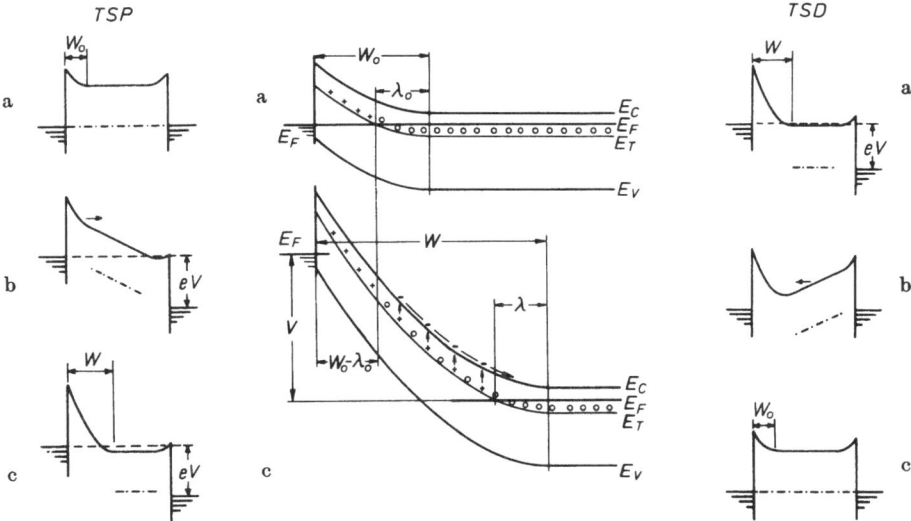

Fig. 3.75. *Left:* Energy-band diagrams of a metal–insulator–metal system during TSP (**a**) in thermal equilibrium before cooling and poling; (**b**) just before heating with poling voltage applied at LNT; and (**c**) in steady-state after TSP.

Middle: Enlarged diagrams of the depletion layer near the cathode in stages *a* and *c* of the TSP sequence. The enlarged diagrams show the expansion of the layer when the sample is biased and make it clear why the deep donor centres between $W - \lambda$ and $W_0 - \lambda_0$ are liable to emit an electron when the biased sample is heated.

Right: Energy-band diagrams during TSD (*a*) in steady-state after cooling with a charging voltage; (*b*) at LNT just before TSD; and (*c*) in steady-state after TSD

Combining this relation with (3.9) we note that a plot of T_m vs $E^{1/2}$ should be a straight line. This is indeed the case (for details see [3.24, 468]).

Of two other theories advanced for the description of the TSD of thin films, one postulates *electron hopping* between two sites separated by an *asymmetric* potential barrier. This theory is due to *Adachi* et al. [3.469] who applied it to their results on SiO. Asymmetric two-site hopping is also field dependent and shows a saturation effect, but *not* to the same extent.

The other theory is by *Simmons* et al. [3.17, 67, 166, 470–474], who point out that *Schottky barriers* are formed at the two SiO_2 interfaces owing to differences in the work functions (Fig. 3.75). The electrons therefore have to surmount a high energy barrier to enter the SiO_2. Hence, during a negative charging there will be *no* electron flow from the Al into the SiO_2. It is more likely that electrons trapped in the SiO_2 at *donor levels* above the Fermi level are released and forced to move to the positive pole, leaving the SiO_2 at the forward biased Schottky barrier on the Si side. The emptied donor levels will create a positive space charge near the Schottky barrier on the Al side. During the subsequent TSD this space charge is neutralized by electrons released from donor levels on the Si side. Although this model also shows saturation with charging voltage [3.17,

470], its TSD curves might be expected to be independent of the charging polarity, because the Schottky barriers at the two interfaces will hardly differ.

Simmons and his group have studied the TSP and TSD of Schottky barriers in nearly all their aspects; but the first use of the method must be credited to others [3.111, 475]. In order to extract as much information as possible from these methods they suggest [3.17, 67, 470] performing a series of chargings and dischargings for different forward- and reverse-bias voltages. This technique enables them to change the number of traps investigated, by changing the width of the depletion layer.

They have also discussed the TSD of MOS structures *not* containing heavily doped Si [3.471]. In such structures *surface* traps at the SiO_2–Si interface can be studied with TSP and TSD. They have further investigated the *memory* traps in MNOS devices [3.472]. Recently, they have shown that it is not only possible to measure the carrier emission from traps, but also the *generation* of electron–hole pairs by inversion of the barrier with a strong reverse bias [3.471b, 472a, 473]. They have also paid attention to the theoretical analysis of the experimental data, such as the calculation of the energy spectrum of the traps [3.214, 474]. It should be mentioned, however, that in their theory they have assumed that the width of the depletion layer remains constant during the heating. This is not entirely true; a more general treatment has been given by *Gupta* and by *Ferenczi* et al. [3.476].

3.14.1 Thermally Stimulated Current Measurements on Semiconductors

The creation of a *depletion* region by a Schottky barrier contact is, of course, also essential for thermally stimulated current measurements on (highly doped) *semiconductors*, in which, without a high resistivity region, any detrapping current would be swallowed by the strong conduction current [3.3, 111, 320–323, 326]. Another way of creating a depletion region and so avoiding the conduction current is to perform the measurement on a *reverse-biased p–n junction* [3.3, 107, 111, 475, 477].

A quite different route for studying semiconductors with TSD is filling its trapping sites by the *field effect* [3.478], see Fig. 3.76. Now the only traps that are filled are the ones adjacent to the noncontacting field electrode.

During TSD the trapped carriers transferred to the conduction band are collected at the contacting electrode. To reach this electrode they have to move *along* the length of the semiconductor. The ensuing current can be calculated from the telephone-line equation [3.30, 478]. However, *Katzir* and *Agarwal* et al. [3.30, 479] have questioned whether the detrapping current is in fact revealed in such a measurement. They show that ohmic conduction may seriously interfere.

In a variant of the field-effect method the charging is done in a two-layer structure with an insulating insert (cf. Sect. 3.12). In this arrangement, the transverse detrapping current and ohmic current are observed [3.30, 68].

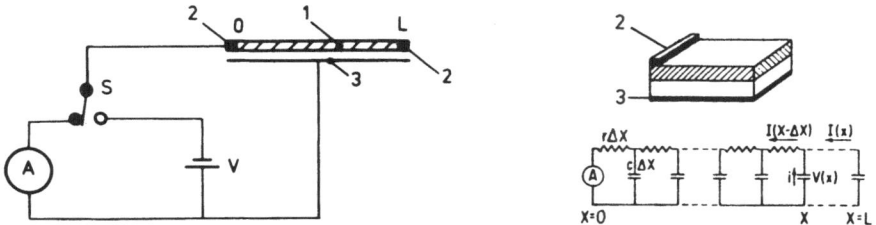

Fig. 3.76. TS capacitor discharge method [3.478]. (*1*) semiconductor under investigation; (*2*) ohmic contacts; (*3*) field electrode. On the right is a schematic representation of the longitudinal discharge currents

3.14.2 Methods Using Thermally Stimulated Admittance and Capacitance

In Sect. 3.4 we stressed that it is often rewarding to compare the TSD results of electrets with those from other independent techniques. Their relation with dielectric measurements was found particularly illuminating (cf. Sect. 3.9). The same holds for semi-insulators and semiconductors; their TSD, too, can be compared most beneficially with *thermally stimulated admittance* ([3.480], see also Ref. 3.3, p. 131) and *thermally stimulated capacitance* measurements [3.107, 477, 481]. Both types of measurements are again conducted on reverse-biased Schottky barriers or *p–n* junctions, the purpose being to suppress ohmic leakage currents by creating a region depleted of free carriers.

Thermally stimulated admittance measurements (TSAdm) actually are dielectric ac measurements that probe the periodic flow of charges to and from the depletion layer when it shrinks and expands under the action of the applied voltage. The charges originate from the recurrent filling and emptying of those traps in the depletion region which, within the time set by the measuring frequency, are able to communicate with the conduction band. We know from Fig. 3.75 that the traps liable to do so are those which cross the Fermi level. We further have from (3.65) that a particular trap will contribute most charge when at a specific temperature its emission rate becomes equal to the reciprocal of the angular measuring frequency.

Hence, if the barrier or junction is warmed linearly from LNT (the temperature of liquid nitrogen) upwards specific peaks will appear in the ac conductance versus temperature spectrum. From the shift of the peaks with frequency the trap depth can be found from an Arrhenius plot, while the trap concentration can be deduced from the height of the peaks. Together with the ac conductance the ac capacitance can be recorded; this quantity only shows stepwise changes, each time (3.65) is fulfilled.

The *thermally stimulated capacitance* measurements (TSCap) advanced by *Sah* [3.481] differ from TSAdm in that they probe the detrapping in the entire depletion region. Another difference is that it is *not* a steady-state measurement, but resembles a TSD *single shot* experiment; i.e., the traps under investigation are loaded and unloaded only once.

To probe the charge state of the traps in the depletion layer TSCap monitors the width of the layer with a high-frequency capacitance bridge, the measured capacitance being

$$C(T) = \varepsilon_0 \varepsilon A / w(T) = \varepsilon_0 \varepsilon A \sqrt{e[N_D + m_i(T)]/2\varepsilon_0 \varepsilon (V_r + V_b)}, \tag{3.115}$$

where A is the electrode area, V_r the reverse bias, V_b the built-in barrier voltage, N_D the concentration of the completely ionized shallow donors, and $m_i(T)$ the concentration of the ionized deep donor traps. The ionization of these deep traps is determined by the temperature during the heating:

$$m_i(T) = m_0 \left[1 - \exp \left(- \int_0^T v(T) dt \right) \right], \tag{3.116}$$

where the electron escape rate $v(T)$ obeys (3.71).

In short, TSCap is performed as follows: The initial charge in the donor traps (or generally the majority-carrier traps) is set by a zero or forward biasing of the barrier (or junction) at LNT. The barrier is next reverse biased and heated so that those filled traps in the depletion region which lie above E_F, get a chance to emit an electron to the conduction band. This thermal emission increases the capacitance, according to (3.115).

Like the charge TSD curves, the capacitance vs T curves exhibit only stepwise changes; they are therefore often differentiated to give a more structured image of the location of the trap levels. The method supplements the thermally stimulated current measurements in that it allows a *direct* discrimination between electron and hole traps, for hole detrapping increases the capacitance, whereas electron detrapping diminishes it. Further details are given in the reviews cited [3.107, 477, 481].

3.14.3 Deep-Level-Transient Spectroscopy

The capacitance measurements on semiconductors can, like dielectric measurements, be done in the *time* domain. The change in capacitance is then registered, upon a step change in bias voltage.

A sophisticated variant of these measurements is the *deep-level-transient spectroscopy* (DLTS) proposed by *Lang* [3.482]. He exposes the reverse-biased barrier or junction to a sequence of voltage pulses repetitively filling the majority-carrier traps. When after each of the pulses the reverse bias is restored, the traps try to empty themselves again, at a rate given by (3.71). This emptying is monitored against time by measurement of the change in high-frequency capacitance. However, instead of recording the whole capacitance transient, *Lang* concentrates on measuring electronically the difference within a certain *time window*. This device allows him to scan the emptying of the traps automatically during a linear heating.

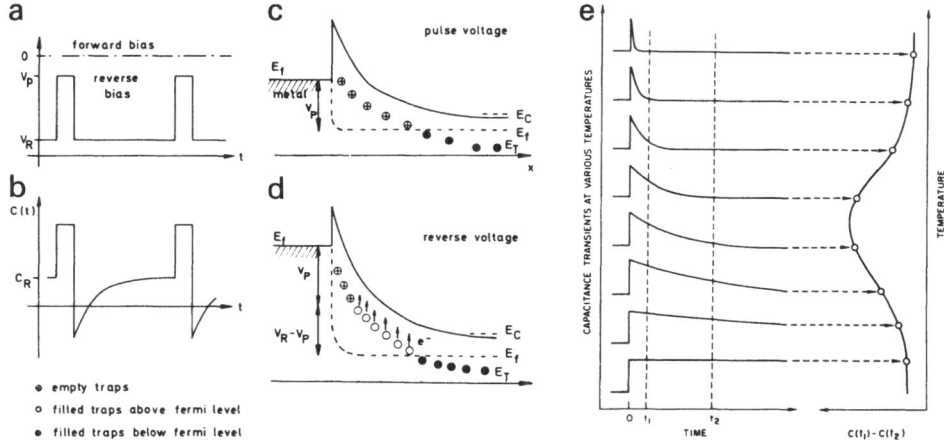

Fig. 3.77a–e. Principle of trap analysis in a Schottky barrier with DLTS (a) time variation of the bias voltage; (b) time variation of the capacitance signal; (c) and (d) energy band bending at the two different bias voltages; (e) illustration of the rate window concept, which shows that the output during a linear temperature ramp corresponds to the difference in capacitance at the sampling times $t_{1,2}$. [3.482, 483]

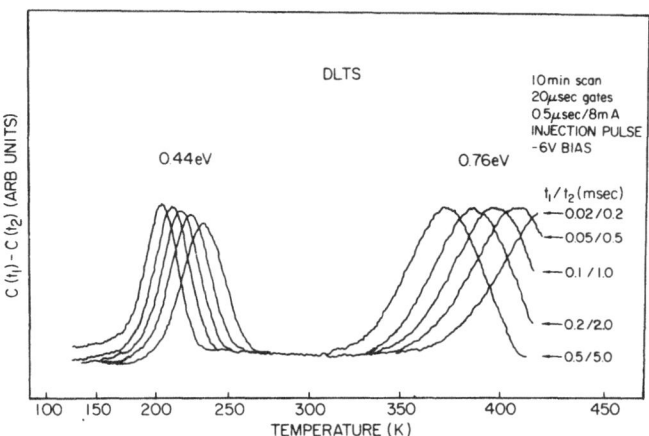

Fig. 3.78. DLTS spectra for two minority-carrier traps (hole traps) in n-GaAs recorded for five different time windows. Both trap concentrations are 1.4×10^{14} cm^{-3}, and the majority-carrier concentration n is 5×10^{15} cm^{-3}. [3.482]

The principle of the method is depicted in Fig. 3.77. The figure also shows that at a certain temperature a peak appears in the $C(t_1) - C(t_2)$ curve. This clearly happens when the trap emission rate falls within the chosen time window.

Just as in TSCap, the *minority*-carrier traps can be investigated as well. For this purpose the minority traps in the barrier are sequentially filled by superposing short forward-bias pulses on the fixed reverse bias. DLTS is further capable of determining the *local* density of the deep traps in the depletion region. A modification particularly suited to this trap-density profiling has recently been proposed [3.483].

A typical DLTS spectrum for GaAs is given in Fig. 3.78. We see that the spectrum is recorded five times by use of five different time windows. The recordings were made in this way for the purpose of easily finding the depths of the two dominant traps, because the depths can be found from an Arrhenius plot in which the reciprocals of the peak temperatures are plotted against the chosen rate window values.

DLTS has become a favorite method for investigating and testing semiconductors and semi-insulators, because it is one of the most powerful techniques for undertaking the spectroscopy of trapping levels in those materials. Compared with TSAdm and TSCap, it has a higher sensitivity, works faster and is more convenient. Moreover, the range of traps it can observe is greater. The instrumentation is, however, more sophisticated, several ingenious electronic circuits can be centred around it. More details are given in [3.482, 484]; the reader should in particular consult the excellent review by *Miller* et al. [3.482b].

Recently, *Martin* et al. [3.323b] have compared DLTS and TSCo spectra in GaAs: Cr, and found that DLTS may help to interpret the trapping levels detected by TSCo.

3.15 Analysis of Charge Detrapping by Other Techniques

3.15.1 Thermoluminescence and Thermally Stimulated Conductivity

Defect levels in insulators are also investigated by other techniques related to TSD. The oldest method using thermal stimulation is the *thermoluminescence* of phosphors [3.2, 4–6, 45, 48]. This method is entirely optical and therefore has the advantage that no electrical contacts are needed. Until recently, most TL research was done on crystalline inorganic materials, but the work is now expanding to other materials, such as polymers [3.44, 485], and even to biomaterials [3.356].

Instead of filling the traps electrically as in TSD, we now fill them at a low temperature by shining *light* on the sample, or by exposing it to *nuclear radiation* (e.g., in radiation dosimetry [3.49, 155], or archaeological dating [3.150, 151]). Light of a photon energy greater than the band-gap energy creates electron–hole pairs which separate and finally settle down at trapping levels between the conduction and valence band.

During the subsequent heating the electrons and holes are gradually released from their traps and radiate light on being captured by opposite carriers in recombination centres (note that nonradiative recombinations may also occur). Measurement of the light emission peaks (or luminescence peaks)

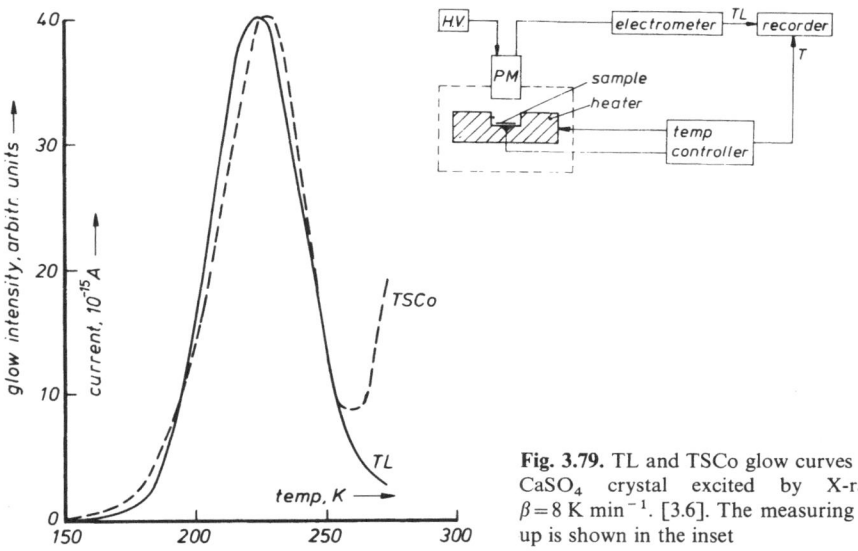

Fig. 3.79. TL and TSCo glow curves of a CaSO$_4$ crystal excited by X-rays; $\beta = 8$ K min^{-1}. [3.6]. The measuring set-up is shown in the inset

with a sensitive photomultiplier as a function of temperature provides particulars about the trapping and recombination centres. Measurement of, in addition, the spectral distribution of the emitted light, yields extra information on the recombination processes. An example of a TL "glow curve" is given in Fig. 3.79, which shows the TL of a CaSO$_4$ crystal excited by X-rays.

The transition processes involved in TL are illustrated in the simple band-gap model of *Schön–Klasens* (Fig. 3.80). One trapping level and one recombination level for electrons are assumed. Electrons released from their traps by the heating diffuse across the conduction band until they meet empty recombination centres, where they are neutralized. The energy freed by this neutralization is emitted as light whose intensity obeys

$$I(t) \propto n(t)h(t) \quad \text{or} \quad I(t) \propto -dh(t)/dt, \tag{3.117}$$

where $n(t)$ is the number of free electrons in the conduction band and $h(t)$ the number of holes in the recombination centres.

The theory of TL is very similar to that described in Sect. 3.10 for TSD, except that a precise book-keeping of the recombination centres should be included in the analysis. To be honest, the TSD theory has borrowed heavily from the older theoretical description of TL. However, the internal variables operating in TL are space *independent*, and so only their time dependence needs to be considered, which is a great simplification. Nevertheless, a proper analysis of TL curves is still complicated, because of the many variables involved [3.2, 4–6, 486], and for an exact solution, the differential equations have to be integrated numerically [3.487].

In setting up the analysis, it is essential to start from a good physical model (electron or hole trapping, degree of filling, mono- or bimolecular recombination, slow or fast retrapping, one or more trapping and recombination levels, a continuous distribution of levels, etc.). For further details the reader is referred to [3.13, 58] and to a critical review by *Bräunlich* et al. [3.488].

The next oldest method is the *thermally stimulated conductivity* method (TSCo) by which the detrapping in photoconductors can be investigated [3.1, 2, 206, 489]. In this method the traps are also filled *optically* (or by nuclear radiation; in particular if the material is either nonphotoconductive or highly insulating, or if it is used for radiation dosimetry, cf. Sect. 4.7. The photoconductor is cooled to well below room temperature, so as to avoid neutralization of the optically generated carriers by ohmic conduction. The sample is illuminated during cooling or at some constant low temperature, sometimes in the presence of a weak *field*, the latter impeding recombination of the positive and negative carriers by keeping them separated.

When the photoconductor is next reheated in the dark with a bias field E, the trapped electrons are gradually released to the conduction band contributing specific peaks to the dark conductivity. When the photoconductor is an n-type one, in which only the detrapped electrons contribute to the conductivity, we have for the TSCo current density

$$i(t) = \mu_0 n(T)E , \qquad (3.118)$$

where μ_0 is the mobility of the electrons in the conduction band.

The TSCo method is applicable not only to photoconductors, but also to semiconductors, p–n junctions, phosphors and other optoelectronic components as well as to *photoelectrets* [3.55, 490]. In photoelectrets a *persistent internal polarization* (PIP) is produced by application of a *poling* field during illumination. The PIP thus generated obviates the need to apply a charge collecting bias during the subsequent heating. As in normal electrets the detrapped carriers are forced to move in the internal field of the photoelectret and so neutralize themselves. Figure 3.81 shows that the peaks generated by this self-neutralization are sharper than those of the TSCo current[25]. Yet, it seems advisable to do both experiments (with zero and nonzero collecting bias during the heating) so as to extract the maximum possible information about the trapping levels [3.492].

An interesting feature of TSCo, illustrated in Fig. 3.82, is that electron and hole traps can be investigated *separately*, by changing the polarity of the bias voltage with respect to the illuminated side of the sample [3.493, 494]. We can also distinguish between electron and hole detrapping by applying a weak *temperature gradient* during the heating and measuring the thermally stimulated thermoelectric power of the excited sample [3.495].

Another interesting variant of conventional TSCo measurements has been described by *Tkach* [3.496]. He determined the thermally stimulated con-

---

25 As *Tkach* [3.491] has pointed out, *Zolotaryov* et al. [3.490] have given a doubtful explanation for the pointedness of the peaks. They attribute it to the fact that under photoelectret conditions $i_m \propto q_0^2$, but we know from Fig. 3.25 that $i_m \propto q_0$, as it is in conventional TSCo.

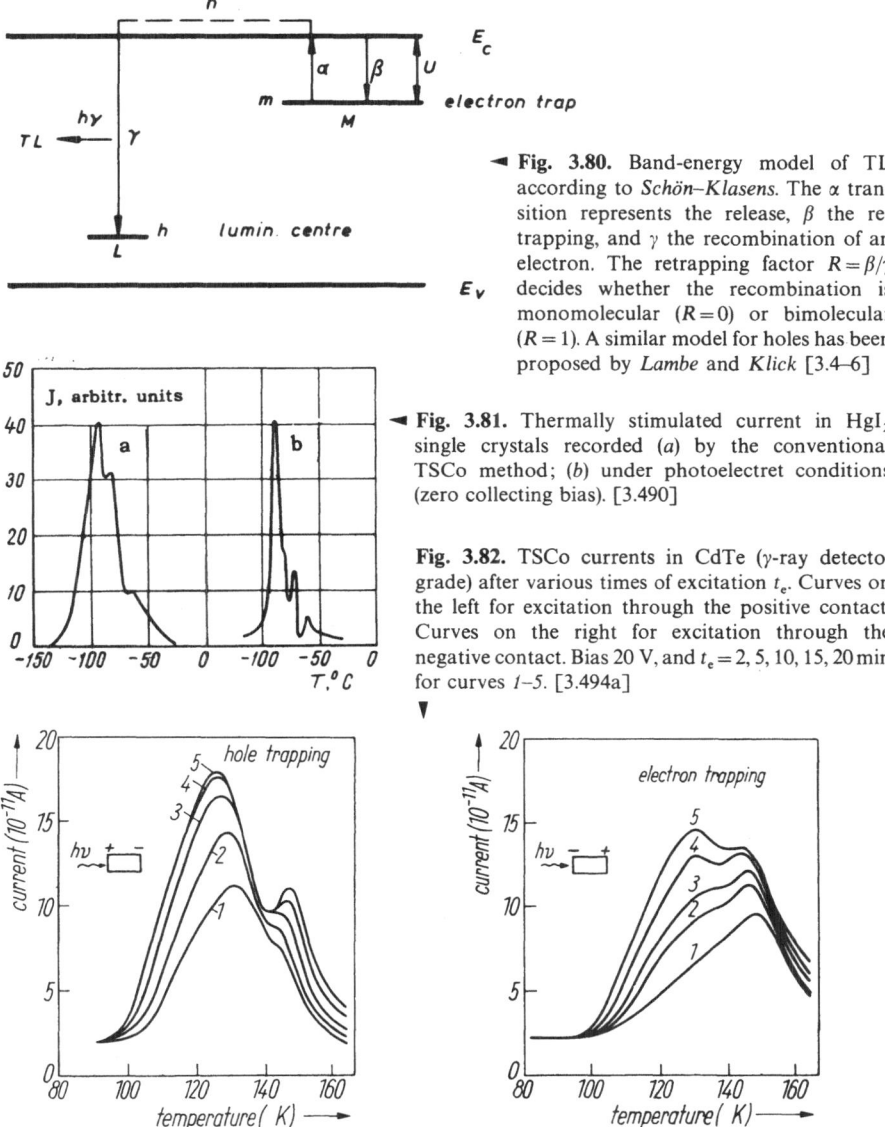

Fig. 3.80. Band-energy model of TL according to *Schön–Klasens*. The α transition represents the release, β the retrapping, and γ the recombination of an electron. The retrapping factor $R = \beta/\gamma$ decides whether the recombination is monomolecular $(R = 0)$ or bimolecular $(R = 1)$. A similar model for holes has been proposed by *Lambe* and *Klick* [3.4–6]

Fig. 3.81. Thermally stimulated current in HgI_2 single crystals recorded (*a*) by the conventional TSCo method; (*b*) under photoelectret conditions (zero collecting bias). [3.490]

Fig. 3.82. TSCo currents in CdTe (γ-ray detector grade) after various times of excitation t_e. Curves on the left for excitation through the positive contact. Curves on the right for excitation through the negative contact. Bias 20 V, and $t_e = 2, 5, 10, 15, 20$ min for curves *1–5*. [3.494a]

ductivity using an *ac bias* whose amplitude he changed. This enabled him to display the same current peak at two different temperatures; from the shift in peak position he could derive the trapping depth. TSCo measurements with an ac instead of a dc bias have also been conducted by *Silverman* et al. [3.497] on an X-ray excited ZnS phosphor.

Returning to Fig. 3.79, which also gives the TSCo results of the X-ray excited $CaSO_4$ crystal, we see that the TSCo peak appears at a higher temperature than the TL peak. As can be proved theoretically TSCo peaks usually do so, because they require more carriers to be emitted into the conduction band before a peak appears [3.499a]. Sometimes, however, no corresponding TL peak is found, namely when the TL proceeds locally *without* free carriers appearing in the conduction band, their transitions from excited states taking place entirely within the forbidden gap (associated donor–acceptor or Prener–Williams recombination [3.4–6]).

It therefore often pays to subject a sample to both experiments, because they complement each other [3.498]; TSCo probes particularly the trapping levels, while TL focusses on the recombination levels. *Chen*, and *Fillard* et al. [3.13, 58, 499] have stressed that a particularly interesting quantity to recorded is the ratio between the TL and TSCo data, i.e., $I_{TL}(T)/I_{TSCo}(T)$, which according to (3.115) and (3.116) equals $h(t)$. By recording this ratio during a run of partial heatings we can deduce whether the recombination is mono- or bimolecular.

The theory of TSCo is again very similar to that of TL and that of TSD (see Sect. 3.10). The usual theories of the technique, however, only take account of the time dependence, space-charge effects being neglected, and so the carriers considered to experience a uniform bias field. This restriction is legitimate only when the traps have been filled in the absence of a field. If the traps have been filled in the presence of a field, the analysis is performed by some authors [3.240, 475] using space-averaged quantities.

Several methods have been proposed for calculating the kinetic parameters from TL and TSCo data [3.2, 12, 13, 58, 199–204, 214c, 489]. They are very much the same as those reviewed in Sect. 3.8 for TSD, except that they are in part focussed on the question whether the kinetics are of first- or second-order [3.13, 204, 500]. Additional measurements at constant temperature may be helpful if there is any ambiguity on this point [3.501].

3.15.2 Optical Charge Detrapping and TSEE

Since traps can be filled optically, they can also be *emptied optically*. Carriers emitted into the conduction band by *photodepopulation* can, as in TSCo, be collected with a bias voltage and measured with an ammeter. Optical charge detrapping has only recently been pursued quantitatively [3.51–53, 463], although it has long been in use for discharging photoconductors [3.1, 55] and photoelectrets [3.54, 133].

Like the thermal scanning in TSD it is most appropriate to scan the optical current release as a function of photon energy by use of an adjustable monochromator. The method has the advantage of rapidity and of leaving the sample intact, because the temperature remains constant. Moreover, deeper traps can be reached (note: $1 \, \mu m \rightarrow 1.24 \, eV$, while $300 \, K \rightarrow 0.025 \, eV$). However, it requires more advanced equipment, and the currents generated are rather weak, because in insulators optical emptying is not very efficient. Results of a

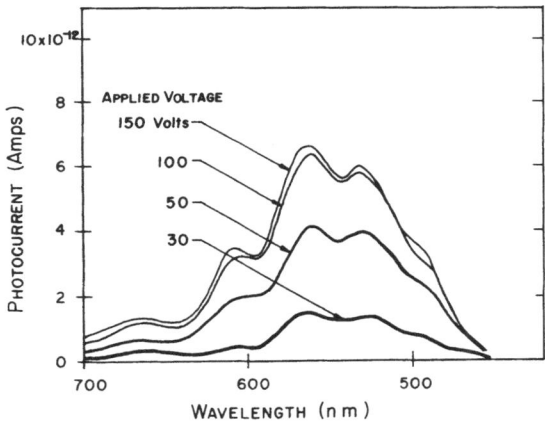

Fig. 3.83. Spectrally resolved photodepopulation response in SiO_2 for various applied voltages. Electrode area 0.18 cm², monochromator sweep rate 0.2 nm s⁻¹, band pass 20 nm. [3.51]

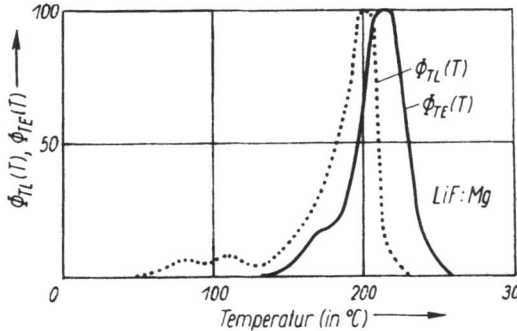

Fig. 3.84. TL and TSEE glow curves of a γ-ray excited and Mg-activated LiF phosphor. [3.155b]

photodepopulation in SiO_2 are given in Fig. 3.83. For more details we refer to [3.51–53, 463]. These articles also discuss the theory which again is very similar to that of Sect. 3.10, except for the fact that the thermal detrapping is replaced with optical detrapping and the isothermal versions of the equations are used.

Finally, we can also investigate the detrapping of a sample by measuring the *electron emission* from its surface [3.45, 49, 57, 155b, 502]. Recalling Fig. 3.4, we see that electrons liberated from their traps during the heating may not only cause conduction or light emission; they may also *leave* the sample if they are near the surface and if their energy is sufficiently high to overcome the electron affinity. This process is called TSEE (*thermally stimulated exoelectron emission*), or OSEE if the electrons are excited optically [3.49, 57, 502]. The measurement is performed by counting of the number of emitted electrons in a vacuum chamber during the temperature ramp. Peaks arise at temperatures often associated with those of TL and TSCo peaks [3.503], see Fig. 3.84. The TSEE technique is mostly used for *radiation dosimetry* [3.49, 57, 155b], but on a much smaller scale than TL.

3.16 Review of Information Obtainable from TSD

TSD has been vital in setting up the criteria that materials must fulfill if reliable technical electrets are to be made from them. Its main task in *electret* research is to arrive at detailed information about deep *traps* in *dielectrics* able to hold charges captive for a prolonged period at room temperature. Such traps arise from structural defects and impurities. TSD has proved to be a useful technique for determining the relevant properties of such defects. Certainly, the main reason for its popularity amongst "trap-level spectroscopists" is that it allows the trapping processes to be unravelled very rapidly. Moreover, it is inexpensive and convenient, and possesses high sensitivity and resolving power. In addition, TSD allows the mean position of the stored charges and their spatial distribution to be found. It also reveals the transport rate (mobility) of charges in electrets, and is valuable in predicting the lifetime of electrets.

Detailed examination of the storage and transport of charges is likewise of significance for several applications of *semi-insulator* and *semiconductor* devices. Present studies in this field concern the trap-level spectroscopy of insulating layers in MOS and thin-film devices, and in the amorphous materials used in switching elements. Methods closely related to TSD, such as thermally stimulated capacitance and admittance, are also productive in this area.

Trapping and detrapping of charges are also required for the proper functioning of *optoelectronic materials* (photoconductors, photoelectrets, electrophotographic layers, and phosphors) and *optoelectronic products* (radiation detectors, lasing diodes, and solar cells). To these materials, variants of TSD – namely, thermally stimulated conductivity and thermoluminescence – are often applied. In these techniques the traps are filled optically, and therefore they provide more information on deeper traps (recombination centres).

TSD has not only advanced our understanding about charge trapping sites, but has also evolved as a powerful tool for identifying and evaluating *dielectric relaxation* mechanisms. As such it is of major importance for the bulk use of dielectrics in capacitors, electrical insulation, and power transmission. All peaks found in current TSD have a counterpart in conventional dielectric loss measurements, except for the ϱ peak, which generally appears in TSD only. In fact, the results of the two techniques are easily interconvertible. Such a conversion shows that TSD data pertain to the ultra-low-frequency range. The resolution of multiple peaks in TSD is therefore excellent; peaks from competing relaxation processes are separated if the pre-exponential factor of these processes differ by as little as a factor of 5, or by as little as 10% in activation energy.

TSD peaks due to native relaxation phenomena have good reproducibility. The ϱ peak, which is due to space charges, and easily affected by absorbed water and extrinsic impurities, is less reproducible. It should be noted that peaks develop fully only when the sample is charged completely by choosing the poling temperature above the highest peak temperature.

TSD allows a discrimination to be made between dipole relaxations arising from either a single relaxation or a distributed relaxation. If the relaxation is distributed, we can find out whether the distribution is one in preexponential factors or one in activation energies by decomposing the composite peak into its elementary components. We can perform this decomposition by activating or deactivating the distributed relaxation in parts using *fractional* polarization or depolarization. TSD also enables us to distinguish between relaxations due to dipoles and space charges. It further provides insight into the behaviour of electrode contacts applied to dielectrics.

TSD can be applied to all kinds of solid materials provided that their ohmic conduction is low enough. Since it scans the relaxations at a slow rate the peaks are observed at the lowest possible temperature, and so the samples need not be overheated. This is of importance for investigations on point defects in crystals and on biological materials.

On a practical scale, TSD is applied in *radiation dosimetry* as well as in the determination of *transition points* (e.g., the T_g points of polymers).

3.17 Conclusions and Prospects

In this chapter we have surveyed the methods currently employed in TSD studies, and given examples of the information obtainable from such studies. All the basic aspects of TSD have been discussed, i.e., its theory, instrumentation and applications, with emphasis on the latest developments. The relatively new field of TSD has proved to be a source of fresh, and potentially useful ideas in electrical materials science. TSD is therefore to be regarded as a fertile field of research for the physicist or chemist interested in solid-state electronics and optics.

The central theme has been the application of TSD in electret research. This research is essential not only for the successful application of existing electrets, but also for the development of new and improved electrets. The method is being used to an increasing extent for routine measurements and pure research on electrets. These activities have in a few years time provided more useful information on the fundamental properties of dipole relaxations, charge trapping and charge transport mechanisms in dielectrics than has been possible in the past by other methods.

TSD research has not been confined to electrets, but has now become widely accepted as an important spectroscopic technique for investigating the dielectric and electrical behavior of solids in general. The drives guiding this development derive mainly from the expansion that is taking place in the field of electronic and optoelectronic devices and microcircuits.

The ultimate goals of TSD studies are to understand the molecular origins of the dielectric and space-charge relaxations observed, and to predict the impact of these relaxations on the electrical properties of a material in a satisfactory manner. This major problem constitutes a challenge to future investigators.

The most promising approach to improving the precise assignment of TSD peaks is to correlate them with results from *complementary* techniques, particularly those devoted to structural characterizations [3.504]. Such "anatomical" studies will demand a concerted effort on the part of investigators skilled, and experienced in a variety of disciplines.

Another problem is that in one and the same temperature range, various decay mechanisms may compete. In such complex situations it may be difficult to find the right physical model for interpreting the experimental results. This is another reason why it would be wrong to analyze TSD data without comparing them with data from other techniques.

The choice of a rigorous theoretical model is not a unique problem of TSD, because many other measurements on condensed matter are not self-contained either. The best thing to do is to sort out the model only after it has been checked for various experimental conditions, and cross-checked with results of other TSD or related experiments. Fortunately, any alteration can be made readily in view of the *short* measuring time. The model can also be refined by simulations on a computer, which is in any case indispensable for the analysis of data with complicated models. However, this poses no serious problem, because most research workers have easy access to a computer.

In particular, we are expecting further progress in the consistent and precise analysis of TSD peaks due to the motion of space charges. The present analysis incorporates a number of idealizations and is only approximate. A crucial advance would be to extend the analysis to multiple traps, so that it agrees better with the actual situation in most materials.

In the future the role played by surface traps and by the electrode contacts (blocking or injecting) may also become clearer. Here a correlation with ESCA techniques for elucidating the structural features of the surface may be helpful [3.505, 506].

We conclude that TSD is passing through a period of intense activity and development, and is rapidly advancing our understanding of electret, electronic, and optoelectronic materials. It is certain that interest will continue, and that a considerable amount of interesting and important information will be dug up.

We can only speculate on what the future has in store for TSD, but it will certainly continue to help in finding the best materials for the best electrets.

Acknowledgements. The author is indebted to P. Th. A. Klaase, P. H. Ong, P. J. Nederveen, G. Roozendaal, and P. J. Droppert for their experimental contributions, and to N. V. Philips' Gloeilampenfabrieken, Eindhoven, for sponsoring part of the work reported. He also thanks G. P. M. Léger for making the most needed corrections to the English, H. E. Koot for her accurate typing of the manuscript, and G. Th. C. Mathlener for drawing some of the figures.

The author is grateful for the patience and understanding shown by his wife and children when he was writing his contribution.

References

An asterisk (∗) preceding an entry denotes a review paper

3.1   R.H.Bube: *Photoconductivity of Solids* (Wiley, New York 1960) Chap. 9

3.2   P.Bräunlich: "Thermoluminescence and Thermally Stimulated Conductivity", in *Thermoluminescence of Geological Materials*, ed. by D.J.McDougall (Academic Press, New York 1968) pp. 61–90

3.3   A.G.Milnes: *Deep Impurities in Semiconductors* (Wiley-Interscience, New York 1973) Chap. 9

3.4   G.F.J.Garlick: Luminescence, in *Handbuch der Physik*, Vol. 26, ed. by S.Flügge (Springer, Berlin, Göttingen, Heidelberg 1958) pp. 1–128

3.5   D.Curie: *Luminescence in Crystals* (Wiley, New York 1963) Chap. 6

3.6   A.Scharmann: "Thermolumineszenz", in *Einführung in die Lumineszenz*, ed. by N.Riehl (Thiemig, München 1970) pp. 182–225

3.7   C.Bucci, R.Fieschi, G.Guidi: Phys. Rev. **148**, 816–823 (1966)

3.8   H.Frei, G.Groetzinger: Phys. Z. **37**, 720–724 (1936)
        O.G. von Altheim: Ann. Phys. (Leipzig) **35**, 417–430 (1939)

3.9   B.Gross, R.J. de Moraes: J. Chem. Phys. **37**, 710–713 (1962)

3.10 A.N.Gubkin, B.N.Matsonashvili: Sov. Phys.-Solid State **4**, 878–884 (1962)

3.11 J. van Turnhout: *Thermally Stimulated Discharge of Polymer Electrets* (Elsevier, Amsterdam 1975)

∗ 3.12 J.Vanderschueren: Bull. Sc. A.I.M. (Liège) **2**, 105–112 (1974); **4**, 291–302 (1974); **4**, 311–319 (1975)

∗ 3.13 R.Chen: J. Mater. Sci. **11**, 1521–1541 (1976)

3.14 L.Brehmer: Faserforsch. Textiltech. **28**, 43–45 (1977)

3.15 T.Takamatsu, E.Fukada: Polym. J. **1**, 101–106 (1970)

3.16 M.M.Perlman: J. Electrochem. Soc. **119**, 892–898 (1972)

3.17 J.G.Simmons, G.W.Taylor: Phys. Rev. B**6**, 4804–4814 (1972)

3.18 P.R.Moran, D.E.Fields: J. Appl. Phys. **45**, 3266–3272 (1974)

3.19 P.Hedvig: *Dielectric Spectroscopy of Polymers* (Hilger, Bristol 1977) pp. 148–155

3.20 J.Vanderschueren: Ph. D. Thesis (Liège 1974)

2.21 V.Adamec, E.Mateova: Polymer **16**, 166–168 (1975)

3.22 S.W.S.McKeever, D.M.Hughes: J. Phys. D**8**, 1520–1529 (1975)

3.23 T.W.Hickmott: J. Appl. Phys. **46**, 2583–2598 (1975)

3.24 J. van Turnhout, A.H. van Rheenen: In *Proc. 4th Int. Conf. on the Physics of Non-Crystalline Solids*, ed. by G.H.Frischat (Trans Tech Publ., Aedermannsdorf 1977) pp. 494–502; in Ref. 3.74, pp. 3–12 (1975)

3.25 J. van Turnhout: In Ref. 3.69, pp. 230–251

3.26 J. van Turnhout: Polym. J. **2**, 173–191 (1971)

3.27 T.Takamatsu, E.Fukada: In Ref. 3.69, pp. 128–140

3.28 Y.Asano, T.Suzuki: Jpn. J. Appl. Phys. **11**, 1134–1146 (1972)

3.29 P.Müller: Phys. Status Solidi A**23**, 165–174 (1974)

3.30 S.C.Agarwal, H.Fritzsche: Phys. Rev. B**10**, 4340–4357 (1974)
        H.Fritzsche, S.Chandra: In Ref. 3.70, pp. 105–117

3.31 A.R. von Hippel (ed.): *Dielectric Materials and Applications* (Wiley, New York 1954)

3.32 J.B.Birks, J.Hart (eds.): *Progress in Dielectrics*, Vols. 1–7 (Heywood, London 1960–1967)

3.33 F.M.Clark: *Insulating Materials for Design and Engineering Practice* (Wiley, New York 1962)

3.34 N.G.McCrum, B.E.Read, G.Williams: *Anelastic and Dielectric Effects in Polymeric Solids* (Wiley, New York 1967)

3.35 H.Fröhlich: *Theory of Dielectrics*, 2nd ed. (Oxford University Press, Oxford 1958)
        V.V.Daniel: *Dielectric Relaxation* (Academic Press, New York 1967)

3.36 N.E.Hill, W.E.Vaughan, A.H.Price, M.Davies: *Dielectric Properties and Molecular Behaviour* (Van Nostrand-Reinhold, London 1969)

* 3.37 R.F.Boyer: Rubber Chem. Technol. **36**, 1303–1421 (1963); J. Polym. Sci. Polym. Symp. **50**, 189–242 (1975); in *Proc. Seminar on "Polymeric Materials"*, ed. by E.Baer, S.V.Radcliffe (American Society of Metals, Metals Park 1975) pp. 277–368
R.F.Boyer (ed.): "Proc. Conf. on 'Transitions and Relaxations in Polymers'" [J. Polym. Sci. C**14** (1966)]

3.38 A.S.Nowick, B.S.Berry: *Anelastic Relaxation in Crystalline Solids* (Academic Press, New York 1972)

3.39 R.P.Kambour, R.E.Robertson: The Mechanical Properties of Plastics", in *Polymer Science*, Vol. 1, ed. by A.D.Jenkins (North-Holland, Amsterdam 1972) Chap. 11

3.40 G.E.Roberts, E.F.T.White: "Relaxation Processes in Amorphous Polymers", in *Physics of Glassy Polymers*, ed. by R.N.Haward (Applied Science Publications, London 1973) Chap. 3

3.41 I.M.Ward: *Mechanical Properties of Solid Polymers* (Wiley-Interscience, London 1971) Chap. 8
L.E.Nielsen: *Mechanical Properties of Polymers and Composites*, Vol. 1 (Dekker, New York 1974) Chap. 4

* 3.42 O.V.Mazurin: J. Non-Cryst. Solids **25**, 130–169 (1977)

3.43 J. van Turnhout, P.H.Ong: In Ref. 3.76, pp. 68–73

3.44 A.Linkens, J.Vanderschueren: In Ref. 3.72, pp. 149–154

* 3.45 A.Scharmann, R.Grasser, M.Böhm: In Ref. 3.72, pp. 1–14

3.46 G.Gira: Z. Elektr. Inf. Energietech. **2**, 186–189 (1972)

3.47 S.Wang: *Solid-State Electronics* (McGraw-Hill, New York 1966) Chap. 11

3.48 B.Ray: *II–VI Compounds* (Pergamon Press, Oxford 1969) Chap. 4

3.49 K.Becker: *Solid-State Dosimetry* (Chemical Rubber Corp., Cleveland 1973)

3.50 R.H.Partridge: "TL in Polymers", in *Radiation Chemistry of Macromolecules*, Vol. 1, ed. by M.Dole (Academic Press, New York 1972) pp. 193–222

3.51 J.H.Thomas, F.J.Feigl: J. Phys. Chem. Solids **33**, 2197–2216 (1972)

2.52 J.D.Brodribb, D.M.Hughes, T.J.Lewis: In Ref. 3.69, pp. 177–187
J.D.Brodribb, D.O'Colmain, D.M.Hughes: J. Phys. D**8**, 856–862 (1975)

3.53 F.J.Feigl, S.R.Butler, D.J.DiMaria, V.J.Kapoor: In Ref. 3.70, pp. 118–134

2.54 V.M.Fridkin, I.S.Zheludev: *Photoelectrets and the Electrophotographic Process* (Consultants Bureau, New York 1961)

3.55 S.M.Ryvkin: *Photoelectric Effects in Semiconductors* (Consultants Bureau, New York 1964) Chap. 10

* 3.56 C.Bowlt: Contemp. Phys. **17**, 461–482 (1976)

3.57 J.A.Ramsey: "Exoelectric Emission", in *Progress in Surface and Membrane Science*, Vol. 11, by D.A.Cadenhead, J.F.Danielli (Academic Press, New York 1976) pp. 117–180

* 3.58 R.Chen: In Ref. 3.72, pp. 15–24

3.59 W.W.Wendlandt: *Thermal Methods of Analysis* (Wiley, New York 1974)

3.60 T.Daniels: *Thermal Analysis* (Kogan Page, London 1973)

* 3.61 R.Capelletti, R.Fieschi: In Ref. 3.69, pp. 1–14

3.62 R.Fieschi, R.Capelletti: In *Proc. 4th Int. Summer School on Point Defects and Their Interaction with Other Lattice Imperfections*, Zakopane 1973, (Polish Scientific Publications, Warsaw 1974) pp. 133–146

* 3.63 R.Capelletti, R.Fieschi: In Ref. 3.71, pp. 131–154

3.64 C.Laj: Radiat. Eff. **4**, 77–83 (1970)
S.Mascarenhas: Radiat. Eff. **4**, 263–269 (1970)
S.Radhakrishna, S.Haridoss: Cryst. Lattice Defects **7**, 191–207 (1978)

3.65 B.T.Kolomiets, V.M.Lyubin: Phys. Status Solidi A**17**, 11–46 (1973)

* 3.66 J.Perret: Bull. Direct. Etud. Rech. (EDF) B No. 2, 27–69 (1974)

* 3.67 J.G.Simmons: In Ref. 3.70, pp. 84–97

* 3.68 P.Müller: In *Amorphous Semiconductors '76*, ed. by I.Kósa Somogyi (Akad. Kiado, Budapest 1977) pp. 199–214

3.69 M.M.Perlman (ed.): *Electrets, Charge Storage and Transport in Dielectrics* (Electrochemical Society, Princeton (1973)

3.70   D.M.Smyth(ed.): *Thermal and Photostimulated Currents in Insulators* (Electrochemical Society, Princeton 1976)
3.71   M.Soares de Campos(ed.): *International Symposium on "Electrets and Dielectrics"* (Acad. Brasil. de Ciências, Rio de Janeiro 1977)
3.72   J.P.Fillard, J. van Turnhout(eds.): *Thermally Stimulated Processes in Solids: New Prospects* (Elsevier, Amsterdam 1977); J. Electrostat. **3**, 1–302 (1977)
3.73   L.Badian (Org.): *Colloquium on "Space Charges in Dielectrics"*, Techn. Univ. of Wroclaw, Sept. (1977); J. Electrostat. **8** (1979)
3.74   Annual Reports *Conf. on "Electrical Insulation and Dielectric Phenomena"* (National Academy of Sciences-National Research Council, Washington, D. C. (from 1974 onwards)
3.75   J. van Turnhout(ed.): J. Electrostat. **1**, 99–163, 209–234 (1975); **2**, 41–57 (1976)
3.76   *IEE Conf. on "Dielectric Materials, Measurements and Applications"*, Cambridge 1975, No. 129 (IEE, Stevenage 1975)
3.77   *Triennially IUPAP Conf. on "Lattice Defects in Ionic Crystals"*, publ. as Suppl. of J. Phys. (Paris) Colloq., 1973/1976
3.78   W.Hoogenstraaten: Philips Res. Rep. **13**, 515–673 (1958)
3.79   H.C.Wright, G.A.Allen: Br. J. Appl. Phys. **1**, 1181–1185 (1966)
3.80   *Proc. 5th Int. Conf. on "Luminescence Dosimetry"*, Sao Paulo, ed. by A.Scharmann (1977)
3.81   *Proc. 5th Int. Symp. on "Exoelectron Emission and Dosimetry"*, Zvikov, Czech., ed. by A.Bochum, A.Scharmann (1976)
3.82   C.T.Moynihan, A.J.Easteal, J.Wilder, J.Tucker: J. Phys. Chem. **78**, 2673–2677 (1974)
3.83   A.J.Kovacs, J.M.Hutchinson, J.J.Aklonis: In *The Structure of Non-Crystalline Solids*, Proc. Symp. Cambridge 1976, ed. by P.H.Gaskell (Francis & Taylor, London 1977) pp. 153–163
3.84   M.Silver: Solid State Commun. **15**, 1785–1787 (1974); in Ref. 3.11, p. 131ff.
3.85   T.Miyamoto, K.Shibahama: J. Appl. Phys. **44**, 5372–5376 (1973)
        R.E.Barker: Pure Appl. Chem. **46**, 157–170 (1974)
3.86   F.Gutmann, L.E.Lyons: *Organic Semiconductors* (Wiley, New York 1967)
3.87   H.J.Wintle: "Theory of Electrical Conduction in Polymers", in *Radiation Chemistry of Macromolecules*, Vol. 1, ed. by M.Dole (Academic Press, New York 1972) Chap. 7
3.88   H.Meier: *Organic Semiconductors – Dark – and Photoconductivity of Organic Solids* (Verlag Chemie, Weinheim 1974); Chimia **27**, 263–274 (1973)
* 3.89   M.Kryszewski: J. Polym. Sci. Polym. Symp. **50**, 359–404 (1975)
        L.Brehmer, M.Kryszewski, C.Ruscher: Faserforsch. Textiltech. **28**, 475–482 (1977)
* 3.90   P.K.C.Pillai, M.Goel: Phys. Status Solidi A**6**, 9–27 (1971)
3.91   J.F.Fowler: Proc. R. Soc. London Ser. A**236**, 464–480 (1956)
*   E.L.Frankovich, B.S.Yakovlev: Int. J. Radiat. Phys. Chem. **6**, 281–296 (1974)
3.92   R.C.Alger: "Radiation Effects in Polymers", in *Physics and Chemistry of the Organic Solid State*, Vol. 2, ed. by D.Fox, M.M.Labes, A.Weissberger (Interscience, New York 1965) Chap. 9
        A.Charlesby: "Radiation Effects in Polymers", in *Polymer Science*, Vol. 2, ed. by A.D.Jenkins (North-Holland, Amsterdam 1972) Chap. 23
        P.Hedvig: "Electrical Conductivity of Irradiated Polymers", in *Radiation Chemistry of Macromolecules*, Vol. 1, ed. by M.Dole (Academic Press, New York 1972) Chap. 8
3.93   G.M.Sessler, J.E.West: Phys. Rev. B**10**, 4488–4491 (1974)
3.94   B.Gross, G.M.Sessler, J.E.West: J. Appl. Phys. **47**, 968–975 (1976)
3.95   N.F.Mott, E.A.Davis: *Electronic Processes in Non-Crystalline Materials* (Clarendon Press, Oxford 1971)
3.96   D.Adler: *Amorphous Semiconductors* (Chemical Rubber Corp., Cleveland 1971)
        J.Tauc(ed.): *Amorphous and Liquid Semiconductors* (Plenum, London 1974)
3.97   D.A.Seanor: "Electrical Properties of Polymers", in *Polymer Science*, Vol. 2, ed. by A.D.Jenkins (North-Holland, Amsterdam 1972) Chap. 17
*   T.J.Lewis: In Ref. 3.74, pp. 533–561 (1976)
        A.E.Owen: J. Non-Cryst. Solids **25**, 372–423 (1977)
3.98   H.Bauser: Kunststoffe **62**, 192–196 (1972)
3.99   T.J.Fabish, C.B.Duke: J. Appl. Phys. **48**, 4256–4266 (1977)
        T.J.Fabish: In Ref. 3.74, pp. 175–188 (1977)
        C.B.Duke, T.J.Fabish: J. Appl. Phys. **49**, 315–321 (1978)

3.100 H.Scher, E.W.Montroll: Phys. Rev. B12, 2455–2477 (1975)
H.Scher: "Theory of Time-Dependent Photoconductivity in Disordered Systems", in *Photoconductivity and Related Phenomena*, ed. by J.Mort, P.M.Pai (Elsevier, Amsterdam 1976) Chap. 3

3.101 A.K.Jonscher, R.M.Hill: "Electrical Conduction in Disordered Nonmetallic Films", in *Physics of Thin Films*, Vol. 8, ed. by G.Hass, M.H.Francombe, R.W.Hoffman (Academic Press, New York 1975) pp. 169–249

3.102 G.Pfister, H.Scher: Phys. Rev. B15, 2062–2083 (1977)

3.103 A.I.Rudenko: J. Non-Cryst. Solids 22, 215–218 (1976)
O.L.Curtis, J.R.Srour: J. Appl. Phys. 48, 3819–3828 (1977)
J.M.Marshall: Philos. Mag. 36, 959–975 (1977)
F.W.Schmidlin: Phys. Rev. B16, 2362–2385 (1977)

3.104 H.K.Henisch: In Ref. 3.72, pp. 233–240

3.105 A.Rose: *Concepts in Photoconductivity and Allied Problems* (Interscience, New York 1963) Chap. 8
J.G.Simmons: "Electrical Conduction Through Thin Insulating Films", in *Handbook of Thin Film Technology*, ed. by L.E.Maissel, R.Glang (McGraw Hill, New York 1970) Chap. 14

3.106 M.A.Lampert, P.Mark: *Current Injection in Solids* (Academic Press, New York 1970)
R.H.Walden: J. Appl. Phys. 43, 1178–1186 (1972)
H.J.Wintle: J. Non-Cryst. Solids 15, 471–486 (1974); Solid-State Electron. 18, 1039–1042 (1975); IEEE Trans. EI-12, 424–428 (1977)

* 3.107 M.G.Buehler, W.E.Phillips: Solid-State Electron. 19, 777–788 (1976)

3.108 G.Dreyfus, J. Lewiner: J. Appl. Phys. 45, 721–726 (1974)

3.109 J. van Turnhout: In *Advances in Static Electricity*, Vol. 1, ed. by W.F. de Geest (Auxilia, Brussels 1971) pp. 56–81

3.110 P.H.Ong, J. van Turnhout: In *Elektrostatische Aufladung* (Verlag Chemie, Weinheim 1974) pp. 105–124

3.111 L.R.Weinberg, H.Schade: J. Appl. Phys. 39, 5149–5151 (1968)

3.112 O.K.Gasanov, V.A.Izvozchikov, V.S.Maidzinskii, V.A.Kozlov, O.A.Timofeev, L.G.Timofeeva: Sov. Phys.-Semicond. 8, 126–127 (1974)

3.113 V.V.Rusakov: Sov. Phys-Semicond. 3, 1581–1582 (1970)

3.114 E.G.Manning, M.A.Littlejohn, J.A.Hutchey, E.M.Oakley: Rev. Sci. Instrum. 43, 324–326 (1972)

3.115 W.L.Paterson: Rev. Sci. Instrum. 46, 196–197 (1975)
A.A.Mills, D.W.Sears, R.Haersey: J. Phys. E10, 51–56 (1977)

3.116 K.Nemeth, P.Sviszt: Exp. Techn. Phys. 21, 443–447 (1973)
K.Inabe: J. Phys. E9, 931–933 (1975)

3.117 R.Szepan: Elektronik 25 (5), 61–62 (1976)
J.Rathlev: Elektronik 26 (8), 64–65 (1977)

3.118 A.Halperin, M.Leibovitz, M.Schlesinger: Rev. Sci. Instrum. 33, 1168–1170 (1962)
P.Müller, J.Teltow: Phys. Status Solidi A12, 471–475 (1972)

3.119 M.Soares de Campos: In Ref. 3.74, pp. 60–67 (1975)

3.120 A.Katzir: J. Phys. E7, 423–424 (1974)

3.121 A.Kessler, R.Pflüger: In Ref. 3.72, pp. 93–97
T.Cairns: Anal. Chem. 48, 266A–280A (1976)

3.122 E.Sacher: J.Macromol. Sci. Phys. B6, 151–165 (1972)

3.123 V.I.Trubin, B.P.Usol'tsev, V.V.Beskrovanov, P.I.Khudaev: Dokl. Akad. Nauk. SSSR 214, 813–814 (1974)
Y.Sawa, D.C.Lee, M.Ieda: Jpn. J. Appl. Phys. 16, 359–364 (1977)
I.Thurzo, D.Barancok, G.Vlasak, J.Doupovec: J. Non-Cryst. Solids 24, 297–299 (1977)
J.P.Crine, D.L.Piron, A.Yelon: In Ref. 3.74, pp. 226–235 (1976)

3.124 S.I.Stupp, S.H.Carr: J. Polym. Sci. Polym Phys. 15, 485–499 (1977)

3.125 J.C.Monpagens, D.G.Chatain, C.Lacabanne, P.G.Gautier: J. Polym. Sci. Polym. Phys. 15, 767–772 (1977)

J.C.Monpagens, D.G.Chatain, C.Lacabanne: In Ref. 3.72, pp. 87–92
C.Lacabanne, D.G.Chatain, J.C.Monpagens: J. Macromol. Sci. Phys. B13, 537–552 (1977)
3.126 M.Oshiki, E.Fukada: Jpn. J. Appl. Phys. 15, 43–52 (1976)
3.127 B.Ai, P.Destruel, H.T.Giam, R.Loussier: Phys. Rev. Lett. 34, 84–85 (1977)
       B.Ai, H.T.Giam, P.Destruel: In Ref. 3.72, pp. 83–85
3.128 S.Radhakrishna, S.Haridoss: Phys. Status Solidi A41, 649–652 (1977)
3.129 J.van Turnhout: In Ref. 3.73
3.130 R.E.Collins: J. Appl. Phys. 47, 4804–4808 (1976); Rev. Sci. Instrum. 48, 83–91 (1977)
3.131 P.Laurenceau, G.Dreyfus, J.Lewiner: Phys. Rev. Lett. 38, 46–49 (1977)
       D.Darmon, P.Laurenceau, G.Dreyfus, J.Lewiner: In Ref. 3.73
3.132 G.M.Sessler, J.E.West, D.A.Berkley, G.Morgenstern: Phys. Rev. Lett. 38, 368–371 (1977)
3.133 V.M.Fridkin: The Physics of the Electrophotographic Process (Focal Press, London 1973) p.
       161 ff
3.134 P.H.Ong, J.van Turnhout: In Ref. 3.69, pp. 213–229
3.135 J.Vanderschueren: J. Polym. Sci. Polym. Phys. 15, 873–880 (1977)
       J.Vanderschueren, A.Linkens: In Ref. 3.72, pp. 155–161
3.136 D.Chatain, C.Lacabanne, J.C.Monpagens: Makromol. Chem. 178, 583–593 (1977); In Ref.
       3.72, pp. 87–92
       P.Fischer, P.Röhl: J. Polym. Sci. Polym. Phys. 14, 531–554 (1976)
3.137 Ch.Ponevski, C.Solunov: J.Polym. Sci. Polym. Phys. 13, 1467–1479 (1975)
3.138 R.A.Creswell, M.M.Perlmann: J. Appl. Phys. 41, 2365–2375 (1970)
3.139 M.Zielinski, M.Kryszewski: In Ref. 3.72, pp. 69–81
3.140 P.H.Ong, J.van Turnhout, J.Booij, G.H.Douma: In Proc. 4th Conf. on Static Electrification,
       ed. by A.R.Blythe (Institute of Physics, London 1975) pp. 188–201
3.141 M.Davies, P.J.Hains, G.Williams: J. Chem. Soc. Faraday Trans. 2 69, 1785–1792 (1973)
3.142 L.R.Weisberg: In Trace Characterization – Chemical and Physical, ed. by W.W.Meinke,
       B.F.Scribner (NBS Monogr. 100, 1967) pp. 39–68
3.143 E.Motyl: Ph. D. Thesis, Wroclaw (1975)
3.144 S.Unger, M.M.Perlman: Phys. Rev. B6, 3973–3981 (1972)
3.145 A.Callens, L.Eersels, R.de Batist: J. Mater. Sci. 12, 1361–1375 (1977)
3.146 P.Hedvig: J. Polym. Sci. Polym. Symp. 42, 1271–1274 (1973)
3.147 J.van Turnhout, P.T.A.Klaase, P.H.Ong, L.C.E.Struik: In Ref. 3.72, pp. 171–179
       L.C.E.Struik, P.T.A.Klaase, P.H.Ong, D.T.F.Pals, J.van Turnhout: In Ref. 3.74, pp. 443–452
       (1976)
3.148 L.C.E.Struik: Polym. Eng. Sci. 17, 165–173 (1977); Physical Aging in Amorphous Polymers
       and Other Materials (Elsevier, Amsterdam 1978)
3.149 F.S.W.Hwang, J.H.Fremlin: Archaeometry 12, 67–71 (1970)
3.150 M.J.Aitken: Physics and Archaelogy, 2nd ed. (Clarendon Press, Oxford 1974) Chap. 3
       M.J.Aitken, S.J.Fleming: "TL dosimetry in Archaeological Dating", in Topics in Radiation
       Dosimetry, Suppl. 1, ed. by F.H.Altix (Academic Press, New York 1972) pp. 1–78
   *   M.A.Seeley: J. Archaeol. Sci. 2, 17–43 (1975)
   * 3.151 D.W.Zimmermann: In Ref. 3.72, pp. 257–268
       I.K.Bailiff, S.G.E.Bowman, S.F.Mobbs, M.J.Aitken: In Ref. 3.72, pp. 269–280
       A.G.Wintle: In Ref. 3.72, pp. 281–288
3.152 P.R.Moran, E.B.Podgorsak, G.E.Fuller, G.D.Fullerton: Med. Phys. 1, 155–160 (1974)
       G.D.Fullerton, J.R.Cameron, M.R.Mayhugh, P.R.Moran: in Biomedical Dosimetry, IAEA-
       SM 193/49 (Int. Atom. Energy Agency, Vienna 1975) pp. 249–265
3.153 N.W.Ramsey: In Ref. 3.76, pp. 60–63
3.154 B.Thomas, J.Conway, M.W.Harper: J. Phys. D10, 55–63 (1977)
3.155 J.R.Cameron, N.Suntharalingam, G.N.Kenney: Thermoluminescence Dosimetry (Wisconsin
       University Press, Madison 1968)
       M.Frank, W.Stolz: Festkörperdosimetrie ionisierender Strahlung (Teubner, Leipzig 1969) pp.
       63–139
3.156 T.Nakajima: J. Appl. Phys. 48, 4880–4885 (1978)

3.157 A. Brun, P. Dansas: J. Phys. C7, 2593–2602 (1974)

3.158 A. S. Nowick: "Defect mobilities in Ionic Crystals Containing Aliovalent Ions", in *Point Defects in Solids*, Vol. 1, ed. by J. H. Crawford, L. M. Slifkin (Plenum, New York 1972) Chap. 3

3.159 B. K. P. Scaife: J. Phys. D7, L171–L173 (1974); J. Phys. D8, L72–L73 (1975)

3.160 J. van Turnhout: J. Phys. D8, L68–L71 (1975)

3.161 B. Gross: J. Phys. D8, L127–L128; In Ref. 3.72, pp. 43–51

3.162 J. P. Fillard, J. Gasiot: Phys. Status Solidi A32, K85–K88 (1975); In Ref. 3.72, pp. 37–42

3.163 J. D. Ferry: *Viscoelastic Properties of Polymers*, 2nd ed. (Wiley, New York 1970)
       F. Bueche: *Physical Properties of Polymers* (Interscience, New York 1962) Chap. 4

3.164 C. Lacabanne, D. Chatain: J. Polym. Sci. Polym. Phys. 11, 2315–2328 (1973)

3.165 W. Squire: J. Comput. Phys. 6, 152–153 (1970)
       W. L. Paterson: J. Comput. Phys. 7, 187–190 (1971)
       M. Balarin: J. Therm. Anal. 12, 169–177 (1977)
       G. Varhegyi: Thermochim. Acta 25, 201–207 (1978)
       W. J. Cody, H. C. Thacher: Math. Comp. 22, 641–649 (1968) [for a simple approximation from the rational Chebyshev approximations for $E_1(x)$]

3.166 J. G. Simmons, G. W. Taylor: Phys. Rev. B5, 1619–1629 (1972)

3.167 L. Nunes de Oliveira, G. F. Leal Ferreira: Nuovo Cimento B23, 385–391 (1974)
       V. Harasta: Fyz. Cas. 19, 232–235 (1969)

3.168 M. Gevers: Philips Res. Rep. 1, 279–313 (1946)

3.169 K. S. Cole, R. H. Cole: J. Chem. Phys. 9, 341–351 (1941)

3.170 R. M. Fuoss, J. G. Kirkwood: J. Am. Chem. Soc. 63, 385–394 (1941)

3.171 B. Gross: *Mathematical Structure of the Theories of Viscoelasticity* (Herman, Paris 1953)

3.172 D. W. Davidson, R. H. Cole: J. Chem. Phys. 19, 1484–1490 (1951)

3.173 S. Havriliak, S. Negami: Polymer 8, 161–210 (1967)
       S. Negami: "Dielectric Measurements", in *Treatise on Coatings: Characterization of Coatings – Physical Techniques*, Vol. 2, ed. by R. R. Myers, J. C. Long (Dekker, New York 1976) pp. 1–65

3.174 I. Thurzo, B. Barančok, J. Doupovec, E. Mariani, J. Janci: J. Non-Cryst. Solids 18, 129–136 (1975)

3.175 A. K. Jonscher: In Ref. 3.72, pp. 53–68

3.176 Ch. A. Solunov, Ch. S. Ponevsky: J. Polym. Sci. Polym. Phys. 15, 969–979 (1977)

3.177 A. C. Lilly, R. M. Henderson, P. S. Sharp: J. Appl. Phys. 41, 2001–2006 (1970)
       K. Kojima, A. Meada, M. Ieda: Jpn. J. Appl. Phys. 15, 2457–2458 (1976)

3.178 I. Chen: J. Appl. Phys. 47, 2988–2994 (1976)

3.179 M. E. Borisova, S. N. Koikov, S. F. Morozov: Sov. Phys. J. 6, 104–110 (1974)

*   3.180 M. Zahn: IEEE Trans. EI-12, 176–190 (1977)

3.181 A. Many, G. Rakavy: Phys. Rev. 126, 1980–1988 (1962)

3.182 J. P. Batra, K. K. Kanazawa, H. Seki: J. Appl. Phys. 41, 3416–3422 (1970)

3.183 J. H. Calderwood, B. K. Scaife: Philos. Trans R. Soc. London 269, 217–232 (1971)

2.184 H. J. Wintle: J. Appl. Phys. 42, 4724–4730 (1971)
       G. F. Leal Ferreira: J. Nonmetals 2, 109–112 (1974)

3.185 G. M. Sessler: J. Appl. Phys. 43, 408–411 (1972)

3.186 G. F. Leal Ferreira, B. Gross: J. Nonmetals 1, 129–131 (1973)

3.187 V. Thomée: J. Soc. Ind. Appl. Math. 10, 229–245 (1962)

3.188 R. D. Richtmyer, K. W. Morton: *Difference Methods for Initial Value Problems* (Interscience, New York 1967) pp. 300–306

3.189 B. Wendroff: *Theoretical Numerical Analysis* (Academic Press, New York 1967) p. 183

3.190 B. Gross, K. A. Wright: Phys. Rev. 114, 725–727 (1959)
       B. Gross, P. V. Murphy: Nukleonik 2, 279–285 (1960)

3.191 L. E. Carrano de Almeida, G. L. Leal Ferreira: Rev. Bras. Fis. 5, 349–361 (1975)

3.192 L. Nunes de Oliveira, G. F. Leal Ferreira: J. Electrostat. 1, 371–380 (1975)

3.193 P. C. Camargo, G. F. Leal Ferreira: In Ref. 3.71, pp. 59–66

3.194 L. Badian, A. Skopec, E. Smycz, Cz. Stec: In Ref. 3.73

3.195 K.E.Lonngren: J. Appl. Phys. **48**, 2630–2631 (1977)
3.196 B.Gross, L.Nunes de Oliveira: J. Appl. Phys. **45**, 4724–4729 (1974); in Ref. 3.71, pp. 15–50
3.197 S.Matsuoka, H.Sunaga, R.Tanaka, M.Hagiwara, K.Araki: IEEE Trans. NS-**23**, 1447–1452 (1976)
3.198 R.Coelho, V.K.Agarwal, R.Haug: J. Phys. D**10**, 1943–1950 (1977)
       V.K.Agarwal, R.Coelho: In Ref. 3.73
3.199 C.S.Shalgaonkar, A.V.Narlikar: J. Mater. Sci. **7**, 1465–1471 (1972)
       R.Chen: J. Mater. Sci. **9**, 345–347 (1974)
* 3.200 R.Chen: In Ref. 3.72, pp. 15–24
* 3.201 P.Kivits, H.J.L.Hagebeuk: J. Lumin. **15**, 1–27 (1977)
3.202 D. de Muer: Physica **48**, 1–12 (1970)
3.203 V.Maxia, S.Onnis, A.Rucci: J. Lumin. **3**, 378–388 (1971)
* 3.204 Y.V.G.S.Murti: In *Proceedings of the National Symposium on "Thermoluminescence and Its Applications"* (Reactor Research Center, Madras 1975) pp. 168–182
3.205 H.L.Friedman: "Thermal Aging and Oxidation with Emphasis on Polymers", in *Treatise on Analytical Chemistry*, Vol. 3, ed. by J.M.Kolthoff, P.J.Elving, F.H.Stross (Wiley, New York 1976) pp. 393–584
3.206 T.A.T.Cowell, J.Woods: Br. J. Appl. Phys. **18**, 1045–1051 (1967)
3.207 N.S.Mohan, R.Chen: J. Phys. D**3**, 243–247 (1970)
3.208 F.R.Schwarzl, A.J.Staverman: Physica **18**, 791–798 (1952)
3.209 T.Alfrey, P.Doty: J. Appl. Phys. **16**, 700–712 (1945)
3.210 J.Vanderschueren: Appl. Phys. Lett. **25**, 270–271 (1974)
3.211 M.Kryszewski, M.Zielinski, S.Sapieha: Polymer **17**, 212–216 (1976)
3.212 G.Pfister, M.A.Abkowitz: J. Appl. Phys. **45**, 1001–1008 (1974)
3.213 S.Ikeda, K.Matsuda: Jpn. J. Appl. Phys. **15**, 963–966 (1976)
3.214 J.G.Simmons, G.W.Taylor, M.C.Tam: Phys. Rev. B**7**, 3714–3719 (1973)
       J.Broser, R.Broser-Warminsky: Br. J. Appl. Phys. **4**, 90–94 (1955)
       H.J.Dittfeld, J.Voigt: Phys. Status Solidi **3**, 509–522 (1966)
3.215 T.Hino: J. Appl. Phys. **46**, 1956–1960 (1975)
3.216 H.Gobrecht, D.Hofmann: J. Phys. Chem. Solids **27**, 509–522 (1966)
3.217 N.Saito, K.Okano, S.Iwayanagi, T.Hideshima: "Molecular Motion in Solid-State Polymers", in *Solid State Physics*, Vol. 14, ed. by F.Seitz, D.Turnbull (Academic Press, New York 1963) pp. 343–502
3.218 C.P.Smyth: "Dielectric Phenomena", in *Physics and Chemistry of the Organic Solid State*, Vol. 1, ed. by D.Fox, M.M.Labes, A.Weissberger (Interscience, New York 1963) Chap. 12
       C.J.F.Böttcher, P.Bordewijk: *"Theory of Electric Polarization. Dielectrics in Time-Dependent Fields*, Vol. 2, 2nd rev. ed. (Elsevier, Amsterdam 1978)
3.219 D.W.McCall: "Relaxations in Solid Polymers" in *Molecular Dynamics and Structure of Solids*, ed. by R.S.Carter, J.J.Rush (NBS, Washington, D. C. 1969) pp. 475–537
3.220 M.Davies (ed.): *Dielectric and Related Molecular Processes*, Vols. 1, 2 (Chemical Society, London, Vol. 1 (1972), Vol. 2 (1975)
3.221 G.L.Link, T.G.Parker: "Dielectric Properties of Polymers", in *Polymer Science*, Vol. 2, ed. by A.D.Jenkins (North-Holland, Amsterdam 1972) pp. 1281–1327
3.222 M.E.Baird: *Electrical Properties of Polymeric Materials* (Plastics Institute, London 1973)
3.223 A.M.North: in *Chemical and Biological Applications of Relaxation Spectroscopy*, ed. by E.Wyn-Jones (Proc. NATO Adv. Study Inst., Reidel, Dordrecht 1975) pp. 17–33; "Molecular Motion in Polymers" in *Molecular Behaviour and the Development of Polymeric Materials*, ed. by A.Ledwith, A.M.North (Chapman and Hall, London 1975) Chap. 11
* 3.224 A.K.Jonscher: Thin Solid Films **36**, 1–20 (1976)
3.225 R.Hayakawa, K.Namiki, T.Sakurai, Y.Wada: Rep. Prog. Polym. Phys. Jpn. **19**, 317–319 (1976)
3.226 Y.Aoki, J.O.Brittain: J. Polym. Sci. Polym. Phys. **14**, 1297–1304 (1976)
3.227 S.B.Dev, A.M.North, R.A.Pethrick: Adv. Mol. Relaxation Processes **4**, 159–191 (1972)
3.228 G.E.Johnson, E.W.Anderson, G.L.Link, D.W.McCall: Am. Chem. Soc. Div. Org. Coat. Plast. Chem. Pap. **35**, 404–409 (1975)

3.229 W.P.Harris: In 1966 *Annu. Rep., Conf. "Electr. Insul. Dielectr. Phenom."* (NAS-NRC, Washington 1967) pp. 72–74

3.230 T.Furukawa, E.Fukada: J. Polym. Sci. Polym. Phys. **14**, 1979–2110 (1976)

3.231 A. van Roggen: IEEE Trans. EI-**13**, 57–58 (1978)

3.232 C.T.Morse: J. Phys. E**7**, 657–662 (1974)

3.233 D.Chatain, P.Gautier, C.Lacabanne: Phys. Status Solidi A**15**, 191–198 (1973)

3.234 M.M.Perlman, S.Unger: J. Appl. Phys. **45**, 2389–2393 (1974)

3.235 S.Onnis, A.Rucci: J. Lumin. **6**, 404–413 (1973)

3.236 D.R.Rao: Phys. Status Solidi A**22**, 337–341 (1974)
       R.P.Khare, J.D.Ranade: Phys. Status Solidi A**32**, 221–226 (1975)
       R.P.Khare, R.Nath: Phys. Status Solidi A**44**, 627–631 (1977)

3.237 A.E.Woodward, J.A.Sauer: "Mechanical Relaxation Phenomena" in *Physics and Chemistry of the Organic Solid State*, Vol. 2, ed. by D.Fox, M.M.Labes, A.Weisberger (Interscience, New York 1965) Chap. 7
       J.A.Sauer: J. Polym. Sci. C**32**, 69–122 (1971)
       A.Hiltner, E.Baer: Crit. Rev. Macromol. Sci. **1**, 215–244 (1972)

3.238 J.Heijboer: Br. Polymer J. **1**, 3–14 (1969); Ph. D. Thesis, Leiden (1972); Int. J. Polym. Mater. **6**, 11–37 (1977)

3.239 G.M.Bartenev, Yu.V.Zelenev (eds.): *Relaxation Phenomena in Polymers* (Wiley, New York 1974)

3.240 A.Samoc, M.Samoc, J.Sworakowski: Phys. Status Solidi A**36**, 735–754 (1976)
       A.Samoc, M.Samoc, J.Sworabowski, J.M.Thomas, J.O.Williams: Phys. Status Solidi A**37**, 271–278 (1976)
       A.Samoc, M.Samoc, J.Sworakowski: Phys. Status Solidi A**39**, 337–344 (1977)
       A.Samoc, M.Samoc: In Ref. 3.73

3.241 R.A.Creswell, M.M.Perlman: J. Appl. Phys. **41**, 2365–2375 (1970)

3.242 L.K.Monteith, J.R.Hauser: J. Appl. Phys. **38**, 5355–5365 (1967)

3.243 G.M.Sessler, J.E.West: J. Appl. Phys. **47**, 3480–3484 (1976)
       G.M.Sessler: In Ref. 3.71, pp. 321–335; In Ref. 3.72, pp. 181–185

3.244 I.P.Batra, H.Seki: J. Appl. Phys. **41**, 3409–3415 (1970)

3.245 A.I.Rudenko: Sov. Phys.-Semicond. **5**, 2097–2099 (1972); Sov. Phys.-Semicond. **14**, 2706–2709 (1973)

3.246 L.Nunes de Oliveira, G.F.Leal Ferreira: In Ref. 3.71, pp. 51–58

3.247 E.H.Martin, J.Hirsch: J. Appl. Phys. **43**, 1001–1015 (1972)

3.248 L.Weaver, J.K.Shulte, R.E.Faw: J. Appl. Phys. **48**, 2762–2770 (1977)

3.249 B.Gross, G.M.Sessler, J.E.West: J. Appl. Phys. **46**, 4674–4677 (1975)

3.250 Y.Suzuoki, T.Mizutani, M.Ieda: Jpn. J. Appl. Phys. **15**, 1665–1668 (1976)

3.251 D.M.Taylor: J. Phys. D**9**, 2269–2279 (1976)

3.252 M.M.Perlman, S.Unger: Appl. Phys. Lett. **24**, 579–580 (1974)

3.253 A.Hampe: Colloid Polym. Sci. **62**, 154–160 (1977)

3.254 G.Sawa, D.C.Lee, M.Ieda: In Ref. 3.76, pp. 64–67

3.255 Y.Suzuoki, K.Yasuda, T.Mizutani, M.Ieda: J. Phys. D**10**, 1985–1990 (1977)
       T.Mizutani, Y.Suzuoki, M.Ieda: Electr. Eng. Jpn. **96** (5), 1–8 (1976); J. Appl. Phys. **48**, 2408–2413 (1977)

3.256 J.H.Crawford, L.M.Slifkin: Annu. Rev. Mater. Sci. **1**, 139–164 (1971)

3.257 J.S.Dryden, R.J.Meakins: Discuss. Faraday Soc. **23**, 39–49 (1957)
       R.J.Meakins: In Ref. 3.32c, Vol. 3 (1961) pp. 151–202

3.258 C.Bucci: Phys. Rev. **164**, 1200–1206 (1967)

3.259 M.Wentz, K.Ledjeff, K.Zierold, F.Granzer: In Ref. 3.77, C**7**, 401–406 (1976)

3.260 J.Kiessling, A.Scharmann: Z. Naturforsch. Teil A **31**, 105–107 (1976)

3.261 C.Lacabanne, D.Chatain, R.Madru, M.Maitrot: J. Chem. Phys. **71**, 597–603 (1974)

3.262 V.V.Popov, N.A.Tikhonov, A.G.Tyulyupa: Sov. Phys.–Solid State **18**, 527–528 (1976)

3.263 J.H.Crawford, G.E.Matthews: In Ref. 3.77, C**7**, 297–301 (1976)

3.264 R.Capelletti, F.Fermi, R.Schiavi: In Ref. 3.71, pp. 117–130
3.265 S.Radhakrishna, S.Haridoss: J. Phys. **38**, 841–844 (1977)
3.266 S.Radhakrishna, S.Haridoss: Solid-State Electron. **18**, 1247–1250 (1976)
3.267 I.V.Murin, B.F.Kornev, A.N.Murin: Sov. Phys.–Solid State **17**, 958–959 (1975)
3.268 R.Capelletti, F.Fermi, F.Leoni, E.Okuno: In Ref. 3.71, pp. 105–116
3.269 R.Capelletti, A.Gainotti, L.Pareti: In Ref. 3.70, pp. 66–83
3.270 S.Benci, R.Capelletti, F.Fermi, M.Manfredi: In Ref. 3.77, C 7, 138–144 (1976)
3.271 R.Capelletti, A.Gainotti: In Ref. 3.77, C 7, 316–321 (1976)
3.272 A.M.Hor, P.W.M.Jacobs, K.S.Moodie: Phys. Status Solidi A **38**, 293–298 (1976)
3.273 G.M.Sordi, S.Watanabe: Nuovo Cimento B **25**, 145–158 (1975)
3.274 R.Capelletti, E.Okuno: In Ref. 3.71, pp. 53–65
3.275 V.N.Erofeev, A.G.Melent'ev, E.M.Nadgornyi, G.I.Peresada: Sov. Phys.–Solid State **18**, 1312–1314 (1976)
3.276 J.Prakash, F.Fischer: Phys. Status Solidi A **39**, 499–507 (1977); In Ref. 3.77, C 7, 167–168 (1976)
3.277 S.C.Zilio, M.de Souza: Phys. Status Solidi B **80**, 597–601 (1977)
3.278 P.Varotsos, D.Kostopoulos, S.Mourikis, S.Kouremenou: Solid State Commun. **21**, 831–832 (1977)
3.279 C.O.Tiller, A.C.Lilly, B.C.LaRoy: Phys. Rev. B **8**, 4787–4794 (1973)
3.280 D.L.Kirk, D.C.Phillips, P.L.Pratt: in *Proc. Conf. on "Mass Transport in Non-Metallic Solids"* (British Ceramic Society, Stoke-on-Trent 1971) Publ. no. 19, pp. 105–134
3.281 F.Cusso, J.L.Pascal, F.Jaque: In Ref. 3.72, pp. 125–131
3.282 A.Kessler: In Ref. 3.70, pp. 45–52
3.283 J.Winter, K.Rössler: In Ref. 3.77, C 7, 265–269 (1976)
3.284 M.Siu, M.de Souza: In Ref. 3.71, pp. 377–385
3.285 G.P.Williams, D.Mullis: Phys. Status Solidi A **28**, 539–544 (1975)
3.286 J.H.Crawford, G.E.Matthews: In Ref. 3.77, C 7, 297–301 (1976)
3.287 N.Kristianpoller, Y.Kirsh: Phys. Status Solidi A **21**, 87–94 (1974)
3.288 B.P.M.Lenting, J.A.J.Numan, E.J.Bijvanck, H.W.den Hartog: Phys. Rev. B **14**, 1811–1817 (1976)
3.289 R.Capelletti: In Ref. 3.70, pp. 1–31
3.290 S.Unger, M.M.Perlman: In Ref. 3.69, pp. 37–53
3.291 J.H.Crawford: J. Phys. Chem. **31**, 399–409 (1970)
3.292 S.Unger, M.M.Perlman: Phys. Rev. B **10**, 3692–3696 (1974)
       M.M.Perlman, S.Unger: J. Electrostat. **1**, 231–234 (1975)
3.293 R.Capelletti, R.Fieschi: Cryst. Lattice Defects **1**, 69–81 (1969)
3.294 V.K.Jain: Phys. Status Solidi B **44**, 11–28 (1971)
3.295 R.Muccillo, J.Rolfe: Phys. Status Solidi B **61**, 579–589 (1974)
3.296 E.B.Podgorsak, P.R.Moran: Phys. Rev. B **8**, 3405–3418 (1973)
       E.B.Podgorsak, G.E.Fuller, P.R.Moran: In Ref. 3.69, pp. 172–176
       E.B.Podgorsak: Ph.D. Thesis, Wisconsin University Madison (1973)
3.297 R.Muccillo, A.R.Blak, S.Watanabe: In Ref. 3.71, pp. 357–370
       R.Muccillo, A.R.Blak: J. Nucl. Mater. **61**, 66–70 (1976)
3.298 M.S.Li, M.de Souza, F.Lüty: Phys. Rev. B **7**, 4677–4681 (1973)
3.299 R.Robert, R.Barboza, G.F.L.Ferreira, M.F.de Souza: Phys. Status Solidi B **59**, 335–342 (1973)
3.300 N.de Souza: In Ref. 3.71, pp. 387–394
3.301 E.L.Kitts, M.Ikeya, J.H.Crawford: Phys. Rev. B **8**, 5840–5846 (1973)
3.302 E.L.Kitts, J.H.Crawford: Phys. Rev. B **9**, 5264–5267 (1974)
3.303 V.I.Bugrienko, V.K.Marinchik, V.M.Belous: Sov. Phys.–Solid State **12**, 36–39 (1970)
3.304 I.Kunze, N.Starbov, A.Buroff: Phys. Status Solidi A **16**, K 59–K 62 (1973)
3.305 J.Jiminez, J.P.Fillard, J.Gasiot, L.F.Sanz, J.A.de Saja: In Ref. 3.72, pp. 133–138
3.306 T.A.Roth: In Ref. 3.69, pp. 29–36
3.307 T.A.Roth: J. Appl. Phys. **44**, 1056–1060 (1973)

3.308 H.Gränicher, C.Jaccard, P.Scherrer, A.Steinemann: Discuss. Faraday Soc. **23**, 50–62 (1957)
3.309 A.von Hippel: J. Chem. Phys. **54**, 145–149 (1971)
3.310 J.B.Hasted: *Aqueous Solutions* (Chapman and Hall, London 1973) Chap. 4
3.311 S.Mascarenhas: In *Proc. Int. Symp. on "Physics of Ice"*, ed. by N.Riehl, B.Bullemer, H.Engelhardt (Plenum, New York 1969) pp. 483–491
3.312 A.Jeneveau, P.Sixou, P.Dansas: Phys. Kondens. Mater. **14**, 252–264 (1972)
3.313 G.P.Johari, S.J.Jones: J. Chem. Phys. **62**, 4213–4223 (1975)
3.314 M.I.El-Azab, C.H.Champness: Appl. Phys. Lett. **31**, 295–297 (1977)
       G.Guillard, J.Fornazero, M.Maitrot, D.Chatain, C.Lacabanne: J. Appl. Phys. **48**, 3428–3433 (1977)
* 3.315 B.T.Kolomiets, V.M.Ljubin, V.L.Averjanov: Mater. Res. Bull. **5**, 655–664 (1970)
       B.T.Kolomiets: In *Proc. 5th Int. Conf. on Amorphous and Liquid Semiconductors*, ed. by J.Stuke, W.Brenig (Taylor and Francis, London 1974) pp. 189–201
3.316 R.A.Street, A.D.Yoffe: Thin Solid Films **11**, 161–174 (1972)
3.317 T.Botila, A.Vancu, M.Lazarescu, L.Vescan, Gr.Ioanid, St.Sladaru: Thin Solid Films **12**, 223–226 (1972)
3.318 I.Thurzo, D.Barancok, E.Mariani: Phys. Status Solidi A **22**, K149–K151 (1974)
       I.Thurzo, M.Pavlikova, E.Mariani: Czech. J. Phys. B **25**, 1279–1284 (1975)
3.319 A.M.Andriesh, S.D.Shutov, V.G.Abashkin, M.R.Chernii: Sov. Phys.–Semicond. **8**, 1254–1257 (1975)
3.320 G.M.Martin, E.Fogarassy, E.Fabre: J. Appl. Phys. **47**, 264–266 (1976)
3.321 B.L.Smith: In *Proc. Conf. on Metal-Semiconductor Contacts*, ed. by M.Pepper (Institute of Physics, London 1974) Publ. no. 22, pp. 210–217
3.322 E.Fabre, R.N.Bhargava, W.K.Zwicker: J. Electron. Mater. **3**, 409–430 (1974)
3.323 G.M.Martin, J.Hallais, G.Poiblaud: In Ref. 3.72, pp. 223–232
3.324 M.Saji, K.C.Kao: Jpn. J. Appl. Phys. **15**, 1393–1394 (1976)
3.325 V.N.Lobushkin, I.M.Sokolova, V.N.Tairov: Sov. Electrochem. **12**, 387–391 (1976)
3.326 G.Caserta, A.Serra: J. Appl. Phys. **42**, 3778–3785 (1971)
3.327 M.M.Perlman: J. Appl. Phys. **42**, 2645–2652 (1971)
3.328 T.Takamatsu, E.Fukada: Rep. Prog. Polym. Phys. Jpn. **14**, 485–488 (1971)
3.329 K.Hartman, R.N.O'Brien: In Ref. 3.69, pp. 444–454
3.330 D.K.Walker, O.Jefimenko: J. Appl. Phys. **43**, 3459–3464 (1973)
3.331 A.P.Rojindar, M.L.Khare, C.S.Bhatnager: Indian J. Pure Appl. Phys. **12**, 849–850 (1974)
3.332 T.Takamatsu, E.Fukada: In Ref. 3.70, pp. 201–212
       Y.Hoshino, Y.Tokunaga: J. Appl. Phys. **48**, 1456–1460 (1977)
3.333 T.Garofano, T.Corazzari, G.Casaline: Nuovo Cimento B **38**, 133–149 (1977)
3.334 M.Campos, G.Leal Ferreira, S.Mascarenhas: J. Nonmetals **2**, 123–126 (1974)
       A.P.Srivastava, S.R.Agarwal: Indian J. Pure Appl. Phys. **13**, 869–870 (1975)
3.335 P.Devaux, M.Schott: Phys. Status Solidi **20**, 301–309 (1967)
       O.E.Lobanova, E.L.Lutsenko: Sov. Phys.–Semicond. **5**, 295–296 (1971)
       B.Sh.Barkhalov, E.L.Lutsenko: Phys. Status Solidi A **11**, 433–439 (1972)
3.336 P.Dansas, P.Sixou, M.Jaffrain: Mol. Phys. **21**, 225–240 (1971)
3.337 J.I.Lauritzen: J. Chem. Phys. **28**, 118–131 (1958)
3.338 G.Williams: In Ref. 3.220, Vol. 2, Chap. 4
3.339 G.P.Johari, M.Goldstein: J. Chem. Phys. **53**, 2372–2388 (1970)
3.340 P.Dansas, P.Sixou: C.R.Hebd. Sceances Acad. Sci. Ser. B **266**, 459–463 (1968)
3.341 R.K.Chan, G.P.Johari: In Ref. 3.74, pp. 52–59 (1975)
3.342 R.J.Heitz, H.Szwarc: In *Proc. 4th Int. Conf. on "The Physics of Non-Crystalline Solids"*, ed. by G.H.Frischat (Trans Tech Aedermannsdorf 1977) pp. 534–541
3.343 S. van Houten: In *Proc. 3rd European Solid-State Devices* (Institute of Physics, London 1973) Publ. No. 19, pp. 131–157
3.344 G.Meier: In Ref. 3.220, Vol. 2, Chap. 5
3.345 S.Bini, R.Capelletti: In Ref. 3.69, pp. 66–74
3.346 J.Costa Ribeiro: An. Acad. Bras. Cienc. **22**, 325–347 (1950)
3.347 T.Tachibana, T.Takamatsu, E.Fukada: Chem. Lett. 907–910 (1973)

3.348 G.Schwarz: In Ref. 3.220, Vol. 1, Chap. 6
      B. Rosenberg, E.Postow: Semiconductivity in Proteins and Nucleic Acids, in *Experimental Methods in Biophysical Chemistry*, ed. by C.Nicolau (Wiley, New York 1973) Chap. 7
3.349 D. Eley: In *Organic Semiconducting Polymers*, ed. by J.E.Katon (Dekker, New York 1968) Chap. 5
3.350 S. D. Bruck: Polymer **16**, 25–30 (1975)
3.351 M. Reichle, T. Nedetzka, A. Mayer, H. Vogel: J. Phys. Chem. **74**, 2659–2666 (1970)
3.352 D. Chatain, C. Lacabanne, M. Maitrot, G. Seytre, J. F. May: Phys. Status Solidi A **16**, 225–233 (1973)
      J. Guillet, G. Seytre, D. Chatain, C. Lacabanne, J. C. Monpagens: J. Polym. Sci. Polym. Phys. **15**, 541–554 (1977)
3.353 P. K. C. Pillai, M. Goel: Polymer **16**, 5–8 (1975)
3.354 A. Ledwith, K. C. Smith, S. M. Walker: Polymer **19**, 51–56 (1978)
3.355 E. Menefee: In Ref. 3.69, pp. 661–675; in *Proc. "Electrically Mediated Growth Mechanisms in Living Systems"*, ed. by A. R. Liboff, R. A. Rinaldï, publ. as Ann. N. Y. Acad. Sci. **238**, 53–67 (1974)
      J. L. Leveque, J. C. Garson, G. Boudourts: Text. Res. J. **44**, 504–505 (1974)
      J. L. Leveque, J. C. Garson: Biopolymers **16**, 1725–1733 (1977)
3.356 L. G. Augenstein, J. O. Williams: "TL in Biological Materials", in *Experimental Methods in Biophysical Chemistry*, ed. by C. Nicolau (Wiley, New York 1973) Chap. 11
* 3.357 E. Fukada: Adv. Biophys. **6**, 121–155 (1974)
3.358 A. Sharples: "Crystallinity", in *Polymer Science*, Vol. 1, ed. by A. D. Jenkins (North-Holland, Amsterdam 1972) Chap. 4
      J. M. Schultz: *Polymer Materials Science* (Prentice-Hall, Englewood Cliffs, N.J. 1974)
      B. Wunderlich: *Macromolecular Physics*, Vols. 1, 2 (Academic Press, New York 1973, 1976)
      J. H. Magill: "Morphogenesis of Solid Polymer Microstructures", in *Properties of Solid Polymeric Materials*, ed. by J. M. Schultz, Treatise on Materials Science and Technology (Academic Press, New York 1977) **10** A, pp. 1–368
3.359 I. M. Ward (ed.): *Structure and Properties of Oriented Polymers* (Applied Science, London 1975)
* 3.360 R. Hosemann: Crit. Rev. Macromol. Sci. **1**, 351–397 (1972)
3.361 G. M. Bartenev, Yu. V. Zelenev: Polym. Sci.–USSR **14**, 1109–1122 (1972)
3.362 Y. Wada, R. Hayakawa: "Relaxation Processes in Crystalline and Non-Crystalline Phases in Polymers", in *Progress in Polymer Science Japan*, Vol. 3, ed. by S. Okamura, M. Takayanagi (Halsted Press, New York 1972) pp. 215–261
3.363 B. Stoll, W. Pechhold, S. Blasenbrey: Kolloid Z. Z. Polym. **250**, 1111–1130 (1972)
3.364 J. Vanderschueren: Polym. Lett. **10**, 543–548 (1972)
3.365 V. G. Lyudskanov, T. A. Vasil'ev, Yu. V. Zelenev: Polym. Sci. USSR **14**, 180–190 (1972)
3.366 C. Lacabanne, D. Chatain: J. Polym. Sci. Polym. Phys. **11**, 2315–2328 (1973)
3.367 H. Solunov, T. Vassilev: J. Polym. Sci. Polym. Phys. **12**, 1273–1282 (1974)
3.368 L. Lamarre, H. P. Schreiber, M. R. Wertheimer: In Ref. 3.74, pp. 218–225 (1977)
3.369 G. A. Baum: J. Appl. Polym. Sci. **17**, 2855–2866 (1973)
3.370 D. Varma, C. S. Bhatnager: Indian J. Pure Appl. Phys. **14**, 93–95 (1976)
3.371 A. R. McGhie, G. McGibbon, A. Sharples, E. J. Stanley: Polymer **13**, 371–378 (1972)
3.372 J. P. Reardon, P. F. Waters: In Ref. 3.70, pp. 185–200
      P. C. Mehendru, S. Chand, N. L. Pathak: Thin Solid Films **44**, L13–L16 (1977)
3.373 P. Murphy, M. Latour: In Ref. 3.71, pp. 227–238; In Ref. 3.72, pp. 163–169
3.374 K. Ikezaki, M. Hattori, Y. Arimoto: Jpn. J. Appl. Phys. **16**, 863–864 (1977)
3.375 P. K. C. Pillai, K. Jain, V. K. Jain: Phys. Lett. A **39**, 216–218 (1972); Indian J. Pure Appl. Phys. **11**, 597–601 (1973)
3.376 M. A. Abkowitz, G. Pfister: J. Appl. Phys. **46**, 2559–2573 (1975)
3.377 G. Pfister, W. M. Prest, D. J. Luca, M. Abkowitz: Appl. Phys. Lett. **27**, 486–488 (1975)
3.378 A. I. Baise, H. Lee, B. Oh, R. E. Salomon, M. M. Labes: Appl. Phys. Lett. **26**, 428–430 (1975)
3.379 N. Murayama, H. Hashizume: J. Polym. Sci. Polym. Phys. **14**, 989–1003 (1976)
3.380 E. J. Sharp, L. E. Garn: J. Appl. Phys. **29**, 480–482 (1976)

3.381 A.Callens, R.de Batist, L.Eersels: Nuovo Cimento B 33, 434–446 (1976)
3.382 D.Chatain, C.Lacabanne, M.Maitrot: Phys. Status Solidi A 13, 303–311 (1972)
       D.Chatain, P.Gautier, C.Lacabanne: J. Polym. Sci. Polym. Phys. 11, 1631–1640 (1973)
       C.Lacabanne, D.Chatain, J.Guillet, G.Seytre, J.F.May: J. Polym. Sci. Polym. Phys. 13,
       445–453 (1975)
3.383 J.P.Soulier, B.Chabert, J.Chauchard, D.Chatain, C.Lacabanne: J. Chim. Phys. 71, 32–36
       (1974)
       C.Lacabanne, D.Chatain: J. Phys. Chem. 79, 283–287 (1975)
3.384 A.C.Lilly, L.L.Stewart, R.M.Henderson: J. Appl. Phys. 41, 2001–2014 (1970)
3.385 D.X.Donchev: C.R. Acad. Bulg. Sci. 25, 1483–1486 (1972)
3.386 G.A.Lushcheikin, L.I.Voiteshonok: Polym. Sci. USSR A 16, 1581–1587 (1974); Polym. Sci.
       USSR A 17, 497–499 (1975)
3.387 T. Hino, Y.Kitamura: Electr. Eng. Jpn. 95 (1), 24–30 (1975)
3.388 Y.Aoki, J.O.Brittain: J. Appl. Polym. Sci. 20, 2879–2892 (1976)
       J. Polym. Sci. Polym. Phys. 15, 199–210 (1977)
3.389 J.Perret, R.Jocteur, B.Fallou: Rev. Gen. Electr. 81, 757–767 (1972)
       R.Lacoste, B.Ai, Y.Sequi, A.Rahal: In Ref. 3.73
3.390 P.Fischer, P.Röhl: Kolloid Z. Z. Polym. 251, 941–946 (1973)
3.391 M.Matsui, M.Murasaki: In Ref. 3.70, pp. 172–184
3.392 M.Pineri, P.Berticat, E.Marchal: J. Polym. Sci. Polym. Phys. 14, 1325–1336 (1976)
3.393 E.Motyl: In Ref. 3.73
3.394 R.J.Gable, N.V.Vijaylaghavan, R.A.Wallace: J. Polym. Sci. Polym. Chem. 11, 2387–2389
       (1973)
3.395 M.Zielinski, M.Kryszewski, S.Sapieha: In Ref. 3.74
3.396 S.I.Stupp, S.H.Carr: J. Appl. Phys. 46, 4120–4123 (1975)
       R.J.Comstock, S.I.Stupp, S.H.Carr: J. Macromol. Sci. Phys. B 13, 101–115 (1977)
3.397 P.Alexandovich, F.E.Karacz, W.J.Knight: J. Appl. Phys. 47, 4251–4254 (1976)
3.398 S.Takeda: J. Appl. Phys. 47, 5480–5481 (1977)
3.399 K.Jain, A.C.Rastogi, K.L.Chopra: Phys. Status Solidi A 21, 685–692 (1974)
3.400 P.K.C.Pillai, K.Jain, V.K.Jain: Nuovo Cimento B 28, 152–160 (1975)
       P.C.Mehendru, K.Jain, V.K.Chopra, P.Mehendru: J. Phys. D 8, 305–313 (1975)
3.401 V.K.Jain, C.L.Gupta, R.K.Jain, S.K.Agarwal, R.C.Tyagi: Thin Solid Films 30, 245–258
       (1975)
       P.C.Mehendru, K.Jain, P.Mehendru: J. Phys. D 9, 83–88 (1976)
3.402 A.Linkens, J.Vanderschueren, S.H.Chor, J.Gasiot: Eur. Polym. J. 12, 137–146 (1976)
3.403 R.A.Wallace, R.J.Gable: J. Appl. Polym. Sci. 17, 3549–3552 (1973)
3.404 J.L.Crowley, R.A.Wallace, R.H.Bube: J. Polym. Sci. Polym. Phys. 14, 1769–1787 (1976)
3.405 P.Eyerer: Gummi Asbest Kunstst. 25, 1–12 (1972); J. Appl. Polym. Sci. 16, 2461–2483 (1972)
3.406 T.Takamatsu, Y.Nakajima: Rep. Prog. Polym. Phys. Jpn. 17, 391–394 (1974)
3.407 D.L.Shelley, S.F.Huber: In Ref. 3.74, pp. 100–108 (1975)
3.408 S.M.Baturin, G.B.Manelis, A.G.Malent'ev, E.M.Nadgornyi, Yu.A.Ol'khov, V.G.
       Shteinberg: Polym. Sci. USSR A 18, 2808–2814 (1976)
3.409 K.Jain, A.C.Rastogi, K.L.Chopra: Phys. Status Solidi A 20, 167–175 (1973)
3.410 T.Tanaka, S.Hayashi, K.Shibayama: J. Appl. Phys. 48, 3478–3483 (1977)
3.411 M.Kryszewski, J.Ulanski: In Proc. Summer School on "Electrical Properties of Organic
       Solids" (Technical University Wroclaw 1974) pp. 361–371
3.412 K.Schon: PTB Mitt. 3, 152–154 (1972)
3.413 P.K.C.Pillai, P.K.Nair, R.Nath: Polymer 17, 921–922 (1976)
3.414 T.Hashimoto, M.Shiraki, T.Sakai: J. Polym. Sci. Polym. Phys. 13, 2401–2410 (1975)
3.415 J.B.Woodward: J. Electron. Mater. 6, 145–162 (1977)
3.416 P.K.C.Pillai, K.Jain, V.K.Jain: Nuovo Cimento B 11, 339–348 (1972)
       P.C.Mehendru, K.Jain, P.Mehendru: J. Phys. D 10, 729–736 (1977)
3.417 P.Hedvig: In Proc. 4th Symp. on "Radiation Chemistry", ed. by P.Hedvig, R.Schiller
       (Akadémiai Kiadó, Budapest 1977) pp. 1–16

3.418 A.Szymanski, M.Kryszewski: J. Polym. Sci. C**22**, 867–879 (1969)
3.419 M.Kryszewski, H.Kasica, J.Patora, J.Piotrowski: J. Polym. Sci. C**30**, 243–260 (1970)
3.420 K.Amakawa, Y.Inuishi: In Ref. 3.69, pp. 115–127
       F.K.Dolezalek: In *Proc. 3rd Int. Congr. on "Static Electricity"* (Eur. Feder. Chem. Engrs.,
       Grenoble 1977) pp. 5a–5c
       T.Mizutani, M.Ieda: J. Phys. D **11**, 185–191 (1978)
3.421 K.Mazur: Polymer **18**, 409–411 (1977)
3.422 C.Linder, I.F.Miller: J. Polym. Sci. Chem. **11**, 1119–1130 (1973)
3.423 B.Ai, H.Carchano, J.Guastavino, D.Chatain, P.Gautier, C.Lacabanne: Thin Solid Films **21**,
       313–324 (1974)
       J.Tanguy: J. Appl. Phys. **47**, 2792–2799 (1976)
3.424 D.Chatain: Ph.D. Thesis, Toulouse (1974)
3.425 C.Lacabanne: Ph.D. Thesis, Toulouse (1974)
3.426 P.V.Murphy, F.W.Fraim: In Ref. 3.69, pp. 603–621
* 3.427 A.J.Kovacs: Fortschr. Hochpolym. Forsch. **3**, 394–507 (1964)
3.428 S.Kästner, M.Dittmer: Kolloid Z. Z. Polym. **204**, 74–83 (1965)
* 3.429 Y.Wada, R.Hayakawa: Jpn. J. Appl. Phys. **15**, 2041–2057 (1976)
       M.G.Broadhurst (Coord.): *Proc. Piezoelectric and Pyroelectric Symposium*, Workshop
       (NBS, Washington 1975)
3.430 J.B.Enns, R.Simha: J. Macromol. Sci. Phys. B**13**, 11–24 (1977)
3.431 C.A.Bucci, S.C.Riva: J. Phys. Chem. Solids **26**, 363–371 (1965)
3.432 V.I.Bugrienko, N.G.Barda: Sov. Phys.–Solid State **16**, 2061–2062 (1975)
3.433 V.Harasta, I.Thurzo: Fys. Cas. **20**, 148–155 (1970)
3.434 I.Kunze, P.Müller: Phys. Status Solidi A **13**, 197–206 (1972)
3.435 B.Gross: J. Electrochem. Soc. **119**, 855–860 (1972); J. Polym. Sci. Pt. A-2 **10**, 1941–1947
       (1972)
3.436 P.Müller: Phys. Status Solidi A **23**, 393–397 (1974)
3.437 L.Badian, J.Klocek: In Ref. 3.73
3.438 B.Gross: J. Chem. Phys. **17**, 866–872 (1949)
3.439 W.P.G.Swann: J. Franklin Inst. **250**, 219–248 (1950)
       A.N.Gubkin: Sov. Phys.–Tech. Phys. **2**, 1813–1824 (1957)
3.440 R.E.Vossteen: In *Proc. 9th Ann. Meeting IEEE Ind. Appl. Soc.* ES-WED-AMI, 799–810
       (1974)
3.441 I.P.Batra, K.K.Kanazawi, H.Seki: J. Appl. Phys. **41**, 3416–3422 (1970)
3.442 J.Mort, I.Chen: "Physics of Xerographic Photoreceptors", in *Applied Solid State Science*,
       Vol. 5 (Academic Press, New York 1975) pp. 69–149
3.443 R.M.Schaffert: *Electrophotography*, 2nd ed. (Focal Press, London 1975) Chap. 12
3.444 B.Gross: Z. Phys. **107**, 217–234 (1937); Phys. Rev. **57**, 57–59 (1940)
3.445 N.D.Borisenko, F.F.Kodzhespirov, S.N.Pisanko, E.I.Yaroshenko: Sov. Phys.–Semicond. **7**,
       285–286 (1973)
       Yu.N.Nikolaev, M.N.Titov: Sov. Phys.–Semicond. **7**, 465–468 (1973)
3.446 R.G.Vijverberg: "Charging Photoconductive Surfaces", in *Xerography and Related
       Processes*, ed. by J.H.Dessauer, H.E.Clark (Focal Press, London 1965) pp. 201–216
       M.Ieda, G.Sawa, U.Shinohara: Electr. Eng. Jpn. **88** (7), 67–73 (1968)
       R.A.Moreno, B.Gross: J. Appl. Phys. **47**, 3397–3402 (1976)
3.447 K.K.Kanazawa, I.P.Batra: J. Appl. Phys. **43**, 1845–1853 (1972)
       H.J.Wintle: J. Appl. Phys. **43**, 2927–2930 (1972)
3.448 A.I.Rudenko: Sov. Phys.–Semicond. **5**, 2097–2099 (1972)
3.449 L.Nunes de Oliveira, G.F.Leal Ferreira: J. Electrostat. **2**, 187–198 (1976)
3.450 P.W.Chudleigh: J. Appl. Phys. **48**, 4591–4596 (1977)
3.451 K.K.Kanazawa, I.P.Batra, H.J.Wintle: J. Appl. Phys. **43**, 719–720 (1972)
3.452 H.J.Wintle: J. Appl. Phys. **41**, 4004–4007 (1970)
       R.M.Hill: J. Phys. C **8**, 2488–2501 (1975)

3.453 E. A. Baum, T. J. Lewis, R. Toomer: J. Phys. D **10**, 487–497 (1977)
      R. Gerhard: Unpubl. results (see [2.169])
3.454 T. J. Sonnonstine, M. M. Perlman: J. Appl. Phys. **46**, 3975–3981 (1975)
      M. M. Perlman, T. J. Sonnonstine: In Ref. 3.71, pp. 337–355
      M. M. Perlman, T. J. Sonnonstine, J. A. St. Pierre: J. Appl. Phys. **47**, 3122–3126 (1976); In Ref. 3.72, pp. 187–193
3.455 E. A. Baum, T. J. Lewis: In *Proc. 4th Conf. on "Static Electrification"*, ed. by A. R. Blythe (Institute of Physics, London 1975) pp. 130–140
      H. T. M. Haenen: J. Electrostat. **1**, 173–185 (1975)
3.456 G. M. Sessler, J. E. West: In *Proc. 2nd Int. Conf. on Electrophotography*, ed. by D. R. White (Society of Photographic Scientists and Engineers, 1974) pp. 162–166
      J. Feder: J. Appl. Phys. **47**, 1741–1745 (1976)
3.457 J. van Turnhout, C. van Bochove, G. J. van Veldhuizen: Staub. Reinhalt. Luft **36**, 36–39 (1976)
3.458 R. J. Atkinson, R. J. Fleming: J. Phys. D **9**, 2027–2040 (1976); Austr. Telecommun. Res. **10**, 48–56 (1976)
3.459 D. W. Vance: J. Appl. Phys. **42**, 5430–5443 (1971)
  * H. Kiess: RCA Rev. **36**, 667–710 (1975)
3.460 A. S. DeReggi, C. M. Gutmann, F. I. Mopsik, G. T. Davis, M. G. Broadhurst: Phys. Rev. Lett. **40**, 413–416 (1978)
      H. von Seggern: Appl. Phys. Lett. **33**, 134–136 (1978)
3.461 B. Andresz, P. Fischer, P. Röhl: Progr. Colloid Polym. Sci. (Suppl. Colloid Polym. Sci.) **62**, 141–148 (1977)
3.462 T. Hino, F. Kaneko: In Ref. 3.74, pp. 17–24 (1976)
3.463 D. J. Dimaria, F. J. Feigl: Phys. Rev. B **9**, 1874–1883 (1974)
      V. J. Kapoor, F. J. Feigl, S. R. Butler: J. Appl. Phys. **48**, 739–749 (1977)
3.464 S. M. Sze: *Physics of Semiconductors* (Wiley, New York 1969)
3.465 E. J. M. Kendall, J. W. Haslett, F. J. Scholz: In Ref. 3.69, pp. 96–104
3.466 J. C. Manifacier, P. Parot, J. P. Fillard: In Ref. 3.72, pp. 203–212
      J. van Turnhout, A. H. van Rheenen: In Ref. 3.72, pp. 213–221
  * 3.467 B. E. Deal: J. Electrochem. Soc. **121**, 198C–205C (1974)
3.468 M. Zielinski, M. Samoc: J. Phys. D **10**, L105–L107 (1977)
3.469 H. Adachi, Y. Shibata: J. Phys. D **8**, 1120–1132 (1975)
3.470 J. G. Simmons, G. S. Nadkarni: Phys. Rev. B **6**, 4815–4827 (1972)
3.471 J. G. Simmons, G. W. Taylor: Solid-State Electron. **17**, 125–130 (1974)
      H. A. Mar, J. G. Simmons: Phys. Rev. B **11**, 775–783 (1975)
3.472 L. S. Wei, J. G. Simmons: Solid-State Electron. **17**, 591–598 (1974)
      H. A. Mar, J. G. Simmons: IEEE Trans. ED-**24**, 540–546 (1977)
3.473 J. G. Simmons, H. A. Mar: Phys. Rev. B **8**, 3865–3874 (1973)
      H. A. Mar, J. G. Simmons: Solid-State Electron. **17**, 1181–1185 (1974)
3.474 H. A. Mar, J. G. Simmons: Solid-State Electron. **17**, 131–135 (1974)
      J. S. Uranwala, J. G. Simmons, H. A. Mar: Solid-State Electron. **19**, 375–380 (1976)
3.475 L. Heyne: Philips Res. Rep. Suppl. No. **4**, 1–161 (1961); Ph.D. Thesis, Amsterdam (1961)
3.476 H. M. Gupta, R. J. van Overstraeten: J. Phys. C **7**, 3560–3572 (1974)
      H. M. Gupta: J. Phys. C **10**, L429–L431 (1977)
      G. Ferenczi, J. Balazs: In *Proc. Conf. on Metal-Semiconductor Contacts*, ed. by M. Pepper (Institute of Physics, London 1974) Publ. No. 22, pp. 249–254
  * 3.477 M. G. Buehler: Solid-State Electron. **15**, 69–79 (1972); in *Proc. 2nd Int. Symp. on "Semiconductor Silicon"*, ed. by H. R. Huff, R. R. Burgess (Electrochemical Society, Princeton 1973) pp. 549–560
3.478 A. G. Zhdan, V. B. Sandominskii, A. D. Ozheredov: Solid-State Electron. **11**, 505–508 (1968); Sov. Phys.–Semicond. **3**, 1130–1133 (1970)
3.479 A. Katzir, A. Halperin: Solid-State Electron. **15**, 573–575 (1972)
3.480 D. L. Losee: J. Appl. Phys. **46**, 2204–2214 (1975)
      G. Vincent, D. Bois, P. Pinard: J. Appl. Phys. **46**, 5173–5178 (1975)

* 3.481 C.T.Sah: Solid-State Electron. **17**, 975–990 (1976)
  3.482 D.V.Lang: J. Appl. Phys. **45**, 3014–3032 (1974)
*   G.L.Miller, D.V.Lang, L.C.Kimerling: Annu. Rev. Mater. Sci. **7**, 377–448 (1977)
  3.483 H.Lefèvre, M.Schulz: Appl. Phys. **12**, 45–53 (1977); IEEE Trans. ED-**24**, 973–978 (1977)
  3.484 G.L.Miller, J.V.Ramirez, D.A.H.Robinson: J. Appl. Phys. **46**, 2638–2644 (1975)
         M.D.Miller, D.R.Patterson: Rev. Sci. Instrum. **48**, 237–239 (1977)
         J.Guldberg: J. Phys. E **10**, 1016–1018 (1977)
  3.485 L.Y.Zlatkevich: Rubber Chem. Technol. **49**, 178–188 (1976)
         S.Radhakrishna, M.R.K.Murthy: J. Polym. Sci. Polym. Phys. **15**, 987–993 (1977)
         R.J.Fleming, L.F.Pender: In Ref. 3.72, pp. 139–148
  3.486 P.Bräunlich, A.Scharmann: Phys. Status Solidi **18**, 307–316 (1966)
         I.J.Saunders: J. Phys. C **2**, 2181–2198 (1969)
  3.487 M.Böhm, A.Scharmann: Phys. Status Solidi A **5**, 563–570 (1971)
         P.Kelly, M.J.Laubitz, P.Bräunlich: Phys. Rev. B **4**, 1960–1968 (1971)
         D.Shenker, R.Chen: J. Comput. Phys. **10**, 272–283 (1972)
         H.J.L.Hagebeuk, P.Kivits: Physica B **83**, 289–294 (1976)
  3.488 P.Bräunlich, P.Kelly: In Ref. 3.72, pp. 25–36
  3.489 R.R.Haering, E.N.Adams: Phys. Rev. **117**, 451–454 (1960)
         K.H.Nicholas, J.Woods: Br. J. Appl. Phys. **15**, 783–795 (1964)
  3.490 V.F.Zolotaryov, D.G.Semak, D.V.Chepur: Phys. Status Solidi **21**, 437–442 (1967); Sov. Phys.–JETP **25**, 557–559 (1967)
  3.491 Y.Y.Tkach: Sov. Phys.–Semicond. **8**, 167–168 (1974)
  3.492 G.A.Bordovskii, V.G.Boitsov, B.A.Demidov: Sov. Phys.–Semicond. **8**, 1243–1245 (1975); Sov. Phys.–Semicond. **10**, 603–604 (1976)
  3.493 F.S.Sinencio, S.Mascarenhas, B.S.H.Royce: Phys. Lett. A **26**, 70–72 (1967)
  3.494 C.Scharager, J.C.Muller, R.Stuck, P.Siffert: Phys. Status Solidi A **31**, 247–253 (1975)
         R.Stuck, J.C.Muller, J.P.Ponpon, C.Scharager, C.Schwab, P.Siffert: J. Appl. Phys. **47**, 1545–1548 (1976)
  3.495 H.C.Wright, R.E.Hunt, G.A.Adler: Solid-State Electron. **10**, 633–639 (1967)
  3.496 Y.Y.Tkach: Sov. Phys.–Semicond. **6**, 451–452 (1972)
  3.497 A.Silverman, H.Kallmann, B.Kramer: Phys. Status Solidi A **16**, 401–411 (1973)
* 3.498 V.K.Mathur: In *Proc. Natl. Symp. on "Thermoluminescence and Its Applications"* (Reactor Research Center, Madras 1975) pp. 234–246
  3.499 R.Chen: J. Appl. Phys. **42**, 5899–5901 (1971)
         J.P.Fillard, J.Gasiot, M.de Murcia: In Ref. 3.73, pp. 99–104
         J.Gasiot, J.P.Fillard: J. Appl. Phys. **48**, 3171–3172 (1977)
  3.500 S.A.Rabie: Phys. Rev. B **14**, 2569–2578 (1976)
  3.501 M.E.Haine, R.E.Carley-Read: J. Phys. D **1**, 1257–1269 (1968)
* 3.502 E.Huster: Naturwissenschaften **64**, 448–460 (1977)
  3.503 P.Kelly: Phys. Rev. **13**, 749–751 (1972)
  3.504 S.Bhoraskar, R.Abe: Jpn. J. Appl. Phys. **15**, 1471–1478 (1976)
         M.Legrand, G.Dreyfus, J.Lewiner: J. Phys. (Paris) Lett. **38**, L439–L440 (1977)
         A.Edgar: J. Phys. E **10**, 1261–1264 (1977)
  3.505 D.Briggs, D.M.Brewis, M.B.Konieczko: Eur. Polym. J. **14**, 1–4 (1978)
  3.506 D.Briggs (ed.): *Handbook X-Ray and Ultraviolet Photoelectron Spectroscopy* (Heyden, London 1977)
*   N.Treitz: J. Phys. E **10**, 573–585 (1977)

4. Radiation-Induced Charge Storage and Polarization Effects

By B. Gross

With 30 Figures

Soon after the discovery of X-rays and radioactivity numerous authors reported the effects of the new radiations on dielectrics. Still in the last century (1896) *Thomson* and *McClelland* [4.1] found that irradiation with X-rays increased the conductivity of paraffin; in 1903 *Becquerel* [4.2] found the same effect using radium as a radiation source. *Joffé* in 1906 [4.3], studying the radiation-induced conductivity in quartz, found that the effects persisted for a considerable time after termination of the irradiation and thus confirmed the existence of a delayed conductivity. *Roentgen* in 1921 [4.4] investigated the behavior of X-ray irradiated rock salt. *Nasledow* and *Scharawsky* [4.5] found in 1929 that the induced conductivity of ceresin wax irradiated with X-rays increased proportional to the square root of the exposure rate; they explained this behavior by assuming that carrier generation and recombination in irradiated dielectrics follow the same laws as in gases.

Roos in 1929 [4.6], *Seidl* in the 1930s [4.7], and *Scislowski* [4.8] found that dielectric absorption in many solid dielectrics, in particular paraffin wax, amber, and Rochelle salt, was strongly increased by irradiation with X-rays and gamma rays. Since dielectric absorption always implies in charge and/or polarization storage, this proved that these effects can be enhanced by irradiation. After the war, irradiation affecting the performance of solid dielectrics became of increasing importance in connection with the development of modern radiation technology and the widespread use of insulating materials in nuclear installations and other ambients with strong radiation fields.

Charge storage effects are observed in most instances when solid dielectrics are exposed to penetrating radiation. They can be due to microscopic structural defects, macroscopic heterogeneity, electrode polarization, and dipole formation and orientation. Charges can be shot into dielectrics by means of charged particle injection, and even by photon irradiation. Charges, once generated, dissipate slowly after the termination of radiation exposure, and in many cases persist over extremely long periods of time. Samples which are electrically neutral on a macroscopic scale, i.e., which contain no excess space charge and no polarization, might contain equal concentrations of separately trapped positive and negative carriers which at any time can be thermally activated and give rise to postirradiation conductivity.

New concepts on carrier excitation, trapping, and recombination developed by *Rose* in the early 1950s [4.9] allowed *Fowler* in 1956 [4.10] to develop a general self-consistent theory which covers radiation-induced conductivity, postirradiation currents, temperature effects, and thermal activation, which provides a working model for most recent work. A list of references on electrical irradiation effects can be found in the bibliographies of [4.11–13]. The following review emphasizes analytical and diagnostic methods and practical aspects related to radiation-induced conductivity, charge storage, radiation dosimetry, and electret formation.

4.1 Radiation-Induced Conductivity

4.1.1 Band-Gap Model

Charge storage and transport properties of irradiated dielectrics were discussed generally in terms of the theory of localized states developed for band-gap photoconductors by *Rose* [4.9] and successfully extended to dielectrics by *Fowler* [4.10]. The amorphous and microcrystalline dielectrics of concern here possess a short-range order. This allows the band-gap model (Fig. 4.1) to be applied over microscopic distances. Although some aspects of these theories might be controversial they still serve as a most useful working model covering a wide range of observations.

The lack of long-range order and the presence of impurity centers and structural disorder implies the appearance of localized states in the gap between conduction and valence band. Carriers in one of these states are considered to be bound. Electrons in shallow states, or shallow traps, are in thermal equilibrium with electrons in the conduction band and are likely to be thermally re-excited into that band. Electrons in ground states are more likely to capture a free hole and recombine with it than to be thermally re-excited into the conduction band; the recombination time might, however, be quite long. Thus ground states act as deep traps and recombination centers. A ground state which is neutral when occupied by an electron and positive when unoccupied is a donor state; an acceptor state is neutral when unoccupied by an electron and negatively charged when occupied.

Shallow traps and ground states are separated by a demarcation line which is the steady-state Fermi level under excitation. Electrons whose energy is that of the Fermi level have approximately the same probability of being excited into the conduction band or to recombine. The position of the Fermi level depends on the concentration of free electrons; if the latter increases due to increased excitation by radiation it moves toward the band edge of the conduction band. The situation for holes is analogous to that for electrons. As Fig. 4.1 shows, the ground states are situated between the Fermi level for holes and that for electrons.

Excitation by radiation generates an equal number of holes and electrons. The electrons are lifted into the conduction band and subsequently trapped in a shallow state or lost by falling into a ground state and eventual recombination. The inverse process occurs with the holes. Free electrons in the conduction band and free holes in the valence band can travel under the action of an electric field

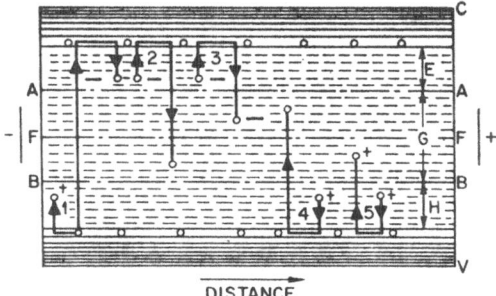

DISTANCE

Fig. 4.1. Band-gap model and trapping events. C: conduction band. V: valence band. F–F: dark Fermi level. A–A: electron Fermi level (under photoexcitation). B–B: hole Fermi level (under excitation). E: shallow electron traps. H: shallow hole traps. G: ground states (recombination centers, deep traps). 1: photoexcitation of molecule; hole is captured in neutral shallow trap; electron is lifted into conduction band and captured in shallow electron trap. 2: shallow trapped electron is thermally activated into conduction band and recombines in ground state. 3: shallow trapped electron, thermally activated into conduction band, is captured by deep trap. 4: shallow trapped hole receives electron from valence band and recombines in ground state. 5: shallow trapped hole receives electron from valence band and is captured in deep trap. The figure shows carrier movement under an applied electrical field [4.14]

and thus give rise to a conduction current. Their mobility in general is different. Charge transport might take place also by other processes such as hopping conduction, i.e., transfer of electrons between densely spaced localized states, field-assisted thermionic emission (Poole–Frenkel effect), and tunneling of electrons into the conduction band under the influence of high electric stress [4.14].

A theory of photoconductivity in insulators with an energy gap containing many discrete trap levels, with a random distribution in the energy densities and the electron and hole capture cross sections for the traps has been developed by *Simmons* and *Taylor* [4.15].

An approach to the problem of induced photoconductivity and charge transport, based on random walk theory, has been developed by *Scher* and *Montroll* [4.16]. A simplified treatment has been given by *Leal Ferreira* [4.17].

4.1.2 Radiation-Induced Conductivity

The conductivity of a dielectric exposed to ionizing radiation increases from a usually quite low dark value to a higher value which depends on the dose rate $\dot{D}$. The relation between the *steady-state radiation-induced conductivity* (RIC), g, and $\dot{D}$ follows for most dielectrics, in particular polymers, a power law [4.10]:

$$g = g_0 (\dot{D}/\dot{D}_0)^\Delta, \quad 0.5 \leqq \Delta \leqq 1, \tag{4.1}$$

where $\dot{D}_0$ is a reference value taken as 1 rad/s and g_0 the conductivity at that value. This relationship can be explained with the assumption of a continuous distribution of traps with energy. A uniform distribution gives $\Delta = 1$ and an exponential distribution gives $0.5 < \Delta < 1$. A single trapping level considered as the limiting case of an extremely steep distribution, gives $\Delta = 0.5$. For pulses in the ns range, whose duration is below the trapping time, g has been found to vary linearly with $\dot{D}$ [4.18].

After termination of the irradiation, the conductivity does not disappear instantaneously. Within a very short time (assumed to be less than 1 μs), it falls off to a fraction κg of its steady state value; subsequently a *delayed radiation-induced conductivity* (DRIC) persists which is given by an expression of the form [4.10],

$$g(t) = \kappa g (1 + t/\tau^*)^{-1}. \tag{4.2}$$

Here, τ^* is a time constant which depends on $g(0)$ and is usually smaller than 1 min.

The fast initial decay of g is due to a delay in reaching equilibrium between detrapping and recombination of carriers; subsequently carriers recombine at the same rate as they are released from traps. If the hyperbolic function for $g(t)$ was valid for infinite times, then the number of carriers initially would have had to be infinite. Thus for long times $g(t)$ must decrease faster than given by (4.2). Nevertheless, the delayed conductivity can be responsible for a considerable amount of charge transport.

The distance s_b over which a carrier drifts in the direction of the electric field before it disappears by recombination is called the *Schubweg*. If the Schubweg is small compared with the thickness of the sample, then the net volume charge density in the bulk of the material can be assumed to be zero. Charge neutrality does not exist in the presence of extrinsic excess charge due to injection caused by irradiation with charged particle beams. In this situation a proportion of carriers captured by ground states cannot recombine due to lack of partners.

Carriers, which at one temperature are trapped in ground states, and thus are considered to be deeply trapped, might have to be considered to be in shallow traps at a higher temperature.

Usually a single relaxation time is attributed to each energy level. This assumption might not necessarily be valid; dielectric relaxation measurements show that a single activation energy might give rise to a spectrum of relaxation times [4.19] extending over several decades.

Schottky barrier formation and other polarization effects can be significant when measurements on thin films are carried out with blocking electrodes, which are used in most cases (cf. Sect. 4.2.1).

4.1.3 Radiation Quantities

The *absorbed dose D* is defined as the mean energy imparted by ionizing radiation to the matter in a volume element divided by the mass of the matter in the volume element; $1\,\text{rad} = 10^{-2}\,\text{J kg}^{-1} = 10^2\,\text{erg g}^{-1} = 6.2 \times 10^{13}\,\text{eV g}^{-1}$ [4.20]. The new SI unit is the gray (Gy), and $1\,\text{Gy} = 1\,\text{J kg}^{-1} = 100\,\text{rad}$.

The *average absorbed dose rate* in a dielectric of area a [cm^2] and mass density δ [g cm^{-3}] irradiated normally to its surface with an electron beam of average range $\bar{r}$ [cm], accelerating voltage V [V] and injection current I_i [A] in the irradiated region $a\bar{r}$ is approximately given by

$$\dot{D} = I_i V \times 10^5 / a\bar{r}\delta \text{ rad s}^{-1}. \qquad (4.3)$$

If the dielectric is irradiated with gamma or X-rays with a linear attenuation coefficient μ_1 [cm^{-1}], $\bar{r}$ in (4.3) is replaced by $1/\mu_1$.

The *energy fluence* ψ is defined by dE_{fl}/da, where dE_{fl} is the sum of the energy of all the particles which enter a sphere of cross-sectional area da [4.20]. Exposure of a material to a unidirectional beam of radiation gives rise to an energy fluence ψ at the irradiated surface which for 1 R of ^{60}Co γ radiation of 1.26 MeV is 3.6×10^3 erg cm^{-2}.

The *energy flux density* or intensity I is given by $I = \dot{\psi}$.

The *exposure X* in roentgen (R) is defined by dQ/dm where dQ is the absolute value of the total charge of the ions of one sign produced in air when all the electrons (negatrons and positrons) liberated by photons in a volume element of air having mass dm are completely stopped in air [4.20], and 1 R = 2.58 $\times 10^{-4}$ C kg^{-1}. When photon energies lie above a few MeV or below a few keV exposure is difficult to measure [4.20]. The *exposure to energy fluence conversion factor* is $\psi/X = 86.9 (\mu_{en}/\delta)_{air}^{-1}$ erg cm^{-2} R^{-1} [4.23], where μ_{en} is the linear energy transfer coefficient.

The absorbed dose for a given exposure depends on the characteristics of the radiation and the properties of the irradiated material. Therefore no unique correspondence between the roentgen and rad units exists. The *energy for the generation of one free ion pair* by photons and electrons in air of NTP is 33.7 eV. From the definition of the roentgen it then follows that for air the conversion factor is 0.869 rad/R. The mean value of the absorbed dose in a material irradiated normal to its surface by a monoenergetic photon beam of energy fluence ψ is approximately given by

$$D = 0.869 \frac{(\mu_{en}/\delta)}{(\mu_{en}/\delta)_{air}} \frac{[1 - \exp(-\mu_l s)]}{\mu_l s} X, \qquad (4.4)$$

where μ_l and μ_{en} are respectively the linear attenuation coefficient and the linear energy transfer coefficient of the material; s is the penetration depth. For values of $\mu_l s$ much less than one, this gives the *absorbed-dose-to-exposure conversion* factor $D/X = 0.869 (\mu_{en}/\delta)/(\mu_{en}/\delta)_{air}$. For a gamma ray of energy of 1.26 MeV, which corresponds to ^{60}Co radiation, $(\mu_{en}/\delta)_{air} = 2.68 \times 10^{-2}$ cm^2 g^{-1} [4.20].

The *carrier production rate* p can be calculated when the energy G for the generation of a free electron–hole pair is known. Since 1 rad = $6.2 \times 10^{13} \delta$ eV cm^{-3}, a dose rate of 1 rad/s gives $6.2 \times 10^{13} \delta/G$ carriers per cm^3 and second. For a typical polymer, δ is of the order of 2 g cm^{-3}, while G

might vary between 100 and 1000eV [4.24]. This then gives approximately $p \simeq 1 \times 10^{11} \, \text{cm}^{-3} \text{s}^{-1}$ per rad/s, using the higher value for G.

The definition of μ_l and μ_{en} used in this section is as follows: The total collision cross section per atom, multiplied by the number of atoms per cm^3 of absorber, is the linear attenuation coefficient per centimeter of travel in the absorber. The total cross section of nonelastic collisions per atom, when multiplied by the number of atoms per cm^3, is the linear energy transfer coefficient. A collision is nonelastic when it results in the generation of a photoelectron, Compton electron, or electron–positron pair.

4.2 General Features of Excess-Charge Transport

For an understanding of the current curves observed during the release of space-charge layers it would be desirable to have a satisfactory analytical theory of the movement of excess charge in dielectrics. This theory, leading to nonlinear differential equations, is difficult; in addition the physical situation in most cases is complex and inadequately known. Thus interpretation frequently remains ambiguous. Here we shall limit ourselves to listing some general aspects, referring to other chapters in this book and to the primary literature for further information.

4.2.1 Electrode Effects

Dielectric effects depend strongly on the behavior of the electrodes. The usual vacuum-deposited metal electrode is a *blocking electrode* at low and intermediate field strengths. It prevents transfer of charge carriers from the electrode into the dielectric while it might accept carriers from the dielectric.

If the dielectric does not contain charge carriers the steady-state current in a dielectric bounded by blocking electrodes is zero. If the dielectric does contain carriers of both polarities whose mobilities are significantly different, electrode polarization due to the *formation of a Schottky layer* occurs at the electrode whose polarity is the same as that of the more mobile carriers [4.25]. The situation can be described in simple terms if the carriers of one polarity, say the electrons, have zero mobility and no further free-carrier generation occurs. An applied field removes the positive carriers from the anode (at $x = 0$). Since the electrode does not supply positive carriers, a negative space charge ϱ of thickness s_s is built up between $x = 0$ and $x = s_s$. After the space-charge layer has been fully developed, the voltage V initially applied along the entire length of the sample has contracted across s_s and the current has become zero. The extension of the layer is given by $s_s = (2\varepsilon V/\varrho)^{1/2}$ and is independent of the applied field. As an example, if the contraction of the immobile (trapped) carriers is $\varrho = 1.6 \times 10^{-4} \, \text{C cm}^{-3}$, $\varepsilon = 2 \times 10^{-13} \, \text{F cm}^{-1}$, and $V = 100 \, \text{V}$, one has $s_s = 5 \times 10^{-4} \, \text{cm}$, which is small compared to the thickness of many samples used in irradiation experiments.

Thus the use of blocking electrodes can prevent the complete removal of the mobile carriers from the dielectric whatever their mobility. The formation and dissipation of a Schottky barrier in a dielectric having an intrinsic conductivity was treated in [4.26]. *Goodman* and *Rose* [4.27] have discussed the double extraction of uniformly radiation-generated electron–hole pairs from insulators with blocking electrodes and have shown how this also leads to the formation of depletion layers in the electrode regions and how it affects the relation between current and dose rate.

A nonconducting interface of finite thickness between dielectric and electrode may produce *barrier polarization*. Formation of a double layer of positive and negative carriers, of molecular dimensions, can, however, not be distinguished by external measurements from charge neutralization at the electrode, and does not generate a blocking effect.

Ohmic (or *neutral*) *electrodes* supply charge carriers when the field strength at the electrode surface is finite. Under irradiation a blocking electrode may become ohmic.

An *injecting electrode*, whose behavior is similar to a glow cathode, spontaneously injects carriers into the dielectric at zero surface field. The space-charge cloud in front of the electrode eventually reduces the emission to zero unless it is dissipated by an externally applied field. The surface condition for an injecting electrode at $x = 0$ is $E(0, t) = 0$ and $\varrho(0) = \pm \infty$, the polarity depending on that of the injected carrier [4.28–30]. This classification is, of course, an idealization; in practice, electrode behavior is more complex [4.31]. The theory of the formation of space-charge layers near injecting electrodes has been dealt with by *Popescu* and *Henisch* for relaxation semiconductors [4.32] and semi-insulators [4.33].

Not all electrodes need to be metallic or electrolytic. An electron beam penetrating part of the way into a dielectric generates a *virtual electrode* [4.34–36]. It produces an electron–hole plasma within the irradiated region from which carriers can move through the end plane of the beam into the nonirradiated region.

4.2.2 The Zero-Field Theorem

Important conclusions can be drawn from a theorem first stated by *Lindmayer* [4.37] and derived from first principles by *Gross* and *Perlman* [4.38]. Assuming a plane-parallel arrangement, we consider the movement of charge along a constant-field plane $x(t)$ defined by $E[x(t)] = \mathrm{const}$, $dE/dt = 0$. Thus

$$I(t) = I_c[x(t), t] + a\varepsilon\partial E[x(t), t]/\partial t, \tag{4.5}$$

where I_c is the conduction current, and a the electrode area. Since $\partial E/\partial t = dE/dt - (\partial E/\partial x)(dx/dt)$, it follows from Poisson's equation that $I = I_c - a\varrho dx/dt$. A zero field plane is defined by the condition $E[x^*(t), t] = 0$. For any such plane where diffusion can be neglected and no other field-independent currents exist, one has $I_c[x^*(t), t] = 0$ and

$$I(t) = - a\varrho(x^*, t)\, dx^*/dt. \tag{4.6}$$

The current is, therefore, given by the charge-density at the zero-field plane times the velocity of displacement of this plane; x^* can be obtained from the equation

$$\varepsilon V/s + \int_0^s \varrho(x, t)(1 - x/s)\,dx = - \int_0^{x^*} \varrho(x, t)\,dx, \tag{4.7}$$

where V is the applied voltage.

Equation (4.6) includes the displacement current and makes no assumption about the space and time dependence of the charge distribution. In the presence of a persistent polarization P_p, i.e., a component of the dielectric volume polarization that lags the field E, a term $adP_p(x^*, t)/dt$ has to be added in (4.6) [4.39]. Field-independent currents can be currents due to the injection of fast particles, in particular, electrons, and photon-radiation driven currents (Compton and photo-Compton currents) [4.38].

If $V = 0$ (short circuit) and $\varrho(x, t) = \varrho_a(x)\varrho_b(t)$, then $x^* = $ const and therefore $I = 0$. This shows that the detrapping of an arbitrary carrier distribution from a single trapping level gives no external short-circuit current [4.41]. The assumption that carriers initially trapped to the "left" of the zero-field plane drift to the "left" electrode and those initially trapped to the "right" of the zero-field plane drift to the "right" electrode, by the same token, leads to zero external current [4.41].

A stationary zero-field plane is compatible with a finite external current only if it coincides with an injecting electrode where $E(x^*) = 0$ and $\varrho(x^*) = \pm \infty$.

In open circuit all zero-field planes remain stationary because $I = 0$. If a dielectric has been charged by radiation (in short circuit or biased by an applied voltage) and subsequently is open-circuited, all zero-field planes remain stationary. Since $I_c(x^*) = 0$, these planes behave like impermeable barriers for charge transport. Charge contained in a region situated between a zero-field plane and an electrode or between two such planes can decay only by conduction within each region.

4.2.3 Ultimate Charge

The current generated by the thermally stimulated depolarization of a persistent volume polarization P_p is a *total differential* [4.41, 42]. Therefore the *ultimate released charge* $\int_0^\infty I[t, T(t)]\,dt$ is the same for all heating functions. Therefore a *principle of charge invariance* is valid [4.42]. The current is *not* a total differential for a space-charge polarization [4.43] or for a Wagner-type layer dielectric

[4.44, 45]. Therefore in these cases the ultimate values of the released charge are different for different heating rates. Frequently the differences are small [4.46]. Therefore the principle of charge invariance might remain approximately valid. The invariance of the ultimate charge is not obvious because current–time curves for different heating rates might differ strongly from each other.

The principle is strictly valid if charge transfer between one of the electrodes and the dielectric is prevented. This can be achieved by inserting a thin air gap between the dielectric and one of the electrodes which prevents any charge transfer between the dielectric and the electrode as long as the interface field is moderate [4.47]. Then the whole charge contained in the dielectric is transferred to the contacting electrode. The technique is used for open-circuit thermally stimulated current (TSC) measurements. A nonconducting insert is also generated by heating the dielectric in a strong temperature gradient, with one surface kept at a sufficiently low temperature, at which conductivity and carrier mobility are low.

4.2.4 Carrier Mobilities

The drift velocity of free carriers in an applied field is $v = \mu_0 E$ where μ_0 is the *free mobility*, i.e., the mobility of electrons in the conduction band or of holes in the valence band. In thermal equilibrium the relation between the free and trapped carrier concentrations n and m of either polarity is constant and $n = \theta m$. This allows one to introduce a *trap-modulated mobility* μ defined by

$$\mu = \mu_0 \varrho_f / \varrho = \mu_0 \theta / (1 + \theta); \varrho = \varrho_f + \varrho_t, \tag{4.8}$$

where $\varrho_f = en$ and $\varrho_t = em$ are, respectively, the free and trapped charge densities. A relation of this form, with a different value of θ, applies separately to holes and electrons. With the introduction of μ, Maxwell's equations can be written in their conventional form [4.29].

The free mobility is a material constant which is not sensitive to temperature variations. Observed values for many insulators vary between 10^{-3} and $1 \, \mathrm{cm^2 \, V^{-1} \, s^{-1}}$. The trap-modulated mobility depends strongly on temperature. For a single electron trapping level one has $n/m = (N/M) \exp(-U/kT)$ where N is the degeneracy (number of vacancies) of the conduction level, M the number of traps, and U the activation energy. This leads for a single trapping level to the expression $\mu = \mu_0 [1 + (M/N) \exp(U/kT)]^{-1}$ [4.48]. For high temperatures, one has $\mu \simeq \mu_0$, and for low temperatures, $\mu \simeq \mu_0 N/M \exp(-U/kT)$. With $\mu_0 = 10^{-3} \, \mathrm{cm^2 \, V^{-1} \, s^{-1}}$, $N \simeq 10^{19} \, \mathrm{cm^{-3}}$, $M \simeq 10^{17} \, \mathrm{cm^{-3}}$, $U \sim 0.5 \, \mathrm{eV}$, and $kT = 2.6 \times 10^{-2} \, \mathrm{eV}$ (room temperature), one has $\mu \sim 5 \times 10^{-10} \, \mathrm{cm^2 \, V^{-1} \, s^{-1}}$ which is of the order of magnitude reported for many polymers [4.49]. Under most experimental conditions the exponential term predominates.

An approximate lower limit for the *free mobility* can be obtained from *Langevin's relation* [4.50, 51] $b/\mu_0 = e/\varepsilon$ where b is the recombination coefficient [cm^3 s^{-1}] between carriers of opposite polarity. The recombination coefficient is inversely proportional to the average molecular thermal velocity $\bar{v}_t$ and directly proportional to the molecular cross section. The latter is equal to or greater than the Coulomb cross section $\bar{a} \simeq 7 \times 10^{-17}$ cm^2. At room temperature one has $\bar{v}_t = 10^7$ cm s^{-1}. This gives $b \gtrsim \bar{a}/\bar{v}_t = 7 \times 10^{-10}$ cm^3 s^{-1}. The dielectric constant of most polymers and many other good insulators is of the order of 2×10^{-13} F cm^{-1}. Substitution of these values into Langevin's equation gives $\mu_0 \gtrsim 10^{-3}$ cm^2 V^{-1} s^{-1}.

An approximate value for the *trap-modulated mobility* in terms of the radiation-induced conductivity can also be obtained using *Fowler's* recombination law [4.10]. The radiation-induced mobility might be written $g = en\mu_0 = e(n+m)\mu$. The free-carrier production rate p and conductivity g are related by the *recombination law* $p = bn(n+m)$ [4.52]; this relation is based on the postulate of charge neutrality (under steady-state conditions) and the assumption that recombination occurs only between free carriers of one polarity and trapped carriers of the opposite polarity. Combining this equation with Langevin's equation, one obtains $\mu = g^2/ep\varepsilon$.

Intuitively, *Langevin's relation* can be obtained as follows. The lifetime of a free carrier in the conduction band is $\tau_0 = 1/bm$, m being the concentration of trapped carriers of the opposite polarity. The "relaxation" time constant $\tau = \varepsilon/g$ is greater than τ_0 by a factor m/n since the loss of carriers by recombination is compensated by the lifting of detrapped carriers into the conduction band. Introducing the expression for g, one has $\tau = (1/bm)(m/n) = \varepsilon/e\mu_0 n$ or $b/\mu_0 = e/\varepsilon$.

Fowler's recombination law has been derived for medium excitation [4.9, 10]. The concentration of ground states n_g containing electrons is great compared with that of those containing holes; holes generated by irradiation are immediately captured in ground states, and the number of ground states filled by such holes is great compared with that of those already filled without irradiation, i.e., in the dark. Then charge neutrality demands that the steady-state concentration p_g of holes in ground states is the same as the concentration of free and shallow-trapped electrons, that is $p_g = n + m$. To recombine, a trapped electron must first be lifted into the conduction band. This gives $p = bn(n+m)$. Similar reasoning can be applied to a situation where the holes are the mobile carriers.

Since the trap-modulated mobility depends on g and p, it must also depend on the dose rate $\dot{D}$. If the RIC is given by (4.1), one has $\mu = \mu(\dot{D}_0) \times (\dot{D}/\dot{D}_0)^{2\Delta-1}$. Therefore μ increases with $\dot{D}$ but at a lower rate than the RIC [4.52]. The values and methods of determination of mobilities remain still a controversial problem.

The *dark conductivity* of good insulators at room temperature is of the order of 10^{-20} (Ω cm)$^{-1}$ or less; it follows a similar activation law to that of the mobility.

The component of the drift velocity of the free carriers in the direction of the applied field can be written $v = \mu_0 E = s_b/\tau_0$. Here τ_0 is the free carrier lifetime, i.e., the total time spent by a carrier in the conduction band before recombination; s_b is the *Schubweg* or drift distance in the direction of the field of the free carriers. As long as $s_b/s \ll 1$, one has $g = ep(\mu_0\tau_0)$. Thus the *mobility–lifetime product* is given by $\mu_0\tau_0 = g/ep$, and $\mu_0 = g/ep\tau_0$. The recombination lifetime is also given by

$\tau_0 \simeq 1/bm$, where m is the occupancy level of traps filled with carriers of the opposite polarity. Thus one finds $m \simeq 1/\tau_0 b$ and $n = m\mu/\mu_0$.

The theory is based on the assumption that the *recombination lifetime* is *small compared with the carrier transit time* and therefore the Schubweg is small compared with the sample thickness. This assumption can be checked by a measurement of the radiation-induced current I. The movement of a carrier with charge e over a distance s_b normal to the electrodes of a capacitor with plane-parallel geometry gives an external charge es_b/s [4.53]. If the carrier generation rate is p over a volume as, one has $I = sa\,eps_b/s$ and $s_b/s = I/asep$. This relation allows one to estimate s_b when the carrier generation rate p is approximately known.

These relations show that a considerable amount of information about radiation properties of irradiated materials can be obtained from experimentally determined values of the steady-state RIC and total current as functions of irradiation dose rate $\dot{D}$. Uncertainty remains about the energy needed for generation of a free carrier pair.

4.2.5 Transport Equation

Frequently one considers the transport of excess charge of one polarity in a medium which has a conductivity g due to thermally stimulated or radiation-excited carriers. If the conduction current due to the transport of excess charge and that due to the conductivity g can be superposed without interaction [4.46, 54], the net conduction current for planar geometry is given by [4.22, 46]

$$I_c = agE + a\mu\varrho E, \tag{4.9}$$

where again $\varrho = \varrho_f + \varrho_t$; with the use of Poisson's equation $\varepsilon\partial E/\partial x = \varrho$ one has

$$I_c = agE + a(\mu\varepsilon/2)\partial E^2/\partial x. \tag{4.10}$$

Stringent conditions for the validity of (4.9) have been established [4.55].

In particular cases the charge decay can be calculated analytically. Differentiation of (4.10) with respect to x gives, with the aid of Poisson's equation and the equation of continuity,

$$\partial I_c/\partial x = -a\partial\varrho/\partial t = a\mu\varrho^2/\varepsilon + ag\varrho/\varepsilon + aE[(\partial g/\partial x + \varrho\partial\mu/\partial x + \mu\partial\varrho/\partial x)]. \tag{4.11}$$

In open circuit, with $I = 0$, (4.6) shows that $(d\varrho/dt)_{x^*} = (\partial\varrho/\partial t)_{x^*}$. The last term in (4.11) disappears at x^* because $E(x^*) = 0$. Then (4.11) is an ordinary differential equation for ϱ. If g and μ do not depend on t, although they might depend on x, the solution is

$$\varrho[x^*, t] = \frac{(g/\mu)\varrho_0}{(\varrho_0 + g/\mu)\exp(tg/\varepsilon) - \varrho_0}, \tag{4.12}$$

where $\varrho_0 = \varrho(x^*, 0)$ and g and μ are taken at x^*. The results can be generalized to include the case where g and/or μ are functions of t.

Equation (4.12) gives the charge decay in an open-circuit sample at the zero-field plane for an arbitrary charge distribution. Expansion for $g \to 0$ gives $\varrho(x^*, t) = \varrho_0 [1 + (\mu/\varepsilon)\varrho_0 t]^{-1}$.

For a *box distribution*, ϱ is independent of x within a region $x_a(t) < x < x_b(t)$ and zero outside this region. If in addition g and μ are independent of x the last term in (4.11) disappears everywhere within the box and $\partial\varrho/\partial t = d\varrho/dt$ in the same interval. The resulting differential equation is the same as that which led to (4.12). Therefore as long as neither edge of the distribution has reached one of the electrodes, $\varrho(x, t)$ is given by (4.12) for $x_a < x < x_b$, and zero outside, in short circuit as well as in open circuit.

The box distribution is important because it is a rigorous solution of the basic equations [4.56], and because an arbitrary distribution always tends to approach a box distribution within a time usually short compared to the transit time [4.57]. After both the leading and the trailing edge of a box distribution have reached the corresponding electrodes the external current is zero. When the conductivity g is finite, a space-charge distribution may have been completely dissipated before any of its edges has reached an electrode.

The thermally activated dissipation of an excess space charge is described by (4.9) with g and μ being functions of temperature and thus of time. If the temperature dependence of both parameters is the same one can write $\mu = \mu^* f[T(t)]$ and $g = g^* f[T(t)]$ where μ^* and g^* now are independent of temperature. Introducing (4.9) into the current equation (4.5), dividing by $f[T(t)]$, and introducing a reduced time defined by $dt' = f[T(t)]dt$, one obtains for the reduced current $I^*(t') = I(t)/f[T(t)]$ an equation in t' which is formally identical with the original equation for $I(t)$, but with μ and g replaced by μ^* and g^*. Thus, when the solution of the isothermal problem is known, the solution of the nonisothermal problem can be obtained by a *transformation of variables and amplitudes*. It also follows that under these conditions, and always when the term gE can be neglected, the principle of charge invariance applies rigorously [4.57].

The case of transport of excess charge of one polarity in a shorted dielectric, without deep trapping, has been discussed by *Camargo* and *Ferreira* [4.57] under the assumption that the conduction current is given by $\mu\varrho E$, with $g = 0$. The charge distribution at zero time was assumed as given (initial condition). For this case an *analytic solution* could be obtained.

A *numerical* treatment of the problem of unipolar conduction for general initial and boundary conditions has been given by *Zahn* et al. [4.57].

The theory based on (4.8) does not consider *deep trapping* in ground states. If this is to be considered, equations become more complex. If the material contains a space charge of one polarity only and has no intrinsic or induced conductivity, the conduction current is $I_c = \mu_0 \varrho_f E$; Poisson's equation is written $\varepsilon \partial E/\partial x = \varrho_f + \varrho_t$. An additional differential equation is needed which gives the relation between ϱ_f and ϱ_t. If the material contains carriers of both polarities, recombination terms have to be included. Rate equations of this type have been discussed by *van Lint* [4.18] and applied to the discussion of instantaneous and delayed conductivity induced by radiation pulses of high intensity (10^7–10^{10} rad s^{-1}) and short duration (in the nanosecond range).

4.2.6 "Floating" Charge Layer

In a *homogeneous* short-circuited dielectric, charge migrates predominantly to the nearest electrode. This can be proved for the case of a "floating" excess charge layer (i.e., a layer which touches neither electrode) of total charge q [4.58]. When the conductivity term gE can be neglected and as long as the charge distribution does not reach one or the other electrode, the current increases exponentially as $I(t) = I(0)\exp(t/\tau)$ where $1/\tau = q\mu/a\varepsilon s$ and

$$I(0) = (q/\tau)\{[\bar{r}(0)/s] - 1/2\}, \tag{4.13}$$

where $\bar{r}(0)$ is the initial depth of the centroid of the charge distribution. Therefore if $\bar{r}(0)/s > 1/2$ the current is positive; if $\bar{r}(0)/s < 1/2$ it is negative. *Charge moves to the nearest electrode.* The time constant is inversely proportional to mobility and total charge. Neither of these results depends on the shape of the charge distribution. For any distribution, whose centroid is halfway between electrodes, the current is zero at least until one of the edges reaches one of the electrodes.

During the thermally stimulated release of a floating charge layer one has $\mu \simeq \mu(T_0)\exp(-U/kT)$ and $\tau \simeq \tau(T_0)\exp[U/kT(t)]$. If the heating rate $\beta = dT/dt$ is constant and heating starts from a temperature T_0 one has [4.58]

$$I(T) = I(T_0)\exp\left[-\frac{U}{k}\left(\frac{1}{T} - \frac{1}{T_0}\right) + \int_{T_0}^{T}\exp\left\{-\frac{U}{k}\left(\frac{1}{T} - \frac{1}{T_0}\right)\right\}dT/\beta\tau(T_0)\right] \tag{4.14}$$

Equation (4.14) has the same form as an approximate expression derived for the thermal release of space charge trapped near one of the electrodes [4.59] or for that of a box distribution [4.60]. It shows that the *initial current in a homogeneous dielectric* is independent of the shape of the charge distribution as long as no part of the charge reaches one of the electrodes. Equation (4.14) is also identical with the expression for thermally-stimulated currents caused by the decay of a dipole polarization [4.61]. Therefore the initial current slope does not discriminate between different types of charge and polarization decay.

The long-lasting DRIC in dielectrics irradiated with nonpenetrating electron beams generates a *heterogeneity* in the samples which obliges one to revise some previous views, as pointed out by *Sessler* and *West* [4.62].

4.2.7 Diffusion

The previous discussion is valid if diffusion can be neglected. Otherwise the diffusion current $aD_F\partial\varrho_f/\partial x$ has to be included in the transport equation, D_F being the free diffusion coefficient relating to the free carriers, so that

$$I_c = agE + a\mu_0\varrho_f E - aD_F \partial\varrho_f/\partial x . \tag{4.15}$$

An approximate value for the *ratio f between the diffusion term and the field-induced transport term* can be obtained [4.63] by means of *Einstein's relation* $D_F = \mu_0(k/e)T$ cm$^2 \cdot$s^{-1} where $k/e = 8.6 \times 10^{-5}$ V is Boltzmann's constant. Then one has $f = kT(\partial\varrho_f/\partial x)/\varrho_f E$. Consider a space-charge layer of thickness d and charge per unit area σ_f. Putting $\partial\varrho_f/\partial x = \sigma_f$ and $\varrho_f \simeq \sigma_f d$ one has $f \simeq kT/Ed$ within the charge layer. As an example we might consider a charge layer of thickness $d = 10^{-4}$ cm. Since at room temperature, $kT/e = 2.6 \times 10^{-2}$ V one finds $f = 2.6 \times 10^2/E$. Thus, provided E exceeds the order of 10^3 V$\cdot$cm, the contribution of diffusion to the local current becomes small.

The external current for a sample which is shorted or is biased by a constant applied voltage is [4.38]

$$I = \frac{1}{s} \int_0^s I_c dx . \tag{4.16}$$

The contribution of the diffusion current to the external current is therefore given by

$$I_D = -a(D_F/s)[\varrho_f(s, t) - \varrho_f(0, t)] . \tag{4.17}$$

For a floating charge distribution (which does not touch the electrodes), $I_D = 0$; the same happens if the space charge densities in both electrode regions are the same.

4.3 Electron Beam Charging

It has been known for a considerable time that injection of electrons into many dielectrics by means of electron beams leads to the formation of negative space-charge layers [4.64]. The presence of these charges might be indicated by breakdown effects followed by the appearance of a Lichtenberg figure whose location corresponds to the depth of the space charge [4.22, 65], by thermal activation [4.64, 66], and by the electret effect [4.67] which has been found to be particularly persistent in many fluorocarbons. It was this last observation that lends a special practical importance to this method of electret formation.

4.3.1 Range–Energy Relations for Electrons

Electrons passing through matter undergo frequent elastic collisions which cause range straggling. Therefore the number transmission curve, or integral range distribution curve, is S shaped, as schematically shown in Fig. 4.2, inset. The

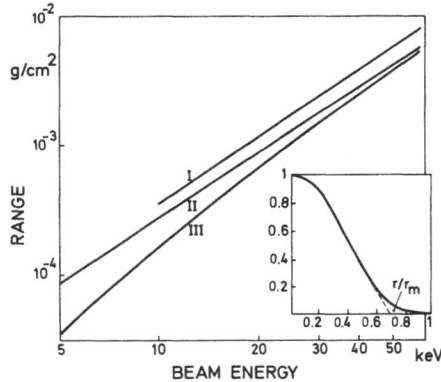

Fig. 4.2. Range–energy relations for 50 keV electrons – theoretical I: csda range for Teflon (*Pages* et al. [4.70]). II: Practical range for air and other light materials (*Gledhill* [4.69]). III: Practical range (*Katz–Penfold* [4.68]). *Inset*: Transmission curve for 0.05 MeV electrons in Al [4.71]

practical range r_p is obtained by extrapolation of the linear part of the absorption curve. For elements beyond ^{5}Be and energies up to 10 MeV it is given by the *Katz–Penfold* relation $r_p/E_0 = 0.412 E_0^n$, where $n = 0.265 - 0.0954 \log_e E_0$, with E_0 given in MeV and r_p in g cm^{-2} [4.68]. For values below 0.05 MeV this relation gives values which are too low. A better approximation then is given by the *Gledhill relation*

$$\log r_p = -5.100 + 1.358 \log E_0 + 0.215 \log^2 E_0 - 0.043 \log^3 E_0 ,$$

with r_p in mg cm^{-2} and E_0 in keV [4.69]. While this relation was developed for measurements in air, it holds for other elements with low atomic number [4.69]. The *average range* is defined by the relation

$$\bar{r}_v = \int_0^{r_m} r(x)dx/r_m ,$$

where r_m is the *maximum range*, i.e., the depth where the absorption curve becomes indistinguishable from the background, electrons becoming thermalized. The latter is approximately equal to the *range in the continuous-slowing-down approximation*,

$$r_{csda} = \int_0^{E_0} (dE/ds)^{-1}dE .$$

This is the path length which a particle would travel in the course of slowing down, in an unbounded medium, from initial to zero energy, if its rate of energy loss along the entire track length were always equal to the mean rate of energy loss. It differs from r_p and r_m which are defined with reference to transmission through a plane-parallel absorber. Extensive tables of r_{csda} are available [4.70].

Backscattering occurs at the surface of incidence of the electron beam. The *backscattering coefficient* increases with increasing atomic number of the irradiated material and with decreasing energy of the electrons [4.71]. For Al and

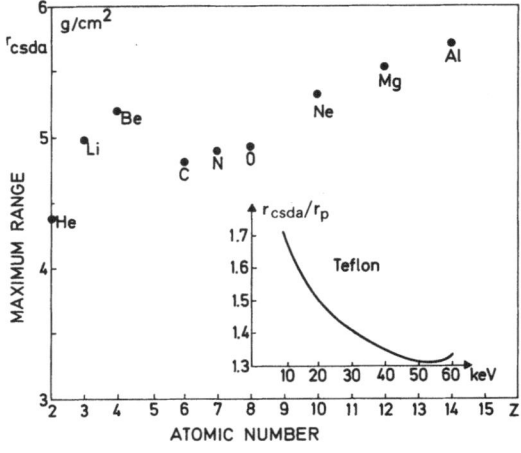

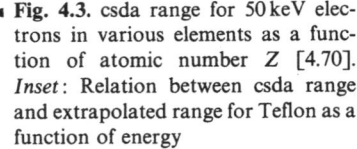

◄ **Fig. 4.3.** csda range for 50 keV electrons in various elements as a function of atomic number Z [4.70]. *Inset*: Relation between csda range and extrapolated range for Teflon as a function of energy

Fig. 4.4. Range relations for electrons – experimental. I: Number transmission curve for 0.159 MeV electrons in Al [4.75]. II: Number transmission curves for 1 MeV electrons in Al [4.71]. III: Charge transmission curve for 2 MeV electrons in plexiglas (Gross-Wright [4.72]). IV: Energy deposition (ionization) curve for 2 MeV electrons in Al [4.73]. V: Energy deposition (dose rate) curve for 2 MeV electrons in polyethylene [4.74]. VI: Charge deposition curve for 3 MeV electrons in plexiglas (*Gross-Wright* [4.72]). VII: Charge deposition curve for 2 MeV electrons in polyethylene [4.74]

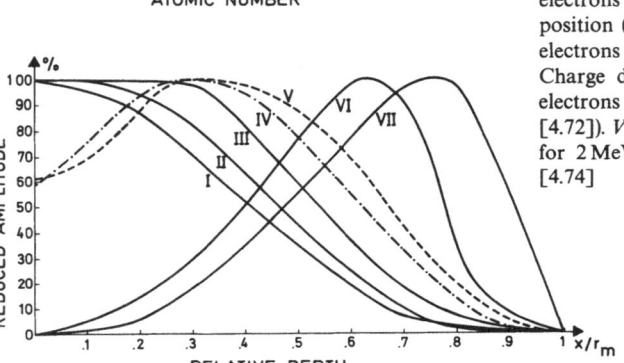

electrons of less than 0.1 MeV, about 13% of the incoming electrons are backscattered.

The *charge transmission* curve reflects the number transmission curve. The *charge deposition* curve reflects the derivative of the number transmission (or differential range distribution) curve. Both can be measured directly [4.72]. The *energy deposition* curve can be measured by ionization [4.73] or by dosimetric methods [4.74]. Transmission curves for many materials have been discussed in detail by *Berger* and *Seltzer* [4.71].

Figure 4.2 shows theoretical values for the extrapolated range r_p and for r_{csda}, confirming the inadequacy of the *Katz–Penfold* relation in the low energy range. Figure 4.3 shows theoretical values of the csda range for elements up to aluminum and an electron energy of 50 keV. Values increase with increasing atomic number Z. The lowest value, of 1.98×10^{-3} g cm^{-2}, is found for hydrogen, in view of the low mass of the hydrogen nucleus. The inset of the figure shows the relation between r_{csda} and r_p as a function of electron energy for Teflon. Hydrogenated materials, in particular many polymers, also show low values of r_{csda}. Table 4.1 gives numerical values of r_{csda} for light materials. Finally, Fig. 4.4 shows

Table 4.1. Ranges of 50 keV electrons in light materials in the continuous slowing down approximation [4.70]

Material	r_{csda} [g cm^{-2}]
Hydrogen H	1.98×10^{-3}
Polyethylene $(CH_2)_n$	3.315×10^{-3}
Ethylene C_2H_4	3.99×10^{-3}
Water H_2O	4.21×10^{-3}
Polystyrene $(C_8H_8O_2)_n$	4.33×10^{-3}
Lucite $(C_5H_8O_2)_n$	4.35×10^{-3}
Stilbene $C_{14}H_{12}$	4.39×10^{-3}
Anthracene $C_{14}H_{10}$	4.461×10^{-3}
Air	4.922×10^{-3}
Teflon $(C_2F_4)_n$	5.26×10^{-3}
Aluminium Al	5.714×10^{-3}
Standard emulsion	6.950×10^{-3}

characteristic absorption, charge deposition, and energy deposition (dose-depth) curves measured by different methods.

4.3.2 Charge Diagnostics with the Split Faraday Cup

To understand the processes associated with trapping, detrapping, and storage of charges generated by electron injection it is desirable to investigate charge buildup and decay *during* as well as *after* irradiation. This can be done with an arrangement first used by *Spear* [4.76] and now termed the "split Faraday cup" (Fig. 4.5) [4.77]. A film of the dielectric to be investigated is sandwiched between two electrodes, usually vacuum-evaporated Al layers. The electrodes are thin enough (up to 500 Å) not to absorb any significant fraction of the incident electron beam. Both electrodes are *monitored separately*. The sample is irradiated over the front electrode by an electron beam whose range can be varied from a very small value until it becomes equal to or exceeds the thickness of the sample. An auxiliary "catcher" electrode might be placed behind the sample during irradiation for measurement of transmitted primary electrons or it might be placed in front of the sample after irradiation for measurement of electron emission by the sample.

In the *current mode* each electrode is connected to ground or to a low impedance voltage source through meters whose impedances are low compared to that of the sample. As long as no primary electrons are transmitted through the rear electrode, one has under steady-state and transient conditions

$$I_i = I_1 + I_2 , \tag{4.18}$$

where I_i is the injection (or radiation-driven) current, I_1 the current flowing from the front electrode to ground, and I_2 the current flowing from the rear electrode to ground. During irradiation, all currents are nonzero; in the absence of irradiation, $I_i = 0$ and $I_1 = -I_2$. The injection current is that part of the beam current I_b which

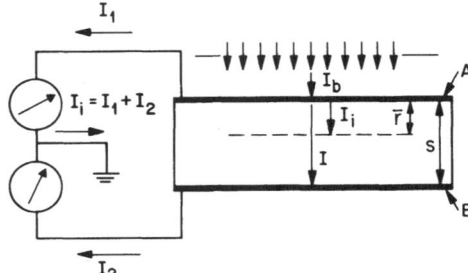

Fig. 4.5. Split Faraday cup. A: front electrode. B: rear electrode. I_b: beam current. I_i: injection current. I_1: front current. I_2: rear current. I: dielectric current. $\bar{r}$: centroid of charge distribution. s: sample thickness [4.77]

actually enters the front electrode. It differs from I_b by the backscatter I_{sc}, and $I_i = I_b - I_{sc}$. The method allows one to simultaneously determine I_i and the dielectric current $I(t) = I_2(t)$ without the need of a separate measurement of I_i.

For the *voltage mode* one has two alternatives.

(I) The *rear* electrode is connected to ground through a current meter, while the voltage of the floating front electrode is monitored. This circuit allows one to measure directly the average electron range while $I_2 = I_i$.

(II) The *front* electrode is connected to ground through a current meter and the voltage of the floating rear electrode is monitored while $I_1 = I_i$ allowing measurement of RIC. Charge diagnostics by these methods are discussed in [4.78].

4.3.3 The Threshold Effect

Numerous authors have investigated the steady-state currents which are induced in short-circuited or externally biased dielectrics by electron beams whose extrapolated range is smaller than the sample thickness. Experiments cover a variety of materials such as silica, MgF_2, and mica [4.79], As_2S_3 and Al_2O_3 [4.80], Sb_2, mica, and Pyrex [4.76], ZnS [4.81], fused natural quartz [4.82], SiO_2 [4.83], and Teflon [4.84]. Results show that significant steady-state currents are induced only when the energy of the incident beam exceeds a threshold value which is only weakly dependent of the applied bias and corresponds to a maximum range considerably smaller than the sample thickness.

Figure 4.6 shows representative results obtained by *Spear* [4.76] for a mica specimen irradiated with electrons of energies up to 45 keV. It gives currents as a function of electron energy for applied fields up to 1.05×10^6 V cm^{-1}. Positive voltage means that the rear electrode is positively biased. The relation between beam energy V_b in kV and extrapolated electron range r_p in cm was found to follow closely the quadratic *Thomson-Widdington* law [4.85], $V_b^2 = c \delta r_p$, with $c = 5.5 \times 10^5$ kV^2 g^{-1} cm^2 for mica ($\delta = 3$ g cm^{-3}). The extrapolated range reaches half the sample thickness for a beam energy of 19 kV, and it becomes equal to the sample thickness for a beam energy of 27 keV. Using a separate catcher electrode placed behind the sample, *Spear* could show that electron transmission through the rear electrode sets in only after $r_p \gtrless s$, while steady-state currents through the sample begin to flow before this happens.

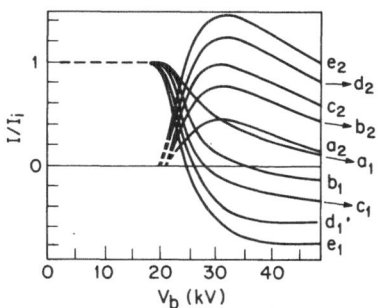

Fig. 4.6. Threshold effect, experimental: Space-charge-limited steady-state currents through 4.3×10^{-4} cm thick electron-irradiated mica foil ($\delta = 3\,\mathrm{g\,cm^{-3}}$) as a function of beam energy for different applied fields. Beam current density 7×10^{-8} A cm^{-2}; applied field: a–a: 0 V; b–b: 2.8×10^5 V cm^{-1}; c–c: 4.7×10^5 V cm^{-1}. d–d: 8.1×10^5 V cm^{-1}; e–e: 1.05×10^6 V cm^{-1}. Rear electrode biased positively. Subscript 1 indicates front currents, subscript 2 indicates rear currents. [4.76]

The absence of a steady-state current for $r_p < s$ is what one would expect at first sight. All dielectrics mentioned are good insulators at room temperature. As long as $r_p < s$, the nonirradiated region (NIR) should be blocking and the steady-state current be zero, assuming that the dark conductivity is small compared with the radiation-induced conductivity. Within the irradiation region (IR) the electron beam generates an approximately uniform RIC; in the steady-state, charge is released through the front electrode at the same rate as it is injected.

The experimental evidence proves however that this simple picture is valid only for low electron penetration. For electron energies above the threshold value the NIR is no longer blocking. The effect was interpreted by *Spear* [4.76] as being due to the superposition of the internal space-charge field and the externally applied field; at some level the space-charge field should change sign reducing currents in the IR and increasing them in the NIR. Conduction currents through the NIR would be due to carriers drawn from the IR into the NIR by the electric field. The theory of space-charge-limited currents provides a quantitative interpretation for the effect.

4.3.4 The Electron Beam as a Virtual Electrode

The use of low energy electron beams for creation of a *virtual electrode* close to the electrode of incidence was suggested by several authors [4.83, 86]. The IR represents a deposit of positive and negative carriers. Since carriers of either sign might be drawn into the NIR depending on the polarity and the value of the applied voltage, the virtual electrode might function as an anode as well as a cathode. This electrode effect is important in the practice of electret production by charge injection.

The present discussion follows a treatment by *Nunes de Oliveira* and *Gross* [4.36]. It is assumed that the IR has a uniform induced conductivity and that no space charge exists between $0 \leqq x < \bar{r}$ while a negative planar charge might exist at $x = \bar{r}$ which is taken as the depth of the electrode. In short circuit, the electron-irradiated dielectric contains only negative excess charges. The zero-field plane is located at the depth $\bar{r}_+$ of the virtual electrode. Thus one has $E(\bar{r}_+) = 0$, where $E(\bar{r}_+) = \lim_{\varepsilon \to 0} E(\bar{r} + \varepsilon)$. To give a finite injection current $\mu \varrho E$, the space-charge density at $\bar{r}_+$ must be infinite. Therefore one has the boundary conditions for an injecting cathode (Sect. 4.2.1). At the depth $\bar{r}$ the electric field changes discontinuously from a finite value in the IR to zero. Therefore a negative planar charge of density σ is generated by the electron beam.

When a positive voltage V is applied to the rear electrode, the value of the field E_a in the IR is decreased because in this region the applied field is in opposition to the space-charge field. The field E_b in the NIR, where internal and applied fields have the same direction, is increased, thus σ is decreased accordingly.

When V is increased, a critical positive voltage V_c is reached where the external and space-charge fields in the IR are equal and opposite and the ohmic current component disappears. Then $I = I_i$ and $E_a = 0$; the discontinuity of the field at $x = \bar{r}$, and with it the planar charge, have disappeared.

If V is increased beyond V_c, the external field exceeds everywhere the space-charge field and no zero-field plane exists in the dielectric. The field at $x = \bar{r}$ remains continuous, so $E(\bar{r}) = E(\bar{r}_+)$; σ remains 0.

If the sample is biased negatively, i.e., for $V < 0$, the value of the field in the IR is increased while that in the NIR is decreased. Accordingly the value of σ increases. Eventually, at a voltage $V = V_0$, I has dropped to zero because the current component due to the radiation-induced conductivity exactly balances the injection current; σ has reached its peak value.

With increasing negative bias the current in the IR becomes more positive. Positive charge arriving at $x = \bar{r}$ is not transferred to the NIR but used up for compensation of the planar negative charge at $x = \bar{r}$. The current I and the space charge in the NIR remain zero; the field in the NIR is constant.

If V becomes still more negative a critical negative voltage V_c' is reached where the negative planar charge is fully compensated and $\sigma = 0$. The dielectric contains no charge at all. The field at $x = \bar{r}$ is continuous and $E_a = E_b$.

For negative voltages beyond V_c' the positive charge reaching the depth $\bar{r}$ exceeds the value required for the neutralization of the continuously injected negative charge. The difference is transferred into the NIR and the current through this region becomes a hole current. The field at $x = \bar{r}$ is continuous and positive as elsewhere in the electric.

Thus the virtual electrode created by the partially penetrating electron beam can behave as a cathode ($V > V_0$) or an anode ($V_c' > V$). It is ohmic for $V > V_c$ and $V_c' > V$, injecting for $V_c > V > V_0$, and blocking for $V_0 > V > V_c'$.

4.3.5 Steady-State Currents as a Function of Electron Range

It has been assumed that in the IR only the injection plus ohmic current term and in the NIR only the transport term need to be considered [4.36]. This gives for the steady-state current in the irradiated dielectric the equations

$$I = agE + I_i \qquad\qquad 0 \leqq x < \bar{r}, \tag{4.19}$$

$$I = \pm a(\mu\varepsilon/2)dE^2/dx \qquad \bar{r}_+ \leqq x \leqq s, \tag{4.20}$$

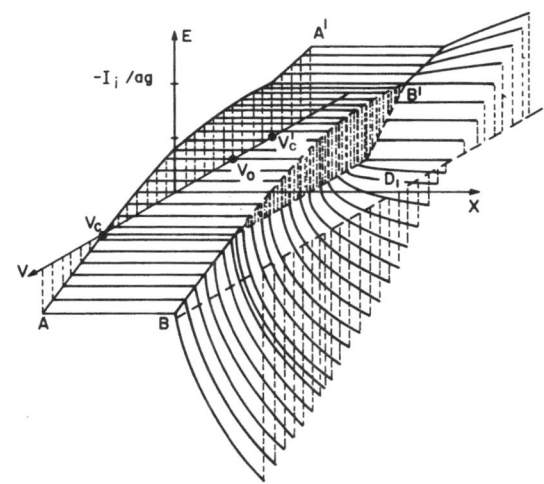

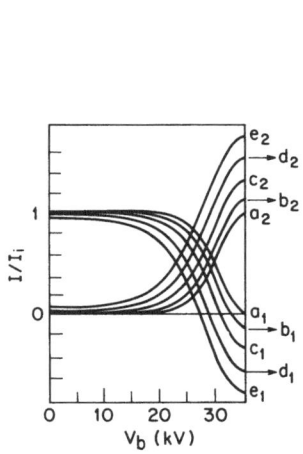

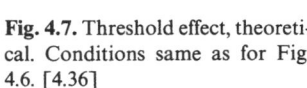

Fig. 4.7. Threshold effect, theoretical. Conditions same as for Fig. 4.6. [4.36]

Fig. 4.8. The virtual electrode: field strength as a function of depth for various values of electrode bias V; rear electrode positive for $V>0$. Beam energy is constant giving relative range $\bar{r}/s = 0.6$. Irradiated region (constant field) is limited by curves $A-A'$ and $B-B'$. Currents through nonirradiated region: hole transport for $V'_c>V$; zero current for $V_0>V>V'_c$; electron transport for $V>V_0$. Planar negative charge at $\bar{r}$, indicated by shaded area, generated only for $V_c>V>V'_c$. Space charge (E increasing) positive for $V'_c>V$ and negative (E decreasing) for $V>V_0$. Virtual electrode is: injecting for $V_c>V_0$ $[dE(\bar{r}_+)/dx = -\infty]$; blocking for $V_0>V>V'_c (I=0)$; ohmic for $V'_0>V$ and $V>V_0 [dE(\bar{r}_+)/dx$ finite]. The curve $A-A'$ is a measure for the current $E/ag = I - I_i$. [4.36]

where the positive sign refers to hole transport and the negative sign to electron transport. The double sign in [4.20] is necessary because the current through the NIR has the same direction for hole and electron transport, while the space charge changes sign.

The boundary conditions at $\bar{r}$ have been discussed in the preceeding section. For the field one has the condition $\int_0^s E dx = -V$; the negative sign means that positive values of V corresponds to a positive bias of the rear electrode. Integration is straightforward.

The material parameters can be determined from the experimental curves shown in Fig. 4.6. The current reaches a peak value for $\bar{r}=s$; the corresponding relation between I and V allows one to determine g. The voltage V_0 for which I_2 becomes zero is given by *Child's law* [4.29]

$$V_0 = \tfrac{8}{9}(-I_i/a\mu\varepsilon)^{1/2}(s-\bar{r})^{3/2} . \tag{4.21}$$

The negative sign appears because I_i is an electron current and is thus negative. This relation allows one to determine μ. V_c is determined as the voltage for which I_1 changes sign. The beam energy V_b which gives a maximum of the current corresponds to $\bar{r} = s$ and $r_p = (V_b/c\delta)^2$; this allows one to determine $\bar{r}/r_p$.

An analysis of the curves for mica (Fig. 4.6) with this method [4.36] gave $\mu = 7.1 \times 10^{-11}$ cm^2 V^{-1} s^{-1} for electrons, $g = 4.8 \times 10^{-13}$ Ω^{-1} cm^{-1} at a dose rate of 1.8×10^6 rad s^{-1}, and $\bar{r}/r_p = 0.6$, values which might be expected under the conditions of the experiment. Theoretical curves calculated with these values are shown in Fig. 4.7. The agreement between the theoretical and experimental curves is quite satisfactory.

The different modes of operation of the partially irradiated dielectric are illustrated for mica by Fig. 4.8 which gives the electric field as a function of depth for various values of the applied voltage and $\bar{r}/r_p = 0.6$. The field in the IR always constant; in the NIR it is constant for voltages between V_0 and V_c'; for $V < V_c'$ it increases and for $V > V_0$ it decreases with x. A discontinuity of the field at $\bar{r}$ and the corresponding planar charge (indicated by the dashed area) are found between V_c and V_c'. The NIR contains a positive space charge for $V < V_c'$ and a negative space charge for $V > V_0$. The curve $A - A'$ connecting the field values at the electrode of incidence shows the behavior of the current agE_a. The current is zero for voltages between V_0 and V_c', positive for $V < V_c'$ (hole transport) and negative for $V > V_0$ (electron transport).

Since adequate experimental data for negative rear bias are not available for mica, mobility for holes has been assumed to be the same as for electrons. This might not be so; therefore curves for hole transport ($V < V_c'$) are only qualitatively valid. The theory neglects *deep trapping in the* NIR which might be responsible for the steep takeoff of the currents at the threshold energy.

Aris et al. [4.87] and *Beckley* et al. [4.87] later developed independently a similar model and successfully applied it to tantalum oxide and polyethylene terephthalate.

4.3.6 Field and Charge Profiles in the Irradiated Region

For electron energies below the threshold value, the steady-state current can be assumed to be zero for all practical purposes. Equation (4.19) then gives $E_a = -I_i/ag$. This relation ignores the transport term $\mu\varrho E$ in the IR and the space charge in this region, as did the previous treatment. To obtain an approximate condition for the validity of this assumption one supposes that the negative excess charge of the irradiated dielectric is smeared out uniformly over the region between the surface of incidence and the extrapolated range of the beam [4.84]. This gives $\varrho = (I_i/a)\tau/\bar{r}$ where τ is the time constant necessary to reach equilibrium. If $\mu\varrho \ll g$, τ is approximately given by $\tau = \varepsilon/g$. Thus the mobility term can be neglected when $\mu \ll a\bar{r}g^2/I_i\varepsilon$. As a numerical example, assuming representative values $\varepsilon = 2 \times 10^{-13}$ F cm^{-1}, $\bar{r} = 10^{-3}$ cm, $I_i/a = 10^{-8}$ A cm^{-2}, and $g = 2 \times 10^{-13}$ Ω^{-1} cm^{-1}, one finds $\mu \ll 2 \times 10^{-8}$ cm^2 V^{-1} s^{-1}. This condition is satisfied for the experimental situation found in the experiments described above.

The theory of space-charge-limited short-circuit currents through the IR including the drift term $\mu \varrho E$, has been treated in detail [4.34] under the assumptions that the NIR is blocking and that the RIC and I_i are constant for $0 \leq x < \bar{r}$. Transient currents following the start of irradiation were calculated by the method of characteristics [4.29]. For times smaller than the transit time t_λ necessary for the first receding front x_f of the injected electrons to reach the front electrode one has $\varrho = 0$ in $0 \leq x \leq x_f$. An analytical solution can be obtained for $t < t_\lambda$. Curves for $\bar{r}/s = 1/4$, $g = 2 \times 10^{-13}$ Ω^{-1} cm^{-1}, $\varepsilon = 1.9 \times 10^{-13}$ F cm^{-1}, and various values of μ between 5×10^{-7} and 10^{-10} cm^2 V^{-1} s^{-1} show that the exponential decay curve obtained for $\mu \varrho \ll g$ is a good approximation for $\mu < 5 \times 10^{-9}$ cm^2 V^{-1} s^{-1}, and thus confirm the estimate above given for the effect of the drift term $\mu \varrho E$. The assumption that the NIR contains no space charge is, however, found to be not entirely satisfactory. Indeed, the steady-state space-charge profile in the IR depends strongly on μ; even for a mobility value as low as 1×10^{-10} cm^2 V^{-1} s^{-1}, a diffuse charge distribution extending from the virtual electrode toward the front electrode covers about 6% of the IR. Analytic expressions are obtained for steady-state values of the field and charge profiles.

Several authors [4.74, 88] have *numerically* integrated the general transport equation

$$I(t) = I_i(x) + ag(x)E + a\varepsilon \partial E(x, t)/\partial t$$

under the assumption that the NIR is blocking, but that I_i and g might be arbitrary functions of depth. An *analytical* solution has been given by *Gross* and *Ferreira* [4.88]. The most detailed results so far have been given by *Matsuoka* et al. [4.74] who have used experimental values for the current-depth profile of $I_i(x)$ and the conductivity-depth profile $g(x)$, as discussed in Sect. 4.3.1. Once the current has been determined, the field as a function of time and depth is obtained by a further analytic integration of the transport equation. The space-charge density then follows as the derivative of E. Figure 4.9a shows the space-charge-density profile for low-density polyethylene irradiated with 1 MeV electrons and an injection current density of 1.5×10^{-7} A cm^{-2} after 10 s. As expected, most of the charge is concentrated in a thin layer toward the end of the IR. In addition there is a small charge concentration near the electrode of incidence. This is connected with the behavior of $g(x)$ which initially increases with depth. Figure 4.9b shows charge profiles for irradiation with electron beams of different energies, but constant current density. The total charge in the dielectric increases because the absorbed dose rate and therefore the conductivity decrease, as follows from (4.3).

A computer simulation of charge dynamics in electron-irradiated polymer foils, based on the general transport equation and realistic data on $\dot{D}(x)$, $I_i(x)$, and $g(x)$, has now been given by *Berkley* [4.74].

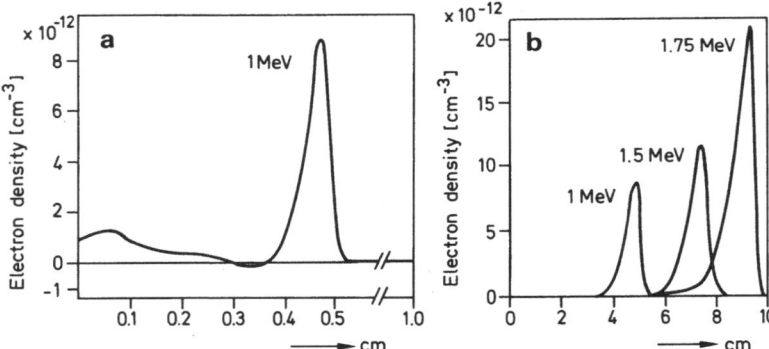

Fig. 4.9a, b. Theoretical charge distribution for electrons in polyethylene [4.74]. Injection current density 1.5×10^{-7} A cm^{-2}; irradiation time 10 s. (a) 1 MeV electrons; (b) 1 MeV, 1.5 MeV, and 1.75 MeV electrons

Transient currents for externally biased samples depend on whether the voltage is applied before or after the inset of irradiation. If the sample shows dielectric absorption and the voltage is applied first, subsequent irradiation leads to a release of the absorbed charge and gives "anomalous" discharge currents which can be used for charge diagnostics. No such currents are expected when the voltage is applied after the onset of irradiation; this case has now been studied. The "anomalous" transient currents of prepoled (10^5–10^6 V cm^{-1}) electron-irradiated polyethylene–terephthalate films observed by *Beckley* et al. [4.86] can possibly be explained by radiation-induced charge release.

An alternative interpretation in terms of generation of low-energy X-rays has also been proposed [4.86]. When two samples were placed in the path of an electron beam unable to penetrate the first one, transient currents were observed during irradiation in the second sample; they were attributed to X-rays generated in the first one [4.89]. For SiO$_2$ it has been estimated that there is an envelope of several micrometers beyond the effective electron range where the secondary, i.e., X-ray generated radiation is above 5% of its peak value, X-ray generation efficiency being estimated as 0.1% [4.90]. There is no indication so far that the X-ray component needs to be taken into account in steady-state theory. However, it has now been shown that the most important factor is the finite build-up time of RIC (B. Gross, J. West, R. Faria, unpublished).

4.4 Charge Diagnostics by Transient Analysis

4.4.1 Equivalent Circuit Model and Circuit Equations

When the energy of the electron beam is sufficiently low that the NIR can be considered blocking, the irradiated dielectric can be represented by the *equivalent circuit* shown in Fig. 4.10 [4.84, 91]. The RIC is represented by the leakage resistor

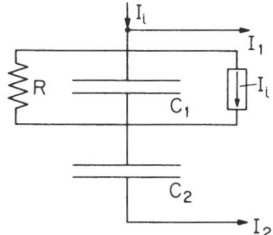

Fig. **4.10.** Equivalent circuit for electron-irradiated dielectric; nonirradiated region assumed to be blocking. R: resistor representing radiation-induced conductivity in irradiated region; C_1 capacity of IR; I_i current generator representing injection current. C_2 capacity of nonirradiated region. [4.84]

$R = \bar{r}/ag$, the capacitances of the IR and the NIR, respectively by the capacitors $C_1 = a\varepsilon/\bar{r}$ and $C_2 = a\varepsilon/(s - \bar{r})$, and the injection current by the current generator I_i. The sum of the partial voltages is $V_1 + V_2 = V$. The differential equation for this circuit is

$$I_2 = C_2 dV_2/dt = C_1 dV_1/dt + V_1/R + I_i. \tag{4.22}$$

In the *steady-state*, charge flows out at the same rate as it is injected, $I_2 = 0$, and thus $agV_1/\bar{r} = -I_i$. The corresponding stored charge is

$$q_\infty = (s/\bar{r}) \int_0^\infty I_2 dt = I_i\tau. \tag{4.23}$$

Since q_∞ and $\bar{r}$ can be measured (see Sect. 4.4.6), Eq. (4.23) allows definition of an empirical time constant $\bar{\tau}$ [4.84, 91]. The circuit diagram gives explicitly

$$\tau = R(C_1 + C_2) = (\varepsilon/g)/(1 - \bar{r}/s). \tag{4.24}$$

In the steady state, the stored charge, represented by a planar charge σ at depth $\bar{r}$, adjusts itself so as to maintain a constant field in the IR which keeps the leakage current through the IR always equal and opposite to the injection current. Without external bias the charge is always negative, since for electron injection I_i is negative. A negative front bias gives a still higher negative charge; a positive bias reduces the value of the negative charge and eventually generates a positive charge in the dielectric [4.92]. With $I_i + V_1/R = 0$ one finds the steady-state value

$$\sigma/\varepsilon = (V + sI_i/ag)(s - \bar{r})^{-1}, \tag{4.25}$$

and $\sigma = 0$ for $V_0 = -\bar{r}I_i s/ag$, $\sigma > 0$ for $V > V_0$, and $\sigma < 0$ for $V < V_0$. These results are simplified versions of the relations obtained from space-charge-limited current theory. They confirm the possibility of obtaining positive as well as negative charges in electron-irradiated dielectrics.

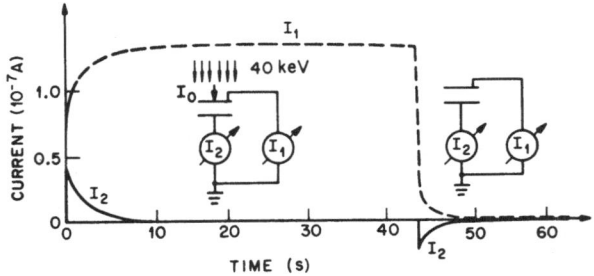

Fig. 4.11. Short-circuit charging and discharging currents for 1 mil Teflon FEP foils irradiated with 40 keV electrons; irradiated area 12.6 cm² [4.84]

If I_i and V are constant, the current $I_2 = dQ/dt$ can be determined from the equation

$$dI_2/dt + I_2/\tau = 0 , \qquad (4.26)$$

which follows from (4.22).

During irradiation the RIC reaches its equilibrium value within a time short compared to the time constant of the circuit; the current decays exponentially

$$I_2 = I_2(0) \exp(-t/\tau), \qquad 0 < t \leqq t_0 . \qquad (4.27)$$

$I_2(0)$ depends on the external bias and for zero bias (short circuit) equals $(\bar{r}/s)I_i$; the time dependence of positive and negative transient currents is the same.

After irradiation the time-dependent DRIC $g(t)$ is given by (4.2). The driving force for the short-circuit discharge current is the space-charge field; since this decreases when stored charge is released, I_2 decreases faster than it would under a constant external bias. Substitution of $g(t)$ from (4.2) into (4.24) and (4.22) and integration shows that I_2, like $g(t)$, is given by a hyperbolic decay function, with an exponent greater than unity [4.84].

4.4.2 Short-Circuit Charging and Discharging Currents

Measurements with 25 μm thick polyfluoroethylene propylene (Teflon FEP, type A) samples [4.78, 84] confirm the predictions of the theory (Fig. 4.11). During irradiation with a scanned electron beam whose average range is about 1/3 of the sample thickness, I_2 decreases from $I_i \bar{r}/s$ to zero, while I_1 increases from $(1 - \bar{r}/s)I_i$ to I_i. The small discharge current after termination of the irradiation is due to the DRIC.

Measurements [4.92] show that application of a positive bias of 200 V to the front electrode produces a positive charging current and therefore leads to storage of positive charge. Negative biasing of the front electrode with 200 V produces a negative charging current which considerably exceeds the short-circuit current. For the same absolute value of the applied voltage negative currents are, as expected, always greater than positive currents (cf. Sect. 4.6.3).

4.4.3 Radiation-Induced Discharge

Conductivity measurements in irradiated and biased dielectrics frequently are disturbed by the buildup of internal space charges opposing the externally applied field [4.93]. These charges can be released by periodically removing the external field and exposing the shorted sample to pulses of penetrating radiation. The technique apparently was first used in electron bombardment studies of diamond [4.94]. It has been applied for clearing samples in measurements of mobilities of carriers drifting in an applied field [4.48, 95]. An investigation has been made of the speed with which pulsed irradiation with a penetrating electron beam removes space charge that previously had built up under an applied field in materials such as As_2S_3, ZnS, and doped ZnS [4.96]. Charge storage and removal by repeated pulsed radiation in short circuit in tantalum capacitors has been studied by measuring charge released in successive pulses [4.97].

The arrangement of the split Faraday cup allows one to make a quantitative study of radiation-induced discharge and depolarization. Referring to Fig. 4.12, the sample is irradiated in open circuit during time $t_0 > \tau$; the stored charge is $q_a = I_i t_0$. Subsequently the sample is short-circuited, while irradiation continues. When the steady state has been reached, the stored charge has decreased to $q_b = I_i \tau$. The charge $\Delta Q = (\bar{r}/s)(q_a - q_b)$ is released and a discharge current is recorded in the external circuit; its time constant is τ. Renewed irradiation in short circuit produces no further discharge current.

The *mechanism of the effect* is not only removal of the stored excess charge but includes charge compensation by induction. This view is supported by the possibility of stepwise discharge, as shown in Fig. 4.13. A sample, previously charged in open circuit with a beam energy eV_1 is reirradiated in short circuit with a beam of energy $eV_2 < eV_1$. Neglecting the DRIC, the low energy beam cannot remove the original charge because it does not reach the depth where it has been stored. It can only partially compensate this charge by allowing charge of the opposite, i.e., positive polarity to build up in the dielectric. Complete compensation or removal is achieved by a second irradiation with a beam of energy eV_1. Subsequent irradiation with higher energy does not produce further discharge currents.

4.4.4 Steady-State Radiation-Induced Conductivity

Measurement of the time constant τ (4.84) of the transient effects, in particular of the radiation-induced discharge currents allows one to determine g at different dose rates. The determination of $\bar{r}$, which is also needed, is discussed in Sect. 4.4.6. Results [4.84] give for Teflon FEP $\Delta = 0.7$ and $g_0 = 1.7 \times 10^{-16} \ \Omega^{-1} \text{cm}^{-1}$, measured at an average dose rate of 2.2×10^4 rad s^{-1} [cf. (3.1)]. The value of g_0 is in approximate agreement with results obtained under irradiation with 2 MeV gamma rays [4.98] (Fig. 4.14).

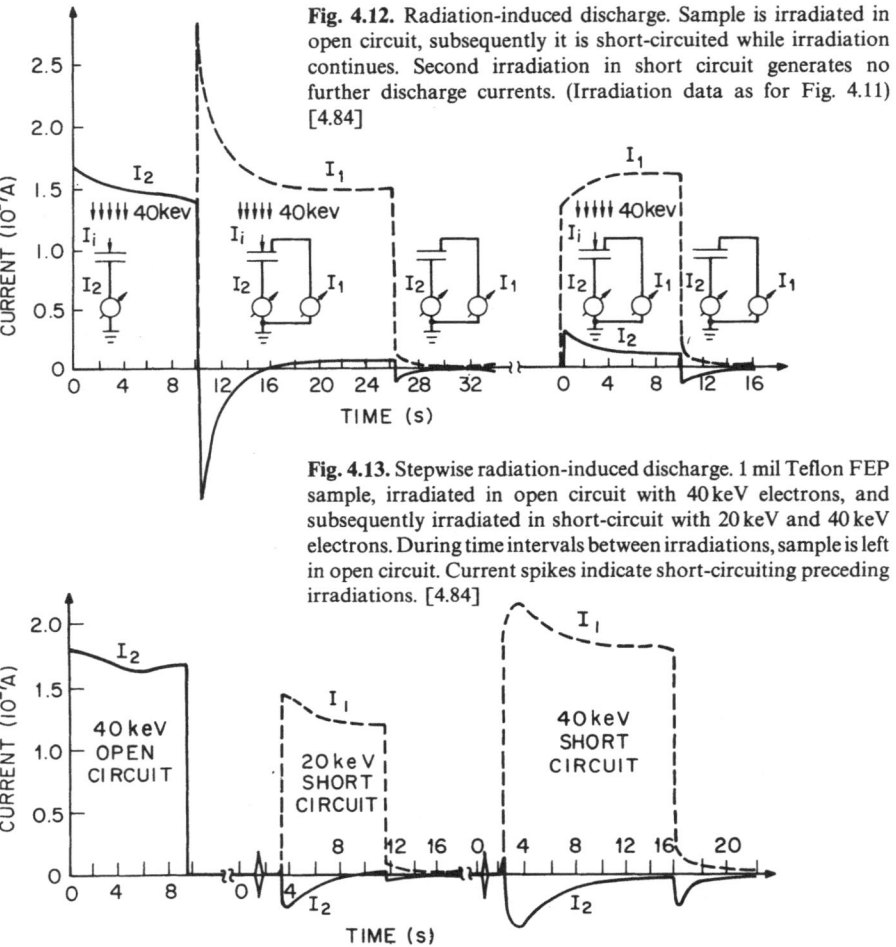

Fig. 4.12. Radiation-induced discharge. Sample is irradiated in open circuit, subsequently it is short-circuited while irradiation continues. Second irradiation in short circuit generates no further discharge currents. (Irradiation data as for Fig. 4.11) [4.84]

Fig. 4.13. Stepwise radiation-induced discharge. 1 mil Teflon FEP sample, irradiated in open circuit with 40 keV electrons, and subsequently irradiated in short-circuit with 20 keV and 40 keV electrons. During time intervals between irradiations, sample is left in open circuit. Current spikes indicate short-circuiting preceding irradiations. [4.84]

4.4.5 Delayed Radiation-Induced Conductivity

The delayed conductivity in practice depends on the dose rate and on the dose received during irradiation before short circuit. This is illustrated in Fig. 4.15 which gives the DRIC for a virgin sample and a preirradiated sample, and a theoretical curve (4.2) calculated with $\tau^* = 1$ s and $\kappa = 1/3$ [4.84]. The latter gives a good average between the two experimental curves. The DRIC is of importance for the understanding of the charge storage properties of dielectrics and, in particular, for the interpretation of TSC curves. After irradiation the dielectric is a heterogeneous medium, with the IR having quite different properties from those of the NIR. In view of the slow decrease of DRIC this difference persists for a long time; possibly the two regions will never be the same as before irradiation.

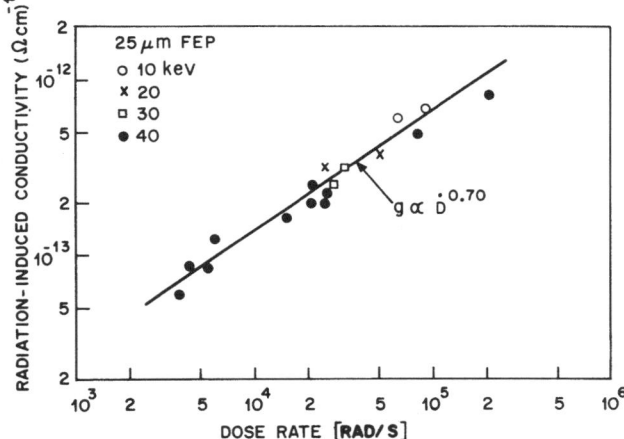

Fig. 4.14. Radiation-induced steady-state conductivity as a function of dose rate, determined from time constants of radiation-induced discharge currents. [4.84]

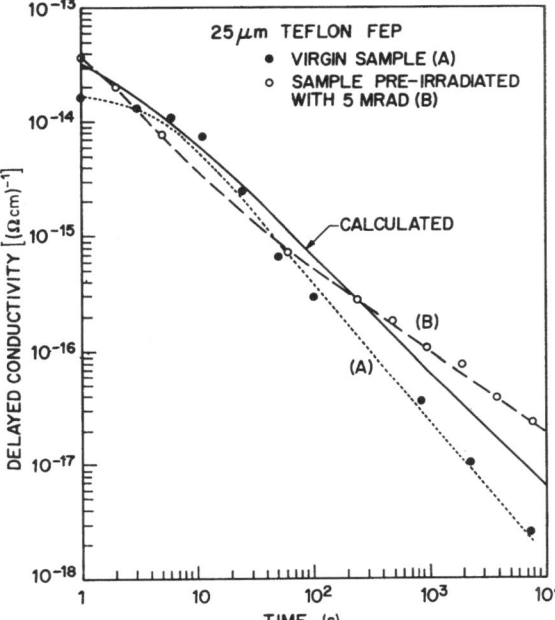

Fig. 4.15. Delayed radiation-induced conductivity as a function of time after irradiation. Sample A: not irradiated prior to irradiation. Sample B: pre-irradiated in short circuit with 5.2 Mrad. Theoretical curve: calculated with $\tau^* = 1$ s and $\kappa = 1/3$. Decay for pre-irradiated sample is faster than for virgin sample. [4.84]

Measurements of short-circuit currents from Mylar (polyethylene terephthalate) films irradiated with a nonpenetrating electron beam and left in open circuit for times between 30 and 1600 s after termination of the irradiation [4.99] gave also a hyperbolic decay over various decades. Results were interpreted in terms of detrapping and drift of injected carriers. Since the presence of the DRIC and the resulting heterogeneity of the irradiated material were not taken into account, interpretation of the results remains controversial.

4.4.6 Charge Centroid

The average depth $\bar{r}$ of the excess charge, or *charge centroid*, is defined by

$$\bar{r} = \int_0^s x\varrho(x,t)dx \bigg/ \int_0^s \varrho(x,t)dx .$$

Under conditions of deep trapping, $\bar{r}$ becomes practically constant after a steady state has been reached under irradiation or after radiation has terminated.

The value of $\bar{r}$ for *irradiation in open circuit* can be measured in the split Faraday cup by an induction method where charges rather than currents are recorded [4.78, 84]. Front and rear electrodes are connected, respectively, to capacitors C_1 and C_2 whose capacitances are much greater than that of the sample. Capacitor voltages are monitored by high impedance voltmeters. In this experiment, the stray capacitance between front electrode and ground must be kept as small as possible because during irradiation it represents a capacitive shunt and lowers the potential of the front electrode [4.100].

The experiment (Fig. 4.16) is carried out in two phases. In phase A, the sample is irradiated in open circuit during a time t_0. Excluding breakdown or leakage to the rear electrode, the total injected charge $q = I_i t_0$ is stored in the sample and represented as a planar charge layer at depth $\bar{r}$. The rear capacitor records the charge $Q = q$. The current is a conduction current $(I = I_i)$ in $0 < x < \bar{r}$ and a displacement current $(I = dq/dt)$ in $\bar{r} < x < s$. During the irradiation period, the front field is $E_a = 0$ and the rear field increases from 0 to $E_b = q/a\varepsilon$.

In phase B, radiation is terminated, and the sample is short-circuited over C_1. After the short circuit, the front field has increased by $\Delta E_a = (q/a\varepsilon)(1 - \bar{r}/s)$ and the rear field has decreased by the same amount. Accordingly the charge of the front capacitor has increased to $Q_1 = q(1 - \bar{r}/s)$ and the charge of the rear capacitor has decreased to $Q_2 = q - q(1 - \bar{r}/s) = q\bar{r}/s$. This gives the depth $\bar{r}/s = Q_2/(Q_1 + Q_2)$.

The duration of the open circuit between the termination of the irradiation and the short-circuiting affects the time functions $Q_1(t)$ and $Q_2(t)$, but not their initial values from which $\bar{r}/s$ is calculated. After a prolonged open circuit, the DRIC is already small and the capacitor charges change little; this is not so after a short open circuit where Q_2 decreases with time and Q_1 increases due to charge leakage to the front electrode. The leakage current is given by $dq/dt = -dQ_2/dt$ and allows calculation of the DRIC. Characteristic records of measurements with Teflon samples are shown in Fig. 4.16 for short (0.5 s) and long (40 min) open-circuit times.

The value of $\bar{r}$ for open-circuit conditions is not necessarily equal to that for short-circuit conditions. In the first case the internal field is limited to the NIR and the IR is field-free; in the second case a field does exist in the IR which generates charge transport to the electrode of incidence, as discussed in Sect. 4.3.6. The initial range for *irradiation in short circuit* can be obtained from the initial values of the electrode currents [4.77] as

$$\bar{r}/s = [1 + I_1(0)/I_2(0)]^{-1} . \tag{4.28}$$

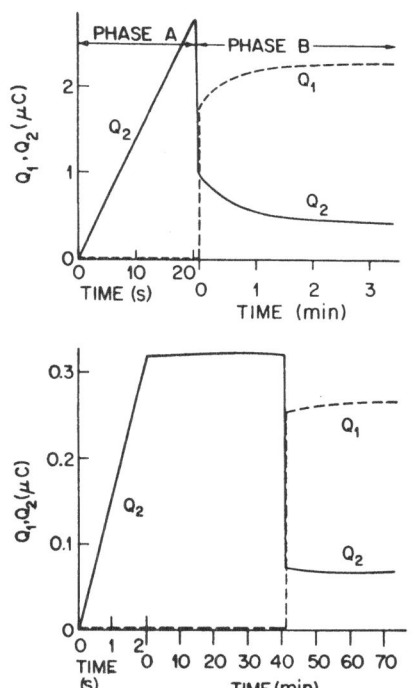

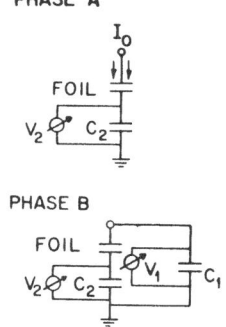

Fig. 4.16. Determination of centroid of charge distribution by induction method: 1 mil Teflon FEP foils, irradiated with diffuse beam of monoenergetic electrons. Phase A: Build-up of charge of rear electrode during open-circuit irradiation. Phase B: Charge on both electrodes in short-circuit. *Top*: Short-circuit performed immediately after termination of irradiation. *Bottom*: Sample remains in open-circuit for 40 min between termination of irradiation and short-circuiting. Irradiation with 40 keV electrons. [4.84].

The *open-circuit front voltage* is $V(t) = (q/C_s)(1 - \bar{r}/s)$. Monitoring V and $q = I_i t$ one obtains therefore $\bar{r}$ as a function of time during irradiation. Measurements with these methods have shown that $\bar{r}$ increases with irradiation time. The effect becomes significant for high electron energies [4.100]. Results are shown in Fig. 2.17. Similar results were obtained with samples not metallized on the irradiated surface [4.101]. C_s is the sample capacitance.

An extension of these methods allows one to determine the *spatial distribution of charges in dielectrics* [4.102]. A monoenergetic electron beam, incident under shorted conditions through the front electrode, creates a virtual electrode. If the dielectric carries a space-charge distribution $\varrho(x)$, it can be shown that

$$\varrho(\bar{r}) = -\frac{d^2(\bar{r}Q_2/a)}{d\bar{r}^2} \tag{4.29}$$

where $\bar{r}$ is the depth of the virtual electrode and Q_2 the induced charge on the rear electrode. The charge density at any depth in the dielectric can thus be be found by sweeping the virtual electrode through the dielectric by means of changing the energy of the beam. In practice the energy and thus the depth of the electrode is changed in steps. The differential expression (4.29) then becomes a difference equation, and $\varrho(\bar{r})$ is determined by measuring the charge released in successive steps. The method involves making a calibration run in which an initially uncharged sample is irradiated with the same beam energies under the same conditions as in the measuring run. The method works well under conditions or for materials where the hole mobility is sufficiently small.

After the successful demonstration of charge injection from some electrodes into dielectrics this effect is now frequently believed to be responsible for almost all cases of dielectric absorption under a dc field. But most metallic electrodes are nonohmic and noninjecting unless the field strength exceeds a critical, rather high value. Therefore in many cases the polarization of the dielectric might be due to formation of a charge depletion layer rather than an injection layer in front of either electrode. The polarity of such a depletion layer is that of a heterocharge, while that of an injection layer is that of a homocharge. Charge diagnostics by means of an electron beam reveals unambiguously the polarity of the charge residing in the dielectric and thus can rule out one or the other hypothesis.

4.4.7 Transit Time Effect and Determination of Mobility

Mobility values in low-mobility solids have been determined by the time-of-flight method [4.103]. A sheet of carriers is injected at one side of the sample and drifted through the bulk to the rear electrode under an applied dc bias. Light beams and laser beams [4.104] have been used to generate the necessary carrier concentration in the dielectric near the electrode of incidence. In the study of carrier mobilities in polymers, an electron beam has also been used to generate a virtual electrode [4.105].

Generally the current traces from laser and electron beams are very similar. Measurements give the value of the shallow-trap modulated mobility in polymers. If the deep-trapping time is sufficiently greater than the transit time, arrival of the first front of carriers at the rear electrode is indicated in the current–time curve by a peak or knee, whose position gives the transit time t_λ, as seen for injecting electrodes [4.28]. The shallow-trap modulated mobility is then given approximately by the relation $\mu = (s - \bar{r})t_\lambda E$ where E is the applied field.

The carrier concentration generated in the dielectric under irradiation with laser or electron beams is higher than for irradiation with light. Therefore it becomes necessary to take into account the bulk recombination of electrons and holes, as has been emphasized by *Chen* [4.104].

Time-of-flight measurements might also be interpreted in terms of a model of hopping transport based on the theory of random walk [4.16, 104]. Here the transit time indicates the arrival of the peak of the carrier distribution, but the concept of carrier mobility has to be abandoned.

In a typical electron beam experiment (*Gross* et al.) [4.105] a 2.5×10^{-3} cm thick Teflon foil was exposed to an electron pulse of 0.5–1 s duration penetrating a few μm into the sample biased by a voltage giving a field of the order of 2×10^5 V cm^{-1}.

The external current is given by

$$I(t) = \frac{1}{s} \int_0^s I_c dx = \frac{1}{s} \int_0^{\bar{r}} (I_c)_1 dx + \frac{1}{s} \int_{\bar{r}}^s (I_c)_2 dx, \tag{4.30}$$

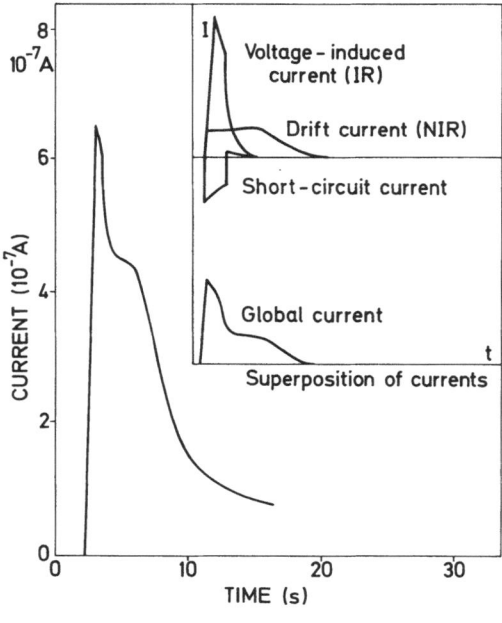

Fig. 4.17. Transit time effect. 25 µm Teflon foil. Electron beam energy 20 keV, applied voltage of front electrode +600 V pulse duration 1 s. Superposition of short-circuit and field-induced currents (schematical)

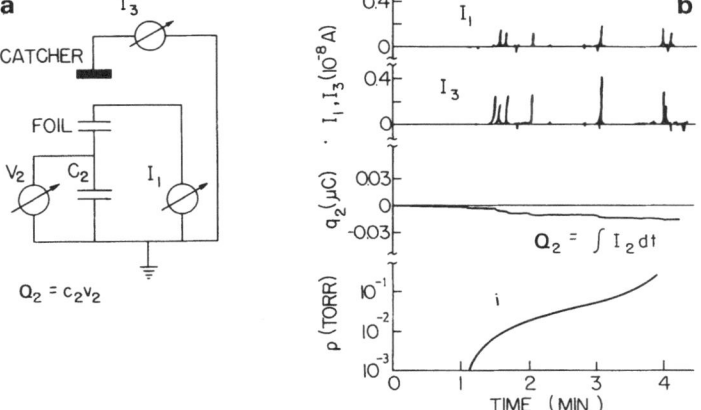

Fig. 4.18. (a) Schematic set-up for measurement of electron emission from 25 µm Teflon TFE previously charged with an electron beam. The capacitive circuit $(C_2 V_2)$ connected to the rear electrode integrates over the current I_1 to the front electrode and the current I_3 to the catcher electrode. Thus Q_2 is the amount of charge lost by the dielectric due to spontaneous current bursts. (b) Typical experimental record, giving emission current I_3 to catcher, front current I_1, rear charge Q_2, and pressure. [4.106]

where $(I_c)_1$ is the conduction current in the IR and $(I_c)_2$ the conduction current in the NIR. If $\bar{r} \ll s$ and the applied voltage sufficiently high, the voltage drop across the IR is small compared to that across the NIR. Then the measured current is approximately given by the superposition of the of the current observed for a blocking NIR and the space-charge-limited current for an injecting electrode. The

Table 4.2. Irradiation parameters of Teflon at dose rate of 100 rad s^{-1} and field of 10^4 V cm^{-1}

	Quantity	Definition	Value
δ	Density	Measured	2.1 g cm^{-3}
ε	Dielectric constant	Measured	1.8×10^{-13} F cm^{-1}
$\dot{D}$	Absorbed dose rate	Measured	10^2 rad s^{-1} = 1.3×10^{16} eV cm^{-3} s^{-1}
σ	Induced conductivity	Measured	$1.7 \times 10^{-15}\,\Omega^{-1}$ cm^{-1}
	Specific induced conductivity	$\sigma/\dot{D}$	$1.7 \times 10^{-17}\,\Omega^{-1}$ cm^{-1} (rad s$^{-1})^{-1}$
G	Energy for generation of free carrier pair		$< 10^3$ eV
μ	Trap-modulated mobility	Measured	5×10^{-10} cm^2 V^{-1} s^{-1}
p	Generation rate of carriers	$p = \dot{D}/G$	10^{13} cm^{-3} s^{-1}
s_b	Schubweg (Field-induced carrier drift distance)	$s_b = i/ep = \sigma E/ep$	10^{-5} cm
m	Trapped carrier concentration	$m = \sigma/e\mu$	10^{13} cm^{-3}
$t*$	Build-up time for trapped carrier population	$t* \gtrsim m/p$	1 s
b	Recombination coefficient	$b \gtrsim a v_t$ $a = 7 \times 10^{-17}$ cm^2 Coulomb cross section $v_t = 10^7$ cm s^{-1} Thermal velocity	10^{-9} cm^3 s^{-1}
n	Free carrier concentration	$n = p/\alpha(m+n)$ $p = bn(m+n)$; Fowler's recombination law	10^9 cm^{-3}
μ_0	Free carrier mobility	$\mu_0 = \mu(m+n)/n$ $\mu_0 = b\varepsilon/e$; Langevin's relation	10^{-5} cm^2 V^{-1} s^{-1} 10^{-3} cm^2 V^{-1} s^{-1}
τ_0	Recombination lifetime	$\tau_0 = s_b/\mu_0 E$ $\tau_0 = 1/\alpha m$	10^{-4} s 10^{-4} s
D_F	Free diffusion coefficient	$D_F = \mu_0(kT/e)$; Einstein's relation	10^{-5} cm^2 s^{-1}
D_T	Trap-modulated diffusion coefficient	$D_T = D_F \times n/(n+m)$ $kT/e = 2.6 \times 10^{-2}$ V	10^{-9} cm^2 s^{-1}

The values of μ, s_b, m, n, $t*$, τ_0, D refer to holes; i is the induced current density.
The values listed in this table allow the consistency of the model to be assessed. They could have been adjusted by using a bigger value for the recombination coefficient. This might, however, be misleading in view of (a) the uncertainty of the experimental value of σ/μ, and (b) the neglect of deep trapping inherent in Fowler's recombination law and the concept of the trap-modulated mobility.

first is given by the sum of the short-circuit current (as observed without bias) and the field-induced current. This superposition of currents is schematically shown in the inset of Fig. 4.17, while the result of a typical measurement for Teflon is shown in the main part of Fig. 4.17.

For Teflon, hole mobilities of the order of 10^{-9} cm^2 V^{-1} s^{-1} were found. Mobilities for electrons are too low to be easily determined by this method.

4.4.8 Breakdown Effects

When the pressure is brought up from 10^{-5} to 10^{-1} Torr on a sample metallized on both surfaces and charged by a scanned electron beam, internal discharges seen

Table 4.3. Electron-beam methods with the split Faraday cup

Determination of charge centroid $\bar{r}/s = \int_0^s x\varrho(x, t)\, dx / \int_0^s s\varrho(x, t)\, dx$		[4.100]

Irradiation in short-circuit	Measured: initial front and rear currents	$\bar{r}/s = I_2(0)/[I_1(0) + I_2(0)]$
Irradiation in open-circuit	Measured: front voltage during irradiation; rear charge during irradiation	$V(t) = (q_2/C_s)(1 - \bar{r}/s);\quad C_s = a\varepsilon/s.$
Open-short circuit	Measured: front and rear charge after short circuit	$\bar{r}/s = Q_2/(Q_1 + Q_2)$

Determination of profile of stored charge $\varrho(x)$	[4.102]

Short-circuit irradiation of charged dielectric with different energies (range sweeping)	Measured: ultimate charge	$\varrho(\bar{r}) = -(d^2/d\bar{r}^2)(\bar{r}Q_2/a)$

Induced conductivity σ	[4.84]

Short-circuid irradiation of uncharged dielectric	Measured: beam injection current and rear charge (NIR assumed blocking)	$\sigma = \varepsilon/\tau$ $\tau = Q_2/I_i$

Trap-modulated mobility μ_+	

Irradiation under high applied voltage with different beam energies	Measured: steady-state currents as functions of applied voltage and range	Obtain from Child's law $V_0 = (8/9)(-I_i/a\mu\varepsilon)^{1\,2}(s - \bar{r})^{3/2}$	[4.36]
Irradiation under high applied voltage	Measured: transit time	$\mu_+ \simeq (s - \bar{r})/t_\lambda E$	[4.105]

as current spikes occur between the charge layer and the front electrode [4.106] as shown in Fig. 4.18. They are not limited to the dielectric; negative charge is also ejected through the front electrode. Charge emission can be monitored by placing a catcher electrode in front of the sample. Similar effects observed by other authors [4.107] have been attributed to exoelectron emission and used to investigate charge profiles due to ion implantation. Continued irradiation in open circuit induces breakdown to the rear electrode and represents one of several "electrodeless" methods [4.108] for breakdown studies.

During irradiation of dielectric samples with a nonmetallized front surface by electron beams of high current density, spontaneous high-frequency current transients have been observed [4.109]. They were explained by electron emission from the surface.

Table 4.4. Methods of analysis for electron irradiation of dielectrics

Model	Equation	Solution
Box model; NIR blocking.	$I(t) = [I_i + a\sigma E(x,t)] \cdot [1 - \theta(x - \bar{r})]$ $+ a\varepsilon \dot{E}(x,t)$	Analytical [4.84]
Range straggling.	$I(t) = I_i(x) + a\sigma(x)E(x,t) + a\varepsilon \dot{E}(x,t)$	Analytical [4.88] (*Gross, Ferreira*) Numerical [4.74.88]
Box model including SCLC in IR. NIR blocking.	$I(t) = [I_i + a\sigma E(x,t) + a\mu\varrho E(x,t)]$ $\cdot [1 - \theta(x - \bar{r})] + a\dot{E}(x,t)$	Analytical [4.34]
Box model for IR. SCLC in NIR	$I(t) = [I_i + a\sigma E(x,t)] \cdot [1 - \theta(x - \bar{r})]$ $+ a\mu\varrho E(x,t)\theta(x - \bar{r}) + a\varepsilon \dot{E}(x,t)$	Analytical [4.36,87]
Range straggling. Beam injection current disregarded; short irradiation burst.	$I(t) = [\mu_+\varrho_+(x,t) + \mu_-\varrho_-(x,t)]E(x,t) + \varepsilon\dot{E}(x,t)$ $\partial\varrho_\pm(x,t)/\partial t = -(\partial/\partial x)[\mu_\pm(x,t)\varrho_\pm(x,t)E(x,t)]$ $\mp b\varrho_+(x,t)\varrho_-(x,t)$	Numerical [4.104] (*Chen*)

Note: SCLC: Space-charge-limited current
   IR: Irratiated region
   NIR: Nonirradiated region
   $\theta(x)$: Unit step function
   $\varrho(x,t)$: Space-charge density (excess charge)
   Short circuit: $\int_0^s E(x,t)dx = 0$
   Open circuit $\Big\}$: $I(t) = 0$
   Steady state $\Big\}$

4.4.9 Radiation Hardening and Pressure-Activated Charge Release

After a Teflon sample has been irradiated in vacuum by a scanned electron beam
with a high dose, of the order of 50 Mrad, the RIC is found to have decreased by at
least an order of magnitude [4.106]. The resistance of Teflon remains, however,
high only as long as the dielectric is kept in a vacuum, as can be seen when a
radiation-hardened and charged sample is brought up from low to high pressure.
After the pressure has gone through the breakdown zone and reached 1 Torr,
charge flows out toward the front electrode and when atmospheric pressure has
been reached the sample is soon completely discharged. The pressure-activated
current curve shows several peaks and appears similar to a TSC curve. Radiation
hardening at atmospheric pressure by beta irradiation has been claimed to be a
general effect and attributed to the generation of additional traps by the radiation
[3.110].

4.4.10 Summary of Electron Beam Data and Methods

Tables 4.2–4 summarize the data and methods discussed in Sects. 4.3 and 4.4.

4.5 Gamma-Beam Charging

4.5.1 Photo-Compton Current

Dielectrics can be charged up by irradiation with a directional beam of X-rays or gamma rays. Photons in the energy range up to about 0.3 MeV interact with matter preferentially by the photoelectric effect while photons above this range interact by the Compton effect. The energy lost by the primary photon beam is thus transferred into the energy of the photoelectrons and the Compton electrons. These secondary electrons are scattered preferentially in the forward direction, their angular distribution becoming steeper with increasing photon energy. The electron flux accompanying the photon flux constitutes an electronic current, the *photo-Compton current*, which is generated in any material exposed to a photon beam [4.111].

The current can be measured in the simple device schematically shown in Fig. 4.19a. This is a dielectric covered by a conducting paint which encapsulates a piece of heavy metal. The conducting paint is connected to ground and acts as the external electrode. The dielectric is the scatterer where the Compton current is generated. The metal acts as the measuring electrode; it has also the task of absorbing the photon radiation. The measured current is given by the difference between the current generated in the front scatterer (between the surface of incidence and the measuring electrode) and the rear scatterer (between the measuring electrode and the rear surface). The latter can be neglected if the photon radiation is completely absorbed in the metal [4.111].

The Compton current I_C like the conduction current agE, is a "local" current. The measured current for planar geometry and complete absorption of the photons in the electrode is $I = s^{-1} \int_0^s [I_C(x) + ag(x)E(x,t)]dx$. Under the assumption that g can be replaced by an average constant value $\bar{g}$, one has $I = s^{-1} \int_0^s I_C(x)dx$, independent of the induced conductivity. The system is supposed to operate in short circuit. The maximum range of the secondary electrons is small compared with the absorption length $1/\mu_1$ of the photons. For a front scatterer whose thickness is several times the electron range, but still small compared with $1/\mu_1$, one has $I = I_C$.

In thick scatterers, the *absorption of the photon beam* has to be considered. Assuming secondary electron equilibrium in the bulk of the material, the electron flux is directly proportional to the photon flux, and the current follows the same exponential absorption law as the photon radiation. Thus one has $I_C(x) = I_C(0) \exp(-\mu_1 x)$. Integration then gives $I = (I_C(0)/\mu_1 s)[1 - \mu_1 s]$ for the value of the measured current [4.111].

An approximate *relation between the exposure rate and the Compton current density* has been obtained under the assumption of planar geometry [4.111]. For low density materials and photon energies of about 1 MeV one obtains approximately 2×10^{-12} A cm^{-2} per R s^{-1} or C cm^{-2} R^{-1}. For materials with

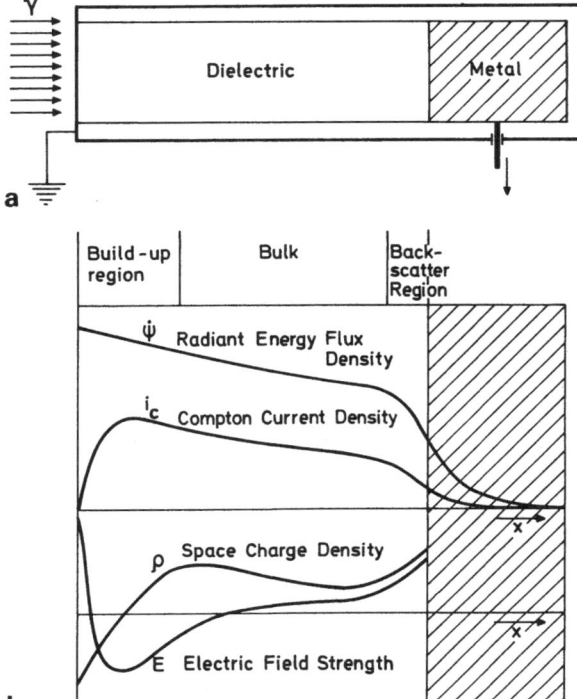

Fig. 4.19. (a) System for measurement of photo-Compton current. (b) Electron current, space-charge density, and internal field as functions of depth. Bulk and interface effects are shown (schematical)

higher atomic number it is necessary to include backscattering and consider departure from linear electron transport; for low photon energies the elementary theory also fails. A detailed theory has, however, been developed whose results are given in tables covering numerous elements and compounds and photon energies between 0.01 and 20 MeV [4.112]. Figure 4.20 gives curves for ethylene, Teflon, Si, and Pb. Ordinates give the relation between the number n_e of forward electrons per number $n_{h\nu}$ of photons. For a strictly monoenergetic photon beam one has $n_{h\nu} = \psi/h\nu$, and $(I_C/a) = e(n_e/n_{h\nu})\,\psi/h\nu$.

4.5.2 Space-Charge Formation

The divergence of the electron current generates space charge. Neglecting the induced conductivity, one has $a\partial\varrho(x, t)/\partial t = -\partial I_C/\partial x$; this gives a linear increase of space charge with time which is valid as long as the internal field is low and all carriers are trapped at the end of their range. If the induced conductivity is included, but carrier detrapping is still neglected, one has [4.22]

$$\varrho(x, t) = -(\varepsilon/g)(\partial I_C/\partial x)\,[1 - \exp(-tg/\varepsilon)]\,a^{-1}.$$

Figure 4.19b shows schematically the behavior of I_C, and the space-charge field E for small irradiation times.

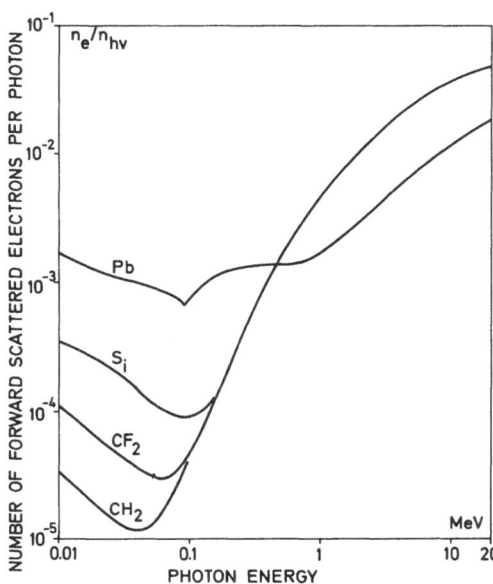

Fig. 4.20. Photo-Compton current production as a function of photon energy. Ordinates give the number of forward scattered electrons per photon for the energies shown as abscissae. Curves are shown for ethylene, Teflon, silicium and lead. [4.112]

The *bulk charges* which in this way can be generated in thick scatterers as well as the corresponding internal fields can be extremely high. It has been found that the high-resistive glass used for hot-cell windows can be completely shattered by internal electric breakdown which sets in after prolonged exposure to gamma radiation [4.113]. Short of spontaneous breakdown, the presence of the space charge can be put into evidence, and its value estimated, by triggering breakdown of the irradiated dielectric. For this purpose the charged dielectric is placed on a grounded table and a point electrode is pressed against its free surface; the electrode is grounded through a capacitor with a large capacitance C connected in parallel with a high-impedance voltmeter [4.22, 65]. Field enhancement at the tip of the point electrode triggers breakdown; the value of the released charge is given by CV, V being the capacitor voltage. Figure 4.21 shows results of experiments in which 10 cm high-resistive glass cubes were exposed to the beam of a ^{60}Co gamma source at an exposure rate of 0.52 MR h^{-1}. Each experimental point corresponds to a different exposure run. *Spontaneous breakdown* sets in after a total exposure of 30 MR giving a maximum released charge of 3×10^{-7} C cm^{-2} [4.113].

Charge buildup occurs also *at the surface of incidence.* The unfiltered radiation emerging from most gamma sources and high-energy X-ray machines is not accompanied by its equilibrium secondary electron flux. The electron component reaches its equilibrium value only after the photon radiation has penetrated into the irradiated material over a distance equal to the maximum electron range. Within this transition zone more electrons are scattered forward than enter the material. Therefore an electron depletion or positive charge layer is generated [4.114]. Information about the presence of this layer and the charge distribution

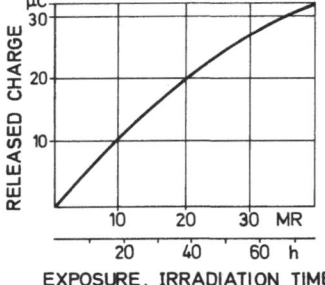

Fig. 4.21. Charge released by triggered breakdown for high-resistive glass as a function of irradiation time and absorbed dose *Hardtke* and *Ferguson* [4.113]

in the bulk of the material could be obtained by a combination of thermally activated currents and the sectioning method [4.114].

The opposite effect, i.e., the formation of an electron excess layer, occurs in the *transition zone* where the radiation goes from a material with low atomic number to one with high atomic number. The number of backscattered electrons from the latter exceeds that of forward scattered electrons in the former. Therefore in front of the material of high atomic number, in particular the measuring electrode, additional negative charge is accumulated. These interface effects and the behavior of multilayer structures are being actively investigated [4.115, 116], and it is believed that internal fields might also reach breakdown values. These effects are also shown schematically in Fig. 4.19.

4.5.3 Compton Diodes for Dosimetry

The measurement of the intensity of Compton currents allows one to determine exposure and exposure rate of the photon radiation generating the currents. Since the currents are "radiation driven", no external power source is requested. The sensitivity of such self-powered dosimeters is low, but together with the their extremely fast response and the absence of saturation this represents an advantage for the measurement of radiation pulses of extreme intensity and very fast rise time (in the nanosecond range). Compton diodes for pulse diagnostics have found widespread application [4.117].

4.6 Thermally Activated Processes

4.6.1 Thermally Stimulated Currents and Voltages

Carrier detrapping is a thermally activated process. The possibility of lifting carriers into the conduction band by a temperature increase suggests the application of TSC methods for the investigation of the characteristics of trapped excess charge. If a the space-charge field provides the driving force, external poling is not required. Thus short-circuit currents or open-circuit voltages are observed when an electron-beam-charged dielectric is heated; current peaks or voltage plateaus, observed for constant heating rate, can give information about nature and energy distribution of the traps in the dielectric [4.118].

Experimental results [4.46, 119–122] have revealed the presence of a complex peak structure for irradiated dielectrics. Interpretation is, however, less straightforward than might appear at first sight. Experiments have shown significant differences between the results of short-circuit and of open-circuit experiments [4.62, 120].

The *short-circuit* TSC gives the average value of the conduction current in the dielectric, i.e., $I = (1/s) \int_0^s I_c \, dx$. The measurement of the *open-circuit voltage* of a dielectric sample with a nonmetallized surface is carried out by means of a "sensing" capacitor C_e, placed at a distance from the sample. The derivative of the capacitor voltage V gives the open-circuit current $I = C_e \, dV/dt$. For V to be the undistorted open-circuit voltage, C_e must be very small compared with the sample capacitance C_s. Then one has for the open-circuit TSC

$$I = [C_e/(C_e + C_s)] (1/s) \int_0^s I_c \, dx.$$

The values of open-circuit and short-circuit TSC are, of course, different [4.45, 123]. A detailed discussion of short- and open-circuit TSC, also called open-circuit TSD, is given in Chap. 3.

The influence of circuit conditions (open or short circuit), storage time t_d between irradiation and thermal activation (up to 500 days), and storage temperature (25 or 100 °C) has been investigated for Teflon FEP foils by *Sessler* and *West* [4.62]. Samples 25 µm thick and metallized on one surface were irradiated through the nonmetallized face with a 20 keV electron beam giving an average dose rate of 4×10^4 rad s^{-1}. Open-circuit measurements were performed with a sensing electrode placed at a distance of 0.4 cm. For short-circuit measurements, a pressed-on electrode was used. Some results are shown in Fig. 4.22. Analogous results are observed if an electrode is vacuum deposited to the free surface after charging but before the TSC measurement. The open-circuit currents, flowing to the rear electrode, are opposite to the short-circuit currents, and exhibit a peak at a higher temperature than the latter. This behavior is also observed with a 40 keV electron beam whose range exceeds half the sample thickness, showing that under these conditions the charge does not flow to the nearest electrode.

This behavior has been explained by the *anisotropy of the irradiated dielectric* due to the presence of the DRIC in the IR [4.62]. The current observed in open circuit is due exclusively to the detrapping of primary injected charge since the NIR (through which this charge is forced) contains neither an induced nor an intrinsic conductivity; the high temperature of the open-circuit peak shows that the injected electrons are deeply trapped. The short-circuit current is due to compensation of the injected and trapped electronic charge by temperature-

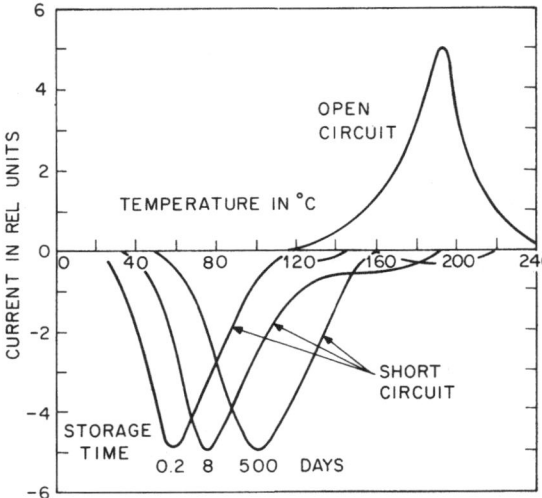

Fig. 4.22. Peak shift: 25 μm thick Teflon FEP films aluminized on one surface, irradiated through nonmetallized surface with 20 keV electrons, charge fluence $3 \times 10^{-8}\,\mathrm{C\,cm^{-2}}$. Short-circuit TSC for storage times $t_d = 2, 8, 500$ days; storage temperature 25 °C. Open-circuit TSC is independent of storage time. Heating rate 1 °C min^{-1}. Amplitudes have been normalized; actual values decrease. [4.62]

activated secondary holes. The secondary carriers, responsible for the delayed conductivity, populate shallow as well as deep traps. The number of the more shallow trapped carriers decreases with storage time by detrapping and recombination; the rate of detrapping increases with increasing temperature. Therefore the temperature necessary to compensate the injected charge by activating a sufficient number of trapped carriers increases with storage time. These effects have been discussed in detail by *Sessler* and *West* [4.62].

4.6.2 Peak Shift Effects

The conductivity in the irradiated region is $g = en\mu_0$. The free carrier density decreases with time as $n = \kappa n(0)/(1 + t/\tau^*)$ where $n(0)$ is its value at the end of the irradiation. After a storage time $t_d \gg \tau^*$ one has $n = \kappa n(0)\tau^*/t_d$. The duration of the TSC experiment starting at time t_d is small compared with t_d. It will be assumed that the trapped carrier reservoir is not significantly depleted, that the detrapping rate prevails over the recombination rate, and that thermal equilibrium exists. Then, for a single trapping level, n increases exponentially with temperature

$$n[T(t_h), t_d] = \kappa n(0)(\tau^*/t_d)\exp[(U/kT_0) - (U/kT)], \qquad (4.31)$$

where T_0 is the storage temperature and $t_h = t - t_d$ is the heating time. The conductivity $g = e\mu_0 n[T(t_h), t_d]$ varies accordingly and the current as a function of time follows from (4.22) and (4.26). Integration gives $I_2 = -Ag(t_h)Q$, where $A = (1 - \bar{r}/s)/\varepsilon$. The current reaches a maximum at a time for which $dI_2/dt_h = 0$, giving the condition $gI_2 + Qdg/dt = 0$. Assuming a linear heating rate $T = T_0 + \beta t_h$ and considering g as a function of T by means of (4.31), one has for T_p the condition

$$g^{-1}(T_p)(dg/dt)_{T_p} = Ag(T_p).$$

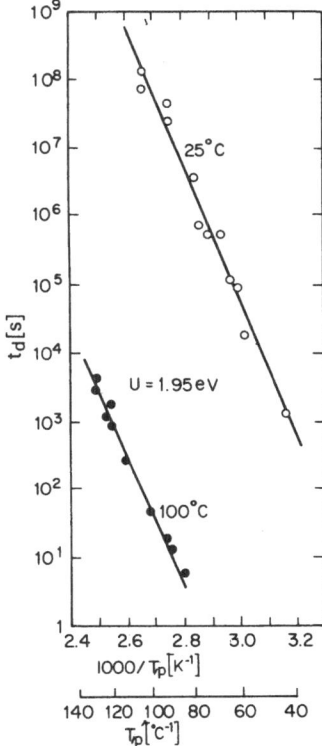

Fig. 4.23. Peak shift: determination of activation energy. Storage time as a function of inverse peak temperature of short-circuit TSC, for storage temperatures of 25 and 100 °C. Samples and irradiation as for Fig. 4.22. [4.62]

Since μ_0 is practically independent of T, substitution of the expression for $g(T)$ gives [4.62]

$$t_d = BT_p^2 \exp[(U/kT_0) - (U/kT_p)], \tag{4.32}$$

where $B = Ae\mu_0\kappa n(0)\tau^* k/U\beta$. Since the temperature dependence of the pre-exponential factor is weak, one obtains a linear relation between $\ln t_d$ and U/kT_p. Curves for different storage temperature T_0' and T_0'' are parallels whose vertical distance is given by $\exp(U/kT_0') - \exp(U/kT_0'')$.

Experimental results are in agreement with the theory. Figure 4.23 gives curves for t_d vs $1/T_p$ calculated from the curves of Fig. 4.22 for the two storage temperatures of 25 and 100 °C. Both curves give an activation energy of 1.95 eV, in agreement with results obtained by other methods [4.123, 124]. Extrapolation of the straight lines of Fig. 4.23 to the temperature of the open-circuit peak (Fig. 4.22) yields storage times of 10^{13} and 10^7 s for 25 and 100 °C respectively. These times would be necessary in each case for the DRIC to decay to the intrinsic conductivity of the material. The corresponding conductivity values are $10^{-27}\,\Omega^{-1}\,cm^{-1}$ and $10^{-20}\,\Omega^{-1}\,cm^{-1}$. The former value is, however, not realistic since other processes, in particular ambient atomic radiations, induce conductivities greater by several orders of magnitude. Time constants for open-circuit decay of Teflon electrets [4.124] are also of the order of 10^{13} (extrapolated) and 10^7 s at 25 and 100 °C. In both cases t ie injected carriers have to drift through a substantial fraction of the thickness of the sample.

Measurements with samples whose surface of incidence was metallized after charging and annealing but prior to the short-circuit TSC measurement have revealed an additional effect. It has been found that annealing induces a *shift of population*, transferring carriers from shallow traps into deep traps. This effect interferes to a certain degree with the results described above. Shift of populations by annealing has also been observed in Teflon FEP thermoelectrets [4.125].

The dependence of peak temperature on temperature during electron injection observed by *Perlman* and *Unger* [4.121] might similarly be explained by the theory given above. The presence of DRIC in a fraction of the irradiated material and its dependence on time, as evident from these results, can also be of importance for the interpretation of short-circuit isothermal and nonisothermal, i.e., thermally stimulated, charge decay. Some analyses in which dielectric properties of irradiated dielectrics have been assumed to be isotropic and time-independent might have to be re-examined.

The long-lasting delayed conductivity in the irradiated region represents a hazard for electron-charged polymer foils. If the free surface accidentally comes into contact with a grounded material, partial compensation of the stored charge might occur. To eliminate this hazard, a "*heat-sealing*" process [4.126] has been proposed. The irradiated dielectric is annealed in open circuit at a temperature high enough to lead to a reduction of the electron–hole densities in the irradiated region by detrapping and recombination, yet not high enough to lead to a significant reduction and loss of excess charge by conduction through the nonirradiated region.

4.6.3 Positive and Negative Charge Storage by Electron Injection

If the dielectric is irradiated with one electrode grounded and the other kept at a constant negative or positive potential, conduction in the irradiated region leads to additional charge storage and allows positive as well as negative charge storage.

The behavior of such samples *charged under an external bias* can be investigated by the same TSC methods used for nonbiased samples [4.92]. Thermograms for positively and negatively biased electron-irradiated Teflon samples (Fig. 4.24) give evidence for storage of positive as well as negative charges. TSC short-circuit currents for positively biased samples have the opposite direction than for negatively biased ones; their amplitudes for comparable bias voltages are smaller since during charging the positive conduction current must first compensate the negative injection current before positive excess charge is generated. Otherwise currents are similar. Both currents show a predominant low-temperature peak called the α peak, followed by a current reversal and smaller peaks at higher temperatures. The α peak appears to be associated with a

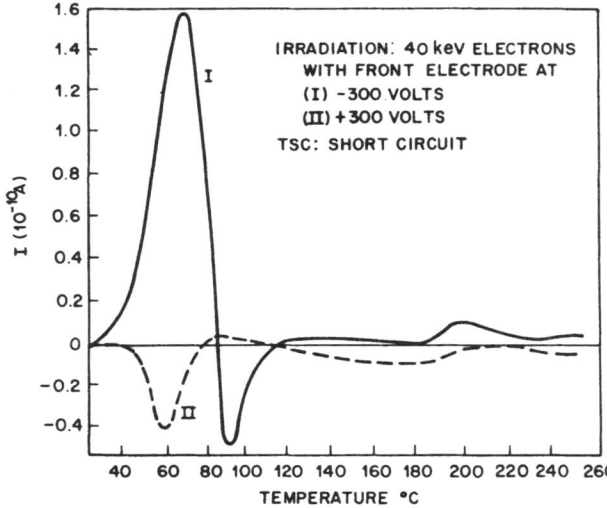

Fig. 4.24. Positive and negative charging by electron injection into externally biased sample. Front bias: Curve *I*, − 300 V. Curve *II*, +300 V. Beam energy 40 keV, beam current density $1.2 \times 10^{-8}\,A\,cm^{-2}$, 25 μm Teflon FEP sample. Heating rate $1\,°C\,min^{-1}$. [4.92]

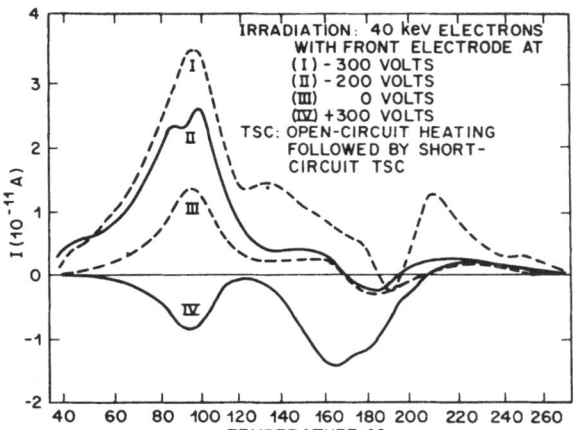

Fig. 4.25. Partial annealing of positively and negatively charged Teflon foils. Annealing temperature 100 °C, annealing time 60 s. Heating rate $1\,°C\,min^{-1}$. [4.92]

region of shallow traps for holes and electrons. The current reversals are believed to be due to self-polarization of the dielectric as will be discussed below.

The field of the trapped primary electrons provides the driving force for the observed currents. Detrapping of this charge is not easily measured in the short-circuit mode because the charge is compensated by thermally activated shallow-trapped secondary holes before the temperature has reached the region where many primary electrons are detrapped. To observe detrapping of the primary charge in short circuit, the number of shallow-trapped carriers in the IR must first be reduced. This is done by temporarily heating the sample in open circuit to a temperature corresponding to the temperature of the α peak, thus activating these carriers and allowing them to recombine.

The α peaks of positive and negative samples annealed in this way at 100–110 °C are indeed considerably lower (Fig. 4.25), while an enhanced peak structure appears at temperatures around 190 °C which corresponds to the ϱ peak [4.46]. The existence of a deep-trap region for electrons has become evident from the previously discussed open-circuit measurements. The existence of such a region for holes is now suggested by the observation of the ϱ peak for positively charged samples.

4.6.4 Conductivity Glow Curves

The interpretation of some aspects of the short-circuit currents of electron-charged dielectrics in terms of the DRIC rather than by the conventionally assumed detrapping and drift of excess charge hinges on the presence of trapped secondary carriers in numbers greatly exceeding that of the injected primaries. A direct proof of the existence of these carriers is obtained by measurements of conductivity glow curves of samples which contain no net charge anywhere in the dielectric.

A *conductivity glow curve* is obtained by heating a dielectric with an externally applied voltage and measuring the external current. If the dielectric has not been irradiated, and the electrodes are not injecting, any measured current is due to the intrinsic (dark) conductivity. If the dielectric has previously been irradiated by fully penetrating gamma or X-rays, a transient current is observed during heating due to the thermal release of the deep-trapped carriers that remain from the previous irradiation.

Samples which have been irradiated with electrons whose average range is considerably in excess of the sample thickness, contain an essentially uniform concentration of trapped electrons and holes without an excess charge. Conductivity glow curves for such samples are due to detrapping and drift of these carriers, and behave in the same way as for gamma-irradiated samples. This equivalence has been confirmed by measurements on 25 μm Teflon foils irradiated with 75 keV electrons or by ^{60}Co gamma rays at the same total dose [4.92], as shown in Fig. 4.26.

Charge neutrality and controlled values of dose rate are nevertheless easier to obtain with gamma rays than with electrons. Investigation of the characteristics of conductivity glow curves therefore are carried out conveniently by irradiation with gamma rays.

Conductivity glow curves have been reported for a variety of irradiated materials. *Wiseal* and *Willard* [4.127] measured the conductivity of glassy 3-methylpentane, gamma-preirradiated at 77 K, as a function of dose, dose rate, temperature, applied voltage, post-gamma irradiation, and nature and concentration of additives. A current peak at 77 K was attributed to the combined mobility of weakly trapped positive and negative charges decaying in approximately 30 min and other peaks at 83 and 102 K to trapped electrons. A current inversion between these peaks, with the current flowing against the applied field,

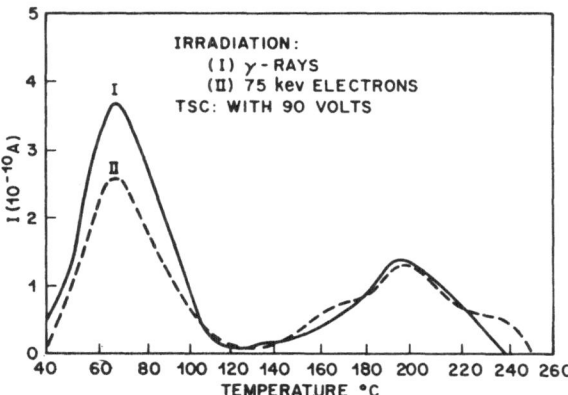

IRRADIATION:
(I) γ - RAYS
(Ⅱ) 75 keV ELECTRONS
TSC: WITH 90 VOLTS

Fig. 4.26. Conductivity glow curves of samples irradiated with gamma rays and penetrating electrons. Samples: 25 μm Teflon FEP foils. Curve I: Irradiation with ^{60}Co gamma rays, absorbed dose 30 krad, dose rate 170 rad s^{-1}. Curve II: Irradiation with 75 keV electrons, irradiation time 170 s, electron current density 1.6×10^{-8} A cm^{-2}. [4.92]

suggested randomization by thermal agitation of a radiation-produced polarization generated at 77 K. *Yahagi* and *Shinohara* [4.128] found for polyethylene, gamma-irradiated at 77 K, a single peak which for low-density material occurred at about 200 K and for high-density material at 250 K. Further investigations were carried out on polyethylene by *Suzuki* et al. [4.128], and by *Nakamura* et al. [4.128]. The latter found two peaks around 60 °C and one near the melting point. The former were attributed to the presence of unsaturated bonds and the latter to oxidation in addition the existence of crystallites. *Blake* et al. [4.129], investigating the same material pre-irradiated at 90 K, found a sharp increase of current of about two orders of magnitude at 250 K followed by a multitude of smaller peaks. Other materials, such as polycrystalline Al_2O_3 [4.130] and MgO [4.131] show a similar behavior. *Pollard* et al. [4.132] investigated the delayed RIC and conductivity glow curves of single crystals of MgO irradiated with relativistic electrons. They found that prolonged annealing of the irradiated samples at 83 °C caused fading of the low-temperature (130 °C) peak but did not affect the high-temperature (230 °C) peak. *Shono* et al. [4.133] found a very complex peak structure in electron-irradiated ZnS which depended on irradiation temperature. Further references, in particular of papers published by Soviet authors, are found in a review article by *Frankevich* and *Yakolev* [4.134]. A detailed investigation of the properties of polyethylene irradiated with X-rays has been performed by *Ramsay* [4.135].

Conductivity glow peaks can be caused by exposure to neutrons as well as by gamma radiation, as has been proved by irradiation of KBr and NaCl crystals with ^{60}Co gamma rays and $^{7}Li(p, n)^{7}$Be neutrons. Exposure to gamma rays at 10 K followed by heating in a field of 10^3 V cm^{-1} gives a predominant peak at 34 K which is absent for neutron irradiation while exposure to neutrons at 77 K gives a predominant peak at 110–120 K which is absent for gammas. This peak structure has been explained by formation of H centers by gammas and of Br$^-$ interstitials by neutrons [4.136].

Table 4.5 summarizes temperature-activated processes in irradiated dielectrics.

Table 4.5. Thermally activated processes in irradiated dielectrics

A) Thermally-stimulated currents in electron-charged dielectrics (no external bias)

Charging	Measurement	Analysis	Ref.
Dielectric *electron-charged in open circuit*; subsequently left in open circuit at room temperature or at higher temperature (annealing).	Dielectric *heated in short-circuit*. Direction of initial current corresponds to transport of holes away from front electrode (or transport of electrons toward front electrode), as consequence of residual electron–hole plasma in IR. At high temperature a current reversal occurs which is caused by drift of released primary electrons to rear electrode.	Current peaks at different temperatures indicate trapping levels. In Teflon, two peaks are observed; the high-temperature peak is attributed to electron detrapping.	[4.46, 118–122]
Dielectric *electron-charged in open circuit*; subsequently left in open circuit at room temperature or at higher temperature (annealing)	Dielectric *heated in open circuit*. Primary charge moves through NIR to rear electrode, in view of absence of field in IR and absence of electron–hole plasma in NIR. Thus only detrapping of primary electrons observed.	Current peaks at different temperatures indicate trapping levels for primary electrons. Frequently only one peak observed (as in Teflon)	[4.62, 123]
Dielectric *electron-charged under applied positive or negative voltage.* Subsequently treated as above.	Dielectric *heated in short circuit*. Depending on polarity of bias, positive or negative current peaks are observed. A peak in one direction usually followed by peak in opposite direction. (Results refer only to Teflon)	Enhancement of negative peak shows increased electron storage as consequence of induced conductivity. Positive peak demonstrates existence of deep hole trap and possibility of storing positive excess charge. Current reversals are mostly due to self-polarization of dielectric during heating.	[4.92]

B) Conductivity glow curves of gamma – irradiated dielectrics (with external bias)

Charging	Measurement	Analysis	Ref.
Dielectric *irradiated by gamma or X-rays or fully penetrating electron-beam.*	Dielectric *heated under external bias.*	Current peaks indicate trapping levels for secondary electrons and holes (thermal activation of trapped carrier population)	[4.92, 127–136]

C) Peak-shift effects (results for Teflon)

Dielectric *electron-charged* in open circuit. *Stored* in open circuit *at room temperature* for different periods of time.	Dielectric *heated in short-circuit*. Temperature of low-temperature peak measured as a function of storage time.	With increasing storage time peak shifts to higher temperatures. Explained by decrease of delayed radiation-induced conductivity in IR during storage. Accordingly the time constant for conduction through IR increases.	[4.62]
Dielectric *electron-charged* in open circuit. *Stored* in open circuit *at increased temperature* for different periods of time.	Dielectric *heated in short circuit*. Peak temperatures and amplitudes measured as functions of storage time.	Low-temperature peak: temperature increases and amplitude decreases with storage time. High-temperature peak: temperature remains constant, amplitude increases with storage time (*population shift*).	[4.62, 121]
Dielectric *irradiated by gamma rays* in short circuit. Stored in short circuit.	Dielectric *heated under applied bias* until first peak is reached, then quenched (while still biased). Process of *heating and quenching repeated* several times.	First peak decreases in amplitude and shifts toward higher temperatures. Attributed to distribution of activation energies (trapping levels). Heating empties increasingly deeper traps until total carrier population is exhausted.	[4.92]

D) Temperature-gradient effect

Dielectric *irradiated by gamma rays in short circuit*.	Dielectric *heated* in short circuit *under temperature gradient*.	External current observed. Possible causes are temperature-activated diffusion or space-charge field due to Compton current generated charges at interfaces between different materials.	[4.143–147]

E) Radioelectret

Dielectric *irradiated by gamma radiation with external voltage* applied during and for some time after irradiation.	Dielectric *heated in short circuit*.	External discharge current observed, exhibiting only low-temperature peak. Possible cause: Schottky layer (barrier) polarization associated with different mobilities of electrons and holes.	[4.92]

4.6.5 Conductivity Glow Curves for Teflon

Investigations with Teflon [4.92] have shown that the field and dose dependence of the low-temperature peak are different from those for the high-temperature peak. The same applies with regard to storage conditions.

The α peak shows only slight *fading* when the sample during storage is kept in open circuit. However, it decreases considerably if the sample is stored with a voltage applied across its terminals and the field during the TSC measurement has the same direction as during room-temperature storage. Since the Schubweg of the carriers is believed to be small compared to the thickness of the sample, removal of carriers by the applied field appears unlikely. A more likely explanation might be polarization of the biased sample due to the formation of a Schottky barrier (Sect. 4.2.1).

The temperature of the α peak increases with increasing *storage time*. This peak shift has to be distinguished from that for short-circuited samples [4.62]; it indicates the existence of a distribution of the activation energies of trapped carriers. The shallow-trapped carriers are released earlier than the more deeply trapped ones. Thus the decrease of the peak amplitude goes together with an increase of the peak temperature. An analogous effect in TSL (thermally stimulated luminescence) has been found in gamma-irradiated cobalt chelate [4.137].

A *peak shift* indicative of the existence of a distribution can also be induced by a repeated heating and cooling cycle. The irradiated sample is heated until the first peak is reached, then cooled fast to room temperature, and heated again. The second peak occurs at a higher temperature than before, but has a lower amplitude. The process can be repeated until the region of shallow traps is depleted. In this way the peak could be moved from 67 °C up to 97 °C [4.92]. A method of partial heating has also been used by van Turnhout [4.138] to prove the existence of a distribution of relaxation times for the dipole polarization of PMMA thermoelectrets.

The possibility of inducing a peak shift by repeated heating might have a bearing on the peak-cleaning technique routinely used in TSC measurements which also employs a heating and cooling cycle and where a similar effect might occur when one is dealing with a distributed polarization or trapping effect [4.139].

For Teflon none of these peak-shift effects have been observed for the high-temperature region. The kinetics of carrier trapping, detrapping, and transport for the α peak is likely to be different from that for the ϱ peak.

Reproducibility of glow curve measurements is not always satisfactory. This might be due to spurious charges residing in the samples before irradiation and to external conditions. It has been found [4.140] that adsorbed and absorbed water can induce a peak at 295 °C in PTFE foils with evaporated gold electrodes. The peak can be eliminated by prolonged heating at 320 °C but gradually reappears over a period of weeks.

Measurements of conductivity glow curves can be distorted by *Schottky layer formation*. Direct evidence for the appearance of a Schottky type polarization in gamma-irradiated Teflon foils and its effect on measured currents could be obtained by *Gross* et al. [4.141].

The field which is applied during the measurement of a conductivity glow curve generates a polarization of the dielectric; the polarization can be "frozen-in" by cooling while the field remains applied. The cooling is initiated shortly after the current has reached its peak value. The process is quite similar to the formation of a thermoelectret.

Analysis of the charge distribution within such an electret by the method of electron injection has shown that a strong negative space charge resides in the region adjacent to the electrode which during poling was the anode. The sign of the charge (negative) excludes the possibility of attributing it to injection from the (positive) electrode. Rather it is a hole depletion layer which was formed because the electrode is blocking.

4.6.6 Short-Circuit TSC in a Temperature Gradient

The anisotropy necessary for the production of discharge currents can be generated by a temperature gradient. Trapped carrier populations in dielectrics irradiated by penetrating electrons, X-rays, and gamma rays have been detected by heating the shorted dielectric while a temperature difference is maintained between opposing electrodes. The conductivity of the dielectric matrix in which the charges are embedded, the mobility of the carriers, and their detrapping are increased in the region of higher temperature, and a discharge current might ensue if hole and electron mobilities are different or if space-charge build-up has occurred during irradiation in the electrode regions due to the interface effect. This effect occurs at the boundary between media of different attenuation and energy transfer coefficients for the radiation concerned [4.142].

Evidence for the *generation of currents and voltages by a thermal gradient* was obtained in experiments where samples of borosilicate glass pre-irradiated by a dose of 4 Mrad of ^{60}Co gamma rays were heated in short circuit in a temperature gradient reaching $23\,°C\,cm^{-1}$ [4.143]. Short-circuit currents of up to $3 \times 10^{-13}\,A\,cm^{-2}$ and open-circuit voltages of 10 V and more were measured. Since voltages of this order can only be due to an electrostatic effect, a conventional thermoelectric effect could be ruled out. This view is supported by the observation that the field necessary to suppress the currents completely appears more closely related to the dose rate than to the temperature gradient [4.144]. Details of the effect have been investigated in various glasses [4.145–147] and polymers [4.148–151].

4.6.7 Radiation-Stimulated Polarization (Radioelectret)

Irradiation of dielectrics by fully penetrating ionizing radiations can induce charge storage and polarization effects when an external polarizing field is applied

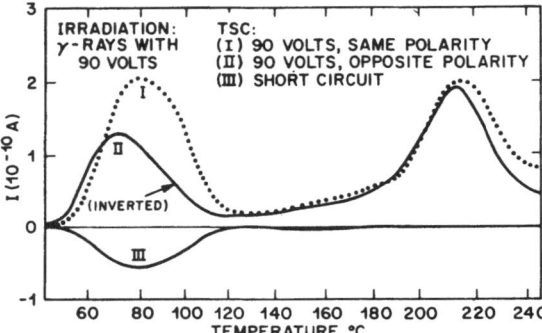

Fig. 4.27. Radiation-stimulated polarization: 25 μm Teflon FEP foils irradiated with ^{60}Co gamma rays, absorbed dose 30 krad. Curve *I*: Conductivity glow curve (bias voltage +90 V), sample irradiated under an applied voltage of +90 V. Polarity of the voltage during heating the same as during irradiation. Curves *II*: Conductivity glow curve (bias voltage −90 V), sample irradiated under an applied voltage of +90 V. Curve *III*: Short-circuit TSC of sample irradiated under an applied voltage of +90 V. [4.92]

during and after irradiation. One instance of radiation polarization is the already mentioned *enhancement of dielectric absorption* which in many dielectrics is considerable. It has been shown [4.152, 153] that the "radioelectret" state could be produced in 0.2 cm thick carnauba wax samples by simultaneous irradiation with beta rays from a ^{90}Sr source and application of an external field of $5 \, \mathrm{kV \, cm^{-1}}$.

Measurements of isothermal decay currents at room temperature and thermally stimulated currents gave polarization charges up to $4 \times 10^{-8} \, \mathrm{C \, cm^{-2}}$ at a mean dose of 8 Mrad. Similar results were obtained with Teflon. Short-circuit TSC thermograms for Teflon (heating rate 1 °C/min) showed a single peak at about 80 °C. Analysis of the initial rise of the TSC curve gave an activation energy of 0.9 eV. Polymethylmethacrylate, polyvinylacetate, polystyrene, polyethylene, and nylon did not form radioelectrets under the same conditions. For nonbiased samples the effect was negligible.

Radioelectrets have also been formed from Teflon foils by irradiation with 30 krad ^{60}Co gamma rays and the simultaneous application of a polarizing field of $35 \, \mathrm{kV \, cm^{-1}}$ [3.92]. Short-circuit thermograms of the irradiated samples show the α peak which for nonpolarized samples is observed only with an external voltage applied during heating. No current is observed in the region of the ϱ peak. The short-circuit thermograms for polarized samples (radioelectrets) were compared with conductivity glow curves of such samples, the applied voltage during polarization being the same as the voltage during heating. When the sample terminals during polarization were connected to the same poles of the voltage source as during the glow curve measurement, current amplitudes decreased because the internal (polarization) and the external fields are in opposition; otherwise they increased (Fig. 4.27). (See also Sect. 4.6.4.)

The polarization effect might explain the *current reversals* observed in short-circuit TSC of electron-charged samples [4.92] (cf. Fig. 4.24). When the temperature approaches the α peak, the secondary carriers populating the shallow traps are thermally activated while the deeply trapped fraction of the field-generating injected charge remains constant. The situation is analogous to that of the radiation-polarization experiment with the difference that this time the carriers are thermally activated, not radiation excited, and that the internal field of the injected charge replaces the external polarizing field. Transport of thermally activated carriers in the direction of the field eventually results in a radioelectret polarization. With increasing temperature the excess charge also decays and the *self-induced polarization* decreases due to further detrapping and randomizing of carriers as envisioned by *Wiseal* and *Willard* [4.127]. The corresponding depolarization current flows in a direction opposite to that of the main component of the TSC. Therefore over a certain temperature interval between the α peak and the ϱ peak the depolarization current prevails and causes a reversal of the external current.

4.6.8 Polarization Effects in Ionic Solids

Radiation-induced polarization effects have also been observed in *ionic solids*, in particular high-purity lithium fluoride [4.154] and calcium fluoride [4.155–158]. LiF irradiated with 100 R X-rays of 35 keV effective energy at a temperature of 130 K in the presence of an applied field of 6.7 kV cm^{-1} and subsequently heated at a rate of 30–40 °C/min shows current peaks at 160 and 180 K. The lower peak was tentatively associated with the dipole moment generated by net charge separation of the V_K center hole and its charge-compensating associated trapped electron. This peak is also observed in Mg-doped LiF.

Polarization effects are found in CaF_2, which forms a thermoelectret with a characteristic depolarization peak at about 130 °C [4.159]. Effects can formally be described by second-order kinetics. When the polarizing field is below 2.5 kV cm^{-1}, the charge released in short-circuit TSC increases linearly with the field; for higher fields, it increases as the square root of the field. Characteristically the released charge for a sample polarized by 10^4 V cm^{-1} at 190 °C is 5×10^{-12} C cm^{-2} per V cm^{-1} corresponding to a dielectric constant of 5×10^{-12} F cm^{-1}. The conductivity is ionic with an activation energy of 1.3 eV. The square-root relation between Q and V is characteristic for the formation of a Schottky layer as has been found for naphthalene crystal electrets [4.160]. Electrode polarization might have occurred also at the copper-foil electrode pressed into contact with one crystal face, the other electrode being a painted-on silver or copper surface layer.

Irradiation effects for CaF_2 have been investigated in detail by *Podgorsak* [4.155] and *Podgorsak* and *Moran* [4.158].

Conductivity glow curves (Fig. 4.28): Samples are irradiated at 77 K with 100 R X-rays (average energy 35 keV), and simultaneously polarized by an external field

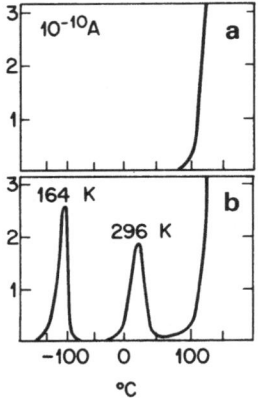

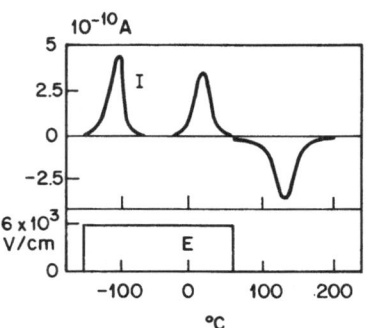

Fig. 4.28a, b. CaF$_2$ conductivity glow curves; heating rate 0.5 °C s^{-1}, applied field 6 × 10^3 V cm^{-1}, sample thickness 0.08 cm, copper-foil biased electrode of 1 cm^2 area; copper or silver painted sensing electrode of 1/3 cm^2 area. (a) Nonirradiated sample. (b) Sample irradiated at 77 K by 100 R, 35 keV$_{eff}$ X-rays. Field applied after irradiation. [4.158]

Fig. 4.29. CaF$_2$ conductivity glow curve followed by short-circuit TSC, for sample irradiated with 20 R at 77 K. External field applied after irradiation over temperature interval covering 164 and 296 K peaks. [4.158]

of 6 × 10^3 V cm^{-1}. After termination of the irradiation they are heated at a rate of 0.5 °C s^{-1} while the external field remains applied. Thermograms show two peaks at 136 and 296 K and a strong current increase above room temperature; the latter is due to the ionic conductivity associated with the thermoelectret state.

If irradiation is performed in short or open circuit, and the polarizing voltage is applied only within the temperature interval of 100–290 K covering the low temperature peaks, further heating in short circuit gives rise to a peak at about 410 K. This peak has the opposite direction as the low-temperature peaks, and is due to the compensation and/or release of the radiation-induced polarization by ionic conduction (cf. Fig. 4.29).

Short-circuit TSC for polarized samples: The internal field associated with the thermoelectret state can replace the external field and induce self-polarization of the dielectric. For this purpose, the thermoelectret state is generated by polarizing the sample at room temperature, cooling it to 77 K with the field applied, and short-circuiting it. The sample is then irradiated at the low temperature as before. The short-circuit thermogram (Fig. 4.30) of such a sample shows the low-temperature peaks at 136 and 296 K, and in addition a high-temperature peak which has a slightly lower amplitude and slightly higher temperature than the previous thermoelectret peak. The difference between the TSC curves of the nonirradiated and the irradiated thermoelectret samples gives the peak structure of the sample biased between 100 and 290 K (cf. Fig. 4.29).

The *efficiency of the conversion process* can be expressed as the relation between the number of measured carriers and the total amount of energy

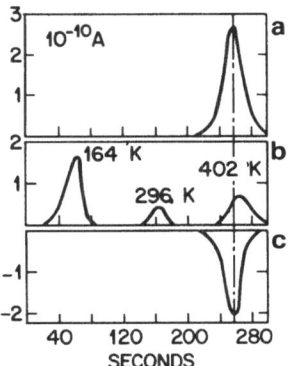

Fig. 4.30a–c. CaF$_2$ thermoelectret irradiation. (**a**) Short-circuit TSC of thermoelectret (no irradiation). (**b**) Short-circuit TSC of thermoelectret irradiated at 77 K with 100 R. (**c**) Difference of thermoelectret peaks of curves (a) and (b) [4.158]

absorbed in the sample. The charge content of the two low-temperature peaks is approximately 7×10^{-9} C or 4.3×10^{10} carriers for a 0.08 cm thick sample with a sensitive area of $1/3$ cm^2 and density of 3.2 g cm^{-3}, prepoled with 3×10^3 V cm^{-1} and exposed to 100 R of 35 keV$_{eff}$ X-rays. The R to rad conversion factor for the low-energy X-rays has been given as 10 rad per R [4.161]. The total amount of energy absorbed in the sensitive volume of the sample is then 5.2×10^{15} eV. Therefore 1 carrier is measured for an absorbed energy of 1.2×10^5 eV.

An approximate value for the energy G for generation of a free carrier pair is 10^3 eV [4.24]. Thus the number of radiation-generated carrier pairs is approximately 5.2×10^{12}. Approximately one out of 1×10^2 radiation generated carriers is measured. Factors which contribute to the high value of G are columnar and geminate recombination. The first occurs along the path of a primary ionizing particle where ionization density is high, and the second occurs between an electron and the ion from which it is produced. Both effects are field dependent and so is G.

Measurements performed with gamma-irradiated carnauba wax [4.162] gave an efficiency orders-of-magnitude lower. Comparison is, however, difficult since the efficiency depends on photon energy and the nature of the internal polarization as well as on the nature of the dielectric. A uniform volume polarization, as it has been observed in carnauba wax, can be expected to have a low efficiency [cf. Sect. 4.7.1].

It is not clear how the internal polarization generates the *low-temperature radiation-activated depolarization currents* in CaF$_2$. A proportion of the X-rays, whose half-length is 0.3 cm in Al, is absorbed in the sample. This can create a space-charge polarization and depth-dependent conductivity. Both effects can contribute to the generation of depolarization currents. A field-independent uniform conductivity, even if it is a function of time, cannot cause the effect, as has been pointed out by *Moran* and *Podgorsak* [4.163]. The same conclusion can be drawn from the zero-field plane theorem. A high efficiency is observed only for the purest materials; addition of impurities which act as recombination centers strongly reduces the measured currents.

4.7 Dosimetry

4.7.1 Dosimetry with Self-Biased Systems

Several methods of radiation dosimetry have been based on processes of interaction between photon radiation and internal polarization and space-charge fields in dielectrics.

The air gap between the electrode and the floating surface of a partially shorted electret behaves like an *ionization chamber* biased by the surface field of the electret. If such a system is irradiated by gamma rays, the surface charge decays exponentially, the time constant depending on the distance between the surface of the electret and the opposing electrode which determines the air-gap capacitance and the air ionization current [4.164].

The surface charge of apparently shorted resin–carnauba wax electrets, measured by an induction plate method, irradiated in short-circuit with gamma rays, also decays exponentially with radiation dose; the reduction is, however, only temporary, the charge recovers within about two weeks [4.165]. The effect presumably is due to an air gap still existing between dielectric and covering metal foil.

A systematic investigation has been carried out by *Fabel* and *Henisch* [4.166] with polymer thermoelectrets, formed at high fields and temperatures. Characteristic values of forming fields and temperatures given were, respectively, $35 \, \text{kV cm}^{-1}$ and $200 \, °\text{C}$ for nylon, $37 \, \text{kV cm}^{-1}$ and $200 \, °\text{C}$ for Mylar, $18 \, \text{kV cm}^{-1}$ and $320 \, °\text{C}$ for Teflon, and $19 \, \text{kV cm}^{-1}$ and $150 \, °\text{C}$ for polystyrene. Recovery effects were found in nylon and to a lesser degree in Mylar, while recovery after read-out was very small in Teflon. Practically all electrets eventually carried homocharges on the measuring surface. Read-out sensitivities down to 1 mR are claimed. Both polystyrene and Teflon were sensitive to neutron irradiation. Since all samples were irradiated in air and wrapped in aluminum foil, charge decay due to a surface effect cannot be ruled out. Similar observations have been reported by *King* [4.167].

A true *volume effect* has first been found in carnauba wax electrets [4.162] irradiated with Mrad doses of ^{60}Co gamma rays; the residual charge was determined as a function of dose by short-circuit TSC measurements. Charge decreased exponentially with dose with a decay constant of 1 Mrad. Typically, 1 Mrad reduced the value of the released charge Q of a sample formed at $5 \, \text{kV cm}^{-1}$ and $70 \, °\text{C}$ from $1.8 \times 10^{-9} \, \text{C cm}^{-2}$ to $0.66 \times 10^{-9} \, \text{C cm}^{-2}$. The low efficiency might be explained by the absence of a macroscopic internal electric field in the wax electret which would be capable of producing charge transport by conduction.

The low radiation sensitivity of materials with a uniform volume polarization is also found for the *pyroelectric effect*. The pyroelectric voltage of tourmaline subjected to uniform heating between 300 and 600 K goes through a maximum [4.168] whose amplitude is decreased after previous radiation with heavy doses of

X-rays. The decrease of the pyroelectric voltage depends on the previously received X-dose and is due to the thermal generation of free carriers excited into traps by the absorbed radiation dose.

The *Compton current* associated with a directional beam of photons generates *space charges* at the interface between a high-Z material (metal) and a low-Z material (dielectric). Unilateral heating of the charged system which produces a temperature gradient, with the metal at the higher temperature, releases the trapped charge. This drifts toward the hot electrode, thus generating a current in the external circuit. If the heating is performed under controlled conditions, the total amount of released charge is proportional to the space charge, which in turn is proportional to absorbed dose. Experimental dosimeter models, made from the borosilicate sealing glass used to make transistor headers consist of a metal–glass–metal sandwich structure with Kovar or Nicoseal electrodes bounded to the glass with glass-to-metal seals [4.169]. When heated, these samples generated currents up to 2.3×10^{-12} A and an integrated output of 37×10^{-12} C after irradiation with a dose of 3.3×10^5 rad of ^{60}Co gamma radiation with slight fading. The use of Compton-current induced polarization by a TSC method has also been proposed [4.170, 171]. This type of Compton dosimetry must, however, be distinguished from that using self-powered Compton diodes (cf. Sect. 4.5.2).

Higher sensitivities can be achieved with the use of CaF_2 *thermoelectrets* [4.155, 159] by analyzing the charge content of the 200 K or the 330 K peak (Sect. 4.6.8). For low exposures, the peak amplitudes grow linearly with dose up to 120 R for the 200 K peak and 75 R for the 330 K peak. Measurements with X-rays of 35 keV average energy gave a radiation response of about 4×10^{-10} C cm^{-2} per R for each peak. Measurements became difficult below 100 mR. Measurements at room temperature are, however, problematic in view of the ionic conductivity of the material. Since the effect apparently is independent of thickness, materials showing acceptable dosimetric response can be fabricated in thin films and used for detection of easily absorbed radiation. A thin-film tritium dosimeter has also been proposed [4.172].

Uniform volume polarization as the driving force for the measurement of radiation-induced conductivity is employed in *ferroelectric radiation detectors*. Irradiation of a poled ferroelectric causes a charge release which within limits varies linearly with absorbed dose. For polycrystalline $Pb(Zr_{0.65}Ti_{0.35})O_3$ + 1 wt. % Nb_2O_5, a sensitivity of 10^{-12} C cm^{-2} per rad can be obtained [4.173, 174]. Response of the system is dynamic, so it can be used for short-time pulse diagnostics, the sensitivity given above having been measured in the pulsed Sandia reactor. The detector responds by virtue of the pyroelectric properties of the material. The absorbed energy causes a slight change of temperature which results in a change of the internal polarization, a corresponding perturbation of the equilibrium field, and the appearance of free charges at the electrodes, the value of released free charge being proportional to the absorbed energy [4.175]. Poled barium titanate has also been used as a detector material [4.176] in the temperature range from 5–30 °C.

An analytical model for the energy balance within a ferroelectric crystal relates ionic spacing to macroscopic quantities such as pyroelectric and piezoelectric coefficients [4.177].

4.7.2 Dosimetry with Externally Biased Systems

Detrapping and recombination of charge carriers is also observed in thermally stimulated luminescence (TSL). Analogies between TSC and TSL have been investigated by many authors [4.129, 137]. Thermoluminescence dosimetry (TSLD) has become one of the most successful modern methods of dosimetry over a wide range of radiation doses and energies. A suitable phosphor, after irradiation, is flashed-out, i.e., heated very fast from room temperature to a temperature of 150–200 °C. Radiation-excited and trapped electrons are temperature-activated and eventually recombine with emission of light quanta. The total light output, measured with a sensitive photomultiplier, is proportional to the absorbed dose within wide limits.

It appears tempting to develop along similar lines a method of *thermo-stimulated-current-dosimetry* (TSCD) in which a uniformly irradiated dielectric is flashed-out and the charge output measured with a current-integrating instrument. While in TSLD no external "driving force" is needed during heating, in TSCD application of an external voltage is used to generate an external current. This might represent a drawback, since it also generates intrinsic conductance currents which eventually mask the radiation-induced current, at least in systems operating above room temperature. On the other hand, TSCD needs for readout only a picoammeter instead of the much more complex photomultiplier system of TSLD, employs a radiation sensor which is insensitive to visible light, and requires no light-proof reader. For dosimetry in radiotherapy, radiobiology, and personnel protection, tissue equivalence is desirable. Thus sensor materials are confined to elements of low atomic number with a very small percentage of higher atomic number elements present as impurities. The method was discussed by *Ramsay* [4.178].

A variety of materials, including polymers, has already been investigated. Useful response was found in MgO, polyethylene, polypropylene, and TPX–R, a methylpentene polymer with a density of $0.83 \ \mathrm{g \ cm^{-3}}$, while high temperature-dependent intrinsic conductivity ruled out TSLD materials like LiF and lithium tetraborate [4.179]. Detectors were shaped as small foils ($10 \times 10 \times 1.5$ mm) to which potentials up to 1 kV were applied for read-out. Painted-on Aquadag layers were used as electrodes; Al electrodes were found to be noisy when evaporated on TPX. For a preliminary assessment of the suitability of various materials, the temperature controller raised the temperature of the sample to 160 °C in 5 s and held it there for about two minutes. Typical peak current densities are $3 \times 10^{-11} \ \mathrm{A \ cm^{-2}}$ with a polarizing field of $2130 \ \mathrm{V \ cm^{-1}}$ after irradiation with 10 rad and a high-temperature static current density of $3 \times 10^{-12} \ \mathrm{A \ cm^{-2}}$. A TSC analysis with linear heating rate of 12 °C/min showed suitable dosimetry peaks for

MgO (140 °C), MgF_2 (110 and 175 °C), and transparent TPX (150 °C), the latter appearing to be the most promising material. Linear dose response over the range 1–1000 rad appears feasible. Other glasses and plastics which appear promising are PTFE (polytetrafluoroethylene) and Spectrosil OH (synthetic fused silica doped with OH) [4.180]. Anomalous dark currents attributed to the painted-on Aquadag electrodes were, however, observed in PTFE which could be annealed but returned after 24 h. It has also been found that the radiation sensitivity of a polymer depends markedly on the grade. Whereas most PTFE samples have a low sensitivity, material made from "Fluon" G 163 granular powder has a considerably improved response [4.181]. A review of the method has been given by *Bowlt* [4.182].

A material which has a sensitivity comparable to that of CaF_2 but shows little fading at room temperature seems to be Al_2O_3, [4.183]. Thermograms of samples irradiated with 35 keV_{eff} X-rays show several useful peaks. A short review article [4.184] describes various TSCD methods and the properties and purification requirements for a variety of materials.

The *measurement of neutron doses* promises to become a particularly important application of the TSC dosimeter. *Ramsay* and *Joesoef* [4.185] have shown that polyethylene, being a highly hydrogenated material, responds to neutron irradiation. The response curve of the thermally liberated charge against dose consists of an initial steep slope up to 150 rad followed by a longer smaller gradient. The system appears suitable for measuring neutron doses between 20 and 10,000 rad in the energy range from 1–14 MeV. The production of a trapped carrier population is believed to be due to the heavily ionizing knock-on protons produced by the neutrons.

So far discussion has been limited to metal–insulator–metal (MIM) structures. Recently the more complex *metal–insulator–semiconductor* (MIS) structures have been investigated for possible use in dosimetry. The principle of measurement is based on the separation of radiation-generated electrons and holes and the preferential trapping of holes. The trapped charge is directly measured by means of a sensing electrode.

The preferred device is the *metal–oxide–semiconductor* (MOS) system, in particular the metal–SiO_2–Si capacitor. Typically the device consists of an Al "gate" electrode in contact with a thin (0.1–1 μm thick) silicon oxide film. A semiconducting silicon film acts as the sensing electrode. The system is biased by an external voltage applied between the gate electrode and an ohmic electrode in contact with the silicon.

A general discussion of the *physics of MOS structures* has been given by several authors [4.186–188]. Electron–hole pairs are created by the primary photon radiation absorbed in the dielectric. The mobility–lifetime product of the holes is much smaller than that of the electrons. For positive gate bias, electrons which are lifted into the conduction band are swept out of the oxide and into the metal plate, while holes are trapped near the SiO_2–Si boundary. This charge can remain in the oxide for weeks or months. For negative bias, the holes are trapped near the SiO_2–gate electrode boundary. Since the readout method depends on the

Table 4.6. Dosimetric effects and methods

Methods	Mechanism	Driving force	Material	Sensitivity	Ref.
Compton diode	Directional beam of photons generates electron current (Photo-Compton current) in irradiated dielectric, resulting in signal in external circuit.	Photon radiation	All dielectrics	10^{-12} A cm^{-2}(R s^{-1})$^{-1}$	[4.111–112]
Compton TSC system	Photo-Compton current generates space charge at metal–dielectric interface. Heating in temperature gradient releases charge toward metal and generates external current.	Space-charge field	Borosilicate glass	10^{-6} C/rad	[4.169]
Ionization chamber	Ionization chamber poled by electret. Collected charge compensates electret surface charge. Charge decay measured by lifting of counterelectrode (induction plate method).	Field of electret		1 mR	[4.164–166]
Conductivity glow curve (TSCD)	Radiation generates electron–hole plasma in dielectric which remains after termination of radiation, with little fading (flash-out) under applied field generates discharge current.	External voltage source	MgO,Al$_2$O$_3$ Spectrosil Teflon	3×10^{-11} A cm^{-2}(10 rad)$^{-1}$ (peak current density)	[4.178–185]

Method	Mechanism	Driving force	Material	Sensitivity	Ref.
Thermoelectret depolarization	Irradiation of thermoelectret reduces internal polarization. Amplitude and charge content of short-circuit TSC changes proportional to exposure.	Polarization of thermoelectret	CaF_2	4×10^{-10} C cm^{-2} R^{-1}	[4.155–159]
Ferroelectric effect	Irradiation of a poled ferroelectric reduces the internal polarization and this generates a displacement current and open-circuit voltage.	Ferroelectric polarization	$Pb(Zn_{0.65} Ti_{0.35})O_3 + 1wt.\%Nb_2O_5$	10^{-12} C cm^{-2} rad^{-1}	[4.173–177]
Pyroelectric effect	Irradiation reduces pyroelectric peak observed during heating. Comparison of peak amplitude of irradiated and and nonirradiated sample gives dose.	Spontaneous internal polarization	Tourmaline	10^5 R	[4.168]
Metal–Oxide–semiconductor system (MOS)	Irradiation of MOS structure under applied field leads to separation of electrons and holes and preferential trapping of holes at oxide–semiconductor interface. Shift of "flat-band voltage" proportional to trapped charge.	External voltage	Al–SiO$_2$–Si	10^{-2} V/rad	[4.187–192]
Electro-mechanical	Movable electrode, balanced by a mechanical force and electret field, moves when field is reduced by irradiation.	Field of electret		Several rad	[4.193]

charges imaged into the silicon, the positive bias method allows maximum sensitivity, though not necessarily highest stability and linearity. Charge buildup and current generation during irradiation with X-rays have been investigated by *Farmer* [4.187]. The hole traps may derive from nonbonding orbitals of an intact Si–O–Si network system or sites containing a broken Si–O bond. The source of the broken bond could be either metallic impurity ions or strain of the network [4.188].

A convenient parameter which can be measured in the device is the gate voltage at which the field-induced inversion current reaches some chosen level. This *threshold* voltage depends on the amount of trapped oxide charge. Irradiation generates additional charge and therefore produces a shift of the threshold voltage. The effect can also be determined by capacitance–voltage measurements. By finding the difference between the measured voltage for a given capacitance, say the "flat-band" capacitance and its "ideal" value, one can determine the total charge that is trapped in surface states as a function of surface potential.

Dosimeters based on these principles have been described by several authors [4.189–191]. Readily available MOS transistors give a sensitivity of 10^{-2} V/rad; an increase of the sensitivity by increasing the thickness of the dielectric could make the device potentially useful in the radiation-safety monitoring field. The maximum dose measurable is controlled by saturation of traps by holes or by field conditions in the oxide. A special circuit allows to extend the region of linear response toward 10^3 rad.

After repeated irradiation an *annealing treatment* can restore the oxide to the unirradiated state for reuse. Heating to 150–200 °C does, however, not lead to a thermal depopulation of holes, which would require a higher temperature, but to a thermally activated injection of electrons from either interface via interface states. This effect is similar to the radiation-induced discharge of electron-charged Teflon foils (Sect. 4.4.3).

Electron trapping has been observed [4.192] in SiO_2 layers implanted with 20 keV aluminum ions. Photoinjection into the oxide resulted in the buildup of a negative space charge which at room temperature proved to be extremely stable. A sample stored in darkness in open circuit for three months showed a change in flat-band voltage of only a few percent. Steady-state and transient analysis methods allowed the spatial distribution of trapped electrons and of traps created by the injected aluminum ions to be determined. The negative space charge could be annealed optically and also thermally at 350 °C. A significant reduction of the concentration of electron traps required, however, prolonged heating to 600 °C.

Table 4.6 summarizes dosimetric effects and methods. Only Compton diodes have so far been used routinely in practical applications.

Acknowledgements. For much help and encouragement received during the preparation of this work the author wishes to thank his friends and colleagues: G. M. Sessler from Technische Hochschule, Darmstadt; J. E. West and D. Berkley from Bell Laboratories; and Sergio Mascarenhas, G. Leal Ferreira, and L. Nunes de Oliveira from the University of São Paulo. Preparation and revision of the typescript has been greatly helped by Mrs. Hildegard Franks and Mrs. Betty Kubli (deceased).

References

4.1   J.J.Thomson, R.McClelland: Proc. Cambridge Philos. Soc. **9**, 15 (1896)
4.2   H.Becquerel: C.R. Acad. Sci. **136**, 1173 (1903)
4.3   A.Joffé: Ann. Phys. (Leipzig) **20**, 946 (1906)
4.4   W.C.Röntgen: Ann. Phys. (Leipzig) **64**, 1 (1921)
4.5   D.Nasledow, W.Scharawsky: Ann. Phys. (Leipzig) **3**, 63 (1929)
4.6   C.Roos: Z. Phys. **36**, 18 (1926)
4.7   F.Seidl: Z. Phys. **73**, 45 (1932); **99**, 695 (1936)
4.8   W.Scislowski: Acta Phys. Pol. **4**, 123 (1935); **6**, 403 (1935); **7**, 27 (1939); **7**, 214 (1939)
4.9   A.Rose: RCA Rev. **12**, 362 (1951); Phys. Rev. **97**, 322, 1538 (1955)
4.10 J.F.Fowler: Proc. R. Soc. London A **236**, 464 (1956)
4.11 B.Gross: *Charge Storage in Solid Dielectrics* (Elsevier, Amsterdam 1964)
4.12 R.K.Thatcher (ed.): TREE (*Transient Radiation Effects in Electronics*) Defense Atomic Support Agency Report DASA 1420-1. Battelle Memorial Inst., August 1967 (AD-824434) J.A.Wall, E.A.Burke, A.R.Frederickson: In *Proceedings of the Spacecraft Charging Technol. Conf.*, ed. by C.P.Pike, R.R.Lovell, Air Force Surveys in Geophysics, No. 364. NASA TMX-73537 (February 1977) p. 569
4.13 B.Gross: *Electrical Irradiation Effects in Solid Dielectrics*. INIS-mf-1235 (1973)
4.14 W.C.Johnson: IEEE Trans. NS-**19**(6), 33 (1972) H.J.Wintle: IEEE Trans. EI-**12**(2), 97 (1977)
4.15 J.G.Simmons, G.W.Taylor: J. Phys. C **7**, 3051 (1974) G.W.Taylor, J.G.Simmons: J. Phys. C **8**, 3360 (1975)
4.16 E.W.Montroll, H.Scher: J. Stat. Phys. **9**, 101 (1973) H.Scher, E.W.Montroll: Phys. Rev. B **12**, 2455 (1975)
4.17 G.F.Leal Ferreira: Phys. Rev. B **16**, 4719 (1977)
4.18 V.A.J.van Lint, J.W.Harrity, T.M.Flanagan: IEEE Trans. NS-**15**(6), 194 (1968) T.J.Ahrens, F.Wooten: IEEE Trans. NS-**23**, 1268 (1976)
4.19 M.G.Broadhurst: In *Dielectric Properties of Polymers*, ed. by F.G.Karasz (Plenum, New York 1972) p. 129
4.20 ICRU (International Commission on Radiological Units and Measurements) Rpt. 19. *Radiation Quantities and Units* (1971) J.H.Hubbell: *Photon Cross Sections, Attenuation Coefficients, and Energy Absorption Coefficients*. NSRDS-NBS 29, August (1969)
4.21 G.J.Hine, G.L.Brownell: *Radiation Dosimetry* (Academic Press, New York 1956) p. 12
4.22 B.Gross: J. Appl. Phys. **36**, 1635 (1965)
4.23 L.Koblinger: Phys. Med. Biol. **19**, 885 (1974) D.W.Anderson: Phys. Med. Biol. **21**, 524 (1976)
4.24 J.Hirsch, E.M.Martin: J. Appl. Phys. **45**, 1008 (1974)
4.25 A.von Hippel, E.P.Gross, J.G.Jelatis, M.Geller: Phys. Rev. **91**, 568 (1953)
4.26 L.Nunes de Oliveira, G.F.Leal Ferreira: Phys. Rev. B **11**, 2311 (1975)
4.27 A.M.Goodman, A.Rose: J. Appl. Phys. **42**, 2823 (1971)
4.28 W.Helfrich, P.Mark: Z. Phys. **166**, 370 (1962) A.Many, G.Rakavy: Phys. Rev. **126**, 1980 (1962)
4.29 M.A.Lampert, P.Mark: *Current Injection in Solids* (Academic Press, New York 1970)
4.30 A.I.Rudenko: J. Non-Cryst. Solids **22**, 215 (1976)
4.31 D.J.Dascalu: J. Appl. Phys. **44**, 3609 (1973)
4.32 C.Popescu, H.K.Henisch: Phys. Rev. B **11**, 1563 (1975)
4.33 C.Popescu, H.K.Henisch: Phys. Rev. B **14**, 517 (1976)
4.34 B.Gross, L.Nunes de Oliveira: J. Appl. Phys. **45**, 4724 (1974)
4.35 B.Gross: Electrostatics **1**, 125 (1975)
4.36 L.Nunes de Oliveira, B.Gross: J. Appl. Phys. **46**, 3132 (1975)
4.37 J.Lindmayer: J. Appl. Phys. **36**, 196 (1965)
4.38 B.Gross, M.M.Perlman: J. Appl. Phys. **43**, 853 (1972)

4.39 G.Dreyfus, J.Lewiner: Phys. Rev. B **8**, 3032 (1973)
4.40 B.Gross, G.Leal Ferreira, L.Nunes de Oliveira, G.Dreyfus, J.Lewiner: Phys. Rev. B **9**, 5318 (1974)
4.41 B.Gross: J. Phys. D **7**, L 103 (1974)
4.42 B.Gross: J. Electrochem. Soc. **115**, 376 (1968)
4.43 B.Gross, R.J.de Moraes: J. Chem. Phys. **37**, 710 (1962)
4.44 B.Gross: J. Polym. Sci. Polym. Phys. Ed. **10**, 1941 (1972)
4.45 B.Gross: J. Electrochem. Soc. **119**, 855 (1972)
4.46 J.van Turnhout: *Thermally Stimulated Discharge of Polymer Electrets* (Elsevier, Amsterdam 1975)
4.47 B.Gross: J. Appl. Phys. **43**, 2449 (1972)
4.48 W.E.Spear: J. Non-Cryst. Solids **1**, 197 (1969)
4.49 H.J.Wintle: J. Appl. Phys. **43**, 2927 (1972)
4.50 P.Langevin: Ann. Chim. Phys. **28**, 289 (1903); **28**, 433 (1903)
4.51 G.A.Ausman, F.B.McLean: Appl. Phys. Lett. **26**, 173 (1975)
4.52 B.Gross: Solid State Commun. **15**, 1655 (1974)
4.53 W.Shockley: J. Appl. Phys. **9**, 635 (1938)
4.54 K.K.Kanazawa, I.P.Batra, H.J.Wintle: J. Appl. Phys. **43**, 719 (1972)
4.55 B.Gross, L.Nunes de Oliveira: In *Electr. Dielectr. Symp.* 1975, ed. by M.Campos (Academia Brasil Ciencias, Rio de Janeiro 1977) p. 15
      L.Nunes de Oliveira, G.F.Leal Ferreira: J. Electrostatics **1**, 371 (1975); **2**, 187 (1976); **2**, 249 (1976/1977)
4.56 J.H.Calderwood, B.K.Scaife: Philos. Trans. R. Soc. London Ser. A **269**, 217 (1971)
      G.F.Leal Ferreira, B.Gross: Rev. Bras. Fis. **2**, 205 (1972)
      G.F.Leal Ferreira: J. Non-Metals **2**, 109 (1974)
4.57 L.E.Carrano de Almeida, G.F.Leal Ferreira: Rev. Bras. Fis. **5**, 349 (1975)
      P.C.de Camargo, G.F.Leal Ferreira: Rev. Bras. Fis. **6**, 231 (1976)
      P.C.Camargo, G.F.Leal Ferreira: In *Electrets and Dielectr.* Symp. 1975, ed. by M.Campos (Academia Brasil Ciencias, Rio de Janeiro 1977) p. 59
      M.Zahn, C.F.Tsang, S.C.Pao: J. Appl. Phys. **45**, 2432 (1974)
      M.Zahn, S.H.Pao: J. Electrostatics **1**, 235 (1975)
4.58 G.F.Leal Ferreira, B.Gross: J. Non-Metals **1**, 129 (1973)
      G.F.Leal Ferreira, B.Gross: In *Electrets*, ed. by M.M.Perlman (Electrochemical Society, Princeton, N.J. 1973) p. 252
4.59 R.A.Creswell, M.M.Perlman: J. Appl. Phys. **41**, 2365 (1970)
4.60 H.J.Wintle: J. Appl. Phys. **43**, 2927 (1972)
4.61 C.Bucci, R.Fieschi: Phys. Rev. Lett. **12**, 16 (1964)
      C.Bucci, R Fieschi, G.Guidi: Phys. Rev. **148**, 816 (1966)
4.62 G.M.Sessler, J.E.West: Phys. Rev. B **10**, 4488 (1974); J. Appl. Phys. **50**, 3328 (1979)
4.63 R.B.Schilling, H.Schachter: J. Appl. Phys. **38**, 841 (1967)
4.64 B.Gross: Phys. Rev. **107**, 368 (1957)
4.65 B.Gross: J. Polym. Sci. **27**, 135 (1958)
      J.E.Rauch, A.Andrews: IEEE Trans. NS-**13**(6), 109 (1966)
      J.Furuta, E.Hiraoka, S.Okamoto: J. Appl. Phys. **37**, 1873 (1966)
      G.Brown: J. Appl. Phys. **38**, 3904 (1967)
      F.Dekoupolos, E.Marx: Electrotech. Z. A **88**, 617 (1967)
4.66 P.V.Murphy, S.Costa Ribeiro: J. Appl. Phys. **34**, 2061 (1963)
4.67 G.M.Sessler, J.E.West: J. Polym. Sci. Polym. Lett. Ed. **7**, 367 (1969)
4.68 L.Katz, A.S.Penfold: Rev. Mod. Phys. **24**, 28 (1952)
4.69 J.A.Gledhill: J. Phys. A **6**, 1420 (1973)
      T.Matsukawa, R.Shimizu, K.Harada, T.Kato: J. Appl. Phys. **45**, 733 (1974)
4.70 M.J.Berger, S.M.Seltzer: *Tables of Energy Losses and Ranges of Electrons and Positrons* (NASA SP 3L12, Washington, D.C. 1964)
      L.Pages, E.Bertel, H.Joffre, L.Sklavenitis: At. Data **4**, 1 (1972)
4.71 S.M.Seltzer, M.J.Berger: Nucl. Instrum. Methods **119**, 157 (1974)

4.72   B.Gross, K.A.Wright: Phys. Rev. **114**, 725 (1959)
        T.Tabata, R.Ito, S.Okabe: Phys. Rev. B **3**, 570 (1971)
        T.Tabata, R.Ito, Y.Fuyita: J. Appl. Phys. **42**, 3361 (1971)
4.73   J.G.Trump, K.A.Wright, A.M.Clarke: J. Appl. Phys. **21**, 345 (1950)
4.74   S.Matsuoka, H.Sunaga, R.Tanaka, H.Hagiwara, K.Araki: IEEE Trans. NS-**23**, 1447 (1976)
        D.A.Berkley: J. Appl. Phys. **50**, 3447 (1979)
        A.R.Frederickson: *Electric Fields in Irradiated Dielectrics*. In: 1978 Spacecraft Charging
        Conf. (US Air Force Acad. Colorado Springs Co., 1 Nov. 1978)
4.75   H.Seliger: Phys. Rev. **100**, 1029 (1955)
4.76   W.E.Spear: Proc. Phys. Soc. B **68**, 991 (1955)
4.77   B.Gross, J.Dow, S.V.Nablo: J. Appl. Phys. **44**, 2459 (1973)
4.78   B.Gross, G.M.Sessler, J.E.West: Appl. Phys. Lett. **22**, 315 (1973)
        G.M.Sessler, J.E.West: J. Electrostatics **1**, 111 (1975)
4.79   L.Pensak: Phys. Rev. **75**, 472 (1949)
        C.Bowlt, W.Ehrenberg: J. Phys. C **2**, 159 (1969)
4.80   F.Ansbacher, W.Ehrenberg: Nature (London) **164**, 144 (1949); Proc. Phys. Soc. London
        A **64**, 362 (1951)
        P.M.Adams, D.P.L.Jones, E.F.Taylor, C.N.W.Litting: J. Phys. D **7**, L165 (1974)
4.81   W.Ehrenberg, N.J.Hidden: J. Phys. Chem. Solids **23**, 1135 (1962)
4.82   W.Ehrenberg, V.B.Gutan, L.K.Vodopyanov: Br. J. Appl. Phys. **17**, 63 (1966)
4.83   M.A.Lampert, W.C.Johnson, W.E.Bottoms: *Study of Electronic Transport and Breakdown
        in Thin Films*, Rpt. AFCRL-TR-73-0263, Dept. Electr. Eng., Princeton University (1973)
        p. 24
4.84   B.Gross, G.M.Sessler, J.E.West: J. Appl. Phys. **45**, 2841 (1974)
4.85   R.Widdington: Proc. R. Soc. London A **89**, 554 (1914)
4.86   L.M.Beckley, T.J.Lewis, D.M.Taylor: Solid State Commun. **10**, 550 (1972)
        O.B.Evdokimov, N.P.Tubalov: Sov. Phys.–Solid State **15**, 421 (1973)
4.87   F.C.Aris, P.M.Davies, T.J.Lewis: J. Phys. C **9**, 797 (1976)
        L.M.Beckley, T.J.Lewis, D.M.Taylor: J. Phys. D **9**, 1355 (1976)
        D.M.Taylor: J. Phys. D **9**, 2269 (1976)
4.88   B.L.Beers: IEEE Trans. NS-**24**, 2429 (1977)
        B.Gross, G.F.Leal Ferreira: J. Appl. Phys. **50**, 1506 (1979)
4.89   W.G.Trodden, R.D.Jenkins: GEC J. **32**, 85 (1965)
4.90   A.G.Thomas, S.R.Butler, J.I.Goldstein, P.D.Parry: IEEE Trans. NS-**21**(4), 14 (1974)
4.91   B.Gross, G.M.Sessler, J.E.West: Annu. Rep. Conf. El. Ins., 1973 (Natl. Acad. Sci. and Natl.
        Res. Counc., Washington 1974) p. 457
4.92   B.Gross, G.M.Sessler, J.E.West: J. Appl. Phys. **47**, 968 (1976)
4.93   D.M.Compton, G.T.Cheney, P.A.Poll: J. Appl. Phys. **36**, 2434 (1965)
4.94   K.G.McKay: Phys. Rev. **74**, 1606 (1948); **77**, 816 (1950)
4.95   D.J.Gibbons, W.E.Spear: J. Phys. Chem. Solids **27**, 1917 (1966)
4.96   D.J.Gibbons: J. Phys. D **7**, 433 (1974)
4.97   T.M.Flanagan, E.E.Leadon, J.F.Colwell: IEEE Trans. NS-**21**(6), 378 (1974)
4.98   R.A.Meyer, F.L.Bouquet, R.S.Alger: J. Appl. Phys. **27**, 1012 (1956)
4.99   L.K.Monteith: J. Appl. Phys. **37**, 2633 (1966)
        L.K.Monteith, J.R.Hauser: J. Appl. Phys. **38**, 5355 (1967)
4.100   B.Gross, G.M.Sessler, J.E.West: J. Appl. Phys. **48**, 4303 (1977)
4.101   G.M.Sessler: J. Appl. Phys. **43**, 405, 408 (1972)
4.102   G.M.Sessler, J.E.West, D.A.Berkley, G.Morgenstern: Phys. Rev. B **38**, 368 (1977)
4.103   W.Spear: J. Non-Cryst. Solids **1**, 197 (1969)
        W.D.Gill: J. Appl. Phys. **43**, 5033 (1972)
4.104   A.C.Papadakis: J. Phys. Chem. Solids **28**, 641 (1967)
        D.J.Gibbons, A.C.Papadakis: J. Phys. Chem. Solids **29**, 115 (1968)
        M.Abkowitz, H.Scher: Philos. Mag. **35**, 1585 (1978)
        G.Pfister, H.Scher: Phys. Rev. B **15**, 2062 (1977)
        G.Pfister: Phys. Bev. B **16**, 3676 (1977)
        R.C.Hughes: J. Chem. Phys. **58**, 2212 (1973)
        I.Chen: J. Appl. Phys. **49**, 1162 (1978)

4.105 K. Hayashi, K. Yoshino, Y. Inuishi: Jpn. J. Appl. Phys. **14**, 39 (1975)
       K. Yoshino, J. Kyokane, T. Nishitani, Y. Inuishi: J. Appl. Phys. **49**, 4849 (1978)
       B. Gross, G. M. Sessler, H. von Seggern, J. E. West: Appl. Phys. Lett. **34**, 555 (1979)
4.106 B. Gross, G. M. Sessler, J. E. West: Appl. Phys. Lett. **24**, 315 (1974)
       B. Gross, G. M. Sessler, J. E. West: Ann. Rep. Conf. El. Insulation 1974 (Natl. Acad. Sci.
       and Natl. Res. Council, Washington, D.C. 1975) p. 654
4.107 J. Drenkhan, H. Gross, H. Glaefeke: Phys. Stat. Sol. a **2**, K 51 (1970)
       M. Schmidt, H. Glaefeke, J. Drenckhan, W. Wild: Radiat. Eff. **17**, 185 (1973)
4.108 R. Williams, M. H. Woods: J. Appl. Phys. **44**, 1026 (1973)
4.109 A. Watson, J. Dow: J. Appl. Phys. **39**, 5935 (1968)
4.110 J. H. Coleman, D. Bohm: J. Appl. Phys. **24**, 497 (1953)
       J. H. Coleman: *Treatment of Electrically Insulating Materials Subjected to Ionizing Radiation
       and Apparatus for Measuring Such Radiation*, British Patent 735847 (August 1955)
4.111 B. Gross: Z. Phys. **155**, 479 (1959), Z. Angew. Phys. **30**, 323 (1971); IEEE Trans. NS-**25**, 1048
       (1978)
4.112 C. J. MacCallum, T. A. Dellin: J. Appl. Phys. **44**, 1878 (1973)
       T. A. Dellin: *Handbook of Photo-Compton Current Data*, Rpt. SCL-RR-720086 (1972)
       T. A. Dellin, J. C. MacCallum: J. Appl. Phys. **46**, 2924 (1975); IEEE Trans. NS-**23**, 1844 (1976)
4.113 V. Culler: Proc. 7th Hot Lab. Equip. Conf., Cleveland, Ohio (Am. Nucl. Soc. Hinsdale,
       Illinois 1959) p. 369
       F. C. Hardtke, K. R. Ferguson: Proc. 7th Hot Lab. Equip. Conf., New York (Am. Nucl. Soc.,
       Hinsdale, Illinois 1963) p. 369
4.114 P. V. Murphy, B. Gross: J. Appl. Phys. **35**, 171 (1964)
4.115 F. C. Kooi, M. Kusnezov: IEEE Trans. NS-**20** (6), 97 (1973)
4.116 A. R. Frederickson: IEEE Trans. NS-**22**, 2556 (1975); NS-**23**, 1867 (1976)
4.117 T. R. Fewell: *Compton Diodes: Theory and Development for Radiation Detectors*, Rpt.
       SC-DR-0118 (1972)
       M. M. Conrad: *Mod. 3 Family of Production Compton Diodes*, Rpt. SLA-73-0979 (1972)
4.118 S. Mascarenhas: Radiat. Eff. **4**, 263 (1970)
4.119 G. M. Sessler, J. E. West: Appl. Phys. Lett. **17**, 507 (1970)
4.120 G. M. Sessler, J. E. West: Annu. Rep. Conf. Electr. Insul., 1970 (Natl. Acad. Sci., and Natl.
       Res. Counc., Washington, D.C. 1971) p. 8
       G. M. Sessler, J. E. West: In *Electrets*, ed. by M. M. Perlman (Electrochemical Society,
       Princeton, N.J. 1973) p. 292
4.121 M. M. Perlman, S. Unger: J. Phys. D **5**, 2115 (1972); Appl. Phys. Lett. **24**, 579 (1974)
4.122 S. Nakamura, G. Sawa, M. Ieda: Jpn. J. Appl. Phys. **18**, 917 (1979)
4.123 M. M. Perlman: J. Electrochem. Soc. **119**, 892 (1972)
4.124 F. W. Chudleigh, R. E. Collins, G. D. Hancock: Appl. Phys. Lett. **23**, 211 (1973)
4.125 K. Ikezaki, M. Hattori, Y. Arimoto: Jpn. J. Appl. Phys. **16**, 863 (1977)
4.126 B. Gross, G. M. Sessler, J. E. West: J. Appl. Phys. **46**, 4674 (1975)
4.127 B. Wiseal, J. E. Willard: J. Chem. Phys. **46**, 4387 (1967)
4.128 K. Yahagi, K. Shinohara: J. Appl. Phys. **37**, 310 (1966)
       Y. Suzuki, T. Mizutami, M. Ieda: Jpn. J. Appl. Phys. **16**, 1929 (1977)
       S. Nakamura, G. Sawa, M. Ieda: Jpn. J. Appl. Phys. **16**, 2165 (1977)
4.129 A. E. Blake, A. Charlesby, K. J. Randle: J. Phys. D **7**, 759 (1974)
4.130 P. S. Pickard, M. V. Davis: J. Appl. Phys. **41**, 2636 (1970)
4.131 W. C. Mallard, J. H. Crawford, Jr.: J. Appl. Phys. **43**, 2060 (1972)
4.132 J. H. Pollard, D. L. Bowler, M. A. Pomerantz: J. Phys. Chem. Solids **26**, 1325 (1965)
4.133 Y. Shono, T. Yoshida, T. Oka: Annu. Rep. Radiat. Cent. Osaka Prefect. **14**, 37 (1973)
4.134 E. L. Frankevich, B. S. Yakolev: Int. J. Radiat. Phys. Chem. **6**, 281 (1974)
4.135 N. W. Ramsay: *Post-Irradiation X-Ray Induced Conductivity in Insulators*, Ph.D. Thesis,
       University of London (1972)
4.136 P. R. Hanley: *Thermally Stimulated Conductivity of Irradiated Alkali Halides*, Ph.D. Thesis,
       Cornell University Ithaca, New York (1969), University Microfilm No. 70-5987

4.137 A.S.Charlesby, S.Gupta, S.Sarup: Int. J. Radiat. Biol. **5**, 141 (1973)
4.138 J. van Turnhout, P.H.Ong: In Proc. IEE Conf. Dielectr. Mater. Measure. Appl., Cambridge (1975) Publ. No. 129 (IEE, Stevenage 1975) pp. 68–74
4.139 M.M.Perlman, S.Unger: In *Electrets*, ed. by M.M.Perlman (Electrochemical Society, Princeton, N.J. 1973) p. 105
4.140 M.R.Grinter, C.Bowlt: J. Phys. D **9**, L61 (1976)
4.141 B.Gross, J.E.West, D.A.Berkley: Annu. Rep. Conf. Electr. Ins., 1978 (Natl. Acad. Sci., Washington, D.C. 1978) p. 163
4.142 P.V.Murphy, B.Gross: J. Appl. Phys. **35**, 171 (1964)
       A.R.Frederickson: IEEE Trans. NS-**22**, 2556 (1975)
4.143 B.Gross: Phys. Rev. **110**, 337 (1958)
4.144 P.H.Murphy, F.E.Hoecker: Radiat. Res. **46**, 1, (1971)
4.145 T.M.Proctor: Phys. Rev. Lett. **3**, 575 (1959); Phys. Rev. **116**, 1436 (1959)
4.146 F.Hardtke: Phys. Rev. Lett. **9**, 339 (1962); J. Chem. Phys. **42**, 3000 (1965)
4.147 J.F.Barnes, F.E.Hoecker, L.Kevan: Radiat. Res. **40**, 235 (1969)
4.148 E.L.Frankevich, V.L.Talroze: Sov. Phys.–Solid State **3**, 131 (1961)
4.149 T.Nomura, K.Yamamoto: Jpn. J. Appl. Phys. **10**, 971 (1971)
4.150 P.H.Murphy, F.E.Hoecker: J. Appl. Phys. **42**, 4094 (1971)
4.151 J.H.Ranicar, R.J.Fleming: J. Polym. Sci. Polym. Phys. Ed. **10**, 1979 (1972)
4.152 P.V.Murphy: J. Phys. Chem. Solids **24**, 329 (1963)
4.153 P.V.Murphy, S.Costa Ribeira, F.Milanez, R.J.de Moraes: J. Chem. Phys. **38**, 2400 (1963)
4.154 D.E.Fields, P.R.Moran: Phys. Rev. Lett. **29**, 721 (1972)
4.155 E.B.Podgorsak: *Radiation and Impurity Induced Thermally Activated Charge Transport in Calcium Fluoride*, Ph.D. Thesis, Wisconsin University (1973) p. 195
4.156 E.B.Podgorsak, G.E.Fuller, P.R.Moran: In *Electrets*, ed. by M.M.Perlman (Electrochemical Society, Princeton, N.J. 1973) p. 172
4.157 E.B.Podgorsak, P.R.Moran: Phys. Rev. Lett. **30**, 926 (1973)
4.158 E.B.Podgorsak, P.R.Moran: Appl. Phys. Lett. **24**, 580 (1974)
4.159 E.B.Podgorsak, P.R.Moran: Phys. Rev. B **8**, 3405 (1973)
4.160 M.Campos, S.Mascarenhas, G.Leal Ferreira: Phys. Rev. Lett. **27**, 1514 (1971)
4.161 E.B.Podgorsak, P.R.Moran: Science **179**, 380 (1973)
4.162 B.Gross, R.J.de Moraes: Phys. Rev. **126**, 930 (1962)
4.163 P.R.Moran, E.B.Podgorsak: Phys. Lett. A **44**, 237 (1973)
4.164 O.A.Myazdrikov: At. Energ. **1**, 64 (1960)
       H.Bauser, W.Runge: Health Phys. **34**, 97 (1978)
4.165 J.L.Wolfson, J.C.Dyment: Health Phys. **7**, 36 (1961)
4.166 G.W.Fabel, H.K.Henisch: Phys. Status Solidi a **6**, 535 (1971)
4.167 M.D.A.King: "Dosimetry by Charge Decay of Polymer Thermoelectrets", Rpt. RL-74-2, RHEL, Chilton, U.K. (1972)
4.168 D.N.Bose, H.K.Henisch, J.M.Toole: Solid State Electron. **12**, 65 (1969)
       J.M.Toole, H.K.Henisch: Solid State Electron. **11**, 743 (1968)
4.169 J.P.Mitchell, D.G.DeNure: IEEE Trans. NS-**20**(5), 67 (1973)
4.170 P.R.Moran: "Disclosure to USAEC: Compton-Effect TAP/TAD Dosimeter", Rpt. COO-1105-204, Wisconsin University (1974)
4.171 P.R.Moran: "Summary of Progress in Solid State Dosimetry", Rpt. COO-1105-215, Wisconsin University (1974)
4.172 P.R.Moran: "Thin Film Tritium Dosimetry", Rpt. COO-1105-205, Wisconsin University (1974)
4.173 D.L.Hester, D.D.Glower, L.J.Overton: IEEE Trans. NS-**11**(6), 145 (1964)
4.174 D.Miller, P.Schlosser, J.Burt, D.Glower, J.McNeilly: IEEE Trans. NS-**14**(6), 245 (1967)
4.175 P.A.Schlosser, D.W.Miller, D.D.Glower: Int. J. Nondestr. Test. **2**, 19 (1970)
4.176 G.M.Panchenkov, D.A.Kraushanskii, V.M.Pluzhnikov, B.P.Glazunov, A.A.Avrorov, M.A.Ositov: "Use of Ferroelectric Crystals as Detectors of Gamma Radiation", in *Dozimetriya i Radiatsionnye Protsessy Dozimetricheskikh Sistemakh*, ed. by L.M.Blaunshtein (Izdatel'stvo Uzbekskoi SSR, Taschkent 1972) p. 140

4.177 J.T.Klopcic, D.L.Swanson: "An Analytical Model of the Ferroelectric Radiation Detector", Rpt. BRL-MR-2310, Ballistic Res. Labs. Aberdeen Proving Ground, Md. (1973)

4.178 N.W.Ramsay: "Dose Measurement by Thermally Stimulated Conductivity a New Method of Dosimetry", INIS-mf-524 (1972) paper 22.7

4.179 M.W.Harper, B.Thomas: Phys. Med. Biol. **18**, 409 (1973)
       B.Thomas, J.Conway, M.W.Harper: J. Phys. D **10**, 55 (1977)

4.180 C.Bowlt, D.J.Waggett: Phys. Med. Biol. **10**, 534 (1974)
       J.Conway, M.W.Harper, B.Thomas: J. Phys. D **10**, 1131 (1977)

4.181 M.Grinter, C.Bowlt: J. Phys. D **8**, L159 (1975)

4.182 C.Bowlt: Contemp. Phys. **17**, 461 (1976)

4.183 G.D.Fullerton, J.R.Cameron, M.R.Mayhugh, P.R.Moran: "High Sensitivity Thermo-current Dosimetry (TCD) with Alumina, in *Biomedical Dosimetry*, IAEA (1976) p. 249
       G.D.Fullerton, P.R.Moran: Med. Phys. **1**, 161 (1974)

4.184 P.R.Moran, E.B.Podgorsak, G.E.Fuller, G.D.Fullerton: Med. Phys. **1**, 155 (1974)

4.185 N.W.Ramsay, L.Y.Joesoef: "The Measurement of Neutron Dose by Thermally Stimulated Current in Insulating Materials, in *Biomedical Dosimetry*, IAEA (1976) p. 95

4.186 K.Zaininger, A.G.Holmes-Siedle: RCA Rev. **28**, 208 (1967)

4.187 A.G.Holmes-Siedle: Proc. IEEE **62**, 1196 (1974)
       W.C.Johnson: IEEE Trans. NS-**22**, 2144 (1975)
       J.W.Farmer: "X-Ray Induced Currents and Space-Charge Buildup in MOS Capacitors", Ph.D. Thesis, Kansas University (1974), Univ. Microfilms No. 74-25599
       R.Freeman, A. Holmes-Siedle: IEEE Trans. NS-**25**, 1216 (1978)

4.188 A.G.Holmes-Siedle: Rep. Prog. Phys. **37**, 699 (1974)

4.189 W.J.Poch, A.G.Holmes-Siedle: RCA Eng. **16**(3), 56 (1970)

4.190 A.G.Holmes-Siedle: Nucl. Instrum. Methods **121**, 169 (1974)
       L.Adams, A. Holmes-Siedle: IEEE Trans. NS-**25**, 1607 (1978)

4.191 D.R.Ciarlo, R.Kalibjian, K.Mayeda, T.A.Boster: IEEE Trans. NS-**19**(1), 350 (1972)

4.192 N.M.Johnson, W.C.Johnson, M.A.Lampert: J. Appl. Phys. **46**, 1216 (1975)

4.193 D.Perino, J.Lewiner, G.Dreyfus: IEEE Trans. NS-**25**, 1117 (1978)

5. Piezo- and Pyroelectric Properties

M. G. Broadhurst and G. T. Davis

With 18 Figures

Heaviside [5.1] in 1892 postulated that certain waxes would form permanently polarized dielectrics when allowed to solidify from the molten state in the presence of an electric field. He viewed an electret as the electrical analog of a magnet, that is, as having a frozen-in, relatively long-lived (compared with the observation time) nonequilibrium electric dipole moment. Present popular usage has expanded this concept of an electret to include monopolar dielectrics having a net frozen-in real charge. For example, the commercial electret microphone employs a monopolar polymer film.

Wax and rosin electrets were made and studied by *Eguchi* in the early 1920s [5.2]. By 1927 it was well understood [5.3] that molecules containing permanent electric moments orient in the direction of an electric field when mobile in the liquid state. Upon solidification of the material in the presence of the field, the dipoles lose their mobility while retaining their preferred orientation. The net dipole orientation produces the electret's permanent polarization (net dipole moment per unit volume). It was also recognized that in addition to the electret's moment there were real charges, generally concentrated near the electret surfaces, which were injected during the formation process by field emission, gas breakdown or similar processes.

In 1927 piezoelectricity and pyroelectricity were shown theoretically and experimentally to be properties exhibited by electrets with preferentially ordered dipoles [5.3, 4]. However, these early wax electrets had poor mechanical strength and low sensitivity, and applications for them did not develop. More recently, strong, highly active polymer films, notably poly(vinylidenefluoride), PVDF, poly(vinylfluoride), PVF, and poly(vinylchloride), PVC, have been recognized for their potential value as thermoelectric and electromechanical transducer materials. Already these materials are finding their way into a new technology of polymer transducers. Japanese scientists have been particularly active in the early research and development of these devices with the work of *Fukada* on natural and synthetic polymers [5.5], *Kawai* who pointed out how general the effect is [5.6], *Hayakawa* and *Wada* with their theoretical analyses [5.7, 8], and industrial scientists who are developing films [5.9, 10] and using them for various devices [5.11–13]. Most of the early polymer electret work in the U.S. has focused on using the pyroelectric response for electromagnetic radiation detection [5.14–22].

In the following sections we shall present concepts, models, experimental considerations, results, and implications which have resulted from some of the work on piezoelectric and pyroelectric polymers. This approach is intended to provide the reader with basic physical concepts needed to identify important molecular and material parameters, deduce guidelines for optimizing desirable properties and provide a basis for selecting new applications.

5.1 Thermodynamic Definitions

Piezo- and pyroelectricity are defined in a formal way by thermodynamics [5.23]. The piezoelectric constant d_{mj} is a tensor component given by a second derivative of the Gibbs free energy with respect to the electric field vector E and stress tensor T,

$$d_{mj} = \left[\frac{\partial^2 G(E, T, T)}{\partial E_m \partial T_j} \right]_T . \tag{5.1}$$

We define a material as being piezoelectric if this second derivative has a value large enough to be measurable. A material is pyroelectric if at least one of the components of the pyroelectric coefficient vector p defined as

$$p_m = \left[\frac{\partial^2 G(E, T, T)}{\partial E_m \partial T} \right]_T \tag{5.2}$$

has a value large enough to be measurable. Being second derivatives of free energy, these coefficients have a basis in common with better-known quantities such as coefficients of expansion, compressibility, heat capacity, and dielectric constant, and can therefore be expected to be complex quantities if measured with an alternating stress or show relaxational behavior in the time domain.

The second derivatives in (5.1) and (5.2) can be taken in any order so that we have

$$d_{mj} = (\partial D_m / \partial T_j)_{T, E} = (\partial S_j / \partial E_m)_{T, T} , \tag{5.3}$$

and

$$p_m = (\partial D_m / \partial T)_E = (\partial \Sigma / \partial E_m)_T . \tag{5.4}$$

In the above, $j = 1, ..., 6$; $m = 1, 2, 3$; T is the stress; S is the strain; D is the electric displacement; E is the electric field; T is the temperature; Σ is the entropy.

The d are often called piezoelectric strain constants whereas the piezoelectric stress constants, e, arise from taking S rather than T as the independent variable in (5.1). Two other piezoelectric constants h and g can be defined by taking D and S and D and T as independent variables in (5.1), [5.24].

It is important here to emphasize that the above relationships are based on thermodynamic quantities such as electric field and mechanical stress, whereas experiments are performed using measured quantities such as voltage and force. A derivative at constant force or voltage is not the same as one at constant stress or field. As a consequence, the above equations must be used with care for experimental purposes as discussed in Sect. 5.2.3, and pointed out previously [5.25, 26]. Before relating the quantities defined in (5.3) and (5.4) to measured quantities, it is well to develop a better description of an electret and understand the influence of real and dipolar charges.

5.2 Physical Description of an Electret

5.2.1 Preparation

Consider a slab of polymer which we take to be amorphous, homogeneous, and elastically isotropic. We first evaporate metallic electrodes on both sides to eliminate air gaps between the polymer and metal, and then follow the temperature–voltage–time sequence shown schematically in Fig. 5.1: (I) raise the temperature from room temperature T_r to an elevated temperature which we show here as being above the glass transition temperature T_g; (II) apply a dc voltage Φ resulting in an electric field of several hundred kilovolts per centimeter of slab thickness, s, between the electrodes; (III) while maintaining this voltage, lower the temperature back to T_r. The electret thus formed can be represented schematically as shown in Fig. 5.2.

The above poling procedure typically results in both real charges and charges resulting from molecular dipole distributions, and these will affect the behavior of the electret in different ways. To explain this difference we consider separately the two types of charge.

5.2.2 Real Charges – Monopolar Electrets

In general, real charges do not contribute to the zero field piezo- and pyroelectric response as long as the sample is strained uniformly [Ref. 5.7, Sect. 2.2]. To illustrate this fact, consider the example of Fig. 5.3. A slab of dielectric with uniform permittivity ε, thickness s, and short-circuited contact electrodes contains a layer of trapped positive charges at a distance x, from the bottom electrode. The charge density on the top electrode is σ_s and the charge density of the trapped charge is σ_x. The trapped charge will induce an equal and opposite charge on the two electrodes divided according to the capacitance between the charge and the electrodes. Thus if the permittivity is uniform

$$\sigma_s = -\sigma_x(x/s). \tag{5.5}$$

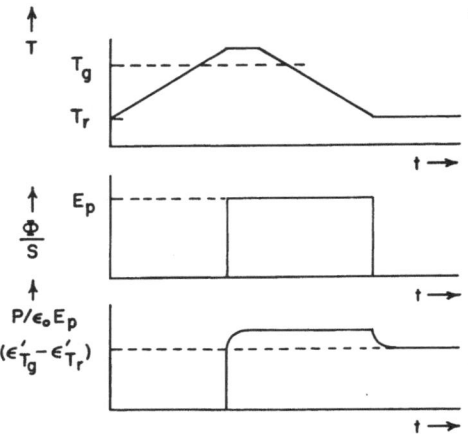

◄ **Fig. 5.1.** A diagram of a poling procedure showing the temperature T, voltage Φ, time t sequence and the resulting polarization P reduced by the permittivity of free space ε_0 and applied field E_p. The remaining frozen-in reduced polarization is the difference between relative permittivity, ε' at the two temperatures T_r and T_g

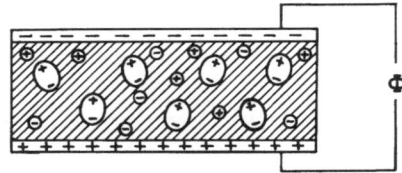

Fig. 5.2. A model of an electret with real charges and preferentially ordered dipolar charges resulting from the applied voltage

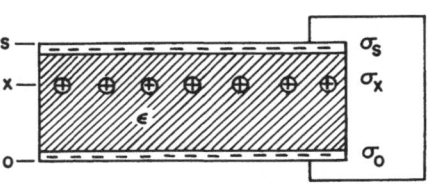

Fig. 5.3. A model of an electret with a sheet of real charge embedded in it and induced charges on the short-circuited electrodes. The σ's are the surface charge densities located at the various positions, o, x, and s

If the material is strained so that the distances x and s change to $x + \Delta x$ and $s + \Delta s$, then σ_s becomes

$$\sigma'_s = -\sigma_x(x/s)[(1 + \Delta x/x)/(1 + \Delta s/s)]. \tag{5.6}$$

For uniform strain, $\Delta x/x = \Delta s/s$, then $\sigma'_s = \sigma_s$ and no charge will flow (at zero field) as a result of pressure and temperature changes which produce the strain.

When the material properties or the resultant strain are nonuniform, the trapped charges can give rise to an electrical response as discussed by *Wada* and *Hayakawa* [5.8]. In their model, a film is heterogeneous in the thickness direction only (x direction in Fig. 5.3). The density of spacecharge $\varrho(x)$ and permittivity $\varepsilon(x)$ are position dependent as are the stress dependences $\alpha_\varepsilon(x) \equiv \partial \ln \varepsilon(x)/\partial X$ and $\alpha_x(x) \equiv \partial \ln x/\partial X$, where X is a mechanical stress or thermal stress (temperature). Considering the film to be a stack of thin layers, (5.5) can be generalized to give the charge on the top electrode of area A_s as

$$Q_s = -\int_0^s \varrho(x)A_s dx \int_0^x dx'/\varepsilon(x') \Big/ \int_0^s dx'/\varepsilon(x'). \tag{5.7}$$

The total charge in a given layer, $\varrho(x)A_s dx$ is assumed constant with stress. In the linear approximation the strained quantities dx' and $\varepsilon(x')$ can be replaced with unstrained quantities $dx_0[1+\alpha_x(x_0)dX]$ and $\varepsilon(x_0)[1+\alpha_\varepsilon(x_0)dX]$. Expanding to first order in the α's,

$$Q_s = -\int_0^s \varrho(x)A_s dx \int_0^x dx_0/\varepsilon \left\{1 + \left[\int_0^x \varepsilon^{-1}(\alpha_x-\alpha_\varepsilon)dX dx_0 \Big/ \int_0^x dx_0/\varepsilon\right]\right.$$
$$\left. - \left[\int_0^s \varepsilon^{-1}(\alpha_x-\alpha_\varepsilon)dX dx_0 \Big/ \int_0^s dx_0/\varepsilon\right]\right\} \Big/ \int_0^s dx_0/\varepsilon. \tag{5.8}$$

The α's and ε are understood to depend on x_0. Subtracting the unstrained Q_{s0} from Q_s and assuming uniform stress dX one obtains the change in charge on the top electrode with stress,

$$A^{-1}\partial Q_s/\partial X = -\int_0^s \varrho(x)\left\{\int_0^x [(\alpha_x-\alpha_\varepsilon)-\langle\alpha_x-\alpha_\varepsilon\rangle]dx_0/\varepsilon(x_0)\right\}dx \Big/ \int_0^s dx_0/\varepsilon(x_0) \tag{5.9}$$

The partial derivative with respect to temperature is at constant mechanical stress and vice versa, and the average of a quantity is defined as

$$\langle A \rangle = \int_0^s A dx_0/\varepsilon(x_0) \Big/ \int_0^s dx_0/\varepsilon(x_0). \tag{5.10}$$

Using the usual formula for integration by parts and the fact that $\int_0^s [(\alpha_x-\alpha_\varepsilon) - \langle\alpha_x-\alpha_\varepsilon\rangle]dx_0/\varepsilon = 0$, the above equation can be written in the form given by *Wada* and *Hayakawa*

$$A^{-1}\partial Q_s/\partial X = \left\langle [(\alpha_x-\alpha_\varepsilon)-\langle\alpha_x-\alpha_\varepsilon\rangle]\left[\int_0^x \varrho(x_0)dx_0\right]\right\rangle. \tag{5.11}$$

In general the response due to trapped charge will be small because of limited charge densities. *Crosnier* et al. found a low level of piezoelectricity in polypropylene and showed a linear relationship between activity and charge density [5.27]. *Ibe* has shown that piezoelectric response from nonpolar polymers can result from bending a specimen which establishes a nonuniform stress in the material and creates a condition where trapped space charge can contribute to the electrode charge when stress is applied [5.28]. To calculate this effect (5.9) must be modified leaving dX inside the integral. The resulting equation will be like that used by *Collins* to describe the electrical response of a charged, nonpolar film to a thermal heat pulse [5.29]. *Collins* used the fact that the thermal stress was nonuniform during the thermal equilibration of the sample to gain information about the distribution of trapped charge in the film. *DeReggi* et al. [5.30] showed that such an experiment can provide several Fourier coefficients of the charge distribution. A related experiment using nonuniform mechanical stress has been proposed by *Laurenceau* et al. as an alternative way of measuring charge distribution in films [5.31].

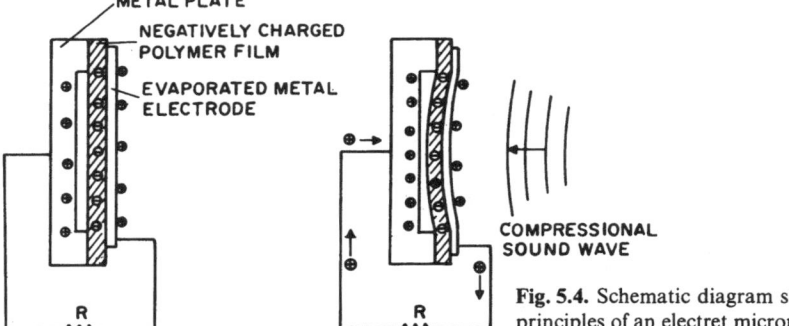

METAL PLATE

NEGATIVELY CHARGED
POLYMER FILM

EVAPORATED METAL
ELECTRODE

COMPRESSIONAL
SOUND WAVE

Fig. 5.4. Schematic diagram showing the principles of an electret microphone

Equation (5.11) suggests the possibility that one can create an artificial piezoelectric from a heterogeneous structure designed to optimize $\varrho(x)$ and the spatial variations in the stress dependence of strain and permittivity. An obvious example is shown in the schematic diagram of Fig. 5.4 of a charged polymer film mounted so that an air gap separates it from a conductive plate. As the charged film moves with respect to the plate (under the influence of sound waves, for example) the plate potential changes and charges flow between it and a conductive contact electrode on the film through an appropriate circuit. This is the operating principle of the electret microphone which typically uses fluorinated ethylene–propylene copolymer [5.32, 33]. This same effect is probably a source of electrical signals generated in a flexible polymer-insulated coaxial cable when it is subjected to mechanical vibrations or pressure changes. Note that the electret microphone system does not undergo uniform strain. That is, the air gap is strained much more than the polymer film and hence the real charges do contribute to current flow in accord with (5.11). It has been proposed that such a mechanism may be responsible for piezoelectricity in polyvinylidene fluoride, as will be discussed in Sect. 5.5.3.

5.2.3 Dipolar Electrets

To illustrate the electret's piezoelectric and pyroelectric behavior consider Figs. 5.5, 6. As the electrically short-circuited electret contracts due to an increase in hydrostatic pressure or a decrease in temperature the metal electrodes move closer to the dipoles and the zero potential is maintained by a flow of charge. This model of a strain sensitive electret is similar to that given by *Adams* [5.3] and accounts for most of the response of piezoelectric and pyroelectric polymers. Note that the model predicts the direction of current flow in terms of the direction of the poling field and also predicts that while the total charge released is proportional to the temperature or pressure change, the current depends on the rate of pressure or temperature change and can be quite large. This effect can be described mathematically as follows.

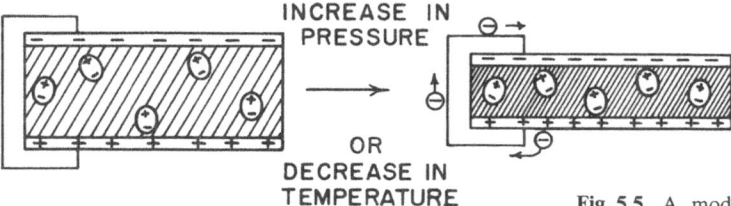

INCREASE IN
PRESSURE

OR

DECREASE IN
TEMPERATURE

Fig. 5.5. A model of a dipolar electret showing the flow of charge resulting from a thickness change due to an increase in pressure or a decrease in temperature

ELECTROMAGNETIC RADIATION

RADIATION ABSORBING
CONDUCTIVE ELECTRODE

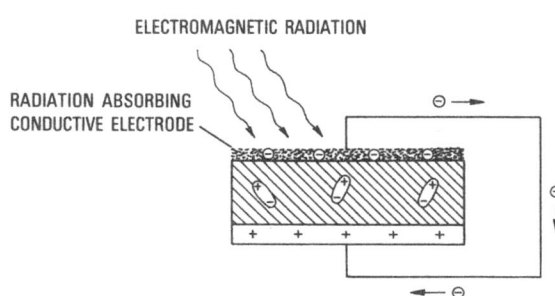

Fig. 5.6. A model showing how the electret of Fig. 5.5 is used for a pyroelectric application. Note the interaction of the film with the radiation takes place in the electrode which in turn acts as a heat bath

In terms of the material's relative permittivity ε' the equilibrium field-induced polarization is given by

$$P = (\varepsilon' - 1)\varepsilon_0 E, \tag{5.12}$$

where ε_0 is the permittivity of vacuum and E is the applied field. At sufficiently high temperatures the material is a dipolar liquid and the field produces a polarization

$$P_L(T) = [\varepsilon'_L(T) - 1]\varepsilon_0 E_P, \tag{5.13}$$

where the subscript L refers to the liquid phase, (T) to the functional dependence on temperature and $E_P = \Phi/s$ is the mean poling field. During poling, the field maintains this polarization while the temperature is lowered enough to immobilize the molecular dipoles. The field is removed and the lost polarization (neglecting volume change) is

$$P_s(T) = [\varepsilon'_s(T) - 1]\varepsilon_0 E_P, \tag{5.14}$$

where the subscripts s refers to the solid phase.

Thus the frozen-in nonequilibrium polarization remaining after removal of the poling field is

$$P_0(T) = [\varepsilon'_L(T_L) - \varepsilon'_s(T)]\varepsilon_0 E_P = \Delta\varepsilon'\varepsilon_0 E_P \tag{5.15}$$

where T_L is the temperature where the material becomes liquid. Equation (5.15) provides a method of calculating the frozen dipole polarization for linear dielectrics from a knowledge of their relative permittivity at the polarization and measuring temperatures. Special consideration of nonlinear dielectrics will be given in Sect. 5.5. In order to calculate the piezoelectric and pyroelectric coefficients from molecular properties one needs to use a more detailed model as shown below.

The polarization (dipole moment per unit volume) is defined as

$$P = N\langle m\rangle / V \tag{5.16}$$

where N is the number of dipoles, V the volume of the electret and $\langle m\rangle$ the mean effective dipole moment in the direction of P. As a model for the electret with preferentially ordered polarizable dipoles of permanent moment μ_0, consider Fig. 5.7. One can use an *Onsager*-type calculation [5.34] to determine the effective moment $\langle m\rangle$ in (5.16) of a representative dipole located in a spherical cavity and oriented at a fixed average angle θ with respect to the direction of overall polarization P. Such a calculation leads to the result [5.35]

$$P_0 = (\varepsilon_\infty + 2)N\mu_0\langle\cos\theta\rangle / 3V, \tag{5.17}$$

where ε_∞ is the high-frequency relative permittivity related to the polarizability through the Clausius–Mosotti relationship, N/V is the number of dipoles per unit volume and P_0 is the frozen-in polarization present at zero applied field. This important equation can be used to calculate the piezo- and pyroelectric coefficients for this model. This calculation is done simply by taking the derivatives of P_0 with respect to pressure and temperature and then expressing the relationship between these derivatives and the defined quantities (5.3) and (5.4).

The relationship between P_0 of (5.17) and electric displacement is given by

$$D = \varepsilon'\varepsilon_0 E + P_0. \tag{5.18}$$

In the simplest case where the short-circuit current is measured while temperature or stress are changed, one obtains

$$\left.\frac{\partial D}{\partial T}\right|_{E=0,T} = \left.\frac{\partial P_0}{\partial T}\right|_{E=0,T} = \left.\frac{\partial(Q/A)}{\partial T}\right|_{E=0,T}, \tag{5.19}$$

and

$$\left.\frac{\partial D}{\partial T}\right|_{E=0,T} = \left.\frac{\partial P_0}{\partial T}\right|_{E=0,T} = \left.\frac{\partial(Q/A)}{\partial T}\right|_{E=0,T} \tag{5.20}$$

where Q/A is the surface charge per unit area of electrode. Here we continue to neglect changes in (Q/A) due to real charges, i.e., uniform strain conditions.

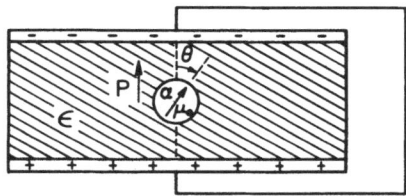

Fig. 5.7. A model for electret containing a representative dipole of moment μ_0, polarizability α, and fixed mean orientation θ with respect to the net polarization P

Generally it is not a change in (Q/A) which is measured but rather a change in Q. Reported values for piezo- and pyroelectric coefficients are thus in error as far as the strict definitions are concerned. In the following we adopt common practice and define experimentally determined piezoelectric and pyroelectric coefficients as

$$d = (1/A)(\partial Q/\partial T)_{E=0,T}, \tag{5.21}$$

and

$$p = (1/A)(\partial Q/\partial T)_{E=0,T}. \tag{5.22}$$

The above distinction becomes particularly significant for polymers where the difference between (5.19) and (5.21) is of the order of magnitude of the terms themselves. Inorganic materials have a much smaller temperature and stress-induced area change and the corresponding difference between (5.19) and (5.21) or (5.20) and (5.22) is small.

Another inconsistency between precise definitions and general practice is sometimes encountered when measurements are reported at voltages considerably greater than zero. Allowing X to represent stress or temperature, the derivative,

$$\frac{\partial D}{\partial X} = \frac{\partial \varepsilon'}{\partial X}\varepsilon_0 E + \varepsilon' \varepsilon_0 \frac{\partial E}{\partial X} + \frac{\partial P_0}{\partial X}, \tag{5.23}$$

has two terms in addition to that in (5.19) and (5.20). The first term involving electrostriction can be large if E is large. From (5.3) and (5.4) this is a legitimate part of p and d which are functions of E. The second term would not appear if E were held constant, but in practice it is the voltage Φ that is held constant and the thickness $s = \Phi/E$ varies with the measurement and gives an electromechanical contribution. Similarly the third term is measured at constant Φ. (Electrostriction and electromechanical contributions are considered in a different way in [5.36].)

To reduce ambiguity, we shall consider measurements made at zero field and for simplicity and a more straightforward comparison of p and d, we shall use hydrostatic pressure as mechanical stress (positive pressure is a negative stress). Without giving the details [5.35], the straightforward differentiation of

LOW TEMPERATURE   HIGH TEMPERATURE

MEAN DIPOLE
MOMENT

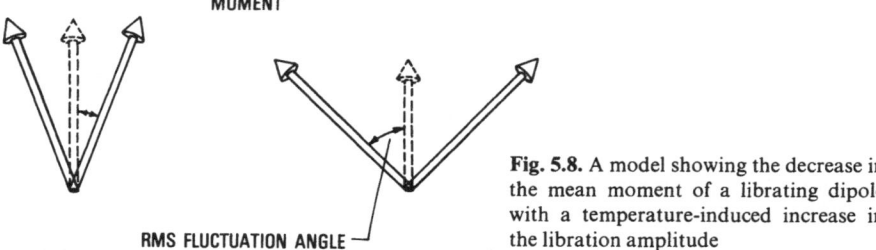

RMS FLUCTUATION ANGLE

Fig. 5.8. A model showing the decrease in the mean moment of a librating dipole with a temperature-induced increase in the libration amplitude

(5.17) to obtain the pressure and temperature derivatives of surface charge in an electret gives

$$A^{-1}(\partial Q/\partial T)_{\mathrm{P}} = -P_0\alpha[\varepsilon_\infty/3 + \phi^2/2\alpha T + \gamma\phi^2], \tag{5.24}$$

$$A^{-1}(\partial Q/\partial p)_{\mathrm{T}} = P_0\beta[\varepsilon_\infty/3 + \gamma\phi^2], \tag{5.25}$$

where $\alpha = (V)^{-1}dV/dT$ is the volume coefficient of thermal expansion, $\beta = -(V)^{-1}dV/dp$ is the volume compressibility, $\gamma = -V\omega^{-1}d\omega/dV$ is a Grüneisen constant for the dipole torsional frequency ω, and ϕ^2 is the mean squared torsional displacement of the dipole fluctuations.

These equations show that for this model most of the piezo- and pyroelectric response comes from volume expansion and its effect on ε_∞. There is an additional contribution from temperature change which can be illustrated with a physical model like that in Fig. 5.8. Although the dipoles have a fixed mean direction, there is always thermal motion whose mean squared amplitude in the simple harmonic approximation is proportional to temperature. Thus increasing the temperature of a dipole reduces the average magnitude of its moment. This effect was the basis of a theory of pyroelectricity in PVDF due to *Aslaksen* [5.37], and accounts for about one-third of the pyroelectricity in PVC and possibly a similar fraction in other polymers. The amplitude of molecular librations is difficult to measure or to predict a priori because of the large number of vibrational modes and molecular conformations contributing. However, one paper gives a value of 10^0 for the root mean squared torsional displacement of polyethylene molecules based on X-ray data [5.38].

5.3 Symmetry and Tensor Components

Crystal symmetry is usually considered in discussions of piezoelectricity. An isotropic amorphous material could not be expected to be either piezo- or pyroelectric at zero field because its response to stress will be the same in all directions. However, if one preferentially aligns molecular dipoles in the

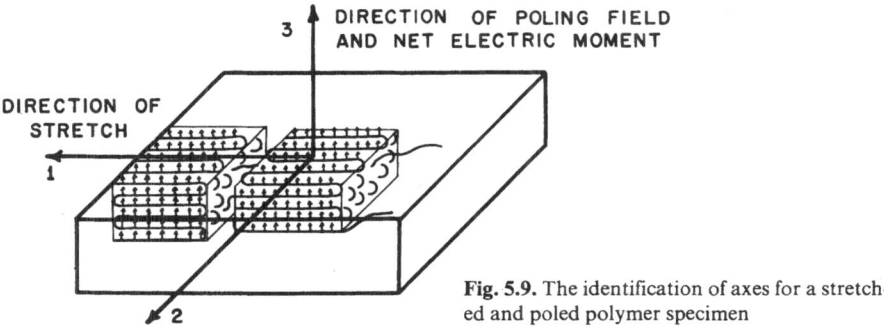

Fig. 5.9. The identification of axes for a stretched and poled polymer specimen

specimen, there is no longer a center of symmetry and the sample will be both piezoelectric and pyroelectric. Polymer films, particularly semicrystalline polymers, are often stretched which preferentially aligns the polymer molecules parallel to the stretch direction and they are then poled to align the dipoles normal to both the stretch direction and the plane of the film. (We consider only those polymers with dipoles normal to the molecular axis.) The result is to remove the isotropy present in the plane of the unstretched film. The axes are usually identified as shown in Fig. 5.9, which depicts a model of a semicrystalline polymer. The expected components of the piezoelectric and pyroelectric tensors for such a specimen and their proper signs are

$$\tilde{d} = \begin{pmatrix} 0 & 0 & 0 & 0 & d_{15}^+ & 0 \\ 0 & 0 & 0 & d_{24}^+ & 0 & 0 \\ d_{31}^+ & d_{32}^+ & d_{33}^- & 0 & 0 & 0 \end{pmatrix}, \tag{5.26}$$

$$\tilde{p} = \begin{pmatrix} 0 \\ 0 \\ p_3^- \end{pmatrix}. \tag{5.27}$$

The assignments can be made by inspection from Fig. 5.9. Stress in the $+3$ direction will increase the sample thickness and thus decrease the electrode charge giving a negative d_{33}. The stress in the 1 and 2 directions will decrease the thickness and increase electrode charge, giving a positive d_{31} and d_{32}. Experimentally, d_{33} is found to be negative [5.17] and d_{31} and d_{32} predominantly positive for PVDF [5.39] and d_{31} is also found to be positive for PVF [5.40]. d was also found to be negative with hydrostatic stress for PVC [5.41]. Remember that we are using the assumption that the electrodes expand with the specimens and that we have adopted (5.21) and (5.22) as our definitions. If we use the proper definitions given by (5.19) and (5.20) stress in all three directions 1, 2, and 3, is expected to increase the volume and decrease the polarization giving negative d_{3j} components.

Unusual effects may be encountered with highly oriented polymers where the Poisson ratio η_{31} for the ratio of the strain in the 3 direction to the strain in the 1 direction when stress is applied in the 1 direction may be greater than 0.5 and η_{32} may be considerably less than 0.3. These values are the usual limits on η for isotropic materials. This point has been discussed by *Sussner* [5.42] and may lead to unusual behavior such as a decrease in volume when stress is applied in the draw direction and positive strain in the draw direction with an increase in hydrostatic pressure.

The shear components result because a positive shear about the 1 axis, T_4, rotates the dipoles into the $+2$ direction and a shear about the 2 axis, T_5, rotates the dipoles into the $+1$ direction. Neither shear causes a change to first order in the moment in the 3 direction. A shear about the 3 axis, T_6, does not change the moment. Because there is no net moment along the 1 and 2 axes, $p_1 = p_2 = 0$, and an increase in temperature produces an increase in volume and decrease in polarization yielding a negative p_3. The $\tilde{d}$ matrix constructed from physical arguments for amorphous polymers is characteristic of C_{2V} symmetry. This symmetry is found for the polar crystal phase of PVDF [5.39], and for polar PVF [5.43]. Poled, unoriented polymers should give $d_{31} = d_{32}$ and $d_{24} = +d_{15}$, characteristics of a piezoelectric matrix with $C_{\infty V}$ symmetry [Ref. 5.5, Sect. IA].

5.4 Structure

5.4.1 General

Using the model discussed above, it is possible to hypothesize four requirements for large piezo- and pyroelectricity in polymers. (I) There must be molecular dipoles present, the higher their moment and concentration the better. (II) There must be some way of aligning the dipoles, the more alignment the better. (III) There must be a way of locking-in the dipole alignment once it is achieved, the more stable the better. (IV) The material should strain with applied stress, the more strain the better (some of the pyroelectric activity need not result from the strain). Evaluating these conditions for a particular polymer requires considerable data on the molecular and bulk structure and properties. In the following discussion we consider in some detail two different types of synthetic polymers–amorphous and semicrystalline. Other types–including the important class of biopolymers which have permanent dipole moments along the molecular axis – will not be considered here. Abbreviations to be used are defined at the beginning of this book.

5.4.2 Amorphous Polymers

Poly(vinyl chloride); (PVC), is an example of an amorphous polymer which can be made piezo- and pyroelectric [5.6, 25, 35, 41, 44]. The repeat unit is polar with an effective dipole moment of 3.6×10^{-30} C m (1.1 D) [5.45]. The usual form of PVC is amorphous because of the nonstereospecific addition of monomer units during polymerization. More stero-regular (syndiotactic) crystallizable PVC can be made and its crystal structure has been determined [5.46]. PVC is an equilibrium liquid above its glass transition temperature (about 80 °C), although thermal decomposition is appreciable above this temperature. Below 80 °C, the kinetics of molecular reorganization are slow enough that a nonequilibrium amorphous solid (glass) is formed. Structural relaxation times of the glass increase rapidly with decreasing temperatures to the order of years at room temperature. Thus, this polymer fits all criteria in Sect. 5.4.1 for piezo- and pyroelectricity. To illustrate the calculation of p and d we can substitute (5.15) for P_0 in (5.24) and (5.25) because the dipoles are small enough that the product of their moment times the field is much less than their thermal energy kT and the polarization is linear with field. For more background see [Ref. 5.47, p. 32]

$$p_y = \Delta\varepsilon'\varepsilon_0 E_p \alpha [\varepsilon_\infty/3 + \phi^2/2\alpha T + \gamma\phi^2], \qquad (5.28)$$

$$d_p = \Delta\varepsilon'\varepsilon_0 E_p \beta [\varepsilon_\infty/3 + \gamma\phi^2]. \qquad (5.29)$$

We use $\Delta\varepsilon' = 10$ and $\varepsilon_\infty = 3$ [5.45], $\alpha = 2.34 \times 10^{-4}/K$, $\beta = 2.58 \times 10^{-10}$ m^2/N [5.41], T = 300 K, and $\phi^2 = 0.07$ rad^2 (from estimate of $\phi = 15°$). The Grüneisen contant is expected to be small because the force constants for dipole rotation are mostly intramolecular and do not depend strongly on volume. Neglecting the small terms in γ, we find $p_y = -0.10$ nC cm^{-2} K^{-1} and $d_p - 0.73$ pC/N when $E_p = 320$ kV cm^{-1}, in good agreement with measured values [5.48]. The subscript p indicates hydrostatic pressure and the subscript y helps to distinguish the symbol for pyroelectric coefficient from that for pressure.

Even if one does not have dielectric data, one can assume $\varepsilon_\infty = 3$ and the quantity $\Delta\varepsilon$ can then be calculated with reasonable confidence from the dipole moment using Onsager's equation [5.45]. *Reddish* [5.45] interpreted the dielectric data on PVC as indicating that the length of the relaxing segments increased as the temperature decreased below T_g and since the $\Delta\varepsilon$ increases linearly with the number of dipoles per rigid unit, large polarizations could be achieved. Unfortunately, we found no enhancement of p and d by lower temperature poling of PVC and suspect the observed effects in dielectric properties are due to space charge.

Since most of the variables in (5.28) and (5.29) will be similar for all polymer glasses, larger coefficients can be sought from polymers with a large dipole moment (p and d will increase as the square of the dipole moment per unit volume) and by increasing the poling field. A possible candidate is polyacrylonitrile (PAN) with a dipole moment greater than 4D. Unfortunately, PAN may have an anomalous liquid phase in which dipole–dipole forces prevent normal polarization [5.49] contrary to criteria (II) in Sect. 5.4.1. In other cases the dipoles may not become immobile at T_g (e.g., polymethylmethacrylate) contrary to criteria (III). A thermally stable, high T_g polar glass may have useful

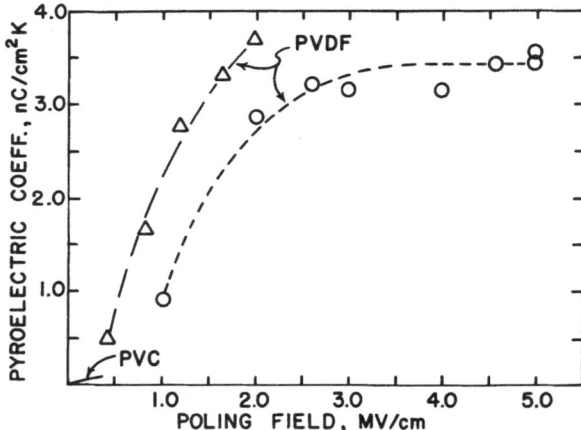

Fig. **5.10.** Pyroelectric coefficient as a function of poling field for the amorphous polymer, PVC, and the semicrystalline polymer, PVDF. PVC data are from [5.48], PVDF data are from [5.19] (△) and [5.114] (○).

high temperature applications, but presently the most sensitive piezo- and pyroelectric polymers are semicrystalline. A comparison of a semicrystalline polymer and PVC is shown in Fig. 5.10.

5.4.3 Semicrystalline Polymers

The most interesting of the semicrystalline polymers are PVDF, PVF and related copolymers. These polymers crystallize much like polyethylene because the fluorines, unlike larger chlorines, are close enough in size to hydrogen so as not to interference with regular packing. Both polymers have head–head and tail–tail defects, where successive repeat units are backwards. Typically, these amount to 5% for PVDF [5.50–52] and 25–32% for PVF [5.50]. A h–h unit in PVDF is immediately followed by a t–t unit [5.50] so that 5% of these defects cancel 10% of the dipole moment of the planar zigzag chain.

The dipole moment of PVF could be quite large in the transplanar conformation if all fluorines were on the same side of the C–C plane (isotactic). For atactic PVF the average moment will be in the C–C plane, perpendicular to the molecular axis and close to 1/2 that of PVDF. However, 30% h–h defects will reduce the net moment of the planar PVF by about 60%, with the result that the net moment of trans PVF is about 20% that of trans PVDF.

Semicrystalline polymers consist of lamellar crystals mixed with amorphous regions. A schematic diagram of a spherulite within an unoriented semicrystalline polymer is shown in Fig. 5.11. Annealing or crystallizing for longer times, at higher temperatures and pressures increases the lamellar thickness and perfection which results in a higher sample density. The crystals grow in the form of spherulites and studies of the morphology of PVDF show that three crystal phases have distinct morphology and can grow simultaneously from the melt or one phase can grow at the expense of another [5.53–55]. A typical molecular weight for these polymers is 10^5 for an extended length of 0.5 μm and

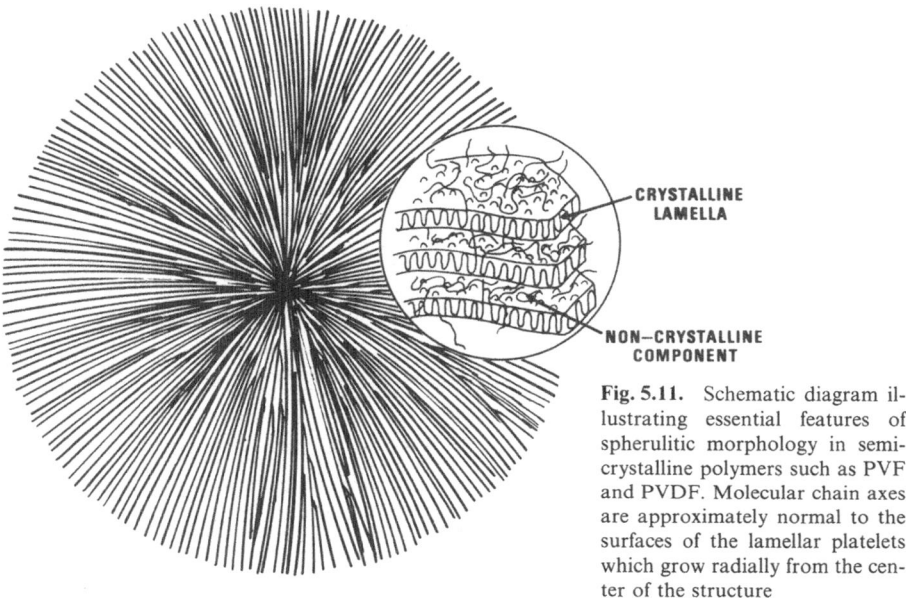

CRYSTALLINE
LAMELLA

NON—CRYSTALLINE
COMPONENT

Fig. 5.11. Schematic diagram illustrating essential features of spherulitic morphology in semicrystalline polymers such as PVF and PVDF. Molecular chain axes are approximately normal to the surfaces of the lamellar platelets which grow radially from the center of the structure

a total of 2000 repeat units. Since the lamellae are of the order of 10^{-8} m thick, a single molecule folds back and forth through the same or different lamellae many times. When stretched to several times the original length, the specimen becomes oriented such that the lamellae are normal to the stretch direction and the molecules are parallel to the stretch direction [5.56, 57]. The polymers currently of greatest interest for piezoelectric applications, PVF and PVDF, are of the order of 50–70% crystalline [5.51, 58, 59].

The amorphous phase is probably mostly confined to layers between the crystal lamellae. The nature of this phase and the degree to which it is oriented and connected to the crystals is a subject of debate. The amorphous phase seems to have normal supercooled liquid properties with a liquid–glass transition region around $-50\,^\circ$C, and a Williams–Landel–Ferry-type dielectric relaxational behavior [5.18, 60, 61]. Broad line NMR [5.60–63], and mechanical relaxation data [5.5, 39, 51, 62, 64] also show a normal liquid–glass relaxation. The magnitude of the associated dielectric dispersion and room temperature relative permittivity increase with amorphous content as expected [5.61, 62].

The dielectric permittivity is quite sensitive to uniaxial and biaxial orientation of the film [5.65], the more orientation the higher the polarizability normal to the draw direction. This effect is typically used to enhance the permittivity of PVDF film used in capacitors and has been attributed to orientation of the liquid material so that rotations about the amorphous molecular axes are more effective in contributing to the polarization. An alternative explanation by *Davies* et al. [5.66] is that since orientation aligns crystal lamellae normal to the draw direction, the liquid–crystal layers are

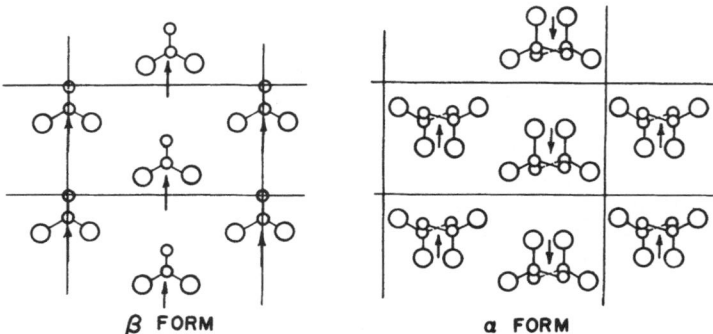

β FORM   α FORM

Fig. 5.12. Projections on a plane normal to the molecular axes of the α and β crystal forms of PVDF. Large circles represent fluorine atoms, small circles represent carbon atoms and hydrogen atoms are omitted

parallel to the applied field giving a mean permittivity larger than in an undrawn specimen where some lamellae are normal to the applied field and their permittivities add in series. From criteria (III) in Sect. 5.4.1 we do not expect the amorphous phase to contribute to piezo- and pyroelectricity in PVDF and PVF. *Murayama* [5.9] has stated the same conclusion.

Three crystal phases have been reported for PVDF. The α phase (alternatively form II) forms most readily upon cooling the molten polymer and assumes a conformation close to trans-gauche–trans-gauche' which then packs in the unit cell to yield an antipolar crystal [5.67, 68]. Mechanical orientation of the α phase at temperatures below 50 °C yields the β phase (form I) which has an extended all-trans (planar zigzag) conformation and packs in the unit cell with the dipole moments of adjacent chains parallel to yield a polar crystal [5.57, 67]. A third form, referred to as γ or form III can be obtained by crystallization from selected solvents such as dimethylsulfoxide or dimethylacetamide [5.69] or by annealing at high temperatures [5.53, 54, 70]. The infrared spectrum of form III closely resembles that of form I so an all-trans conformation was assumed [5.71] to aid in the indexing of X-ray diffraction spacings [5.72]. More recent data may require modifying the present conclusions regarding form III to account for a spacing along the chain which is double that exhibited by the α phase [5.73]. For example, a gtttg'ttt conformation would account for the c axis repeat and the trans sequences would make it vibrationally similar to that of β phase. Projections of the α and β conformations onto a plane normal to the molecular axes are shown in Fig. 5.12. The crystal structure of PVF is the same as the β phase of PVDF [5.43] and mixtures and copolymers of these two monomers tend to cause crystallization into the polar β phase [5.74]. Model calculations have been made for comonomer units and head-to-head defects within PVDF and their effects on the potential energy of chain conformations [5.75] are consistent with experimental observations.

The repeat unit for PVDF, $(-CF_2-CH_2-)_x$, is assumed to have a dipole moment close to that of difluoroethane [5.76] which is $7.56 \times 10^{-28}\,C\,cm$

(2.27 D). In the all-trans conformation (planar zigzag) the component of this moment normal to the long molecular axis is about 6.9×10^{-28} C cm (2.1 D). In the α phase tgtg' conformation the same 2.27 D per repeat unit, using the atomic coordinates of [5.72] yields a dipole moment of $\mu_{\perp}^{\alpha} = 4.03 \times 10^{-28}$ C cm perpendicular to the long axis and $\mu_{\parallel}^{\alpha} = 3.36 \times 10^{-28}$ C cm parallel to the long axis. The antipolar unit cell does not have proper symmetry to yield piezo- and pyroelectric activity. However, recent X-ray data show that the antipolar unit cell can be transformed by a large electric field to give a stable polar modification of the tgtg' conformation which is then both piezo- and pyroelectrically active [5.77]. This finding modifies earlier conclusions that the β form is necessary for activity.

The usual methods of identifying the fraction of crystallinity in a specimen are to compare its density to that of crystal and amorphous densities [5.78] or to compare its heat of fusion to the crystalline heat of fusion [5.78]. The usual measures of crystal phase fractions are X-ray diffraction [5.79] and infrared absorption [5.70] intensity ratios.

5.5 Properties of Semicrystalline Polymers

5.5.1 Crystal Relaxations

In the crystal phases of PVF and PVDF the question of rotational freedom of the dipoles is crucial [Sect. 5.4.1, ii]. Ample dielectric relaxation data exist on PVF [5.80] and α phase PVDF [5.18, 61, 78, 80–83], and mechanical data on PVF [5.64] and α phase PVDF [5.62], and thermally stimulated current (TSC) data on α phase PVDF [5.18] to conclude that a crystal relaxation α_c occurs at about 80 °C at a measuring frequency of 100 Hz. At room temperature the relaxation time τ_c for the α_c relaxation is increased to about 1 s. Log τ_c is linear with $1/T$ and the activation energy is around 100 kJ/mol. The β phase of PVDF is reported to have a mechanical crystal relaxation at 110 °C at 10 Hz [5.39, 62]. Its activation energy has not been determined. Dielectric α_c relaxations in β- and γ-phase PVDF are generally not observed possibly because of rapidly increasing ε' and ε'' with temperature. This behavior is usually attributed to space-charge effects [5.80, 82, 83]. TSC data give strong background currents even from unpoled specimens [5.12, 84]. At lower temperatures current with a broad maximum at 80°, and integrated charge of up to 3μC cm^{-2} is observed [5.21]. *Davies* has reported [5.85] that after removing a large fraction of mobile charges by application of a large DC field (field cleaning) the α_c relaxation in β-phase PVDF was observed at about 140 °C and 10 Hz. The activation energy was about 100 kJ/mol. That the dielectric α_c is a crystal relaxation was shown for α-phase PVDF by the dependence of its amplitude on crystallinity and by observing its disappearance at the melting temperature [5.61]. That it can exist in the β phase is demonstrated by its presence in PVF [5.80], observed following field cleaning. The relaxation probably involves rotation of an entire intralamellar segment with twisting at the lamellae surfaces analogous to the well-documented α_c mechanism in PE [5.86]. Since twisting must be about C–C bonds, these rotations would be restricted by the crystal fields of neighboring molecules. This rigid-rod model is also supported by the dependence of the α_c relaxation parameters on molecular thickness [5.60, 78, 82]. However, some evidence has been reported that the polarization of the α_c relaxation in α-phase PVDF is along the c crystal axis contrary to the usual case [5.87]. Additional direct evidence for electric field induced rotation of molecules in the crystal phase of PVDF is discussed in Sect. 5.5.2.

It is useful to review linear relaxation theory before proceeding to nonlinear effects at high field. The crystal relaxation occurs by rotation of the molecules about their long axes within lamellar crystals either by rigid-rod rotation or twisting [5.86]. To simplify the calculation we assume the crystal lamellae are thin slabs aligned normal to the polymer film as expected in most commonly measured films from uniaxial or biaxial orientation. The net sample polarization P_s arises from the average of the crystal P_c and the liquid P_l polarizations

$$P_s = \Psi P_c + (1 - \Psi)P_l, \tag{5.30}$$

where Ψ is the crystal volume fraction.

Since $P = (\varepsilon' - 1)\varepsilon_0 E$ and the field, E, for these oriented thin lamellae will be the same inside and outside the crystal (tangential component of E continuous across the crystal-liquid interface) we can write the sample permittivity as a simple sum of the crystal ε_c and liquid ε_l permittivities

$$\varepsilon_s = \Psi \varepsilon_c + (1 - \Psi)\varepsilon_l. \tag{5.31}$$

At radio frequencies well above the glass transition temperature the relaxation amplitude of the sample is

$$\Delta \varepsilon_s = \Psi \Delta \varepsilon_c \tag{5.32}$$

where $\Delta \varepsilon$ is the difference between relaxed and unrelaxed values of ε. To relate $\Delta \varepsilon_c$ to microscopic quantities we adopt the familiar 2-site model [5.86] with an important modification to allow for the cooperative effects common to ferroelectric materials.

Assume a molecule in the crystal has its most probable orientation with its dipole moment m_0 at an angle θ with respect to an applied electric field E. A second possible orientation (site 2) is at an angle $\theta + \pi$ and the lattice potential energy of a molecule in site 2 is greater than in site 1 by the energy U as indicated in Fig. 5.13 in which $\theta = 0$. The probability of occupation of site 2 will be $f_2 = C \exp[-(2U + 2m_0 E \cos \theta)/kT]$ and that of site 1 will be $f_1 = C$ where $C = 1/1 + \exp[-(2U + 2m_0 E \cos \theta)/kT]$ is a normalization factor chosen so that $f_1 + f_2 = 1$, and $2U + 2m_0 E \cos \theta$ is the work to move one segment from site 1 to

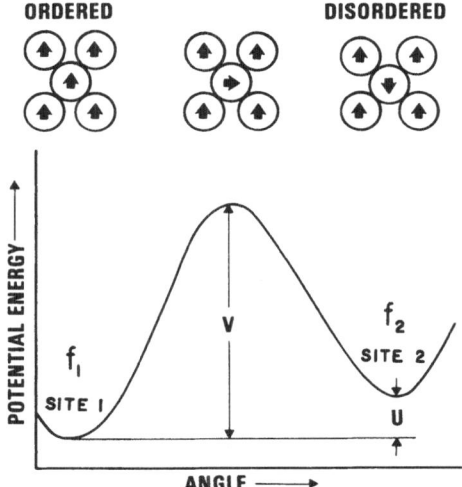

ORDERED   **DISORDERED**

POTENTIAL ENERGY ⟶

f_1 f_2 V SITE 2   SITE I U

ANGLE ⟶

Fig. 5.13. Potential energy diagram for a two site model in which a molecular dipole becomes reoriented with respect to its neighbors under the influence of an electric field

site 2 [5.88]. (The term $2U$ includes the work to move a molecule from site 1 to site 2 plus the work to adjust the energies of the remaining molecules.)

Let m be the apparent dipole moment per molecule for any arbitrary distribution of dipoles between sites 1 and 2 in a given crystallite. For such a crystal, the average moment in the direction of the field will be

$$m \cos\theta = m_0[f_1 \cos\theta + f_2(\cos\theta + \pi)] = m_0(f_1 - f_2)\cos\theta \tag{5.33}$$

whence,

$$2m_0 E(\cos\theta)/kT = -2U/kT + \ln[(1 + m/m_0)/(1 - m/m_0)]. \tag{5.34}$$

The cooperative aspects of this model arise from the requirement that the system normally ordered in site 1 is equivalent to the same system normally ordered in site 2, and that if both sites are equally populated then sites 1 and 2 must have equivalent energy. That is, the energy U must depend on the values of f_1 and f_2. The type of dependence which has been applied to theories of ferromagnetism [5.89], Bragg–Williams order–disorder transitions in alloys [5.90] and ferroelectricity in Rochelle salt [5.91] is

$$U = U_0(f_1 - f_2) = U_0(m/m_0) \tag{5.35}$$

where U_0 is the lattice energy difference between site 2 and 1 when site 1 is completely populated ($f_1 = 1, f_2 = 0$).

Using this result in (5.34) we have an important relationship between the average moment of a dipole in this crystal and the applied field, or

$$2m_0^2 E(\cos\theta)/kT = -2U_0 m/kT + m_0 \ln[(1 + m/m_0)/(1 - m/m_0)]. \tag{5.36}$$

To find $\Delta\varepsilon_s$ we need to calculate the change in sample moment with electric field. The differential of (5.36) at $E=0$ gives

$$(\partial m/\partial E)_{E=0} = 4f_1 f_2 [1 - 2f_1 f_2 \ln(f_1/f_2)/(f_1 - f_2)]^{-1} m_0^2 (\cos\theta)/kT. \qquad (5.37)$$

This result differs from that usually obtained for the 2-site model [5.92] by the additional term in brackets (due to the cooperativity), and the absence of the term $3\varepsilon_l/(2\varepsilon_l + \varepsilon_c)$ due to the assumption of lamellae rather than spheres. It leads in a straightforward way [5.47] to the result

$$\Delta\varepsilon = \psi(N/V) 4f_1 f_2 [1 - 2f_1 f_2 \ln(f_1/f_2)/(f_1 - f_2)]^{-1} \mu_0^2 (\varepsilon_c + 2)^2/18kT \qquad (5.38)$$

where $\mu_0 = 3m_0(\varepsilon_c + 2)^{-1}$ [5.92] is the vacuum moment of the molecular segment involved in the rotation, ε_c the nonrelaxed relative permittivity of the crystal, N/V is the number of crystal dipoles per unit volume, ψ is the volume fraction crystallinity, and the average of $\cos^2\theta$ for single axis rotators with the axes in the plane of the film is $\frac{1}{2}$. The main features of the result are that, for a given f_1, the relaxation amplitude is larger than for the noncooperative (constant U) case tending towards infinity at $U_0/kT = 1$.

Very high dielectric constants are sometimes reported for PVF and PVDF [5.80, 93] but these are usually attributable to space-charge effects which can be greatly reduced by application of high DC fields to the sample [5.80]. It is probable that the enhancement of ε_s near the Curie temperature due to dipole disordering has not yet been observed because the Curie temperatures expected for PVF and PVDF are well above the temperature at which these crystals melt. Equation (5.38) should be replaced by a more general equation for conducting dispersions and random orientation of lamellae when necessary. Note that the amplitude of the crystal and liquid relaxations as measured will appear smaller than if measured in each phase separately because of the factor ψ.

Since the pertinent data are available for PVF, it is instructive to apply the above model to this polymer. A typical all-trans segment within the lamellar crystal can be expected to contain about 40 repeat units. 60% of these are nonpolar because of head-to-head defects and their effective rigid-rod moment (μ_0) will be $16 \times 3.3 \times 10^{-28}$ C cm. The number of such rods per unit volume is $1.8 \times 10^{22}/40$ cm^{-3}, and $T = 350$ K.

The typical commercial films of PVF are highly oriented and have a crystallinity around 50% [5.59]. We can use (5.38) and data from [5.80] which show $\Delta\varepsilon_s \approx 5$ to calculate f_1 and from the definition of f_1 at $E=0$ obtain the difference in crystal energy for disordered and ordered molecules. Equation (5.36) yields a fraction of disordered molecules of 5% (an alternative solution, $f_1 = 50\%$, is not stable [5.88]) when $U_0/kT = 1.67$. That is, even though PVF has a polar, trans-planar crystal structure at a temperature of 350 K approximately 5% of the molecular segments in the crystal lamellae will be oriented with their net moments opposite that of the host lamellae. This is a dynamic disorder such that all segments spend about 5% of the time in the antipolar orientation. The angular frequency with which a segment changes orientation and the potential energy barrier which must be traversed, W, are calculable from the simple Arrhenius expression

$$v = 2v_0 \exp(-W/kT),\qquad\qquad\qquad(5.39)$$

where v_0 is the librational frequency of the segment in the crystal (about 2×10^{12} Hz for β-phase PVDF [5.94]). From dielectric data at 100 Hz and 350 K, $W/kT = 24$ for PVF [5.80]. Both U_0 and W will depend on pressure and temperature decreasing by about 10 % for each 1 % increase in crystal volume [5.95]. Even at room temperature the relaxation time for dynamic disordering of the crystal is of the order of seconds with several percent of the crystalline segments antipolar to the crystal moment at any instant.

Qualitatively similar behavior is expected to occur in α phase PVDF except that nearest neighbors are antipolar in the ordered state and the disordered state consists of several percent polar nearest-neighbor pairs. The absence of published dielectric data on the α_c relaxation in β-phase PVDF makes quantitative analyses of this case unclear, but there is little doubt that the same type of dynamic crystal disordering typical of semicrystalline polymers is present.

It is important to distinguish between the linear response of the dynamic disordering of molecular segments in the PVDF crystal to changes in applied electric fields at the low voltages usually employed for dielectric measurements and the nonlinear response expected at the high voltages used to polarize piezo- and pyroelectric samples. The linear low-field response does not involve permanent changes in polarization while the high-field response leads to ferroelectric reorientation.

5.5.2 Ferroelectricity

PVDF in the polar form is often supposed to be a ferroelectric [5.96, 97], which means that it is not only a polar crystal but that the stable equilibrium crystal polarization can be reoriented with an applied electric field. Direct evidence for a field-induced change in the unit cell orientation in PVDF measured by X-ray pole figures has been reported [5.22, 98]. Molecular orientation measurements using Raman techniques suffer from the dilution effect of the liquid phase [5.99]. The usual hysteresis measurements [5.12, 100] of charge versus field are difficult because of space-charge effects and the results are ambiguous. Measurement of piezoelectric [5.12, 36] and pyroelectric [5.93] response as the field is cycled from large positive to negative values does give a hysteresis loop. A poled PVDF film was shown to require 450 kV cm^{-1} to suppress its piezoelectric response [5.36]. High dielectric constants of the order of 1000 [5.93] are also indicative of ferroelectric switching in PVDF and PVDF-TFE copolymer but since this behavior can often be eliminated by annealing in the

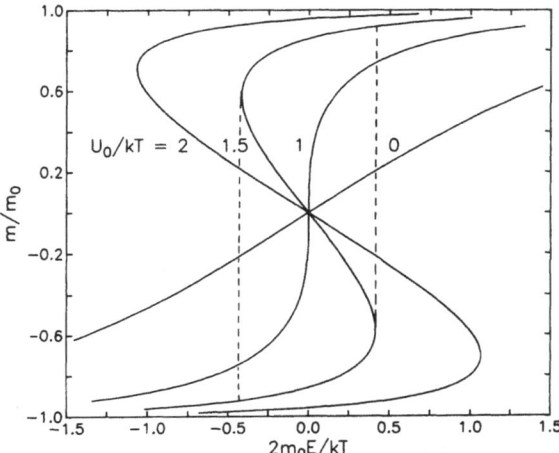

Fig. 5.14. The ratio of the average moment to the actual moment per dipole in the two site model of Fig. 5.13 as a function of the electric field to which it is subjected. Curves are drawn for different values of the energy difference between sites for the perfectly ordered system. The dashed lines correspond to a ferroelectric hysteresis loop for an ideal crystal

presence of a DC field [5.80, 101] space-charge effects rather than ferroelectric switching is probably responsible.

In the work of *Kepler* and *Anderson* [5.98] referred to above, they assumed the molecular segments could rotate about their long axes and occupy any of six orientational positions at $\pi/3$ radians apart. They found that the length of a rotating unit was $40\,CF_2$–CH_2 repeat units which is about the same as the number of units in one molecular segment in the crystal. They also reported that the crystal alignment was partially lost by high temperature annealing. *Fukada* et al. [5.102] used polarized infrared absorptions for symmetric and antisymmetric CF_2 stretch vibrations to show that these dipoles orient in an electric field. Their results indicated that the relationship between piezoelectricity and dipole orientation was not unique. Since they measured the average $\langle\cos^2\theta\rangle$ rather than $\langle\cos\theta\rangle$ (where θ is the angle between the dipole moment and electric field) it is possible that nonuniform polarization which is commonly observed at lower poling voltages [5.19, 30] is an important factor. Infrared measurements of crystal dipole orientation at lower frequencies were also reported to show hysteresis [5.103]. That is, CF_2 dipoles in the crystal do align with an applied field and some of the alignment persists until a strong field is applied in another direction.

Not only does field-induced alignment of dipoles occur but conversions from one nonpolar phase to at least two other polar crystal phases has been demonstrated [5.77, 104–108]. These and other results [5.12, 100, 109] provide strong evidence for ferroelectric switching in the crystal phase of PVDF. It is still possible, however, that not all of the piezoelectric and pyroelectric activity in PVDF is due to oriented dipoles, and the possibility that trapped space charges are an important factor is often suggested [5.10, 16, 110].

According to (5.36) we can calculate the dipole moment for a cooperative 2-site model of PVDF as a function of electric field. The results are shown in Fig. 5.14 for several values of the parameter U_0/kT. We can write the free

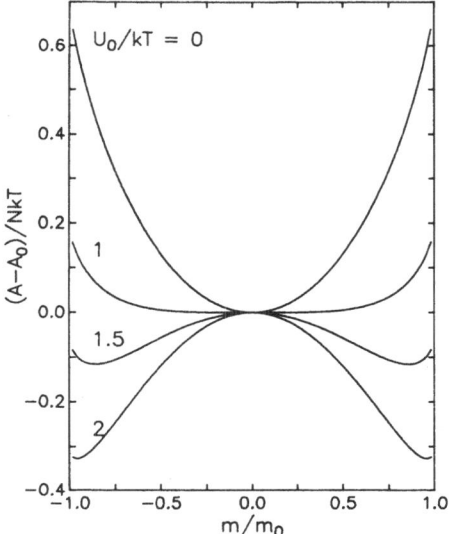

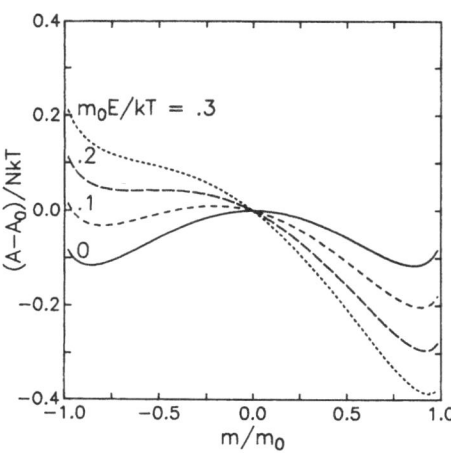

Fig. 5.15. Calculated equilibrium Helmholtz free energy of partially ordered system of dipoles relative to the completely disordered system as a function of the ratio of the average moment to the actual moment per dipole. Curves are drawn for different values of the energy difference between sites as in Fig. 5.14 in the absence of an applied field ($E = 0$)

Fig. 5.16. Calculated Helmholtz free energy of a partially ordered system of dipoles as a function of the ratio of the average moment to the actual moment per molecule. Curves are drawn for different values of an externally applied field and apply to the case when $U_0/kT = 1.5$

energy difference between the partially ordered system and the completely disordered system as,

$$A(m/m_0) - A(0) = -U_0(m/m_0)^2/2 - mE\cos\theta + (kT/2)[(1+m/m_0)\ln(1+m/m_0) + (1-m/m_0)\ln(1-m/m_0)]. \tag{5.40}$$

The extrema of $A - A(0)$ are given by letting $\partial[A - A(0)]/\partial(m/m_0) = 0$ which gives (5.36). The values of $[A - A(0)]/NkT$ for $E = 0$ are given in Fig. 5.15 as a function of m/m_0 for several values of U_0/kT.

The ferroelectricity of the model arises for $U_0/kT > 1$ because of the existence of two minima in the free energy corresponding to positive and negative values for the crystal moment. The system can be switched from $+m$ to $-m$ by application of a sufficiently high electric field in a direction opposite m as illustrated in Fig. 5.16 for the case $U_0/kT = 1.5$. At a value of $m_0E/kT \approx 0.2$, the minimum at $m = -0.95m_0$ is removed and the only remaining minimum is at $m > +0.95m_0$.

Physically then we can describe poling of PVDF in terms of Fig. 5.14 as follows. When a large field is applied to the specimen such that E is considerably greater than its critical value E_c given by the point of infinite slope in Fig. 5.14, the moment of the crystal will be positive. If the field is then decreased beyond $-E_c$, the stable solution for positive m disappears and m switches to a negative value. This procedure of cycling E from large positive to negative values results in a hysteresis curve ideally given, for example, by the dashed path shown on the $U_0/kT = 1.5$ curve of Fig. 5.14.

In the case of PVF we saw that dielectric data yield $U_0/kT = 1.7$ so that $m_0 E_c/kT = 0.3$ and, using the moment given in Sect. 5.5.1, $E_c = 0.3\,\mathrm{MV\,cm^{-1}}$. This number is quantitatively similar to the lowest fields used for poling PVF and PVDF, and gives some indication that a simple cooperative model is realistic.

5.5.3 Space Charges

Space charge measurements are frequently made by measuring the currents generated by heating, at a uniform rate, a specimen with evaporated electrodes which has previously been cooled under an applied field. The thermally stimulated short-circuit currents (TSC) result from dipole reorientation and from the change in the dipole moment of the space-charge distribution [5.111]. Studies on both amorphous and semicrystalline polymers have shown the space charge to be predominantly positive near the negative poling electrode and negative near the positive poling electrode (heterocharged electret). Kerr effect measurements in liquid nitrobenzene with dc fields show uniform space-charge distributions with a net positive charge density of the order of $10^{-8}\,\mathrm{C\,cm^{-3}}$ [5.112]. (For a further discussion of such observations, see Sect. 2.6.3.) Dielectric measurements of PVF and PVDF typically show anomalously high values of ε' at high temperatures and low frequencies [5.80, 83, 113]. This effect is generally attributed to ionic space charges [5.83]. It was shown that the interfacial polarization in solid PVDF is different from the electrode interfacial polarization effects in liquid PVDF and was attributed to space-charge polarization at the liquid-crystal interfaces [5.83]. In TSC measurements it was shown that repeated cycling of PVDF from 25 to 100 °C with an applied field, reduced the space-charge effects [5.18]. Space-charge concentrations in liquid nitrobenzene [5.112], and the space-charge effects seen in dielectric measurements of PVDF [5.80, 83, 113], and PVF [5.80] were also reduced by application of dc fields for several hours. This reduction partially recovers with time after removal of the field. At very high fields ($E > 500\,\mathrm{kV\,cm^{-1}}$) there is a change in the above behavior and current tends to increase with time at constant voltage leading ultimately to breakdown [5.114].

The general behavior of a conducting liquid with dispersed crystals of roughly equal volume is quite complicated [5.115]. EMFs generated by chemical or electronic interactions between a polymer film and the metal electrodes [5.116] are a possible source of anomalous effects such as the extremely high isothermal short-circuit currents observed in vinylidene fluoride polymers at high temperatures [5.117] and variations in current-voltage behavior with electrode metal [5.80, 118]. If, as seems certain, ferroelectric

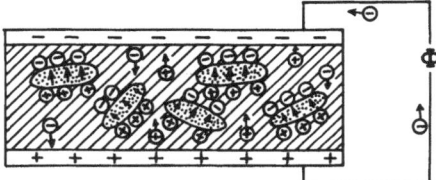

Fig. 5.17. Schematic diagram of interfacial polarization in a semicrystalline polymer in an applied dc field due to charge build up at crystalline surfaces

switching of the polar crystals occurs during the early stages of poling, then continued poling results in a flow of charge, mostly through the more conductive liquid phase regions with positive charges moving toward the negative electrode and negative charges toward the positive electrode. At normal poling temperatures ($\sim 100\,^{\circ}$C) the current is time dependent and bulk interfacial polarization effects are evident [5.83]. The charge carriers could tend to pile up at the crystal surfaces where their normal drift is hindered as shown in Fig. 5.17. Electric fields from oriented crystal dipoles would tend to trap these charges at suitable crystal-liquid interfaces. This situation is analogous to concentrated emulsions of oil and water [5.119]. TSC results show that space charges are released at temperatures higher than the dipole relaxation maxima and these for PVDF are considerably above room temperature [5.111, 120]. As a result, when the specimen is cooled to room temperature the charges are immobilized and remain at the crystal surfaces as in Fig. 5.17. The space charges form a dipole which has the stiffness of the crystal and because of nonuniform strains will produce a piezo- and pyroelectric response [see Sect. 5.2.2]. Note however that the space-charge dipoles if formed from interfacial polarization of the poling-current charges are opposing the molecular dipoles and will reduce the piezo- and pyroelectric response. The observation of a slow poling process reported for PVDF [5.10, 36, 103], whereby the piezoelectricity slowly increases with time, and the observation that space charges are removed by dc fields on a similar time scale [5.18, 80, 83, 112, 113] support the idea that a decrease in space charge reduces the masking effects of space-charge dipoles. Also the partial decay of piezoelectricity with time after removal of the field [5.10] and the partial recovery of space-charge density with time [5.80, 83, 113] may be similarly related to the postulated space-charge dipoles.

Probably the most important role of space charge is in the poling process. It is well known that space charge in a material has associated with it an electric field which will cause the local electric field in the material to be greater or less than the applied field [5.121, 122]. If for example a negative space charge is present in the polymer, the potential difference between the negative electrode and the interior of the polymer will be less than that for an uncharged polymer and correspondingly greater at the positive electrode. Thus, if the polarization due to crystal alignment increases with the local poling field, then the polarization will be greater at the positive electrode for a negatively charged polymer. The distribution of activity in PVDF has been measured and found to be greater at the positive electrode [5.19, 110]. Higher temperatures, longer poling times and higher electric fields reduce this nonuniformity [5.19]. It is surprising that the polarization in PVF tends to be greater at the negative poling electrode [5.123]. Assuming that nonuniform polarization in these materials results from space charge in the bulk polymer, then we conclude that PVDF accumulates a net negative space charge and PVF accumulates a net positive space charge. This charging phenomena is sometimes discussed in terms of a work function difference between the polymer and metal electrodes [5.124–126], which in turn depends on the detailed molecular structure [5.124].

The distortions in local poling fields due to space charge tend to become small at high fields ($>1\,\mathrm{MV\,cm^{-1}}$), and while studies of space-charge effects are necessary for a complete understanding of PVDF, it seems likely that the presence of space charge is not a major factor in the behavior of well-poled PVDF.

5.6 Measurements and Data

Piezo- and pyroelectric measurements on polymer films are usually made by applying tension to a strip of polymer and measuring on opposing electrodes the short-circuit charge or open circuit voltage due to a change in the tension. The stress and strain are measured with a load cell and strain gauge and the stress is usually sinusoidal. This basic technique has been discussed previously [5.5, 7]. Measurements can be made by clamping a sample in a vise in series with a load cell. Noise can be reduced for these measurements by the use of good contact electrodes (e.g., evaporated). A double film sandwich with the high potential electrodes together and shielded by the outer grounded electrodes greatly reduces noise problems. Piezoelectric measurements have also been made by applying a field to the specimen and observing the length and thickness change [5.17], by analyzing the response of a piezoelectric film driven electrically in the neighborhood of its resonant frequencies [5.13] and by applying hydrostatic pressure to the film with He gas [5.41]. Pyroelectric measurements are conveniently made by changing the temperature of the specimen and measuring the charge produced [5.41] correcting for any irreversible effects. Heating and cooling can be done with a Peltier device [5.127] which is noisier than a heater or, for electromagnetic purposes, with optical radiation [5.20]. A very convenient dielectric heating method has recently been demonstrated [5.128]. The accuracy of piezoelectric data is hard to assess since error analyses are seldom mentioned. Piezoelectric and electrostriction data on PVDF as a function of temperature and frequency have been adequately reviewed previously [5.5, 7, 10].

There are many reports in the literature of piezoelectric and pyroelectric response from PVDF as a function of poling conditions. A summary of all of them would be complicated because there are so many parameters, each of which can be varied over a wide range, and conclusions drawn from one set of

Table 5.1

Material	d, pC/N		p, nC cm^{-2}K	
Quartz, d_{11}	2.3	(5.134)[a]		
PZT-4, d_{33}	289	(5.135)	27	(5.132)
BaTio$_3$, d_{33}	190	(5.135)	20	(5.132)
Rochelle Salt, d_{25}	53	(5.134)		
Triglycine Sulfate	50	(5.136)	30	(5.132)
Sr$_{0.5}$Ba$_{0.5}$Nb$_2$O$_6$	95	(5.136)	60	(5.136)
Polyvinylchloride, d_p	0.7	(5.25)	0.1	(5.25)
Polyvinylfluoride	1	(5.6)	1.0	(5.137)
Nylon 11	0.26	(5.138)	0.5	(5.138)
Polyvinylidenefluoride, d_{31}	28	(5.13)	4	(5.19)

[a] Numbers in parentheses refer to references at the end of the chapter.

measurements are not valid for other values of the parameters. For example, early data [5.10] on samples with different contents of form I and form II crystals when poled at 320 kV cm^{-1} and 40 °C showed a variation in piezoelectric response of more than 100-fold, implying a need for the presence of polar form I in order to obtain good piezoelectric response. Subsequent poling [5.13] of oriented form II at slightly higher fields yielded films with piezoelectric coefficients as much as one-half as large as those from form I. More recently [5.105] it has been shown that electric fields in the vicinity of 1 MV cm^{-1} can cause a crystal phase change from antipolar form II to a piezoelectric polar form II and at higher fields, an additional phase change to form I occurs. Again, early data [5.10] showed a strong dependence of electrical response on poling temperature for poling times of 30 min at 1 MV cm^{-1}. More recent data for much longer times (10^3 min) implies that the ultimate response from a given sample depends only upon applied field and that temperature merely affects the rate at which polarization occurs [5.129]. In measurements of pyroelectric response the uniformity of polarization increased significantly when the poling time was increased from 5 to 30 min [5.19].

Polarization versus time measurements for PVDF show two stages according to several authors [5.10, 103, 110, 130]. Some [5.10, 103, 110] report a fast response at less than a second and a slow response at 1–2 h, while others [5.130] report the fast response at 1 min and the slow response at 1 h. More recently it has been shown that for stronger electric fields (in the vicinity of 2 MV cm^{-1}), the polarization occurs within the first few seconds of poling–even at room temperature [5.105, 108, 114]. The nature of the electrodes or the presence of charge carriers can influence the uniformity of polarization over certain ranges of the variables mentioned above. In general, high fields and longer poling times lead to uniform polarization [5.19].

There is disagreement about whether polarization saturates at high temperatures and high fields. *Murayama* et al. [5.9, 10] reported no saturation for the piezoelectric response in PVDF and *Day* et al. [5.19] reported none for the pyroelectric response (although the data for only uniformly poled samples do show saturation effects) [5.19]. Other data show saturation of the piezoelectric response [5.39] in PVDF and in our laboratory both the piezoelectric and pyroelectric response of PVDF and VDF-TFE copolymers saturated as functions of field and temperature [5.48, 114]. Of course, the response also depends upon the orientation of the polymer chains in the film before poling and the direction in which the film is strained while measuring the electrical response. Despite the apparent inconsistencies in the details of the poling process, many reports agree that the maximum value of d_{31} for PVDF which has been obtained so far is 15–20 pC/N [5.13, 56, 109, 114, 131] and the corresponding pyroelectric coefficient is 3–4 nC cm^{2}K. [5.19, 108, 114, 132, 133]. These values are compared with the electrical response from other piezoelectric and pyroelectric materials in Table 5.1. At present there have been few reports of detailed measurements on a single specimen to yield several components of the piezoelectric tensor (Sect. 5.3).

Applications of polymer transducers for reflectivity measurements [5.139], a photocopy process [5.15], radiation detectors [5.20], night vision targets [5.132], intrusion and fire detectors [5.140], hydrophones [5.141], earphones and speakers [5.11], pressure sensors, strain gauges and many more are being developed or investigated, and the list of applications is growing rapidly, as discussed in Chap. 7. Several recent relatively nontechnical publications have stimulated widespread interest in piezoelectric polymer devices [5.142–144].

5.7 Dipole Model Applied to Semicrystalline Polymers

Evidence for the presence of preferentially aligned crystal domains giving PVDF a net dipole moment has been given in Sect. 5.5.2. It is important to know if aligned crystal domains can account for the observed activity. A model similar to that for amorphous polymers (Sect. 5.2.3) has been analyzed [5.145]. In the model the polymer is assumed to consist of crystal lamellae dispersed in an amorphous liquid and oriented approximately normal to the large film surfaces. The molecular segments are preferentially aligned so that their dipole moments are parallel. A typical crystal is assumed to be as shown in Fig. 5.18, with the crystal moment at an angle θ_0 with respect to the film surface and an equilibrium amount of real charge trapped at the crystal surfaces which are normal to the crystal moment. Calculation of the charge dQ appearing at the film surface as a result of changes in temperature dT or hydrostatic stress $-dp$, leads to the following equations for piezo- and pyroelectric response:

$$d_p = \frac{1}{A}\frac{dQ}{dp} = P_0\beta_c[(\varepsilon_c - 1)/3 + \phi_0^2\gamma/2 + \partial(\ln l_c)/\partial(\ln v_c)], \tag{5.41}$$

$$p_y = \frac{1}{A}\frac{dQ}{dT} = -P_0\alpha_c\{(\varepsilon_c - 1)/3 + \phi_0^2[\gamma + (2T\alpha_c)^{-1}]/2 + \partial(\ln l_c)/\partial(\ln v_c)\}, \tag{5.42}$$

FILM   SURFACE

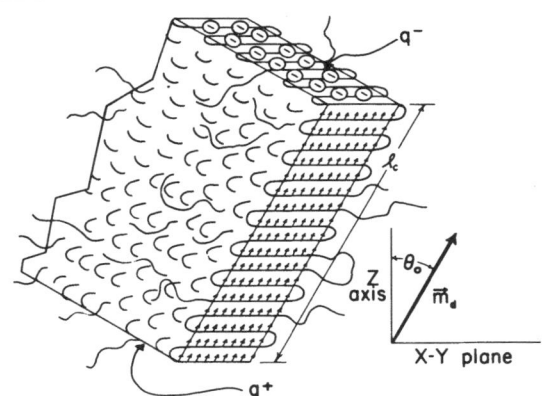

Fig. 5.18. A schematic diagram illustrating dipole alignment within a polar crystal and possible counter charge at the crystal liquid interface. A typical active polymer film consists of an array of such objects with a preferential dipole alignment resulting from the poling procedure

where the polarization from crystal dipoles is

$$P_0 = \Phi(\varepsilon_c + 2)N\mu_0 J_0(\phi_0)\langle\cos\theta_0\rangle/3V_c. \tag{5.43}$$

For the case of no counter charge the crystal length l_c is replaced by the sample thickness l_s. The equations were evaluated using the following experimental values: Temperature, $T = 300$ K, volume expansion coefficients for the crystal $\alpha_c = 1.7 \times 10^{-4}$ K^{-1} [5.146] and sample $\alpha_s = 4.2 \times 10^{-4}$ K^{-1} [5.146]; volume compressibilities for the crystal $\beta_c = 1.1 \times 10^{-10}$ m^2 N^{-1} [5.147], and sample $\beta_s = 2.39 \times 10^{-10}$ m^2 N^{-1} [5.148] crystal relative permittivity, $\varepsilon'_c = 3$ [5.35, 62], volume fraction of crystals, $\psi = 0.5$ [5.149], crystal (vacuum) polarization for Form I, $N\mu_0/V_c = 12 \times 10^{-6}$ C cm^{-2}, $d_p/P_0 = 2 \times 10^{-6}$ cm^2 N^{-1} and $p_y/P_0 = 4 \times 10^{-4}$ K^{-1} [5.105].

The agreement between this calculation and experiment is excellent if one assumes negligible amounts of space charge at crystal-amorphous interfaces. Even with an equilibrium amount of space charge (which gives a contribution opposite that of the dipolar crystals), the model still accounts for about 75 % of the observed activity in PVDF. Apparently, the largest contribution comes from changes in film thickness with changes in temperature or applied stress. This mechanism is very similar to that for amorphous polymers considered in Sect. 5.2.3. Also, the predicted ratio of piezo- to pyroelectric coefficient of $d_p/d_y = 0.005$ K cm^2 N^{-1} is in excellent agreement with experiment and strongly supports the concept that both effects arise from the same basic mechanism, i.e., thermally or mechanically induced dimensional changes in the PVDF film. This type of piezoelectricity is usually called secondary piezoelectricity [5.134]. *Kepler* has analyzed how much of the pyroelectricity in PVDF is associated with the secondary piezoelectricity and has concluded that no more than half of the pyroelectricity in PVDF arises from the mechanism of secondary piezoelectricity [5.150], in contrast to our conclusion above. His results depend on an assumption of isotropic mechanical properties for biaxially or uniaxially drawn films of PVDF. However, Poisson's ratio of semicrystalline polymers, for example, is known to vary greatly with orientation [5.42].

The calculated maximum polarization obtainable with a β phase PVDF crystal is about $22\,\mu C\,cm^{-2}$ (the often quoted value of $13\,\mu C\,cm^{-2}$ is calculated from vacuum moments and does not include the effect of the crystal environment). The corresponding maximum values of d_p (max) and p_y (max) are $44\,pC/N$ and $9\,nC\,cm^{-2}\,K$, respectively. These are more than double the typical values measured for well poled samples where d_p (typical) $= 15\,pC/N$ and p_y (typical) $= 3.5\,nC\,cm^{-2}\,K^{-1}$. This difference between predicted ideal single-crystal values and observed values is similar to the difference between piezo-electric coefficients and polarization of $BaTiO_3$ single crystals and powders [5.151], and can be attributed to inefficiency of the poling process in a poly-crystalline sample.

The model described above applies to any polar phase of PVDF including polar α phase, and β and γ phases. Probably, there will not be a significant difference in activity for various phases, because of compensating effects. For example, increased compressibility of the polar α phase relative to β phase [5.152] crystals will increase p_y and d_p in partial compensation for the decrease in p_y and d_p due to the decreased moment of the α phase repeat unit.

In many applications the electrical response across the film thickness to a change in the uniaxial stress (d_{31} or d_{32}) is of primary concern. Equation (5.41) can be modified by replacing hydrostatic stress ($-p$) with the appropriate stress, T_1 or T_2. Because of the sensitivity of Poisson's ratio to orientation, the change in film thickness will be much greater for a stress in the draw direction T_1 than for a stress transverse to the draw direction, T_2.

Poisson's ratio is also a strong function of the amorphous state of the polymer, being closer to normal values $1/3 < \eta < 1/2$ below T_g (about $-40\,°C$) and being outside this range above T_g. In spite of the apparent importance of η to the basic mechanism of piezo- and pyroelectricity in PVDF (that is, the change in sample thickness) [5.145], there are yet no reported data for η. This is due in part to the difficulty in measuring strains in the thickness of thin films. However, we can estimate from measured values of Youngs modulus $Y_1 \approx Y_2 \approx 2.5 \times 10^9\,N\,m^{-2}$ [5.39, 131] and assumed values of $\eta_{31} = 0.6$ and $\eta_{32} = 0.1$ that $d_{31} \approx 2d_p$ and $d_{32} \approx 0.4d_p$. This large difference between d_{31} and d_{32} has been shown experimentally [5.7, 13, 39] and has been interpreted as evidence for a model where stress along the molecular axis causes dipole alignment and hence an increase in polarization. Here we suggest that differences in the thickness change is the dominant effect.

Other models for piezoelectricity in PVDF include gradual stress-induced melting and crystallization of preferentially aligned crystals [5.153], orientation of dipoles in an anomalous liquid phase [5.154], and increased perfection of the planar zigzag structures [5.13], all by applied stress. Calculations on the polarization kinetics using several models for PVDF have been made [5.155].

5.8 Summary and Conclusions

Some polymers can be made both piezo- and pyroelectric by suitable application of a large electric field. This effect is true piezo- and pyroelectricity rather than electrostriction, conduction, electromechanical effects, or the motion of conductors in the field of space charges. Two distinct types of polymers can be piezoelectric. Amorphous polymers are piezo- and pyroelectric by virtue of a nonequilibium but kinetically stable net dipole orientation in the amorphous phase of the polymer. The semicrystalline polymers are piezoelectric due to alignment of polar, ferroelectric crystals dispersed in the amorphous phase. In both types of polymers, magnitudes of the piezo- and pyroelectric effects are in accord with the expected temperature and pressure dependence of the dipole model. Polarization changes primarily because of dimensional changes of the sample. Space charges embedded in the polymer normally will not produce large piezoelectric and pyroelectric currents. Those embedded near the crystal–liquid interfaces tend to reduce the piezo- and pyroelectricity. Improved orientation of dipoles and reduction of ionic impurities should increase p_y and d_p for PVDF by a factor of two above typical values presently reported. The sensitivity of amorphous and semicrystalline polymers is limited mainly by dipole moment per unit volume and breakdown strength.

Some of the models presented here were developed along with the writing and were used as a framework for the presentation in order to make the chapter more coherent. It is hoped that these ideas, some largely untested, will provide direction and stimulation for further work in this field.

Acknowledgement. Partial support of this work by the Office of Naval Research is gratefully acknowledged.

References

5.1   O. Heaviside: *Electrical Papers*, I (MacMillan, London 1892) p. 488
5.2   M. Eguchi: Philos. Mag. **49**, 178 (1925)
5.3   E. P. Adams: J. Franklin Inst. **204**, 469 (1927)
5.4   A. Meissner, R. Bechmann: Z. Tech. Phys. **9**, 174, 430 (1928)
5.5   E. Fukada: Prog. Polym. Sci., Jpn. **2**, 329 (1971)
5.6   H. Kawai: Jpn. J. Appl. Phys. **8**, 975 (1969)
5.7   R. Hayakawa, Y. Wada: *Advances in Polymer Science*, Vol. XI (Springer, Berlin, Heidelberg, New York 1973) p. 1
5.8   Y. Wada, R. Hayakawa: Jpn. J. Appl. Phys. **15**, 2041 (1976)
5.9   N. Murayama: J. Polym. Sci., Polym. Phys. Ed. **13**, 929 (1975)
5.10   N. Murayama, T. Oikawa, T. Kaito, K. Nakamura: J. Polym. Sci., Polym. Phys. Ed. **13**, 1033 (1975)
5.11   N. Tamura, T. Yamaguchi, T. Oyaba, T. Yoshimi: J. Audio Eng. Soc. **23**, 21 (1975)
5.12   M. Tamura, K. Ogasawara, N. Ono, S. Hagiwara: J. Appl. Phys. **45**, 3768 (1974)
5.13   H. Ohigashi: J. Appl. Phys. **47**, 949 (1976)

5.14 J.H.McFee, J.G.Bergman, Jr., G.R.Crane: Ferroelectrics **3**, 305 (1972)
5.15 J.G.Bergman, G.R.Crane, A.A.Ballman, H.M.O'Bryant, Jr.: Appl. Phys. Lett. **21**, 497 (1972)
5.16 G.Pfister, M.Abkowitz, R.G.Crystal: J. Appl. Phys. **44**, 2064 (1973)
5.17 H.Burkard, G.Pfister: J. Appl. Phys. **45**, 3360 (1974)
5.18 M.Abkowitz, G.Pfister: J. Appl. Phys. **46**, 2559 (1975)
5.19 G.W.Day, C.A.Hamilton, R.L.Peterson, R.J.Phelan, Jr., L.O.Mullen: Appl. Phys. Lett. **24**, 456 (1974)
5.20 R.L.Peterson, G.W.Day, P.M.Gruzensky, R.J.Phelan, Jr.: J. Appl. Phys. **45**, 3296 (1974)
5.21 R.G.Kepler, P.M.Beeson: Bulletin APS, Series II, **19**, 265 (1974)
5.22 R.G.Kepler, E.J.Graeber, P.M.Beeson: Bulletin APS Series II, **20**, 350 (1975)
5.23 D.A.Berlincourt, D.R.Curran, H.Jafee: *Physical Acoustics*, Vol. 1A (Academic Press, New York 1964) p. 183
5.24 T.Furukawa, E.Fukada: Rep. Prog. Polym. Phys. Jpn. **16**, 457 (1973)
5.25 M.G.Broadhurst, W.P.Harris, F.I.Mopsik, C.G.Malmberg: ACS Polymer Preprints **14**, 820 (1973)
5.26 R.A.Anderson, R.G.Kepler: Bull. Am. Phys. Soc. **23**, 379 (1978)
5.27 J.J.Crosnier, F.Micheron, G.Dreyfus, J.Lewiner: J. Appl. Phys. **47**, 4798 (1976)
5.28 T.Ibe: Jpn. J. Appl. Phys. **13**, 197 (1974)
5.29 R.E.Collins: Rev. Sci. Instrum. **48**, 83 (1977)
5.30 A.S.DeReggi, C.M.Guttman, F.I.Mopsik, G.T.Davis, M.G.Broadhurst: Phys. Rev. Lett. **40**, 413 (1978)
5.31 P.Laurenceau, G.Dreyfus, J.Lewiner: Phys. Rev. Lett. **38**, 46 (1977)
5.32 R.E.Collins: Proc. IEEE **34**, 381 (1973)
5.33 G.M.Sessler, J.E.West: J. Acoust. Soc. Am. **53**, 1589 (1973)
5.34 L.Onsager: J. Amer. Chem. Soc. **58**, 1486 (1936)
5.35 F.I.Mopsik, M.G.Broadhurst: J. Appl. Phys. **46**, 4204 (1975)
5.36 M.Oshiki, E.Fukada: J. Matls. Sci. **10**, 1 (1975)
5.37 E.W.Aslaksen: J. Chem. Phys. **57**, 2358 (1972)
5.38 K.Iohara, K.Imada, M.Takayanagi: Polym. J. **3**, 357 (1972)
5.39 E.Fukada, T.Sakurai: Polym. J. **2**, 657 (1971)
5.40 E.Fukada, K.Nishiyama: Jpn. J. Appl. Phys. **11**, 36 (1972)
5.41 M.G.Broadhurst, C.G.Malmberg, F.I.Mopsik, W.P.Harris: In *Electrets, Charge Storage and Transport in Dielectrics*, ed. by M.M.Perlman (The Electrochemical Soc., New York 1973) p. 492
5.42 H.Sussner: Phys. Lett. **58** A, 426 (1976)
5.43 G.Natta, I.W.Bassi, G.Allegra: Atti Accad. Nazl. Lincei Rend., Classe Sci. Fis. Mat. Nat. **31**, 350 (1961)
5.44 J.Cohen, S.Edelman: J. Appl. Phys. **42**, 3072 (1971)
5.45 W.Reddish: J. Polym. Sci., P. C, **14**, 123 (1966)
5.46 C.E.Wilkes, V.L.Folt, S.Krimm: Macromolecules **6**, 235 (1973)
5.47 A.von Hipple: *Dielectric Matls. and Applications* (Technology Press of MIT and Wiley, New York 1954)
5.48 G.T.Davis, M.G.Broadhurst: In *International Symposium on Electrets and Dielectrics*, ed. by M.S.deCampos (Academia Brasileira de Cienacias, Rio de Janeiro 1977) p. 299
5.49 H.G.Olf: North Carolina State University (private communication)
5.50 C.W.Wilson III, E.R.Santes, Jr.: J. Polym. Sci., Pt. C, **8**, 97 (1965)
5.51 M.Gorlitz, R.Minke, W.Trautvetter, G.Weisgerber: Angew. Makromol. Chem. **29/30**, 137 (1973)
5.52 J.P.Stallings, S.G.Howell: Polymer Eng. Sci. **11**, 507 (1971)
5.53 W.M.Prest, Jr., D.J.Luca: J. Appl. Phys. **46**, 4136 (1975)
5.54 W.M.Prest, J., D.J.Luca: J. Appl. Phys. **49**, 5042 (1978)
5.55 S.Osaki, Y.Ishida: J. Polym. Sci., Polym. Phys. Ed. **13**, 1071 (1975)
5.56 R.J.Shuford, A.F.Wilde, J.J.Ricca, G.R.Thomas: Polym. Eng. Sci. **16**, 25 (1976)

5.57   J.B.Lando, H.G.Olf, A.Peterlin: J. Polym. Sci. A-1 4, 941 (1966)
5.58   K.Nakagawa, Y.Ishida: J. Polym. Sci., Polym. Phys. Ed. 11, 2153 (1973)
5.59   J.P.Reardon, P.F.Waters: In *Thermal and Photostimulated Currents in Insulators*, ed. by
       D.M.Smyth (Electrochem. Soc., Princeton, N.J. 1976) p. 185
5.60   N.Sasabe, S.Saito, M.Asahina, H.Kakutani: J. Polym. Sci. A-2 7, 1405 (1969)
5.61   S.Osaki, Y.Ishida: J. Polym. Sci., Polym. Phys. Ed. 12, 1727 (1974)
5.62   H.Kakutani: J. Polym. Sci. A-2 8, 1177 (1970)
5.63   M.Blukis, C.Lewa, S.Letowski, A.Sliwinski: Ultrasonics Int'l. 1977, 474
5.64   E.Fukada, K.Nishiyama: Jpn. J. Appl. Phys. 11, 36 (1972)
5.65   M.E.Baird, P.Blackburn, B.W.Delf: J. Matls. Sci. 10, 1248 (1975)
5.66   G.R.Davies, A.Killey, A.Rushworth, H.Singh: Organic Coatings and Plastics Chemistry 38,
       257 (1978) [Preprints for ACS meeting in Anaheim, CA., Mar. 1978]
5.67   E.L.Galperin, Yu.V.Strogalin, M.P.Mlenik: Vysokomol. Soedin. 7, 933 (1965)
5.68   W.W.Doll, J.B.Lando: J. Macromol. Sci.-Phys. B4, 309 (1970)
5.69   M.Kobayashi, K.Tashiro, H.Tadokoro: Macromolecules 8, 159 (1975)
5.70   S.Osaki, Y.Ishida: J. Polymer Sci., Polym. Phys. Ed. 13, 1071 (1975)
5.71   R.Hasegawa, Y.Tanabe, M.Kobayashi, H.Tadokoro, A.Sawaoka, N.Kawai: J. Polym. Sci.,
       A-2 8, 1073 (1970)
5.72   R.Hasegawa, Y.Takahashi, Y.Chatani, H.Tadokoro: Polymer J. 3, 600 (1972)
5.73   J.B.Lando, M.H.Litt, S.Weinhold: Case Western Reserve Univ., private communication
5.74   G.Natta, G.Allegra, I.W.Bassi, D.Sianese, G.Copoucci, E.Torti: J. Polym. Sci. A 3, 4263
       (1965)
5.75   B.L.Farmer, A.J.Hopfinger, J.B.Lando: J. Appl. Phys. 43, 4293 (1972)
5.76   R.D.Nelson, Jr., D.R.Lide, Jr., A.A.Maryott: *Selected Values of Electric Dipole Moments
       for Molecules in the Gas Phase*. NSRDS-NBSIO, (U.S. Government Printing Office,
       Washington, D.C. 1967)
5.77   G.T.Davis, J.E.McKinney, M.G.Broadhurst, S.C.Roth: J. Appl. Phys. 49, 4998 (1978)
5.78   K.Nakagawa, Y.Ishida: J. Polym. Sci. A-2 11, 1503 (1973)
5.79   L.E.Alexander: *X-Ray Diffraction Methods in Polymer Science* (Wiley, New York 1969)
5.80   S.Osaki, S.Uemura, Y.Ishida: J. Polym. Sci. A-2 9, 585 (1971)
5.81   A.Peterlin, J.Elwell: J. Matls. Sci. 2, 1 (1967)
5.82   S.Yano: J. Polym. Sci. A-2 8, 1057 (1970)
5.83   S.Yano, K.Tadano, K.Aoki, N.Koizumi: J. Polym. Sci., Polym. Phys. Ed. 12, 1875 (1974)
5.84   N.Murayama, H.Hashizume: J. Polym. Sci., Polym. Phys. Ed. 14, 989 (1976)
5.85   G.R.Davies, A.Killey: University of Leeds (private communication from thesis of A.Killey)
5.86   J.D.Hoffman, G.Williams, E.Passaglia: J. Polym. Sci., C 14, 173 (1966)
5.87   Y.Miyamoto, H.Miyaji, K.Asai: Rep. on Prog. Polym. Phys. Jpn. 20, 371 (1977)
5.88   M.G.Broadhurst, G.T.Davis: Ferroelectrics 32, 177 (1981)
5.89   P.Weiss: J. Phys. 6, 661 (1907)
5.90   W.L.Bragg, E.J.Williams: Proc. R. Soc. London A 145, 699 (1934)
5.91   W.P.Mason: Phys. Rev. 72, 854 (1947)
5.92   H.Fröhlich: *Theory of Dielectrics* (Oxford University Press, London 1949)
5.93   P.Buchman: Ferroelectrics 5, 39 (1973)
5.94   M.Kobayashi, K.Tashiro, H.Tadokoro: Macromolecules 8, 163 (1975)
5.95   M.G.Broadhurst, F.I.Mopsik: J.Chem. Phys. 52, 3634 (1970)
5.96   K.Nakamura, Y.Wada: J. Polym. Sci., A-2 9, 161 (1971)
5.97   J.H.McFee, J.G.Bergman, Jr., R.R.Crane: Ferroelectrics 3, 305 (1972)
5.98   R.G.Kepler, R.A.Anderson: J. Appl. Phys. 49, 1232 (1978)
5.99   G.L.Cessac, J.G.Curro: J. Polym. Sci., Polym. Phys. Ed. 12, 695 (1974)
5.100 R.G.Kepler: Organic Coatings and Plastics Chemistry 38, 278 (1978) (Preprints for ACS
       Meeting in Anaheim, CA., March 1978)
5.101 J.C.Hicks, T.E.Jones, J.C.Logan: J. Appl. Phys. 49, 6092 (1978)
5.102 E.Fukada, M.Date, T.Furukawa: Organic Coatings and Plastics Chemistry 38, 262 (1978)
       (Preprints for ACS Meeting in Anaheim, CA., March 1978)

5.103 D. Naegele, D. Y. Yoon: Appl. Phys. Lett. **33**, 132 (1978)

5.104 J. P. Luongo: J. Polym. Sci. A-2 **10**, 1119 (1972)

5.105 J. E. McKinney, G. T. Davis: Organic Coatings and Plastics Chemistry **38**, (Preprints of paper for 175th meeting of ACS in Anaheim, CA., March 1978) p. 271

5.106 D. K. Das Gupta, K. Doughty: Appl. Phys. Lett. **31**, 585 (1977)

5.107 D. K. Das Gupta, K. Doughty: J. Appl. Phys. **49**, 4601 (1978)

5.108 P. D. Southgate: Appl. Phys. Lett. **28**, 250 (1976)

5.109 M. Tamura, S. Hagiwara, S. Matsumoto, N. Ono: J. Appl. Phys. **48**, 513 (1977)

5.110 H. Sussner, K. Dransfeld: J. Polym. Sci., Polym. Phys. Ed. **16**, 529 (1978)

5.111 J. van Turnhout: Polym. J. **2**, 173 (1971)

5.112 E. C. Cassidy, R. E. Hebner, Jr., M. Zahn, R. J. Sojka: IEEE Trans. EI-**9**, 43 (1974)

5.113 S. Uemura: J. Polym. Sci., Polym. Phys. Ed. **10**, 2155 (1972)

5.114 J. M. Kenney: J. Research Natl. Bul. Stds. (US) **84** (6), (1979)

5.115 T. Hanai: Kolloid-Z. **171**, 23 (1960)

5.116 A. K. Vijh: J. Appl. Phys. **49**, 3621 (1978)

5.117 A. I. Baise, H. Lee, B. Oh, R. E. Salomon, M. M. Labes: Appl. Phys. Lett. **26**, 428 (1975)

5.118 H. Sussner, D. Y. Yoon: Organic Coatings and Plastics Chemistry **38**, 331 (1978) (Preprints for ACS Meeting in Anaheim, CA., March 1978)

5.119 T. Hanai, N. Koizumi, T. Saigano, R. Gotoh: Kolloid-Z. **171**, 20 (1960)

5.120 J. van Turnhout: Thesis Leiden (1972), TNO Central Laboratorium Communication, No. 471

5.121 Y. Sakamoto, H. Fukagawa, T. Shikama, K. Kimura, H. Takehana: Fujikura Tech. Review (1977) p. 22

5.122 P. E. Bloomfield, I. Lefkowitz, A. D. Aronoff: Phys. Rev. B**4**, 974 (1971)

5.123 R. J. Phelan, Jr., R. L. Peterson, C. A. Hamilton, G. W. Day: Ferroelectrics **7**, 375 (1974)

5.124 C. B. Duke, T. J. Fabish: J. Appl. Phys. **49**, 315 (1978)

5.125 D. K. Davies: Br. J. Appl. Phys. **2**, 1533 (1969)

5.126 D. K. Davies: J. Phys. D **6**, 1017 (1973)

5.127 A. W. Stephens, A. W. Levine, J. Fech, Jr., T. J. Zrebiec, A. V. Cafiero, A. M. Garofalo: Thin Solid Films **24**, 362 (1974)

5.128 H. Sussner, D. E. Horne, D. Y. Yoon: Bull. Am. Phys. Soc. **23**, 379 (1978)

5.129 W. R. Blevin: Appl. Phys. Lett. **31**, 6 (1977)

5.130 M. Oshiki, E. Fukada: Jpn. J. Appl. Phys. **15**, 43 (1976)

5.131 N. Murayama, K. Nakamura, H. Obara, M. Segewa: Ultrasonics **14**, 15 (1976)

5.132 L. E. Garn, E. J. Sharp: IEEE Trans. PHP-**10**, 28 (1974)

5.133 K. Takahashi, R. E. Salomon, M. M. Labes: Bull. Am. Phys. Soc. **23**, 378 (1978)

5.134 W. G. Cady: *Piezoelectricity* (McGraw Hill, New York 1946)

5.135 D. A. Berlincourt, D. R. Curran, H. Jaffe: In *Physical Acoustics*, Vol. 1, Pt. A (Academic Press, New York 1964) Chap. 3

5.136 S. T. Liu, D. Long: Proc. IEEE **66**, 14 (1978)

5.137 R. J. Phelan, Jr., R. J. Mahler, A. R. Cook: Appl. Phys. Lett. **19**, 337 (1971)

5.138 M. H. Litt, C. Hsu, P. Basu: Tech. Rpt. No. 5 on Office of Naval Research Contract No. N000 14-75-C-0842

5.139 W. R. Blevin, J. Geist: Appl. Opt. **13**, 2212 (1974)

5.140 J. Stern, S. Edelman: Nat'l. Bur. of Stds. (U.S.) Technical News Bulletin **56** (3), 52 (1972)

5.141 J. M. Powers: Natl. Bur. Stds. (U.S.) Interagency Rpt. 75–760 (1975) p. 209

5.142 M. Jacobs: NBS Dimensions **62** (2), 2 (1978); U.S. Dept. of Commerce

5.143 A. L. Robinson: Science **200**, 1371 (1978)

5.144 Mater. Eng., **1978** (5), p⁄ 6

5.145 M. G. Broadhurst, G. T. Davis, J. E. McKinney, R. E. Collins: J. Appl. Phys. **49**, 4992 (1978)

5.146 K. N. Nakagawa, Y. I. Ishida: Kolloid Z. Z. Polym. **251**, 103 (1973)

5.147 B. A. Newman, C. H. Yoon, K. D. Pae: Tech. Rpt. No. 11 under Office of Naval Research Contract N00014-75-C-0540

5.148 W. W. Doll, J. B. Lando: J. Macromol. Sci. Phys. B**2**, 219 (1968)

5.149 K. Nakagawa, Y. Ishida: J. Polym. Sci., Polym. Phys. Ed. **11**, 2153 (1973)
5.150 R. G. Kepler, R. A. Anderson: J. Appl. Phys. **49**, 4490 (1978)
5.151 D. Berlincourt, H. H. A. Krueger: J. Appl. Phys. **30**, 1804 (1959)
5.152 B. A. Newman: Rutgers University (private communication)
5.153 R. G. Kepler, R. A. Anderson: J. Appl. Phys. **49**, 4918 (1978)
5.154 R. Hayakawa, J. Kusuhara, K. Hattori, Y. Wada: Rept. Progr. Polym. Phys. Jpn. **16**, 477 (1975)
5.155 R. E. Salomon, M. M. Labes: Natl. Bur. Stds. (U.S. Interagency Rpt. 75–760 (1975) pp. 199–209

6. Bioelectrets: Electrets in Biomaterials and Biopolymers

S. Mascarenhas

With 14 Figures

Charge and polarization storage via the electret state has now been found in many biomaterials. The importance of the electret effect in these materials has to do with biomedical applications as well as with its possible role in more fundamental biophysical phenomena. As biomaterials, electrets have found interesting applications as antithrombogenic surfaces. Other uses have been mentioned in the literature, such as the stimulation of tissue growth in bone and special artificial membranes. The electret effect has also been found in most biopolymers of importance such as proteins, polysaccharides, and some polynucleotides. Fundamental macromolecules of biology, such as collagen, hemoglobin, DNA, and chitin, not only exhibit the effect, but may have various sources, or, to use a more biological term, "compartments", for polarization and charge storage: dipoles and ions bound to the molecules.

One of the most important aspects of electret research in biophysics is that water bound to biopolymers in the so-called structured form (also called bound water, or biowater) may also be induced into the electret state. Electret investigation techniques were used to study this most important form of water in conjunction with the biopolymer.

The electret state has been considered in various biophysical models as a basis for the understanding of membranes, neural signals, biological memory in regeneration, electrical mediation in tissue growth, and other phenomena. One of the most interesting models claimed to depend on an induced ferroelectric metastable state similar to the electret is Fröhlich's model for coherent longitudinal polarization waves in biological systems. Fröhlich waves have been invoked to explain enzyme action, and recently the electret effect was found in various important enzymes in the solid state such as tripsin, urease, and others. For biomedical applications and in molecular biophysics, the electret concept begins to open new avenues for research, which seem to justify the usage of the term *bioelectrets*.

6.1 Introductory Remarks

It is interesting to observe that the first electret was made with a material of biological origin: carnauba wax, from the carnauba palm tree of Brazil. This was the material originally used by *Eguchi* [6.1] to verify experimentally the theoretical proposal of *Heaviside* [6.2], who coined the name electret. In fact, carnauba wax proved to be, for many years, the main material for electret investigations. The samples prepared by *Eguchi* in 1922 are still electrized and subject to monitoring measurements in the laboratory of Eiichi Fukada in Japan. The other pioneer of electret investigations, *Gross*, also investigated carnauba wax electrets in many of his fundamental papers (listed in [6.3a, b]). Electret research gradually moved to simpler materials like ionic and organic crystals and polymers where fundamental solid-state properties could be correlated with the electret behavior [6.4–8]. More recently, however, the electret effect was studied in materials of biological origin like proteins, and now the picture emerges that the electret effect may in fact be a universal property of biopolymers in general such as polypeptides, polynucleotides, and polysaccharides [6.9a] (see also [6.9b], which may be the first paper on TSD from a protein–hemoglobin).

For biomedical applications, polymers with good biological compatibility (such as teflon) are also considered as biomaterials, and though, strictly, they are not biopolymers, they will be treated as biomaterials in this chapter. In this way we are led to consider the electret properties of artificial polymers such as teflon and polysulfonate films which are of importance for biological or medical applications.

The techniques used to study the electret effect in biomaterials (and biopolymers) are essentially the same as for general electret research [e.g., thermally stimulated depolarization current (TSDC)], and we shall not discuss them in detail here since they are covered in other chapters of the present volume or in the literature [6.7]. Specific changes in these techniques and important details required for electret investigation in biomaterials will be explicitly discussed, however.

Some general observations are nevertheless required in relation to experimental techniques in the special case of biomaterials.

a) Materials of biological origin in general cannot easily be put in single-crystal form. For most biopolymers, for instance, fibers or powder samples are used. Sometimes (as in the case of DNA and cellulose), a film may be obtained. In this case, the orientation of the macromolecule in relation to the film may be investigated by X-rays or optical techniques, and may be important for the interpretation of the effects observed. A typical case is the natural electret effect in keratin found in highly oriented samples of biological origin [6.10], which will be discussed in detail in another section.

b) The purity and origin of samples of a biological nature become a very important parameters electret investigations, and, in general, great attention has to be given to the preparation of samples, preferably with the assistance of biochemists and biologists. Here electret investigations really need to be interdisciplinary.

c) Many biopolymers change their properties by denaturation, hydration (or dehydration), oxidation, and even as a result of exposure to light. Careful attention should be given to all these factors during electret investigations, and again such properties of the materials should be known before detailed measurements are begun.

d) Most biological molecules have been studied intensively in solution, but, in the majority of cases, little is known about the true "solid-state physics" of the material, and the investigator should be careful not to assume that properties investigated in solution apply to solid samples. For instance, the collagen molecule, a triple helix, may denature thermally partially into random coils, or completely, into isolated helices (gelatin). In solution, the denaturation temperature is around 65 °C. However, in the solid state, collagen can be heated above 150 °C with no appreciable denaturation (in vacuum) [6.11].

Biomaterials of nonbiological origin, but good compatibility in vivo, have been used as electrets in biomedical applications. Examples are teflon, dacron, and other polymers. Several biomedical applications of electrets have been proposed. We discuss here mainly electrets in antithrombogenic surfaces [6.12] and artificial membranes [6.13, 14].

Electrets have also been used in biomedicine or bioengineering in different contexts not directly linked to our previous classification. For instance, in medical dosimetry, where high sensitivity do dose is required, a new form of electret dosimeter has been developed recently [6.14] (see Sect. 4.5). The development of transducers with electret films applicable in bioengineering (for example, ultrasound probes or hearing aids) will be discussed in Chap. 7. Obviously these areas also belong to the interface of electret physics and biomedical research or applications. Exotic applications such as the possible use of electret filters for biological ions and free radicals or charged traps for bacteria have been mentioned but are also not discussed here.

Our discussion will be based on themes which we believe are varied enough to give the reader a broad view of the previous research on, and present potential of, electrets in biomaterials and in biophysics.

6.2 General Concepts in Electret Research

Before we describe general results obtained in the investigation of electrets in biomaterials and biophysics, it may be useful to summarize here some general results and concepts on electrets. The reader will find a complete and detailed treatment in the other chapters of this book. In particular, we shall not discuss the action of radiation on biomaterials – a subject of great potential interest in biophysics. For the action of radiation on electrets the reader should consult Gross's chapter.

The general method of polarizing materials may be understood from Figs. 6.1, 2 (see also Sect. 2.2.2). The material is placed in a closed vessel V, where vacuum or a controlled gaseous atmosphere can be established. Metal electrodes (in general, with plane symmetry) are attached to the sample S. A polarizing field E_p is applied during a certain time t, called the polarization time, at a certain

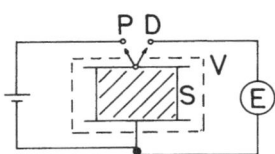

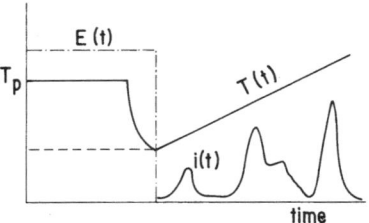

Fig. 6.1. Diagramatic representation of electret polarization and depolarization system

Fig. 6.2. Field (E), temperature (T), and depolarization current (i) as functions of time (t) during a typical experimental run

temperature T, the polarization temperature. The choice of t is made by a previous investigation in which the polarization stored in the material (to be defined below) is measured for several values of t. The minimum time for which the maximum stored polarization is achieved at the chosen polarization temperature T is usually used. The main concept in thermoelectrets is that the relaxation time τ for the decay (or growth) of the polarization P is a function of the temperature $\tau(T)$. $P(T)$ is then said to be thermally activated, and in general, the relaxation time is found to be a function of the form

$$\tau(T) = \tau_0 \exp(A/kT), \tag{6.1}$$

where τ_0 is a constant; A, the activation energy; k, Boltzmann's constant; and T, the absolute temperature. By varying the temperature of the sample, it is thus possible to obtain large variations in $\tau(T)$ depending on A. By lowering the temperature, decay times as long as centuries, for instance, may be obtained. An electret is defined most conveniently in terms of its relaxation time: *a substance is said to be an electret if the decay time of its stored polarization is long in relation to the characteristic time of experiments performed on the material.* If one particular experiment involves time intervals of a few seconds, then a decay time of several minutes will indicate electret behavior. On the other hand, if observations are related to time lapses of days or months, electret decay times will certainly have to be of this order or larger. In biophysics, phenomena may last from fractions of a second to years, and the characteristic electret lifetime or decay time is an important parameter to be measured. We shall see below how this can be done experimentally.

After polarizing the material, the temperature is then lowered (see Fig. 6.2) so that the relaxation time increases and the polarization can be "frozen in" the material. When the lowest temperature has been attained (T_1), the field is switched off, and an electrometer (such as a Keithley 602 or 610 or a Cary vibrating-reed instrument) is put in series with the sample (position D of switch in Fig. 6.1). Preferably, a double-pen recorder is also used so that the current and the temperature of the sample (monitored by a suitable thermocouple) are simultaneously recorded. Alternatively, an $x-y$ recorder may be used if the heating rates are reproducible. The sample is then warmed up most conveniently at a constant heating rate $\beta = dT/dt$ which may vary from a few °C per minute to a few tens of °C per minute. A common rate for most biological substances is about 15 °C min^{-1}. During warm-up, the polarization stored in the material will decay when the temperature is sufficiently high. In the vicinity of this temperature, a current will be observed in the external circuit due in principle to the decay of the polarization plus other conduction components. If we disregard the conductivity for the moment, the external current will be equal to the displacement current, which for unit area is

$$i(T) = dP/dt \, .$$

This current will rise initially and then pass through a maximum at a certain temperature T_p because the polarization stored is finite. Since many sub-polarizations may be present in the material, several such current peaks may appear during depolarization. The resulting current as a function of temperature is called the thermally stimulated current (TSC) or thermally stimulated depolarization current (TSDC). This function contains important information on the electret behavior and it will be one of our objects to describe such curves and the corresponding interpretations for the case of biological materials. Two fundamental questions may be asked after a TSDC is observed for a biological material:

I) *What are the physical sources of charge or polarization storage in the electret, and how can they be identified from the properties of the* TSDC?

II) *How may this storage of charge or polarization be important in biophysical phenomena or for biomedical applications?*

The first question is a question in basic physics, and a complete answer may require the applications of different techniques, such as optical measurements, dielectric constant investigations, variation of the sample parameters (e.g., doping or chemical changes), or EPR. In general, the following sources (or "compartments", to use more biological language) may be responsible for the production of charge or polarization storage:

  a) dipoles of the material or related impurities or defects;
  b) ionic carriers, either as impurities or intrinsic to the material;
  c) electronic carriers (electrons or holes);
  d) structured water bound to the material (mainly in the case of biological macromolecules).

In the case of dipoles, the relevant equations for the thermally stimulated current have been solved (assuming a nonconducting sample) by *Bucci* and *Fieschi* [6.4] and by *Gross* [6.15] in a more general form. *Bucci* and *Fieschi*, assuming a single relaxation time for the dipoles, obtain the current as

$$i(T) = [Np^2 E/3kT_p\tau(T)] \exp\left\{-\int_{T_0}^{T'} dT'/[\beta\tau(T')]\right\}, \tag{6.2}$$

where N is the number of dipoles per unit volume, p the dipole moment, and the other symbols are as defined before. The maximum temperature of the peak T_p will be given by

$$T_p^2 = \beta A \tau(T_p)/k , \tag{6.3}$$

and the total polarization stored P_s as measured by the area under the peak will be

$$P_s = Np^2 E/3kT_p . \tag{6.4}$$

For times at the very beginning of the peak, in the so-called initial rise region, the following approximation is valid:

$$i(T) = i_0 \exp(-A/kT) . \tag{6.5}$$

This is udeful for obtaining the activation energy for an isolated single peak. *Gross* generalized the expressions above for a distribution of relaxation times. Further theoretical developments are discussed in detail by *van Turnhout* [6.7] and in Chap. 3 above, and the reader is directed to these and various other papers [6.3b, 16] on the subject.

In the case of electronic carriers, the charge may be stored in traps which may bind these carriers. The trap distribution in energy and space is then very important. In the case of polymers, especially irradiated ones, where a large number of electrons and holes may be produced by irradiation and subsequently trapped, this will be an important case (see Chap. 4). Ionic carriers, leading to so-called space-charge storage, usually produce peaks in the TSDC at higher temperatures where ionic conductivity begins to appear (around room temperature and above for biological molecules). If the electrodes are blocking (that is, do not allow free passage of carriers between sample and the external circuit), a space charge will build up in the sample, and may be stored if the temperature is low enough so that the electrical conductivity is small. The general equations for depolarization currents arising from space charge have not been deduced in a complete form, because, in general, they lead to nonlinear differential equations in the field.

Recently *Gross* et al. [6.17] have solved some of the space-charge problems either in closed analytical form or by numerical solutions with a computer, but a complete analysis of the experimental curves is still difficult and will not be discussed here. The following observations may be made for the purpose of general interpretations of the TSDC from biological substances:

I) A dipole peak is distinguished from a space-charge peak if it shows the following characteristics:

a) Equations (6.2–4) hold, that is, polarization is a linear function of the applied field and the peak may be fitted by first-order kinetics;

b) the peak temperature generally does not shift with variations of the polarization temperatures (for a single Debye relaxation);

c) the polarization does not depend on the thickness of the sample – essentially because for dipoles (if uniformly distributed), the following equation holds:

$$\text{div} P = 0$$

provided field and polarization temperature are kept constant.

These are just general observations found to apply when simple conditions hold, such as isolated single peaks, or separable overlapping peaks (the separation technique will be discussed below).

II) A space-charge peak may:

a) show a nonlinear dependence of polarization with field (since it saturates at lower polarization levels);

b) not obey (6.2–4)

c) the peak profile and peak temperature may shift with polarization temperature [because the spatial charge distribution (through the carrier mobility among other things) will depend on polarization temperature]; in general, however, the actual shift is hardly recognizable;

d) the polarization may be thickness dependent (for the same field values).

In the case of bound water, the peaks are most easily distinguishable through their variation with degree of hydration as will be shown for many biological materials like proteins and polysaccharides.

It is very important to note that some phenomena related to peaks which are dependent on the nature of the electrodes. Charge or polarization storage in this case may be due to carrier injection from the electrodes, or from Schottky barrier formation. These will be assumed to have been previously investigated – by changing the nature of the electrodes or by interposition of a thin insulating layer (such as a mylar film) to block the junctions. Since these effects are not intrinsic to the material, they will not be considered as true sources of electret action in the sense we are discussing in this chapter, but rather as artifacts to be avoided or to be properly controlled.

Two other aspects of experimental electret work should be mentioned. The first is related to a technique of peak separation, the so-called peak-cleaning method. The second is how the various parameters like T or t may be used for further analysis and peak isolation of the TSDC.

The peak-cleaning method is used when several bands in the TSDC overlap. The overlap and the number of peaks must not be too large if the method is to be useful. Starting, for example, with two ideal overlapping peaks as in Fig. 6.3, after polarization, the sample is brought up to a temperature T smaller than T_2, but larger than T_1. This will "clean" the first peak out of the spectrum. After cooling down and warming up again, a "pure" peak at T_2 will be recorded. The activation energy and profile of peak 2 may then be properly measured. *Perlman* [6.18] discusses criteria and computer programs for analysis of TSDC with peak-cleaning techniques. Some authors in the field have a cautions attitude towards using extensive peak cleaning as a safe technique. They argue that before

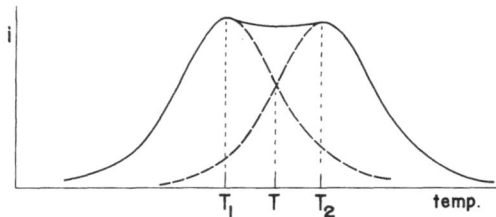

Fig. 6.3. Two overlaping peaks may be separated by warming up only to intermediate temperature T, cooling and rewarming for isolated detection of peak at T_2

knowing the particular kinetics (first, second, or higher order) of the overlapping peaks, peak cleaning may introduce changes in the cleaned peak profile. A good check will thus be to reduce the height of the cleaned peak further (by other successive warming and cooling cycles) to investigate whether peak position or profile will change as a function of the diminishing stored polarization.

Peak analysis may also be done by using a growth curve for the polarization P as a function of polarization time t. Sometimes one peak will grow faster than the other and in this way they can be separated. This is seen in the growth curves of Fig. 6.4. At time t_1, peak 1 is much larger than peak 2. The use of different T is also useful for cleaning out a higher-temperature peak without wiping off a peak occurring at lower temperatures.

We have already mentioned that care must be taken with electrode effects. It is always very important in electret research to be sure that the TSDC changes polarity with a reversal of the polarization field E_p. If asymmetries arise (with a nonoriented sample, such as a polycrystalline material) artifacts must be looked for, and the experimental conditions controlled until satisfactory results are obtained. When using gaseous atmospheres, it is almost certain that "reversed" peaks will appear if no guard rings are used or surface and electrode conditions interfere. These are peaks with the "wrong" sign in relation to the polarization field (homocharge effects). Finally, thermal gradients in the measuring system may induce thermoelectric effects or depolarization effects, resulting from a nonuniform temperature distribution (e.g., some effects found by Gross in certain charged insulators). For this reason, it may be useful to obtain a TSDC with no field applied before extensive measurements are done. Also, the use of a calibrating sample such as a single crystal of divalent doped alkalihalide with a known amount of dipoles is a very useful way of checking the measuring system. Some of the alkali halides have sharp, isolated peaks in their TSDC with well-known parameters like activation energy and T_p. The reader should, however, consult the review by *Fieschi* [6.18] before using these samples, because some precautions must be taken in the thermal treatment of the samples for dispersing the aggregated dipoles.

Finally, in working with biological materials, in most cases a polycrystalline pressed sample will probably be the only form available for experiments. In this situation, Maxwell–Wagner losses may be present. This is certainly one of the main criticisms of the use of polycrystalline samples. Whether peaks are due to Maxwell–Wagner losses rather than to intrinsic dipolar effects can be seen by changing the physical conditions of the sample, such as crystallite size or forming

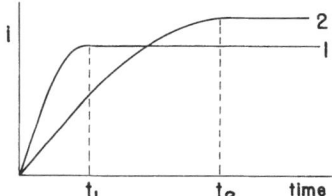

Fig. 6.4. Peaks may show a growth curve of polarization P as a function of time, and separation may be achieved by using different polarization times, such as t_1

pressure. Maxwell–Wagner peaks will shift and change profile or area under these changes. In any case, it is not always simple to separate Maxwell–Wagner losses from intrinsic bulk properties of the material under investigation, and this is a point always to be kept in mind with biological samples.

6.3 Other Dielectric Techniques Complementary to TSDC

To obtain a complete picture of the system under analysis, certain complementary techniques should be used. Of these, mention measurements of dielectric constant and absorption, usually done by an ac bridge method. These techniques are extensively discussed in the literature and will not be examined here. It has been asked why the term electret is introduced if, after all, only an extremely low-frequency dielectric relaxation measurement is performed with TSDC using a dc technique? In principle, this is a valid point. However, for extremely long relaxation times, say of the order of days, the ac bridge technique would be useless, and the electret really merits the new name because a new kind of metastable equilibrium has been attained by the material with important physical implications. Since the fundamental mathematical demonstration given by Gross that electret behavior is not due to a dielectric anomaly, but, in broad terms, can be understood by consistent use of the superposition principle under nonisothermal conditions, the study of electrets has been shown to belong to ordinary dielectric theory. It must also be said that, more recently, new ac bridges going down to extremely low frequencies (sometimes below 0.01 Hz) have been developed. These will probably prove to be very useful in electret research, especially with biological materials. A comparative study of these techniques is given by *van Turnhout* (Ref. 6.3b, p. 97, [6.19]). In the case of alkali halide electrets, a simultaneous analysis of TSDC and dielectric absorption has been done for many systems. For biological materials, a complete analysis and comparison remains to be done for most materials and especially for biopolymers.

Another technique to be used is isothermal polarization and depolarization [6.20]. If one applies a constant dc voltage to the sample, the current rises and then decays to a residual conductivity at constant temperature. By integrating the area below the $i(t)$ function and correcting for the residual conductivity current, the absorbed charge can be measured. This measurement of the stored polarization may be made at several temperatures and under different field conditions. We have applied this technique to investigate the stored polarization in urease, an important enzyme, with interesting results (see Sect. 6.7). Charge and discharge current analysis under isothermal conditions may also be used to identify polarization kinetics by dipoles or by the filling (emptying) of traps by electronic carriers, as was done in the case of naphthalene [6.21].

Piezoelectricity is an important property of many biological molecules, as has been shown by the pioneering work of *Fukada* [6.22]. In principle, it can be shown that an electret will also show piezoelectricity, as has been demonstrated by Gubkin, and, more recently, in a very important series of studies by *Broadhurst* et al. [6.23a] (see also Chap. 5, and [6.22].) *Zimmerman* and colleagues in our laboratory also found an induced piezoelectricity in electretized materials [6.24]. Piezoelectricity has been interpreted as due to degrees of freedom in the macromolecule connected with dipolar relaxation, but a complete analysis is still lacking, and only general correlations have been found for biological materials. In this case, the comparison is made between the TSDC and the temperature-dependent curve of either the real or imaginary part of the piezoelectric constant. Since the piezoelectric constant also depends on elastic properties, the temperature dependence of these complicate the interpretation and the comparison. A definite comparison has been made in the case of the natural electret, keratin [6.24]. At the present stage, these investigations, and the correlations between electret and piezoelectric behavior, are still at an early stage, and though the subject represents a fertile line of research, it will not be further pursued here (see, however, Chap. 5); we will, however, discuss the specific case of the simultaneous use of these techniques with keratin, where some results of interest have been obtained.

6.4 Proteins

Proteins are polymers of amino acids. The 20 natural amino acids divide themselves in hydrophilic and hydrophobic classes. Thus, in general, naturally occurring proteins contain dipolar residues, and can in principle be induced into the electret state. Of course, artificial polypeptides could also be investigated, and, in this case, typically nonpolar compounds like polyglycine will not present dipolar peaks, though they might present other forms of polarization storage. Incidentally, a systematic investigation of artificial polypeptide electrets has not yet been done. The electret effect has been found in many proteins, fibrous and globular, but in the present section we discuss only the particular examples of collagen and gelatin [6.25]. Enzymes are also proteins but, due to their importance in biological phenomena, we discuss this case separately. Also the case of keratin, because it exhibits the so-called native, or natural, electret state (obtained without application of a field, as by the *Costa-Ribeiro effect* [6.26] or by biological growth [6.10, 24]), and because of the simultaneous presence of piezoelectricity, will be discussed in another section.

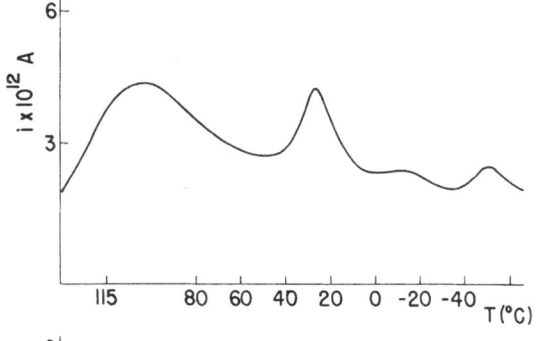

Fig. 6.5. TSDC for gelatin

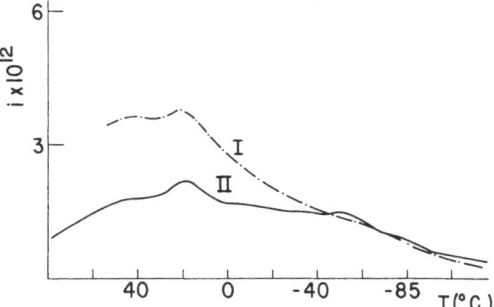

Fig. 6.6. TSDC for collagen (I), and after partial denaturation (II)

Collagen and gelatin are known to possess dipolar groups. The appearance of dipolar orientation components in the TSDC of these materials was thus expected. Such was the case shown by results (Fig. 6.5) for gelatin. Collagen also shows typical electret behavior (Fig. 6.6). Immediately one is struck by the fact that the gelatin spectrum is much richer than that of collagen at lower temperatures. However, if we thermally denature collagen in vacuum in the solid state, its electret spectrum is found to change gradually to that of gelatin (curve II of Fig. 6.6). In fact, the TSDC spectrum has proved to be a very sensitive tool for the characterization of the denaturation state of collagen through the presence of the low-temperature electret bands. The more prominent peak in Fig. 6.5 at higher temperatures was shown to be due to a space-charge effect, and the low-temperature peaks, to a molecular dipole orientation mechanism. This was done, as previously discussed, by the investigation of T_P for the peaks as a function of T_0. If the heating rate is kept constant, a dipole peak will generally not change its position with variable T_0. However, a space-charge peak mechanism for electret behavior will give peaks with T_P often dependent on T_0, as has been shown for several electrets of ionic crystals [6.18], ice [6.27], and polymers [6.15]. We could thus separate in the richer gelatin TSD spectrum the peaks due to molecular dipole orientation from the peak due to space-charge' formation.

Experimental evidence now points to the presence of the electret state in all materials containing dipoles or ions. These are not the only source for electret behavior, for if the protein or any other biopolymer is hydrated, electret behavior can also be demonstrated with so-called bound water, as we discuss below.

6.5 Bound Water (Structured Water or Biowater)

The presence of water bound to proteins and other biopolymers is considered to be of fundamental physical and biological importance. Recently a conference on the subject was organized by the New York Academy of Sciences and the interested reader may consult the volume published [6.28] for further references. The fascinating history of the bound-water problem begins in 1938 with the basic work of *Teller* et al. [6.29]. In the following decades Bull, Pauling, Bernal, Szent-Gyorgyi, and a score of distinguished physicists, physical chemists, biochemists, biologists, and biomedical researchers became interested in the problem. A number of methods have been used to look for these properties and in particular to elucidate the structure of water bound to biopolymers: for example, NMR, infrared spectroscopy, circular dichroism, X-ray diffraction, and dielectric absorption. The particular case of water bound to collagen or gelatin has been studied intensively, and *Berendsen, Grigera,* and *Mascarenhas* have given reviews of the field [6.30].

Studies of water bound to collagen (or gelatin) may be important for the understanding of phenomena ranging from the structure of bound water (like the so-called question of icelike phases) to the interaction of water with the conformation of the macromolecule. Also the influence of hydration on solubility, cross-linking, and mechanorheological properties of collagen and gelatin have been studied by several authors. Recently, a study of dielectric absorption of hydrated collagen has been presented by *Tomaselli* and *Shamos* [6.31]. From results described in [6.5], the TSDC of collagen and gelatin in the dry state could be interpreted in terms of electrical energy storage in two compartments: dipoles and ionic charges. One is thus led to suspect that dipoles of water, if bound in a structured way to the macromolecule, could have energy levels for dipole rotation in the local field of the biopolymer macromolecule. Thus bound water might also exhibit the electret state. Of course we would have to limit ourselves to low hydration levels, where the resistivity of the samples would be high enough for the application of the polarization field. But these lower hydration levels are known to be related to the filling of primary hydration sites when structured binding occurs. If this is so, bound-water energy levels for dipole rotation could be studied in a most direct way. That such is indeed the case is shown by results with gelatin of Fig. 6.7. First one obtains the TSDC of the dry material. In the case of gelatin, this can be done most conveniently, because the sample can be safely warmed up to 150 °C in vacuum without any further denaturation or oxidative pyrolysis. If the sample is then left to equilibrate to a hydration degree of approximately 10 %, the TSDC shows new electret bands. That these bands are due to bound water can be seen in different ways:

a) difference curves of TSDC with succeeding lower hydration, obtainable by continuous pumping of the sample, indicates a difference spectrum where the bands occur;

b) by increasing the hydration degree the same two bands increase;

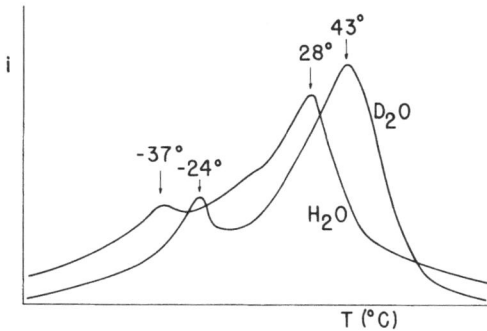

Fig. 6.7. Shift of bound-water peaks in gelatin due to heavy water

c) upon drying, the two bands disappear, and the original gelatin TSDC is obtained; this transformation can be done several times in a reversible way;

d) finally, as we discuss in more detail below, the bands can be shown to shift the maximum temperature if the hydration is done with heavy water (Fig. 6.7).

The cleaning technique to separate bound-water bands from the underlying TSDC was applied, and a quantitative analysis made of the peaks. It was found that the *Bucci–Fieschi* equations could be applied for the second peak where complete separation was possible. The activation energies of the first and second peaks were also obtained with the help of the initial rise technique as will be discussed below. The effect is present in collagen, where a similar set of two bound-water bands were found and with DNA, and many other biopolymers, where the TSDC also changed upon dehydration.

All these results on the electret state of bound water are presented in detail elsewhere [6.9, 25, 32].

We applied (6.2) (whole band analysis) to the higher-temperature bound-water peak. A good fit was found, indicating the first-order nature of the process (Fig. 6.8). For the lower band, the activation energy was determined by (6.5) (initial rise). The values of the activation energies (0.15 and 0.4 eV) suggested the following interpretation: the low-temperature peak is due to a single hydrogen-bonded water molecule, probably attached to polar residues or to the peptide bond. The higher temperature corresponds to multiple hydrogen-bonded water molecules, whose dipole can rotate only by breaking several bonds (corresponding to the 0.4 eV barrier). In view of the fact that the hydrogen bond is known to be cooperative, no suggestion is made for the exact number of linkages, beyond a minimum of the order of 3 bonds (if 0.15 eV is taken as an approximate value for the bond energy). This peak would then correspond to the "structured" bound-water or icelike phase, proposed to be a chainlike structure around the backbone of the polypeptide by several authors. The fact that the peak occurs near 40 °C indicates the presence of water as a structured phase, as required by the existence of a definite energy barrier even at this temperature. On the other hand, since (6.2) was found to be valid, another useful parameter can be obtained: the relaxation frequency ω_0, which can be interpreted in lattice-dynamic terms (as it has been in the case of alkali halide electrets). ω_0 can help in deciding the very important

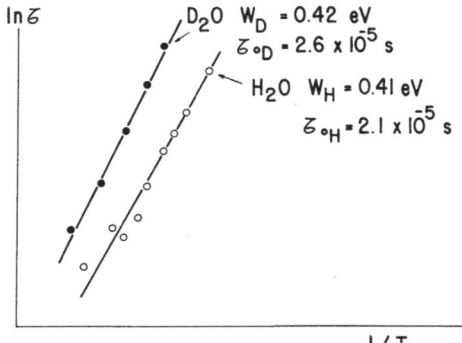

Fig. 6.8. Plot of $\ln X$ versus $1/T$ for bound-water peaks, showing validity of (6.2)

question of whether this bound water is more similar to a liquid or to a solid. The values of ω_0 (10^{-5} s) taken in conjunction with the values of the energy barrier according to the above interpretation lead us to suspect that bound water is more like a solid than a liquid (for which values of ω_0 are many orders of magnitude higher). At the same time, while the value of ω_0 is not identical to that of pure ice, it is of the same order of magnitude as the values determined by dielectric absorption in recent measurements by *G. Gross* for impure ice [6.33]. The electret state of bound water, if this interpretation is correct, may be the first direct experimental evidence for the solidlike phases of bound water structured around biopolymers. We are at present looking for the electret behavior not only of other biopolymers, but of hydrated inorganic salts and the so-called clathrate hydrates extensively investigated by *Jeffrey* [6.34a].

It should be noted that bound water may also be found in a state closely resembling a frozen gas with dipoles whose rotation is constrained. Instead of the long-range order characteristic of a solid, short-range structuring may also lead to a localized energy barrier for the rotation of dipoles. This is similar to what was found by *Onsager* et al. for a natural amorphous-ice electret [6.34b].

6.6 Polysaccharides and Polynucleotides

Besides proteins there are other classes of important biopolymers for which the electret state has been found: polysaccharides and polynucleotides. Again a systematic investigation has not been performed, and we shall take as examples the cases of chitin and cellulose (polysaccharides) and DNA (a polynucleotide).

In the case of DNA, *Fukada* had already found the piezoelectric effect, indicating the existence of polarization storage sources. However, since there are many physicochemical parameters affecting the chemical and structural nature of DNA, a more systematic investigation should be undertaken. The main peak in DNA at high temperatures (50 °C) is of a space-charge nature, and thus a systematic investigation of the salt content and other physicochemical parameters is needed, and is currently in progress at Sao Carlos.

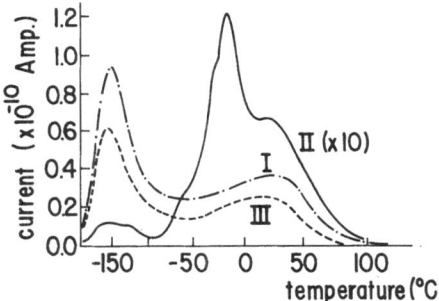

Fig. 6.9. TSDC for alpha-chitin (Dry: *I* and *III*, hum.: *II*)

For the investigation of polysaccharides, chitin will be taken as the main example. Chitin is an important structural material in biology [6.35]. Though there are several structural forms of chitin, we shall refer mainly to alpha-chitin, which can be obtained from lobsters or crab shell. Alpha-chitin forms fibers and its chemical structure is given in [6.35a]. It is seen that OH dipolar groups are present and these in principle may give rise to electret behavior. In fact, Fukada and Sasaki (personal communication) have found piezoelectricity in alpha-chitin, indicating some dipolar degrees of freedom in the material. We report here on results to be published obtained for chitin from crab shells by Fukada, Slaets, Zimmerman, and Mascarenhas to be published. Other results have also been obtained for chitin from the perisarch of tubularia by Mascarenhas, Rakesh, and Liuzzi, and from lobster tail by Mascarenhas, Rakesh, and Glimcher.

The alpha-chitin from crab shells was obtained in the form of a sheet, with the main fiber direction parallel to the sheet, as shown by X-ray diffraction experiments. The chain directions were found to be almost uniformly random within the sheet from the X-ray pattern. The electret state was found to depend also on hydration (curve II in Fig. 6.9), indicating water structuring around the fibers. After successive warm-ups in vacuum, the sample gradually dried (curves I, II), and an intrinsic spectrum characteristic of the material was obtained with two main bands in the TSDC, as shown in Fig. 6.9. The low-temperature peak was shown to be due to dipoles by investigating, as before, the maximum temperature (T_p) shift with T_0. The activation temperature for this peak was measured, and found to be very small (0.1 eV). It is proposed from an analysis of the structure of alpha-chitin (see, for instance, [6.35]) that this peak is due to orientation of the OH dipole which is internally hydrogen-bonded in the molecule. This bond is known to be weak for that particular hydroxyl group, and some of these dipoles would be free to orient under the external field. From the TSDC analysis, approximately 10^{18} dipoles per cm^3 were found to be free. This indicates that alpha-chitin is a strongly knit hydrogen biopolymer, since only a very small fraction of the dipoles are free to orient. The high-temperature peak was found to be due to space charge. As in most cases, the nature of the space charge is unknown, possible sources being protonic carriers or compensating ions.

Following the report by *Mascarenhas* and *Malavolta* [6.35b] on the electret behaviour of cellulose, investigations have been made by on several aspects of this problem. *Talwar* and *Sharma* reported on TSD of cellulose in a study of the nature of the processes responsible for relaxation [Ref. 6.16b, p. 134]. *Pillai* and *Mollah* [Ref. 6.16b, p. 138] studied paper cellulose and the possible influence of water. *Sawatari* [Ref. 6.16b, p. 136] attributed the low-temperature broad peak appearing in cellulose at $-130\,°C$ to molecular motion of the primary hydroxyl group of glucose in the amorphous region of cellulose.

6.7 Enzymes

Fröhlich, in a series of papers [6.36] related to the application of quantum mechanics to biology, proposed that enzyme action could be understood on the basis of its electrical polarization properties. Coherent longitudinal polarization waves coupled to the elastic field of the material constituting the enzyme would induce a metastable ferroelectric state in the system.

It has been shown [6.44] that important enzymes such as trypsin and urease exhibit the electret state, and several peaks are present in their TSDC. Each of these peaks is related to one compartment for charge and polarization storage and could in principle support *Fröhlich* waves. However, the intrinsic existence of the enzyme bioelectret is, of course, independent of the correctness of the *Fröhlich* model, being a direct experiment al fact [6.45].

The TSD for trypsin presents several peaks in the low and ambient temperature region as can be seen from Fig. 6.10. For the case of trypsin it was found by the previous methods that all the peaks below RT are of dipolar nature, while the higher temperature peak is due to space charge. An interesting effect due to bound water in trypsin was also found: hydration enhances the bioelectret behavior of the enzyme, as can be seen from Fig. 6.10 (curve I). This is probably due to the electret behavior of bound or structured water as it has been demonstrated in other biopolymers. Urease and ribonuclease were also investigated and both showed the electret state. Curves for urease and ribonuclease are shown in Figs. 6.11, 12. It is seen that the TSDC for all three enzymes are different, thereby indicating perhaps a certain structure dependence. In the case of urease, it was possible again to investigate the question of hydration and it was seen that hydration enhances the bioelectret behavior of the enzyme. For trypsin and urease, it was possible to show the reversibility of the hydration and dehydration cycle as far as the electret behavior is concerned: upon dehydration in vacuum, the TSDC changed; and upon exposure to humidity, the original TSDC was obtained. This could be done many times with trypsin with good reproducibility. This also indicates that no denaturation during warm-ups occurred for trypsin, and that the bioelectret state is a good monitor of the hydration of the enzyme.

Bioelectret behavior having been demonstrated for trypsin, urease, and ribonuclease, one should consider how it might be important for biophysical

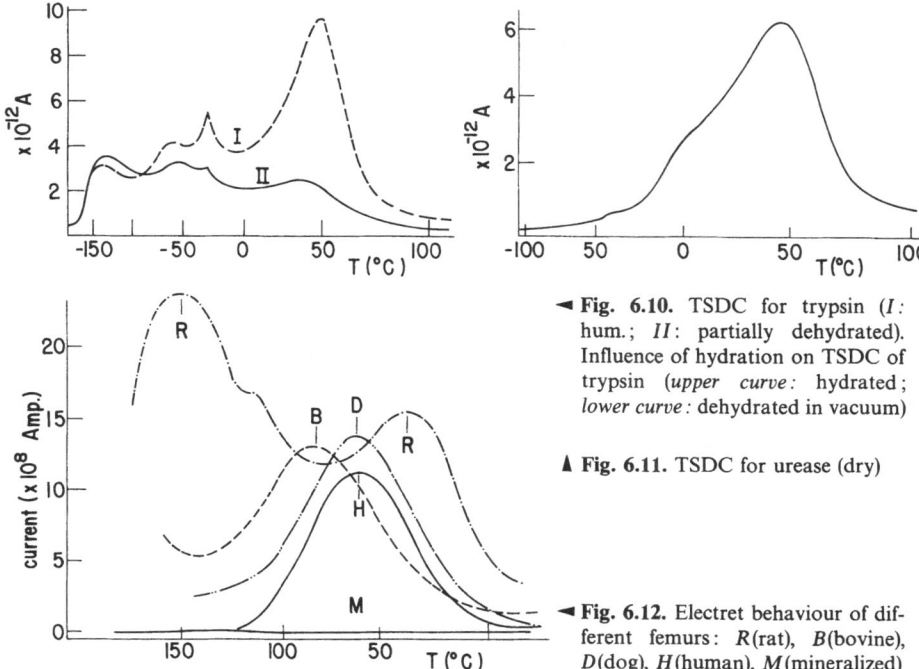

◄ **Fig. 6.10.** TSDC for trypsin (*I*: hum.; *II*: partially dehydrated). Influence of hydration on TSDC of trypsin (*upper curve*: hydrated; *lower curve*: dehydrated in vacuum)

▲ **Fig. 6.11.** TSDC for urease (dry)

◄ **Fig. 6.12.** Electret behaviour of different femurs: *R*(rat), *B*(bovine), *D*(dog), *H*(human), *M*(mineralized)

phenomena? Though it is still too early to draw any firm indication from these results, we would like to offer some conclusions related to the solid-state physics aspects of the experiments, especially in the case of trypsin. It has been demonstrated that there are at least three sources or compartments for charge and polarization storage in trypsin: dipoles, space charge, and bound water. The dipolar part of the electret TSDC is probably due to the dipolar residues which are part of the protein constituting the enzyme. At present, however, it is impossible to make specific assignments on the basis of these results alone. The space-charge peak is probably due to ionic conduction in the molecule. Thus may be related to ions or to proton motion in the molecule. Proton semiconduction has been invoked as a possible mechanism in ice and organic materials. In the case of bound-water peaks, the fact that the peaks decrease with hydration and appear reversibly upon hydration demonstrates the importance of water for the electrical behavior of the enzyme. In the case of trypsin, a fourfold variation in the stored polarization was observed in the hydrated material.

Extrapolating from the results found with collagen and gelatin, and also hemoglobin [6.38], the increase in polarization is due to the electret behavior of hydration shells bound to the protein. In this case, water hydrates the protein in a solid icelike phase, and this phase has different energy levels depending on the orientation of the water dipole, leading to the electret behavior.

Finally, possible relations of these results with Fröhlich's model for the electrical behavior of enzymes should be mentioned. *Fröhlich* proposes that enzymes are capable of presenting longitudinal polarization waves. The fact that several relaxation polarizations, were found for the enzymes investigated suggests that some of these relaxations may sustain Fröhlich waves. Whether the bioelectret behavior of enzymes favors the model first introduced by *Fröhlich* remains to be seen; it is, however, important to observe that enzymes are indeed able to store large amounts of electrical polarization in their structure (amounting sometimes to $10^{-7}\,\mathrm{C\,cm^{-2}}$). Also, theoretical calculations by Luiz Nunes de Oliveira (unpublished) on the field dependence of the Fröhlich model indicates a threshold value for the metastable ferroelectric transition. A detailed experimental analysis for all the individual polarizations in the TSDC of trypsin, for instance, as a function of the field should be done to see whether a threshold field, as predicted by the Fröhlich model, can be observed.

The question may also be asked: the electret state was induced in vitro in the enzyme by external applied fields; how can this happen in vivo? As discussed by *Fröhlich*, the fields provided by ion absorption in the macromolecule may be of the order of $10^4\,\mathrm{V\,cm^{-1}}$. This is easily sufficient to induce the bioelectret state in the enzyme. The fact that the electret peak exists near and above room temperature and relevant ranges such as 30–40 °C also demonstrates that the enzyme would remain an electret at usual in vivo ranges. In fact, the question may be raised about "the electrical denaturation" of enzymes when used above the temperatures of the relevant electret peaks. At Sao Carlos we are presently extending our observations to other enzymes in an exploratory way, to gain familiarity with the general properties of the electret behavior of these important biological substances.

6.8 Thermally Stimulated Pressure and Bound Water

As has been discussed in the previous sections, the bioelectret state depends drastically on the amount of hydration. Bound water with long relaxation times for dielectric relaxation is responsible for the polarization storage in bioelectrets, and the corresponding peaks in the TSD spectrum. Another technique that can be used to investigate hydration effects is Thermally Stimulated Pressure (TSP) [6.37]. Since during the TSD measurements, the sample is continuously heated, the changing temperature also changes the degree of hydrogen. Thus there is one basic inherent difficulty with TSD measurements in hydrated bioelectrets: the hydration may be changing during the experiment. Changes in hydration may also correspond, and they usually do, to changes in conformation of the biological molecule under investigation. The same objection is valid for all hydrated electret materials, biological or not. In order to investigate this problem, we have developed a very simple technique. The partial pressure of the water vapor desorbed from the sample is measured continuously as the sample is warmed. The corresponding pressure P as a function of temperature T can be measured continuously with a normal vacuum thermocouple. Of course the thermocouple must be properly calibrated against absolute values of the pressure.

The water desorption can be monitored with this technique, and if the temperature derivative of the pressure $P(T)$ (i.e., dP/dT) is plotted as a function of T, peaks may be found. This type of spectrum, similar to TSD, is what we call the thermally stimulated pressure (TSP) spectrum. *Celaschi* and *Mascarenhas* have investigated lysozyme using TSD, TSP, as well as thermogravimetric analysis (to measure the change in mass during water desorption). TSP turns out to be more sensitive than thermogravimetric techniques in measuring the kinetics of the dehydration process. It is also much simpler and more convenient. In the case of lysozyme, the very interesting effect was found that bound-water dipoles, previously oriented with an external field, show electric current peaks during desorption. This electrical effect associated with desorption of water dipoles is similar but basically different from the electric effect found by *Onsager* et al. [6.38] in amorphous ice, in which electric potentials and currents were found during the growth and subsequent heating of the samples.

Assuming that TSP spectra could be analyzed in the same way as TSD and that superposition of several peaks could occur, activation energies, the number of water molecules, and other corresponding parameters could be calculated. Details can be found in [6.37].

The other cases in which TSP was used are DNA and RNA bioelectrets. As in the case of lysozyme, the isothermal polarization decay (IPD) is a very convenient means of observing the electret behavior. The reason for this is that, with IPD, the temperature being constant, there is no change in hydration. Making measurements at different temperatures, the behavior of the sample as a function of hydration can be properly investigated. The DNA and RNA cases are very interesting not only because of the importance of these macromolecules in biophysics, but because they were claimed to be ferroelectrics in the literature [6.39]. On the basis of this supposed ferroelectricity, various speculations were made as to the role of electric polarization in memory storage mechanisms [6.40]. It was later shown [6.41] that, due to nonlinear transport mechanisms, DNA could display hysteresis loops when the Sawyer–Tower technique was used to detect ferroelectricity. In the case of RNA, ferroelectricity was similarly reported [6.42], and, as in the case of DNA, it was shown by *Mascarenhas* et al. [6.43] that nonlinearity was also operative. Both DNA and RNA were found to be strong bioelectrets. TSP studies were also made, and the bioelectret state in both DNA and RNA was found to depend strongly on hydration. Since hydration has been found to be fundamental for polarization storage in bioelectrets, the use of IPD together with TSP is essential for the proper study of these systems.

The other very interesting case is that of cellulose and chitin. TSP studies indicate broad compound peaks of water desorption in these materials. The study of cellulose now seems to have attracted a great deal of attention. In investigating its bioelectret behavior, TSP should be applied simultaneously with IPD and TSD.

6.9 Bone, Artificial Biomaterials, and Biomedical Applications

We shall mention here three selected topics, which will serve to demonstrate the potential of electret applications in medicine: antithrombus-formation surfaces, artificial membranes, and the electret state in bones.

Blood-Compatible Electrets

The main problem to be solved in obtaining blood-compatible biomaterials (for use as, for example, cannula, heart valves, and even entire artificial-heart systems) is that of avoiding thrombus formation (blood coagulation). Since the fundamental discovery by Sawyer, and results of Sawyer and Brattain that blood platelets are electrically charged, the idea was pursued that a negatively biased surface would inhibit coagulation.

By using metals, such as magnesium (and *not* electropositive noble metals) which are electronegative, in cannular implants in dogs, it was indeed shown that the concept would work. Unfortunately Mg was also poisonous, and could not be used for a permanent implant. Combining the well-known biological compatibility of teflon with its good electret behavior, the problem was investigated again by *Murphy* and *Marchant* [6.46] and hopeful results were reported. *Mascarenhas* et al. [6.47] following these investigations, also studied electretized cannular implants. We refer the reader to the detailed reports on the subject [6.48].

Membranes

The possible use of electret membranes has been proposed and discussed by several authors both from an experimental and theoretical viewpoint. The main idea here is that electret membranes might show different transport properties as well as different surface properties. With this motivation, *Linder* and *Miller* [6.13] investigated artificial polyelectrolyte membrane electrets. Measurements of polarization as a function of composition temperature, applied field, molecular weight of the polarizing component, and time were made on a series of membranes containing sodium polystyrenesulfonate in matrices of polyvinyl alcohol, polyacrilamide, and polyvinylpyrrolidone. The results indicate that the process depends mainly on interactions between components and not on the intrinsic nature of the components themselves. For detailed results we refer the reader to [6.13] and to other papers on membrane electrets.

Bone

TSD for several types of bone samples are shown in Fig. 6.12. It is seen that a main broad electret band appears in the range 30–100 °C. The effect seems to be a general property of all bone types investigated: bovine, canine, rat, and human femoral or tibial samples. If one measures the area under the current versus time curve, a measurement of the total polarization stored can be obtained. This has been investigated for bones as a function of several parameters like E_p, T, and nature of sample. A saturation polarization was found with fields of the order of $1 \, kV \, mm^{-1}$ for polarizations near 40 °C. In terms of this saturation polarization expressed in coulombs per cm^2, a practical parameter can be obtained as a measure of the capacity for storing polarization. Typical results for bone are of the order of $10^{-8} \, C \, cm^{-2}$, a value comparable to the polarization storage obtainable with good electret materials. Since the results on bone have been published elsewhere (Ref. 6.16a, p. 650; [6.25, 49]), we report here only the more fundamental results. From the curves of Fig. 6.12, it can be seen that the effect is present in samples which have been demineralized but is not present in mineralized samples from which all the protein has been extracted.

The electret state is thus related to the electret behavior of collagen, which we have discussed above. It is interesting to observe that since the denaturation of collagen can be detected by the electret technique, it was possible to show that collagen in the mineralized tissue did not denature even after strong heating or mild chemical treatment. This particular behavior of collagen in bone had been previously demonstrated by *Glimcher* and *Katz* [6.50] using biochemical techniques. Here we find a useful and simple application of the electret investigations. More recently Slaets (personal communication) has found in our laboratory that the saturation behavior of bone samples may depend on hydration.

6.10 Natural Electrets

Menefee [6.10] made the important discovery that the native keratin from porcupine quills presented strong thermocurrents when heated to temperatures at which melting of the alpha helix occurs. At these temperatures, dipole disordering takes place simultaneously, and it is this configurational change that induces the observed currents. Thus the existence of natural electrets has been demonstrated for a material in native conditions. *Menefee* also was able to show that the effect showed the expected symmetry, that is, reversing the orientation of the sample (in relation to its in vivo geometry) reversed the current observed during heating. Also, mounting the sample in such a way that the quill fiber axis was in the plane of the measuring electrodes, the released current was negligible, indicating a strong anisotropy. This was expected, since the dipoles groups are known to lie along the helix axis. For samples with nonoriented dipoles (that is, sample with no true permanent dipole moment, such as nylon), no effect was observed. *Menefee's* results are shown in Fig. 6.13.

Fukada et al. [6.24] have also confirmed *Menefee's* results and extended the observations with simultaneous piezoelectric measurements for the case of horn keratin. *Menefee* has also proposed a simple model for charge separation by helix disordering. The model presupposes an initial polarization P_0 decaying as a result of the helix melting exponentially with a relaxation time τ_1. With the

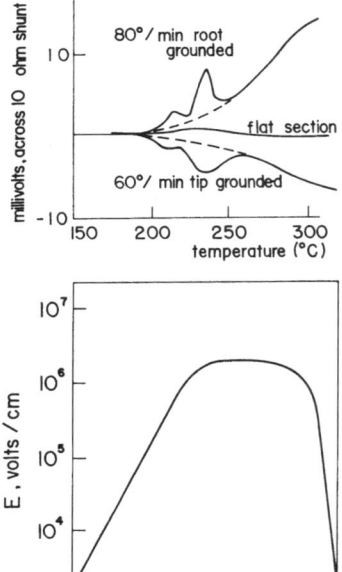

Fig. 6.13. Depolarization of native keratin [6.10]

Fig. 6.14. Field from melting α-helix dipoles [6.10]

capacitance of the system C, and the sample resistance R, the following equation is obtained:

$$(2C - C_0)dV/dt + V/R = (P_0/\tau_1)\exp(-t/\tau_1),\tag{6.6}$$

where C_0 is the vacuum capacitance of the system. With boundary condition $V = 0$ at $t = 0$ and $t = \infty$, one can solve (6.6) for the electric field E:

$$E = [\varrho P_0/(\tau_1 - \tau_2)][\exp(-t/\tau_1) - \exp(-t/\tau_2)],\tag{6.7}$$

where ϱ is the specific resistivity, and τ_2 is given by

$$\tau_2 = \varrho(2\varepsilon - 1)\varepsilon_0.$$

Assuming $P_0 = 3 \times 10^{-6}$ C cm^{-2}, $\tau_2 = 1.7 \times 10^{-4}$ s, and $\tau_1 = 10^{-7}$ s and using (6.7) he obtained the results of Fig. 6.14.

The assumptions above were made for a hypothetical gellike membrane consisting of alpha-helix material undergoing a transition with the relaxation time which has been assumed to apply to such polypeptides.

It is seen that fields as high as 10^6 V cm^{-1} can be calculated from the model. Of course the main point is whether the relaxation time τ_1 used in the calculations corresponds or not to experimental conditions. Nevertheless, the model is very appealing from the point of view of possible biophysical applications, such as the study of phenomena that may occur in membranes or in the generation of signals from sensory receptors, as in the olfactory system.

The possible relation between these natural electrets and their production in vivo invites much speculation as to the role of the electret state in biological processes at the cellular and membrane level.

6.11 Conclusions

We have not tried to cover the entire literature concerning applications of electrets in biology. The tutorial nature of this monograph required a more compact presentation which we hope will have introduced the reader to certain specific but more or less typical problems of the field. The main conclusion is certainly that the field is still in its beginnings and that much valuable fundamental work remains to be done. Another conclusion is that the interdisciplinary nature of the research on these problems imposes limitations on the organization of the investigations. For instance, in bioengineering applications of electrets as biomaterials, the cooperation of biologists or medical specialists is certainly required. Even in the case of more basic investigations, such as the electret properties of biopolymers, the cooperation of biochemists in sample preparation and adequate control of physicochemical conditions is also necessary.

Last, but not least, the use of accessory techniques such as piezoelectricity, X-ray diffraction, and optical measurements as in other areas of solid-state physics will be very important for corroborating and reinforcing possible models and interpretations.

The fact that the electret state has been found for practically all important classes of biopolymer (polypeptides, polysaccharides, polynucleotides) opens wide areas for systematic investigations. The extremely important field of structured-water electrets will certainly attract the attention and interest of many, and we think that it may prove to be one of the most fertile fields for applications in biology. It is also per se a unique method for investigating water structuring around macromolecules. Membrane electrets constitute another field of great potential. In this respect, theoretical semiquantitative calculations such as those given by *Menefee* for signal propagation in a gellike helical polypeptide membrane may be a source of further ideas and investigations.

Natural electrets, as found in keratin, may also be present in vivo in a variety of other biological materials, including bone, tendon, and other tissues. They may also be a general feature of certain biopolymers capable of presenting dipole ordering in their structure. Such may be the case for chitin. In fact, just to mention one example in this field, the possibility exists that the perisarc of tubularia (a hydra), composed of chitin, is such a natural electret and may be involved in important biophysical effects (*Mascarenhas, Shuhan, Liuggi*: unpublished results). The theoretical work is also needed not only to predict possible electret effects in biology but also for a better understanding of basic electret behavior in biological materials, such as the structured-water problem. Also there is the practically untouched field of the action of radiation on the electret behavior of biological molecules, and photoelectret properties, which may be an interesting side field in the new and large area of photobiology.

One of the most important aspects of electret work in general is related to the nonlinear electret. This is particularly important in the case of bioelectrets. With the examples of DNA and RNA, recently investigated by *Mascarenhas* et al. [6.43], and of collagen, by *Povoa* and *Zimmerman* [6.51], it is clear that ionic transport is an essential aspect of charge and polarization storage, inducing highly nonlinear effects. These are further coupled to hydration effects, as discussed previously. *Fröhlich* has called attention to these aspects [6.52]. An important picture that emerges from the bioelectret work is that there are perhaps strong connections between conformation of the biopolymers, compensating charges, hydration, and dipolar orientation. Changing the polarization storage induces changes in conformation that are of paramount importance for the biological action of the molecule. It is in this respect that the nonlinear bioelectret is a very rich concept. Throughout this chapter, we have considered the electret as a microscopic, molecular concept, in which sufficiently polarization or charge is stored locally with a relaxation time long in relation to the phenomena under observation. This concept is important in all our considerations of bioelectrets.

The work described in this chapter leads us to the conclusion that the study of electrets in biomaterials and biophysics is a new and fascinating field of basic and applied research.

Acknowledgements. We would like to thank the Foundation for Research of the State of Sao Paulo, the National Research Council of Brazil, and the NSF for financial assistance. The cooperation of colleagues at the Institute of Physics and Chemistry, Sao Carlos, and stimulating discussions with the late Professor L. Onsager, and with H. Fröhlich, E. Fukada, and M. Kasha are gratefully acknowledged.

References

6.1   M. Eguchi: Philos. Mag. **49**, 178 (1925)
6.2   O. Heaviside: *Electrical Papers*, Vol. 1 (Macmillan, London 1892) p. 488
6.3a   B. Gross: *Charge Storage in Solid Dielectrics* (Elsevier, Amsterdam 1964)
   b   M. S. Campos (ed.): *Rep. Int. Symp. Electrets and Dielectrics* (Brazilian Academy of Sciences, Rio de Janeiro 1977)
6.4   C. Bucci, R. Fieschi: Phys. Rev. Lett. **12**, 16 (1964)
6.5   M. Campos, S. Mascarenhas, G. F. Leal Ferreira: Phys. Rev. Lett. **27**, 1432 (1971)
6.6   F. S. Sinencio, S. Mascarenhas, B. Royce: Phys. Lett. **26** A, 70 (1967)
6.7   J. van Turnhout: *Thermally Stimulated Discharge of Polymer Electrets* (Elsevier, Amsterdam 1975)

6.8   M.Perlman, C.Reedyck: J. Elec. Soc. **115**, 45 (1968)
6.9a S.Mascarenhas: J. Electrostatics **1**, 141 (1975); Ann. NY Acad. Sci. **238**, 36 (1974)
   b T.Reichle, M.Nedeska, A.Mayer: J. Phys. Chem. **74**, 2659 (1970)
6.10 E.Menefee: In *Electrets*, ed. by M.Perlman (Electrochemical Society, New York 1973) p. 661
6.11 I.Yannas: In *Biomedical Physics and Biomaterials Science*, ed. by E.Stanley (MIT Press, Cambridge, Mass. 1973) Chap. 3
6.12 E.Murphy, S.Merchant: In *Electrets*, ed. by M.Perlman (Electrochemical Society, New York 1973) p. 627
6.13 C.Linder, I.Miller: J. Phys. Chem. **76**, 3434 (1972)
6.14 M.Goel, S.Meera, P.Pellai: In *Abstracts Int. Workshop Electrical Charges in Dielectrics*, Kyoto, Japan, ed. by E.Fukada (1978) p. 64
6.15 B.Gross: In *Electrets and Charge Storage Phenomena*, ed. by M.Perlman, L.Baxter (Electrochemical Society, New York 1968); Appl. Opt., Supp. 3: "Electrophotography", 176 (1969)
6.16a M.Perlman, L.Baxter(eds.): *Report Int. Electret Symposium* (Electrochemical Society, New York 1968)
   b E.Fukada(ed.): *Abstracts Int. Workshop Electrical Charges in Dielectrics*, Kyoto, Japan (1978)
6.17 G.F.Leal Ferreira, L.N.Oliveira: Phys. Rev. **11**, 2311 (1975);
      G.F.Leal Ferreira, B.Gross: Rev. Bras. Fisica, **2**, 205 (1972);
      B.Gross et al., Phys. Rev. B**9**, 5318 (1974)
6.18 R.Creswell, M.Perlman: J. Appl. Phys. **41**, 2365 (1970);
      R.Capelletti, R.Fieschi: In *Electrets: Charge Storage and Transport in Dielectrics*, ed. by M.Perlman (Electrochemical Society, Princeton, N.J. 1973) p. 1
6.19 J.Slaets: "Ultralow frequency bridge", M.Sc.Thesis (University of Sao Paulo, Sao Carlos 1976)
6.20 B.Gross: L.F.De Novd: Phys. Rev. **67**, 253 (1945);
      B.Gross: J. Chem. Phys. **17**, 866 (1949)
6.21 M.Campos, S.Mascarenhas, G.F.Leal Ferreira: Phys. Rev. Lett. **27**, 1432 (1968)
6.22 E.Fukada: Prog. Polym. Sci. Jpn. Kodan-Sha **2**, 329 (1971); in *Electrets, Charge Storage, and Transport in Dielectrics*, ed. by M.Perlman (Electrochemical Society, Princeton, N. J. 1973) p. 486; Ann. N. Y. Acad. Sci. **238**, 7 (1974)
6.23a M.G.Broadhurst et al.: In *Electrets*, ed. by M.Perlman (Electrochemical Society, Princeton, N. J. 1973) p. 492
   b R.L.Zimmerman et al.: J. Appl. Polym. Sci. **19**, 1373 (1975)
6.24 E.Fukada, R.Zimmerman, S.Mascarenhas: Biochem. Biophys. Res. Commun. **62**, 415 (1975)
6.25 S.Mascarenhas: Ann. N. Y. Acad. Sci. **238**, 36 (1974)
6.26 J.Costa Ribeiro: An. Acad. Bras. Cienc. **22**, 325 (1950)
      P.Eyerer: Adv. Colloid Interface Sci. **3**, 223 (1972)
6.27 S.Mascarenhas: In *Physics of Ice*, ed. by N.Riehl, R.Bullemer, H.Engelhardt (Plenum, New York 1969) p. 483
6.28 C.Hazlewood(ed.): Ann. NY Acad. Sci. **204**, (1973)
6.29 S.Brunauer, P.Emmett, E.Teller: J. Am. Chem. Soc. **60**, 309 (1938)
6.30 H.Berendsen: In *Biology of the Mouth*, (American Association for the Advancement of Science, Washington, D. C. 1968) p. 145
      R.Grigera: *Introduction to the Biophysics of Water* (Universitaria, Buenos Aires 1976)
      R.Grigera, S.Mascarenhas: Stud. Biophys. (to be published)
6.31 V.Tomaselli, M.Shamos: Biopolymers **12**, 353 (1973)
      T.Ghilardi Neto, R.L.Zimmerman: Biophys. J. **15**, 573 (1975)
6.32 S.Mascarenhas: In [6.16b]
6.33 G.Gross: In *Electrets, Charge Storage and Transport in Dielectrics*, ed. by M.Perlman (Electrochemical Society, Princeton N. J., 1973) p. 560
6.34a G.A.Jeffrey, R.McMullan: Inorg. Chem. **8**, 43 (1967)
   b L.Onsager, D.Staebler, S.Mascarenhas: J. Chem. Phys. **68**, 3823 (1978)
6.35 N.Dweltz: Biochim Biophys. Acta **44**, 416 (1960)
6.36 H.Fröhlich: Int. J. Quantum Chem. **2**, 641 (1968); Phys. Lett. **44**A, 385 (1973); Proc. Nat. Acad. Sci. USA **72**, 4211 (1975)

6.37 S.Celaschi, S.Mascarenhas: Biophys. J. **20**, 273 (1977)

6.38 L.Onsager, D.Staebler, S.Mascarenhas: J. Chem. Phys. **68**, 3823 (1978)

6.39 J.Polonsky, Douzou, C.Sadron: C. R. Acad. Sci. **251**, 976 (1961)

6.40 P.Fong: Bull. G. Acad. Sci. **30**, 13 (1972);
B.T.Mathias: In *Proc. 3rd Conf. from Theoretical Physics to Biology*, Inst. de La Vie, ed. by M.Marois (Karger, Basel 1973)

6.41 C.Brot et al.: J. Chem. Phys. **43**, B603 (1965)

6.42 A.Stanford, A.Lorey: Nature (London) **219**, 1250 (1968)

6.43 S.Quezado: "DNA: Bioelectret, Ferroelectricity, and Non-linearity", M. Sc. Thesis, University of Sao Paulo, Sao Carlos (1978)
S.Mascarenhas et al.: "On the the problem of ferroelectricity and non-linearity of DNA and RNA", in *Riken Symposium on Electrical Properties of Polymers*, Saitama, Japan, ed. by E.Fukada (1978)

6.44 S.Mascarenhas et al.: An. Acad. Bras. Cienc. **51**, 605 (1979)

6.45 S.Mascarenhas: "Bioelectrical Properties in connection with Fröhlich's Theory", in *Conf. Int. Physique Théorique à la Biologie*, ed. by M.Marois (Institut de La Vie, Versailles 1975)

6.46 P.V.Murphy, S.Marchant: In *Electrets, Charge Storage and Transport in Dielectrics*, ed. by M.Perlman (Electrochemical Society, Princeton, N. J. 1973) p. 627

6.47 I.M.Spinelli, S.Mascarenhas, A.Sader: In "Rep. st Latin-Am. Cong. Med. Phys., ed. by S.Watanabe (University of Sao Paulo, 1972) p. 14

6.48 E.Fukada, T.Takamatsu, I.Yasuda: Jpn. J. Appl. Phys. **14**, 2079 (1975)
J.J.Konikoff, J.West: In *Rep. Conf. Electrical Insulation and Dielectric Phenomena* (National Academy of Sciences, Washington, D. C. 1978) p. 304

6.49 S.Haridoss, R.Bullard: Proc. Indian Acad. Sci. **24**, 34 (1978)

6.50 M.Glimcher, E.J.Katz: Ultrastruc. Res. **12**, 705 (1965)

6.51 J.Povoa: "Non-linear transport in collagen", M.Sc. Thesis (University of Sao Paulo, Sao Carlos 1978)

6.52 H.Fröhlich: "Non-linear Dielectric Properties of Biological Materials", in *Rept. Conf. Electrical Insulators and Dielectric Phenomena* (National Academy of Sciences, Washington D. C. 1978) p. 259

7. Applications

G. M. Sessler and J. E. West

With 28 Figures

Permanent electrification effects in dielectrics have been utilized in a wide variety of applications. These reach from the technical areas to the biological and medical fields and are in various states of research, development, or production.

Some of the oldest practical devices of this kind are electret transducers (microphones, earphones, etc.) which were first described in 1928 [7.1]. These transducers, as well as the electret microphones discussed in the following years [7.2, 3] and those used before and during the Second World War [7.4, 5] proved unsatisfactory since they contained wax electrets which have insufficient electrical stability under normal environmental conditions. In 1962, microphones with thin flexible polymer electrets were introduced [7.6] which, upon proper choice of electret materials [7.7] and other improvements [7.8], gained widespread commercial acceptance.

Another application of charge-storage phenomena of great practical importance is in the field of electrophotography. The basic process used in many electrophotographic methods, namely the production of a charge pattern on an appropriate carrier and its development with powders, was already studied in the early 1930s [7.9]. The breakthrough in this field came a few years later when investigations of photoconductive image formation [7.10] led to the development of Xerographic reproduction methods (see [7.11]). Related nonphotographic processes, going back to the same period [7.9], have gained importance in the recording of alphanumeric characters, facsimile, and other information.

More recently introduced electret devices include gas filters, motors, relay switches, optical display systems, and radiation dosimeters. While the now commercialized gas filters use corona-charged electret fibers to capture submicron particles by electrostatic attraction [7.12], the still experimental electret motors employ stators or rotors of charged dielectric [7.13]. The relay type switches utilize the external field of electrets to open or close contacts [7.14]. In radiation dosimeters, the decay of a previously stored charge or the generation of a radiation induced conductivity is employed to measure radiation doses [7.15].

A number of new electret applications are based on the piezoelectric or pyroelectric effects in polarized polymer materials. Of particular importance are electroacoustic transducers utilizing piezoelectric films [7.16, 17] operating in transverse or longitudinal modes. Earphones based on this principle are now in wide practical use and many other transducer applications, including underwater devices, have been suggested. Apart from these, electromechanical transducers based on piezoelectric polymers have also been developed. Of the pyroelectric devices, most of which are still in the experimental stage, light detectors and vidicons deserve particular mention [7.18, 19].

Another group of applications, namely those relating to biophysics, promise to have great future potential. Of interest in this context are attempts to improve the blood compatibility of polymers by negative charge deposition [7.20] and the observation of electret properties of human and other bones and blood-vessel walls [7.21]. It was also shown that foil electrets placed in contact with bones of animals in vivo cause accelerated growth of callus, necessary for fracture healing [7.22]. Moreover, electret bandages put on skin incisions considerably improve the tensile strength of the wound over a given period of time and thus speed the healing process [7.22].

In the present chapter, the emphasis is placed on electret transducers, electret filters, piezoelectric polymer transducers, and pyroelectric devices. The very extensive fields of electrophotography and electrostatic recording have already been thoroughly reviewed in the literature (see [7.11]). It is therefore felt that a relatively brief and selective treatment of these subjects is justified. Finally, electret applications in dosimeters are covered in Chap. 4 (see also Sect. 7.6) while some of the electret applications relating to biophysics are discussed in Chap. 6.

7.1 Electret Transducers

7.1.1 Microphones

Analysis

Electret microphones are electrostatic transducers with a permanently charged solid dielectric. A schematic cross section of such a microphone, consisting of a metallized electret diaphragm and a backplate separated from the diaphragm by an air gap, is shown in Fig. 7.1 where vertical dimensions are enlarged. In real cases, such dimensions are so small that we can neglect effects due to the finite lateral size of the electret. The metal layer and the backplate are connected by a resistor R. The electret charge of density σ_1, considered to be constant, generates an electric field E_1 in the air gap.

If a sound wave impinges on the diaphragm, it experiences a deflection of amplitude $\hat{s}$, changing the air gap thickness s_1. Under open-circuit conditions, the field and the induced charges remain constant and one obtains from (2.22) for the output-voltage amplitude $\hat{V} = -E_1\hat{s}$ the frequency-independent value

$$\hat{V} = \frac{s\hat{s}\sigma_1}{\varepsilon_0(s + \varepsilon s_1)}. \tag{7.1}$$

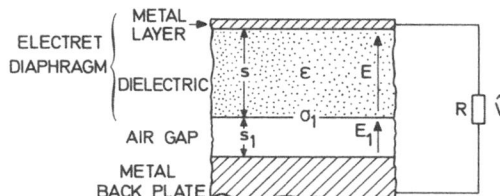

The generated voltage is thus in phase with the deflection of the air layer if σ_1 is positive. If R is finite, the expression in (7.1) is multiplied by $\omega RC / \sqrt{1 + (\omega RC)^2}$, where C is the capacitance of the transducer.

If the restoring forces are due to the elasticity of the air cavities behind the diaphragm (effective thickness s_0) and the tension T of the foil, its displacement $\hat{s}$ due to a sound pressure $\hat{p}$ of a frequency well below resonance will, for negligible losses, be given by [Ref. 7.23, Eqs. (20.10, 23.11)]

$$\hat{s} = \frac{\hat{p}}{(\gamma p_0 / s_0) + (8\pi T / A)}, \tag{7.2}$$

where γ is the specific heat ratio, p_0 the atmospheric pressure, and A the membrane area.

Substituting $\hat{s}$ from (7.2) into (7.1) gives the sensitivity $\varrho_m = \hat{V} / \hat{p}$ of the electret microphone below resonance,

$$\varrho_m = \frac{s \sigma_1}{\varepsilon_0 (s + \varepsilon s_1)[(\gamma p_0 / s_0) + (8\pi T / A)]}. \tag{7.3}$$

If $A \gamma p_0 \gg 8\pi T s_0$, as is frequently the case (see below), one has

$$\varrho_m = \frac{s s_0 \sigma_1}{\varepsilon_0 (s + \varepsilon s_1) \gamma p_0}. \tag{7.4}$$

Then the sensitivity does not depend on area. Values of a few mV/µbar are expected, in agreement with measured data. The resonance frequency $\omega_r = (\gamma p_0 / s_0 M)^{1/2}$ (M is the mass per unit area of the foil) is usually placed at the upper end of the audio range or higher.

Design

The design of an electret microphone is shown in the cutaway drawing of Fig. 7.2. The diaphragm, typically 12 or 25 µm Teflon TFE or FEP with a 500–1000 Å thick deposited metal electrode on one surface, is given a charge of 10–20 nC cm^{-2}. The nonmetallized surface of the foil electret is placed next to a backplate, leaving a shallow air gap whose thickness (10–30 µm) is controlled

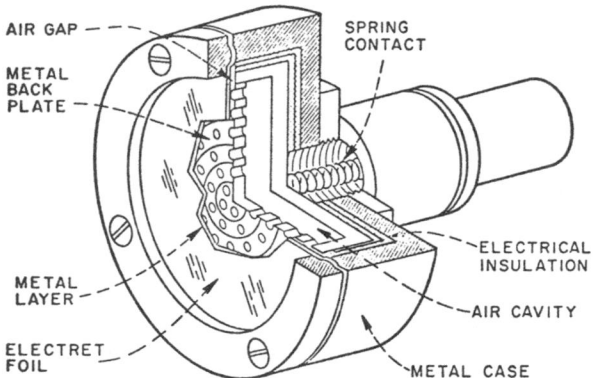

AIR GAP

METAL
BACK
PLATE

SPRING
CONTACT

ELECTRICAL
INSULATION

METAL
LAYER

AIR CAVITY

ELECTRET
FOIL

METAL CASE

Fig. 7.2. Cutaway drawing of foil-electret microphone [7.8]

by ridges or raised points on the backplate surface. The backelectrode is either a metal disk or a metal-coated dielectric having a thermal expansion coefficient about equal to that of the foil. To decrease the stiffness of the air layer and thus improve the microphone sensitivity by lowering the resonance frequency, the air gap is connected to a larger air cavity by means of small holes through the backplate.

The electret microphone differs from the classical electrostatic transducers with solid dielectric [7.24] in the sense that it does not require a dc bias. For the above range of conditions, the sensitivity of the electret microphone corresponds to that of a nonelectret transducer with an external bias of 70–280 V.

The mechanical tension of the foil is generally kept at a relatively low value (about $10 \, \mathrm{N \, m^{-1}}$), so that the restoring force is determined by the air-gap compressibility. Control of the restoring force by the air gap has the advantage that changes in tension due to stress relaxation have only a minor effect on the sensitivity. To achieve better control of the tension, rectangular systems in which the foil is only tensioned in one direction are being used [7.25].

A number of modifications of the design shown in Fig. 7.2 have been implemented of which two deserve particular attention. In one case, the transducer consists of a metallic backplate coated with a layer of Teflon FEP which is permanently charged [7.25]. A thin Mylar diaphragm (typically 3–6 μm thick), metallized on its inner surface, is stretched over the backplate. In this approach, the excellent electrical properties of the Teflon and the good mechanical properties of the Mylar can be used to advantage. Subminiature electret-condenser microphones (size $8 \times 5.6 \, \mathrm{mm}$) have been designed on this principle for use in hearing aids [7.26]. Because of the small diaphragm mass, such a system also provides a 10–20 dB decrease in vibration sensitivity over conventional foil-electret transducers of similar dimensions. In the second modification [7.27], an unmetallized Teflon electret 12–25 μm thick is positioned between a fixed electrode and a metallized polyester film of comparable thickness. Such a two-layer sandwich technique also optimizes the electrical and mechanical properties of the microphone.

Electrical and Mechanical Stability

Many independent studies have been made on charge stability of electrets at elevated temperatures and increased humidity (see Chap. 2) and, more importantly in this context, on the change in sensitivity of electret microphones. Some of the latter investigations were carried on for periods of many years [7.28] at room temperature and others for shorter periods of time at elevated temperatures [7.25, 27, 29–33]. It was found that the electret materials with the best electrical properties are the halocarbon materials Teflon TFE and FEP [7.7, 25, 34–35], and Aclar (CTFE) [7.36].

Figure 7.3 shows the results of a comparative study of microphones with electron-beam charged Teflon FEP foils under extreme environmental conditions and under normal room conditions. The sensitivity variation of both sets of microphones is less than ± 1 dB over the measured time. Similar results have been obtained by other laboratories with some of the data obtained on "aged" microphones subjected to a pretreatment at elevated temperature.

Measuring microphones must maintain constant sensitivity within a fraction of a dB over a wide range of temperatures and humidities. In one study, conventional condenser, ceramic and foil-electret measuring microphones were tested for about one year during which they were exposed to an outdoor environment in Concord, Massachusetts [7.32]. While a substantial sensitivity shift due to water condensation was reported for conventional condenser microphones, both the ceramic and foil-electret transducers performed well under the extreme weather conditions.

From all these tests, it is generally agreed that well-designed unprotected electret microphones can be operated at 50 °C and 95 % relative humidity with a sensitivity loss of only about 1 dB per year. The minimum 3-dB lifetimes in dry and humid air, as estimated from all available sources, is shown in Fig. 7.4. This performance equals or exceeds that of the integrated-circuit preamplifiers used together with these transducers.

The temperature coefficient of the sensitivity of early electret measuring microphones could not be determined due to random instabilities [7.37]. Recent microphones of this kind show temperature coefficients of less than 0.03 dB/°C in the temperature range -10 to $+50$ °C [7.38]. Although this is an excellent figure for an unprotected microphone, it is not quite as good as the value of 0.005 dB/°C for a protected standard condenser microphone.

Another question concerns the mechanical stability of electret microphones, that is the relation between the static deflections of the diaphragm and the electrical field. This problem was investigated using numerical techniques [7.39, 40]. It was found that the analytical results for circular membranes accurately describe the nonlinearity of the observed deflections and also predict the critical stability point. Annular membranes have also been studied using these methods.

Attention has also been given to the question of shaping the backplate-surface profile of electret microphones [7.40]. This is not only important for the

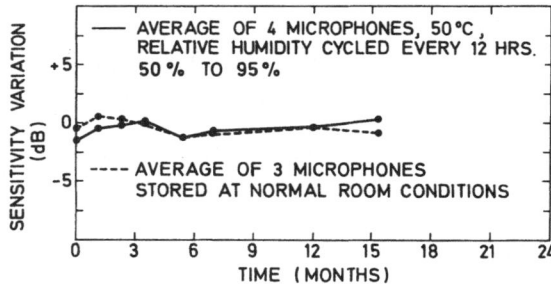

Fig. 7.3. Temporal change of sensitivity of electret microphones with electron-beam charged 25 µm Teflon FEP electrets under different storage conditions [7.8]

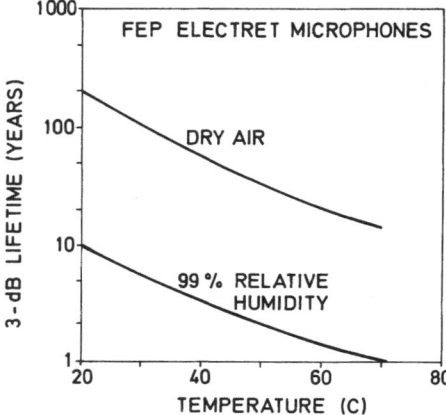

Fig. 7.4. Minimum 3-dB lifetime of FEP electret microphones, estimated from literature values

stability of the sensitivity but also for the sensitivity value as such. By controlling the backplate profile, determined by ridge height and ridge spacing together with tension and thickness of the diaphragm, microphones can be designed whose sensitivities remain constant even if the charge density of the electret decreases. On the other hand, a microphone assembly which does not depend on taut or tensioned membranes has been suggested [7.41]. In many of the newer transducers, however, separation of the electret and diaphragm functions and choice of a high-strength diaphragm eases the mechanical stability problems.

Acoustic Properties

Foil-electret microphones have more desirable features than any other microphone types. Apart from the very wide frequency range, which for various electret transducers can extend down to 10^{-3} Hz and up to hundreds of MHz (see below), they feature a flat frequency response, low harmonic distortion, low vibration sensitivity, good impulse response, insensitivity to magnetic fields and simplicity in design. In addition to these properties, foil-electret microphones have three distinct advantages as compared to conventional condenser microphones: (I) they do not require a dc bias; (II) due to the use of solid dielectrics

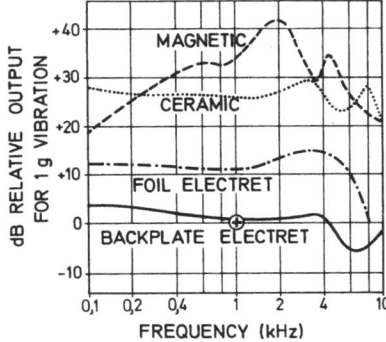

Fig. 7.5. Vibration sensitivities for a variety of microphones without vibration isolation [7.26]

and a shallow air gap, they have a higher capacitance per unit area; and (III) they are insensitive to shorting due to water condensation.

The frequency response of well-designed electret microphones is flat within ±1 dB in the audio-frequency range. Sensitivities of such electret microphones are typically a few mV/μbar. Electret microphones for precision sound level measurements have flat responses up to 30 kHz and sensitivities between 0.1 and 2 mV/μbar [7.32].

At the low end of the frequency range covered by electret transducers, infrasonic microphones were designed [7.42] with lower cutoff frequencies of 10^{-3} Hz. Such corner frequencies can be achieved with a very small pressure-equalization hole, usually an adjustable capillary, and with preamplifier-microphone combinations having a large RC constant. The frequency response of such infrasonic microphones can be made flat up to several kHz.

In the ultrasonic range, foil-electret transducers are being used for exciting and detecting waves up to 200 MHz in liquids and solids [7.43]. In these applications, an external bias is applied to the transducer in addition to its electret bias. The polarity of the external voltage is such that the field in the gap, which determines the sensitivity, is equal to the sum of electret and external-bias fields, while the field in the polymer is equal to the difference of these fields, thus eliminating the possibility of breakdown in the polymer. The sensitivity is raised above the limit achievable with only a single kind of biasing.

The electrical noise [7.31, 44] of an electret microphone–preamplifier combination is primarily due to two sources: the preamplifier input resistor and the field-effect transistor. For well-designed systems with a medium or large size cartridge (diameter 1 cm or more) and a low-noise field-effect transistor (FET), the noise is principally due to the input resistor and corresponds to 15 dB SPL. With extremely low-noise amplifiers, the self-noise of the microphone is also detectable [7.44]. For miniature-size microphones [7.26, 30] however, the FET current noise predominates and the equivalent noise level rises to about 30 dB SPL for a cartridge diameter of 2.5 mm. Because of the larger capacitance per unit area, these noise levels are lower than those of ordinary condenser microphones of comparable size.

The upper limit of the dynamic range [7.6] is typically determined by the distortion of the microphone. For sound pressure levels below 130 dB SPL the distortion is less than 1 %, while it reaches 3 % at 145 dB SPL.

A very desirable property of foil-electret microphones, shown in Fig. 7.5, is the low vibration sensitivity, crucial in applications in which high solid-borne noise is encountered. It can be seen from the figure that the vibration sensitivity of a conventional electret microphone is about 10–30 dB lower than that of other widely used microphones while that of an electret microphone with a backplate-bonded electret is another 10 dB lower [7.26]. The low vibration sensitivity is due to the relatively small mass of the electret-microphone diaphragm [7.35].

Applications

Because of their favorable properties, simplicity, and low cost, electret microphones are being used in many applications, both as research tools and in the commercial market (see [7.8] and references therein). Among the *research applications* are microphones for use in optoacoustic spectroscopy, applied to the detection of air pollution, to the study of reaction kinetics of gases, and to the investigation of optical absorption of solids. The understanding of optoacoustic cells, used in such studies, is now at a very satisfactory stage [7.45]. Because of the favorable noise performance of electret microphones, the detection threshold for air pollutants has been lowered by more than an order of magnitude [7.45]. For example, NO concentrations of 10^7 molecules per cm^3 can now be detected in the atmosphere. Other applications of electret microphones have been reported in aeronautics and shocktube studies, in which the low vibration sensitivity of these transducers is crucial. The wide frequency range of electret transducers, discussed above, made possible their application in infrasonic atmospheric studies and in ultrasonic investigations of liquids, solids and in lowtemperature physics. Such transducers have been used, for example, to measure the first, second, and third sound velocities in He II at 1.2 K [7.45]. In addition, ultrasonic arrays of electret microphones have been applied to acoustic holography (see [7.8]).

Of all *commercial applications* of electret transducers, the high-fidelity electret microphone for amateur, professional, studio, and tape-recorder use is most prominent. Other uses of electret microphones are in sound-level meters, noise dosimeters, and movie cameras, while miniature microphones are customary in hearing aids. The success of the electret microphone in these applications is primarily due to its excellent acoustic quality and low cost. It is noteworthy that, owing to their low vibration sensitivity, foil-electret microphones were the first transducers to be built directly into widely used tape recorders and movie cameras. Although the electret microphone was brought to the market only about 10 years ago, more than 50 million of these transducers are now produced annually all over the world, accounting for over one-half of the entire output of high-fidelity microphones.

In the US and Canadian *telephone system*, foil electret microphones are being used in speakerphones, while Canadian operators' headsets incorporate noise canceling gradient electret transducers (see below).

Many countries are evaluating foil electret microphones as a potential replacement for the universal carbon transmitter (see, e.g. [7.27, 46]). Apart from the already mentioned advantages of electret transducers, their application in telephones is particularly attractive because of low production cost and relatively small power consumption of the preamplifier.

Another interesting application of foil-electret microphones is in *tablets* used for encoding graphic information. Two ultrasonic foil-electret strip microphones are placed perpendicular to each other along two edges of a writing surface. The signals from a pulse-emitting ultrasonic stylus (for example connected to a pencil) is received by both strip microphones. The delay of the received pulse at each microphone is proportional to the *x* and *y* coordinates of the pencil allowing a remote mechanical system to reconstruct its position. Such systems are being offered commercially as computer-input devices and are used experimentally for transmitting handwriting over normal telephone lines [7.47].

7.1.2 Directional Microphones

Directional electroacoustic transducers have many applications ranging from highly specialized acoustical measurements to a variety of general uses. Most directional foil-electret microphones depend on the gradient principle which can be utilized to design transducers with bidirectional, toroidal, and unidirectional patterns of first, second, and higher order [7.48]. While first-order microphones, particularly those with bidirectional and unidirectional patterns, are used extensively, only occasional consideration has been given to second-order microphones. Apart from such gradient systems, dimensional electret microphones (having diameters large compared to the wavelength) are used to obtain directivity in ultrasonic applications.

A first-order gradient microphone is obtained by exposing the diaphragm to the sound field from both sides. The design necessitates a partially open back cavity, while the back cavity in an omnidirectional unit is, with the exception of an air-equalization hole, closed. The restoring forces in the gradient unit are therefore due only to the stiffness and tension of the foil, while in the omnidirectional unit these forces are also due to the stiffness of the air in the cavity. The sensitivities and resonance frequencies are therefore expected to be different for the two microphones. Used as a close-talking microphone the first-order gradient discriminates against distant sound sources.

The response of an electret gradient microphone to air-borne sound is plotted in Fig. 7.6 for three different sound fields [7.35]. These are the nearly spherical field generated by an artifical mouth and far fields generated by a

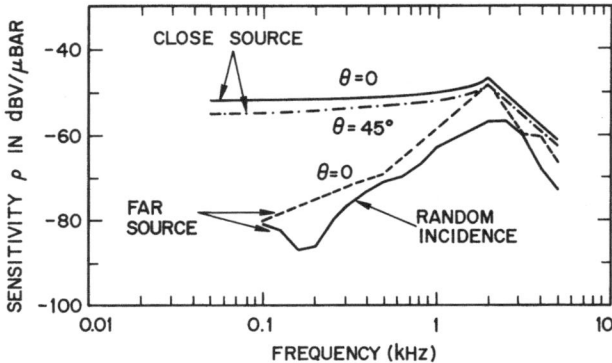

Fig. 7.6. Sensitivity of pressure-gradient electret microphone to sound waves generated by close source (artificial mouth, distance 1.25 cm) and by far source (loudspeaker) at various angles of incidence [7.35]

loudspeaker in an anechoic chamber ($\theta = 0$) and in an office room (random incidence). The low-frequency response is flat if the microphone is positioned close to the artificial mouth and the difference in sensitivity between $\theta = 0$ and $\theta = 45°$ is about 3 dB, as expected. For the far-field positions, the low-frequency response is proportional to frequency and the sensitivity for randomly incident sound is smaller by about 5 dB (the expected value is 4.8 dB) than the sensitivity in the direction $\theta = 0$.

Subminiature directional electret microphones utilizing the Teflon-coated backplate approach (see above) are now being widely utilized [7.49]. The directivity pattern is controlled in a simple manner by the selection of appropriate tubing attached to one side of the diaphragm, as shown in Fig. 7.7 for a cardioid system. Such microphones are now widely used as noise-canceling transducers in hearing aids. Similarly, other electret-gradient microphones are applied in a variety of devices such as operators headsets and aircraft-communication devices [7.50].

The implementation of a second-order toroidal microphone [7.51] is shown in Fig. 7.8. It consists in principle of two second-order transducers arranged at right angles to each other (left part of figure). Each of these second-order transducers is in itself a combination of two first-order transducers. It can be shown that such an array has a toroidal directivity pattern as depicted in the right part of the figure. This microphone was realized by placing a single foil-electret sensor in the center plane of a cylindrical cavity (center part of figure). The sound field is sampled with eight tubes, four going into the upper cavity and four into the lower. Thus, the required "positive" and "negative" sensitivities are realized. The inherent ω^2 dependence of the frequency response is compensated by placing the tube-cavity resonance at the lower end of the desired frequency band. Some of the measured polar patterns are shown in Fig. 7.9. A "distorted" toroidal pattern can be obtained by changing the spacing of the tube openings belonging to one of the second-order transducers. A unidirectional second-order microphone using a foil-electret sensor has also been realized [7.51].

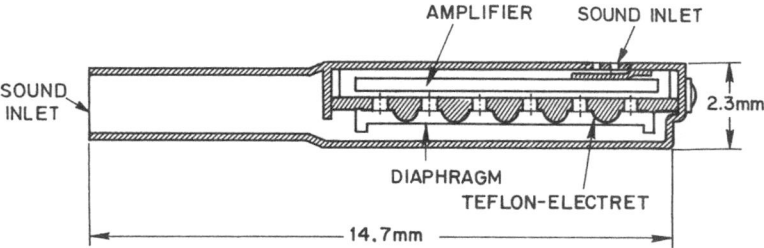

Fig. 7.7. Schematic cross section of a hearing-aid microphone with cardioid characteristic [7.49]

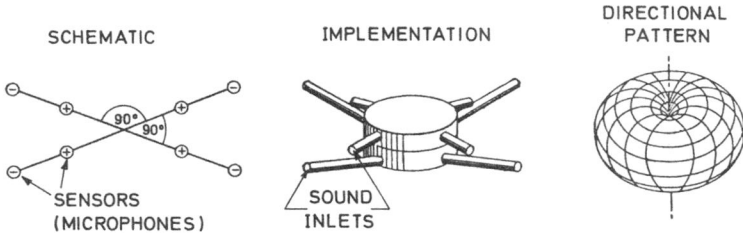

Fig. 7.8. Second-order toroidal microphone (perspective views). *Left part:* Schematic arrangement of sensors. *Center part:* Implementation of microphone, consisting of eight tubes to sample the sound field, and cylindrical cavity subdivided in center by foil-electret microphone. *Right part:* Toroidal directivity pattern. [7.51]

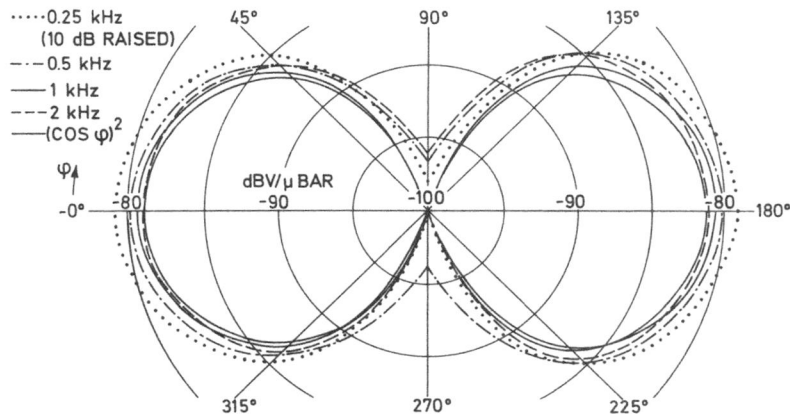

Fig. 7.9. Measured polar pattern of toroidal microphone shown in Fig. 7.8 center. Also shown is calculated $(\cos \varphi)^2$ dependence [7.51]

7.1.3 Headphones and Loudspeakers

Work on externally biased condenser transmitters with solid dielectric [7.24, 52] suggested the use of similar transducers with electret biasing as headphones, loudspeakers, and ultrasonic sources. Although the above microphone design can be utilized in this case, such transducers exhibit nonlinear performance of the diaphragm due to the quadratic dependence of the force F per unit area on the applied ac voltage V. This can be seen from (2.34), which may be rewritten as

$$F = -\tfrac{1}{2}(s\sigma_1 + \varepsilon\varepsilon_0 V)^2/\varepsilon_0(s + \varepsilon s_1)^2 . \tag{7.5}$$

In a well-designed earphone the nonlinearity is, however, not too severe under normal operating conditions. Low-priced systems of this kind are commercially available [7.53].

The nonlinearity can be overcome by use of a push–pull transducer [7.54, 55]. The principle of such a system is shown in Fig. 7.10a. A diaphragm consisting of two electrets, each metallized on one side and with these metal layers in contact, is sandwiched between two perforated metal electrodes, forming a symmetrical system. Application of a signal voltage $\bar{V}$ in antiphase to the two electrodes results in forces F_1 and F_2 on the diaphragm. These forces are given by (7.5) if $\pm\bar{V}$ is substituted for V. The net force $F = F_1 - F_2$ on the diaphragm centered between the electrodes is then

$$F = -2\bar{V}\sigma_1\varepsilon s/(s + \varepsilon s_1)^2 , \tag{7.6}$$

which is linear in $\bar{V}$, indicating the absence of nonlinear distortion.

An interesting modification of the push–pull transducer is possible with the use of nonmetallized monocharge electrets (see Fig. 1.2) [7.8, 55], as schematically shown in Fig. 7.10b. These transducers have the advantage of permitting large electrode–electret separations without reduction in the electric field in the air gap [7.8]. This is possible since, in the absence of a compensation charge on the electret, all the field lines originating in the electret terminate at image charges on the electrodes. With transducers of this kind, electret–electrode spacings of the order of 1 mm are customary. Such transducers allow correspondingly large dynamic electret deflections and are thus suitable for the generation of large-amplitude acoustic signals.

Headphones based on this principle are now commercially available [7.56]. These systems are designed with a low resonance frequency of about 50 Hz and are used in an open-air arrangement providing for wearing comfort. Their frequency response is flat within ±3 dB from 40 Hz to 20 kHz and their harmonic distortion is less than 1 % for a sound pressure level of 104 dB. Because of signal-input requirements, such headphones have to be operated with a transformer.

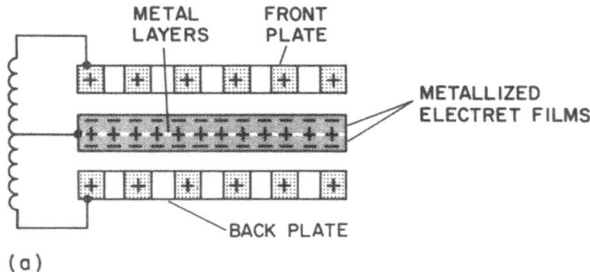

(a)

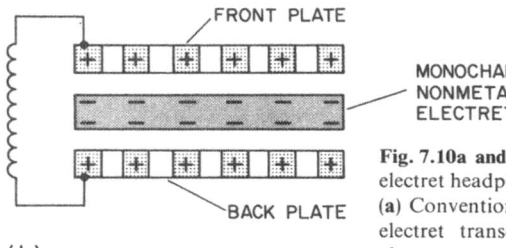

(b)

Fig. 7.10a and b. Schematic cross sections of push–pull electret headphones, showing static charge distributions. (a) Conventional double-electret transducer. (b) Single-electret transducer with nonmetalized monocharged electret

For telephone applications, sound-pressure levels of 90 dB need to be generated at the ear. Without a transformer, signal voltages of about 3 V are necessary to generate such levels with electret headphones [7.57]. This requirement has so far barred use of electret receivers in telephones.

Loudspeakers based on the electret principle have also been designed. As direct radiators, their response is rising up to the resonance frequency of the transducer. To improve the frequency response, horn loading has been used [7.58]. A horn-loaded 6 cm diameter push–pull transducer with electrets deposited on the fixed perforated electrodes has a frequency response within ±3 dB from 1.5 to 20 kHz and relatively high efficiency.

A reversible transducer based on the slot effect [2.5, 6] and particularly suitable as a loudspeaker has been described recently [7.59]. In its simplest implementation it consists of two adjacent and oppositely poled plate capacitors with a joint electret dielectric which is linked to a membrane. Such transducers provide linear transduction and the generated force is independent of location over a wide range of excursions.

7.1.4 Electromechanical Transducers

Apart from its use in electroacoustic transducers, such as microphones and earphones, the electret principle has recently been applied to electromechanical transducers. Examples are phonograph cartridges, touch or key transducers, impact transducers, display devices, and relays.

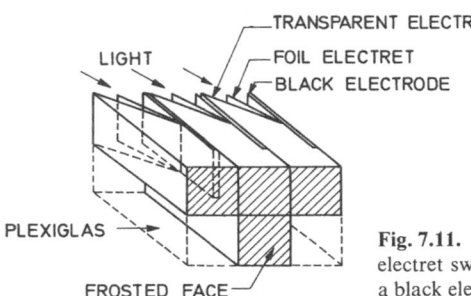

Fig. 7.11. Partial view of optical display using a foil electret switchable between a transparent electrode and a black electrode [7.64]

Two types of phonograph cartridges have been implemented. In one system [7.60], the stylus is coupled by a cavity to an electret microphone. The cavity serves as an acoustic transformer that converts the large-amplitude stylus vibrations to sound waves sensed by the microphones. Such cartridges have favorably low mass, are capable of high trackability and linearity over a wide frequency range, and yet, are immune to magnetic fields. Another design [7.61] consists essentially of a cantilever to which a stylus is attached, and two electrets. The electrets, together with the cantilever, form a pair of electret transducers actuated by the vibrations of the cantilever. Advantages of these cartridges are low intermodulation distortion and a dynamic range of 100 dB. Such cartridges are now commercially available.

Touch-actuated electret transducers similar in design to electret microphones have also been suggested for telephone dialing and data transmission [7.62, 63]. These transducers generate output signals of the order of 10–100 V due to direct or key operation resulting in a deflection of the diaphragm. Because of the small diaphragm deflections, tactile feedback is almost totally absent but can be mechanically provided.

Optical display devices with foil electrets have been designed for use in ambient-illumination, large-size character-display panels [7.64]. The forces between a hinged nonmetallized but opaque foil electret with a charge density of 10^{-10} C cm^{-2} and two widely spaced metal surfaces are utilized to open and close a light emitting window as indicated in Fig. 7.11. The electret which is attracted to one of the metal surfaces is switched in about 50 ms to the other plate by applying a voltage pulse of 500 V opposing the electric field. The drawback of this system is the high switching voltage required. However, the necessary currents are small and only flow during the switching operation.

Bistable electret arrangements have also been used to implement mechanical relays [7.65]. Such devices have power requirements 1000 times smaller than equivalent electromagnetic systems. Thus considerable power conservation over present magnetic transducers is possible.

Impact-sensitive electret transducers similar in design to the above-described microphones have been used for the detection of leaks in space stations [7.66]. Vibration transducers using electrets were suggested for monitoring machine and motor noises [7.67].

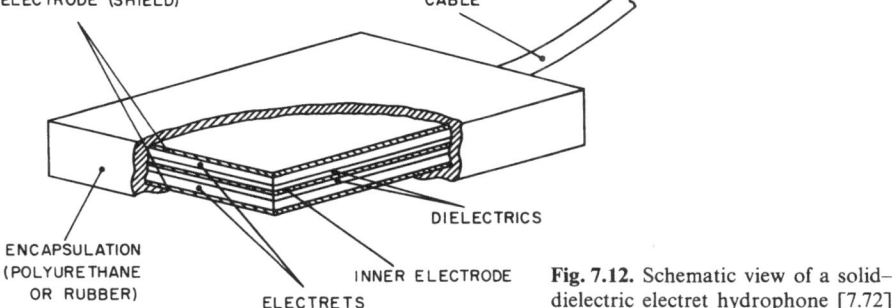

EXTERNAL
ELECTRODE (SHIELD)

COAXIAL
CABLE

DIELECTRICS

ENCAPSULATION
(POLYURETHANE
OR RUBBER)

INNER ELECTRODE

ELECTRETS

Fig. 7.12. Schematic view of a solid–dielectric electret hydrophone [7.72]

Electret line transducers [7.68, 69] consist essentially of a coaxial cable with a charged Teflon dielectric. In such transducers, the center conductor and the shield serve as electrodes. Mechanical excitation resulting in a deformation of the shield at any point along the length of such a cable produces an electrical output signal. Line transducers of this type provide continous sensitivity over the entire length and are in commercial use in security systems.

7.1.5 Underwater Transducers

The use of electrets in underwater transducers is of relatively recent origin. One of the first such systems was the above-discussed line transducer [7.68] which is also used as a hydrophone.

Recently, electret hydrophones based on two different principles have been described [7.70–72]. One system is essentially shaped after transducers for airborne sound [7.71], consisting of a Teflon electret separated from the metal backplate by an air gap. The air gap is maintained by an interposed dielectric mesh or by ridges on the back electrode. Such systems have sensitivities of -170 to -180 dB re 1 V/µPa, considerably larger than those of piezoelectric (PVDF) hydrophones (see Sect. 7.7.2). Drawbacks result from the properties of the air gap which causes an impedance mismatch in water and restricts the use of this transducer to relatively small sound-pressure levels and to shallow waters.

Another electret hydrophone is based on the use of a layered structure between external electrodes, as shown in Fig. 7.12 [7.72]. The inner layers represent a parallel arrangement of two transducers, each consisting of an electret and a dielectric layer of different tensile modulus. No air gaps are used in this system. Sound waves will cause different deformations of these layers and therefore generate an electrical output signal. The sensitivity of this hydrophone is within ± 3 dB from 1 Hz to 15 kHz and amounts to about -200 dB re 1 V/µPa; it is thus lower than that of the air gap system and corresponds to the sensitivity of PVDF transducers. Such hydrophones are, however, applicable to the detection of large-amplitude signals and to operation in a wide range of static pressures.

7.2 Electrophotography

The field of electrophotography comprises a number of different methods which combine photographic and photoelectric techniques to produce a permanent image of an object on paper. Electrophotography is to be distinguished from electrostatic recording which is based on signal, line, or character recording rather than on photographic image production. Electrostatic recording techniques are dealt with in Sect. 7.3.

Following the terminology of *Schaffert* [7.73] we classify the presently most important methods of electrophotography as xerographic, persistent conductivity, and photodielectric techniques. All these methods use charge-storage effects in one way or another. They are therefore within the scope of this chapter and will be described briefly in the following subsections. For a detailed discussion of these as well as a number of other methods the reader is referred to the special literature [7.73–78].

7.2.1 Xerography

The following steps are necessary in xerographic imaging: First, the surface of a photoconductive insulating plate is charged and then exposed to an optical image. Due to photoconductive discharge of the surface, a latent electrostatic image is formed on the plate which is developed with a pigmented powder. Thereafter the image is transferred and finally fused to paper. These steps will be described in somewhat more detail in the following.

The latent-image plate usually consists of a conductive substrate on which a thin layer of photoconductive material such as amorphous selenium or zinc oxide is deposited. The layer combines high dark resistivity (10^{14}–$10^{16}\,\Omega\,cm$) with high photosensitivity (exposure index 1 ASA or higher). To produce a print, the plate is first uniformly charged ("sensitized") in the dark by a corona discharge (see Sect. 2.2.3). Usually positive charge deposition is used for selenium surfaces and negative charge deposition for zinc oxide surfaces. As seen in Fig. 7.13, the plate potential cannot exceed a certain value, the acceptance potential, since the dark resistivity of the photoconductive coating decreases with increasing field and thus with increasing charge. The sensitizing of the plate is generally terminated at a potential somewhat below the acceptance potential.

In the absence of light, the surface charge on the photoconductor is subject to a dark decay. For amorphous selenium, the time constant of this decay depends on charge polarity, charge density, geometry, and deposition method and may vary from a few seconds to many hours. Upon light exposure, the charge decays (see Fig. 7.13) corresponding to the light pattern with a decay rate depending on light intensity and wavelength. Time constants are set to be of the order of a second for the highest exposure. Thus a latent electrostatic image is present in the form of a charge pattern.

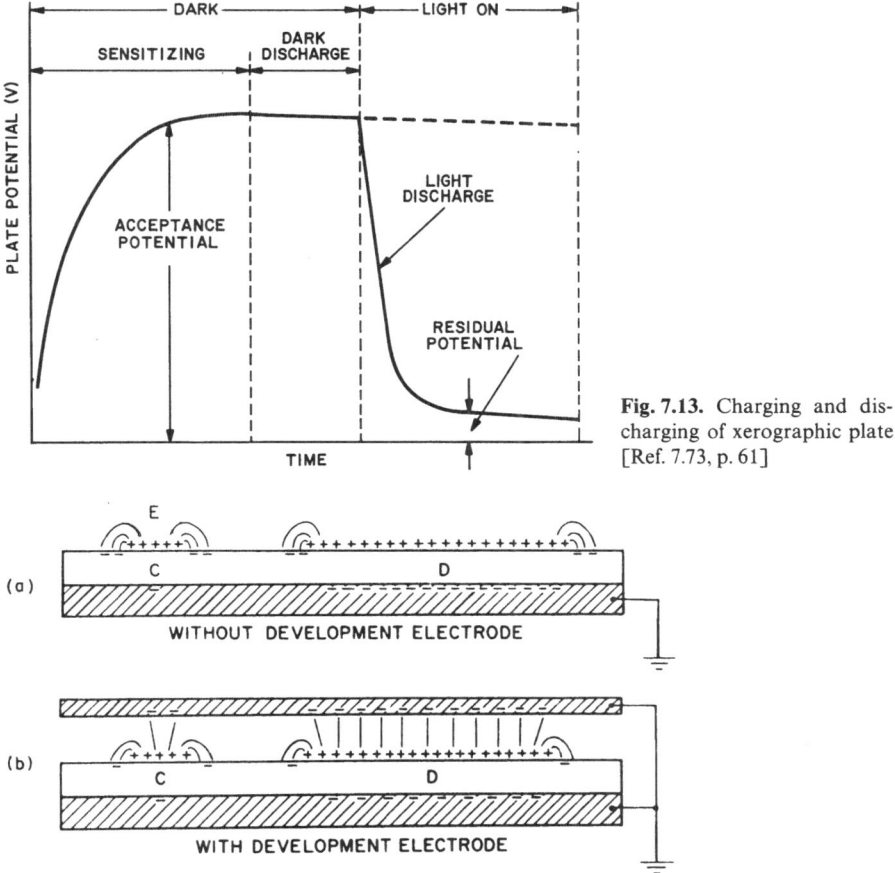

Fig. 7.13. Charging and discharging of xerographic plate [Ref. 7.73, p. 61]

Fig. 7.14a,b. Fields in vicinity of charged area. (a) Without development electrode. (b) With development electrode [Ref. 7.73, p. 35]

The latent image is developed by depositing pigmented particles onto the charged areas of the photoconductor [7.79]. This is most readily achieved in cascade development depending on triboelectrification of a mixture of two different components, namely a relatively coarse carrier and a fine powder (toner). If the toner assumes a polarity opposite to that of the charged surface, it will be attracted by the charged areas. Thus toner deposition is enhanced in the areas of large charge gradients where the field emanating from the surface is strongest, as shown in Fig. 7.14a. To make the deposition more nearly proportional to charge density, a development electrode parallel to, and at a small distance from, the image plane is used (see Fig. 7.14b). A bias may be applied to reduce spurious "background" toner deposition.

The image transfer from the photoconductive plate to paper can be performed electrostatically by placing the paper onto the powdered image and applying a corona discharge of the same polarity as the sensitizing discharge to the rear side of the paper. The charge deposited on the paper attracts the powder sufficiently so that it sticks to the paper upon its removal. An alternate transfer method depends on the use of an adhesive layer on the paper which picks up the powder from the image plate. If fusible toners are used, the image can be fixed on the paper by heating. Other fixing methods include the application of solvent vapors or coatings to the toner image.

The resolution achieved in xerography is determined by the material used for development, since the resolution of the latent image is very high. Typical resolutions are 1–15 lines/mm in cascade development and up to 250 lines/mm in liquid-dispersion development.

Two modifications of xerography have also attained great practical significance. One method, referred to as xeroradiography and used in medical diagnostics and nondestructive testing, depends on X-rays for radiological imaging of an object [7.80]. Apart from this, xeroradiography corresponds closely to xerography. Another method, known as the Electrofax process [7.81], employs photoconductive paper which is not only used for image formation and development but also for fixing of the image. Thus the image-transfer step is avoided.

7.2.2 Persistent-Conductivity Methods

These methods depend on delayed-conductivity effects which occur in some dielectrics after exposure to light or other forms of radiation (see Sect. 4.1). If such a material is exposed to an optical image, a conductivity pattern corresponding to the image is produced which persists in the dark following exposure. One obtains therefore a conductivity pattern (rather than the charge pattern in xerography) which can be used in several ways to generate a permanent print.

One such method consists of the deposition of a charge onto the conductivity image by means of a corona technique. Due to charge decay in the more conductive areas, a charge pattern forms which can be developed by one of the processes used in xerography.

In another method, which is used commercially, the latent image is directly developed by an electrolytic method with the image plate serving as an electrode. Since electrolytic deposition occurs only in the conductive areas, a permanent image is formed on the plate.

7.2.3 Photodielectric Processes

Photodielectric processes differ from xerography in the techniques used for electrostatic-image formation and retention [7.82]. In one such method, illustrated in Fig. 7.15 a photoconductor is pressed against a dielectric surface leaving a small air gap. A voltage applied across this sandwich will cause

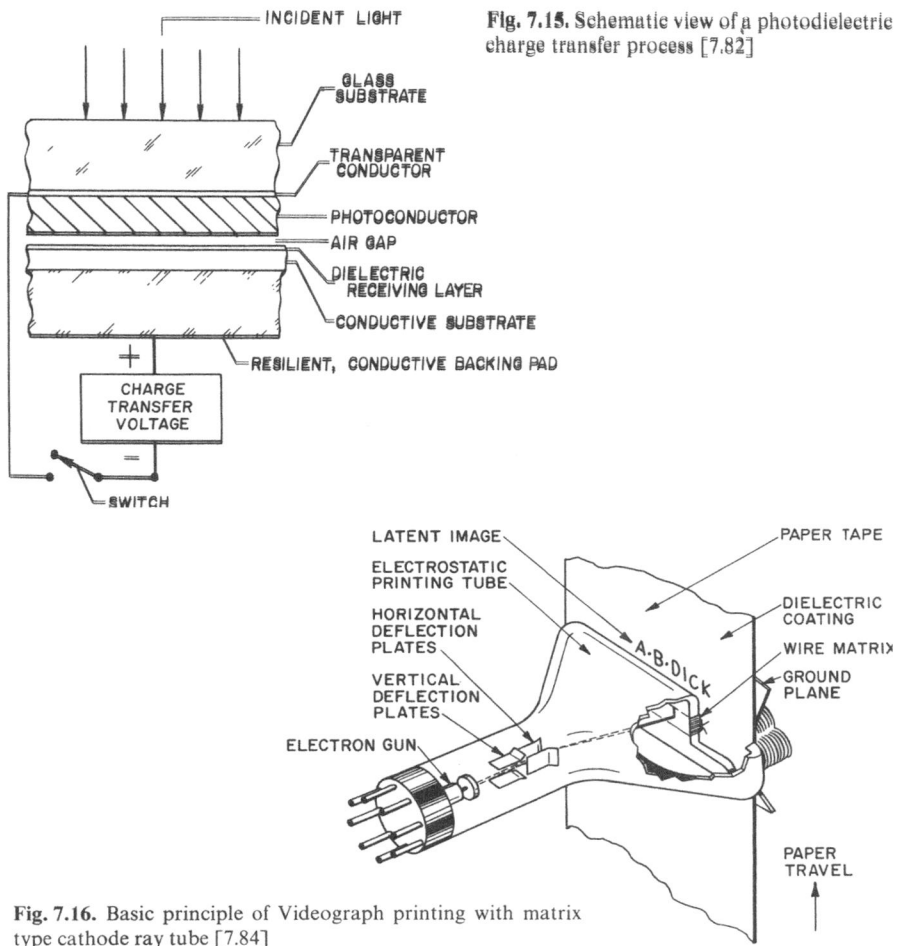

INCIDENT LIGHT

GLASS SUBSTRATE

TRANSPARENT CONDUCTOR

PHOTOCONDUCTOR

AIR GAP

DIELECTRIC RECEIVING LAYER

CONDUCTIVE SUBSTRATE

RESILIENT, CONDUCTIVE BACKING PAD

CHARGE TRANSFER VOLTAGE

SWITCH

Fig. 7.15. Schematic view of a photodielectric charge transfer process [7.82]

LATENT IMAGE

ELECTROSTATIC PRINTING TUBE

HORIZONTAL DEFLECTION PLATES

VERTICAL DEFLECTION PLATES

ELECTRON GUN

A·B·DICK

PAPER TAPE

DIELECTRIC COATING

WIRE MATRIX

GROUND PLANE

PAPER TRAVEL

Fig. 7.16. Basic principle of Videograph printing with matrix type cathode ray tube [7.84]

discharge in areas where the resistivity of the photoconductor has been lowered by projection of an optical image. The resulting charge-deposition pattern on the dielectric can then be developed with a liquid toner. Details of this and other methods are found in the literature [7.73].

7.3 Electrostatic Recording

A variety of electrostatic techniques can be used to record electrical signals, digital information, facsimile, or alpha-numeric characters on carrier materials such as paper or plastic films [7.83]. These techniques differ from the methods described in Sect. 7.2 in the sense that they are not based on photographic imaging. Electrostatic recording processes belong to the category of electrographic methods [7.73] which also encompass electrolytic, magnetic and other techniques which are not of interest in the context of charge-storage phenomena.

One of the most versatile tools to deposit charge patterns onto insulating surfaces is the electron beam, which can be employed in a number of ways in electrostatic recording. In a commercially used system known as the video-graph [7.84], schematically shown in Fig. 7.16, a cathode-ray tube is used which has embedded in its face, and extending through it, a large number of parallel wires insulated from each other. The outside of the face of the cathode-ray tube is contacted by a paper tape coated with a high-resistivity dielectric. The wires conduct the charges of the electron beam to the dielectric where they are deposited. Thus, by controlling modulation and sweep of the electron beam, a latent image consisting of a charge pattern is formed, which may be developed with a powder process as described in Sect. 7.2.1.

Electron beams have also been used experimentally to write charge patterns directly onto dielectrics [7.85, 86], similar to the methods described above [7.9]. In one recent experiment [7.86], an approximately 3×3 mm test pattern was deposited onto a 25 µm thick Teflon film by a computer-controlled 10 keV electron-beam exposure system having an electron-spot diameter of 0.5 µm. The latent image, being a static charge pattern, can be made visible by a number of methods. Examination is possible, for example, with a scanning electron beam by monitoring secondary emission from the sample. To read, yet preserve the pattern, a beam of low current (10^{-12} A) and low energy (1 keV) has to be used. Under these conditions, a resolution of 100 line pairs/mm is achieved. Other reading methods depend on xerographic techniques, such as cascade and liquid (electrophoretic) development, the latter yielding the same resolution as electron-beam reading. The use of Teflon as a substrate ensures retention of the charge pattern over long periods of time (see Sect. 2.6.5).

Other widely used electrostatic-recording techniques employ electrical discharge methods at atmospheric pressure [7.87]. In most of those processes of interest in the context of charge-storage phenomena, a discharge between a pointed electrode or an arrangement of such electrodes and a metal-backed paper sheet with high-resistivity coating is generated (see Fig. 7.17) which deposits a circular dot of charge or a number of such dots on the paper. The size of this dot is controlled by the electrode-paper distance which for high-resolution printing is usually chosen to be of the order of 100 µm or less. For the printing of alpha-numeric characters, a row of print heads consisting of a matrix of wire electrodes are used. By simultaneously pulsing proper combinations of such electrodes in each head, the desired charge patterns along one line on the paper can be generated. Further lines are deposited by advancing the paper and repeating the pulsed-discharge process. Development of the charge pattern is achieved by processes similar to those used in xerography. Discharge methods employing other kinds of print heads are used for facsimile recording and oscillograph recording.

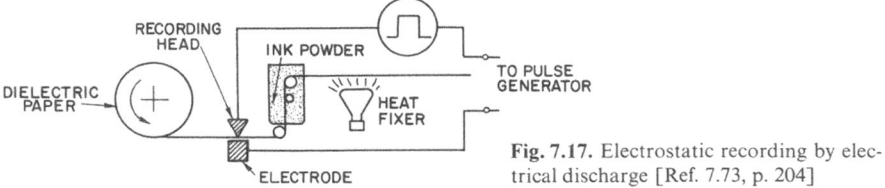

Fig. 7.17. Electrostatic recording by electrical discharge [Ref. 7.73, p. 204]

A somewhat different method has been suggested for the one-dimensional recording of electrical signals on polymer tapes [7.88, 89]. Simultaneous application of a dc voltage through a roller and an ac signal from a stylus results in discharges which deposit onto the tape permanent charges corresponding to the ac signal. The charge pattern can be read nondestructively with a capacitive-probe arrangement.

7.4 Electret Air Filters

The demand for high-efficiency filters which remove submicron particles from polluted gases has led to the development of electret filters consisting of charged polymer fibers [7.12]. The theoretical aspects of particle deposition on such fibers can be adopted from a related analysis valid for externally polarized fibers [7.90].

The electret filters are made from polypropylene film which is first stretched to increase its mechanical strength and then charged by application of a corona discharge. Thereafter, the 9 μm thick film is fibrillated. Due to repulsive forces the fibers spread into a broad web.

The capture of particles by electret fibers depends on Coulomb and induction forces. While the former attract charged particles, the latter act on neutral ones due to the induction of dipoles. Since the operation of the electret filter is based on long-range electrical forces, it can have an open structure and can be made from fibers much thicker, and at much greater spacing, than the particles to be filtered. This results in a low-pressure drop across the filter and high efficiency.

Experimental results are shown in Fig. 7.18 which compares the penetration through, and pressure drop across, an electret filter with the corresponding quantities of a commercial glass filter during loading with submicron sodium chloride aerosol at a filtration velocity of $20\,\text{cm}\,\text{s}^{-1}$. While the penetration through both filters is almost the same, namely about 0.3 %, the pressure drop across the electret filter is 4–20 times less, thus cutting the energy used in the filtration process. The dependence of the penetration on filtration velocity is weak up to velocities of $20\,\text{cm}\,\text{s}^{-1}$. Further experiments show nonelectret filters to have a penetration 80 times less than nonelectret filters of otherwise the same material and structure.

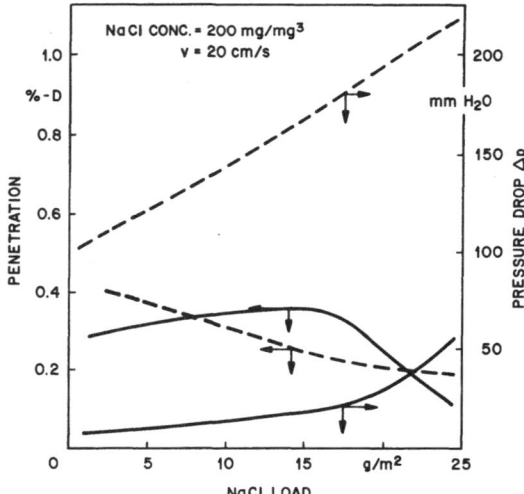

Fig. 7.18. Comparison of penetration-load curve and pressure drop of an electret filter (335 g·m⁻², solid lines) and a commercial glass fiber filter (95 g·m⁻², dashed lines) [7.12]

It is noteworthy that electret filters are not significantly discharged by humid or ionized air or by temporary immersion in water or alcohol. Even exposure to 80 °C for up to 100 days results only in a drop of the efficiency from 99.5 to 92 %. Electret filters are now commercially available.

7.5 Electret Motors and Generators

An experimentally attractive electrostatic device is the corona motor [7.13] which uses voltages up to 2000 V to corona-charge an insulating rotor. Such motors are, however, impractical because of the required high voltage and the strong dependence ·on atmospheric conditions. These limitations were overcome by the use of electrets [7.91, 92].

One such electret motor, shown schematically in Fig. 7.19, utilizes the slot effect [7.91]. In this motor two circular disk electrodes, each subdivided in half by an insulating strip, are placed on opposite sides and close to a circular disk electret of equal diameter. The electret is composed of two oppositely polarized half-disks. When a voltage V is applied to both sets of electrodes, as shown in the figure, the electret experiences a torque[1]

$$T = \frac{(1 + d/t)}{(1 + \varepsilon d/t)^2} \, \sigma (r_0^2 - r_i^2) V , \tag{7.7}$$

where d is the total air gap thickness on both sides of the electret, t is the electret thickness, ε is the dielectric constant of the electret material, σ is the electret's effective surface-charge density, and r_0 and r_i are the outer and inner radii of the electrodes.

---

1 The validity of (7.7) has been discussed recently, see [8.496].

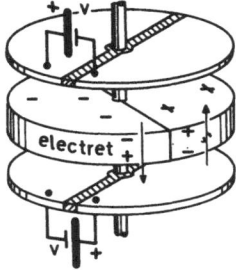

Fig. 7.19. Schematic diagram of slot-effect motor [7.91]

Slot-effect motors appear to be particularly useful in systems where power up to 1 W is needed. They have been used intermittently for periods of over two years without failure. Some weakening of the field of the shielded wax electrets was observed after several hours of operation but the field recovered after a rest period of comparable length. Polymer electrets could lead to improvements in stability and to a reduction in the electret's thickness dimension.

Electret generators, have been suggested [7.93] for use in power engineering and, more recently, for use in tachometers [7.31].

7.6 Electret Dosimeters

Of the many different approaches to radiation dosimetry [7.15], only the methods based on charge-storage phenomena and radiation-induced conductivity in dielectrics are of interest in the present context. For a discussion of these methods, the reader is referred to Chap. 4. Specifically, dosimetry with self-biased systems (dielectrics with an excess charge or a uniform volume polarization present prior to irradiation or with space-charge layers generated by the irradiation) is described in Sect. 4.7.1, while dosimetry with externally biased systems is reviewed in Sect. 4.7.2.

7.7 Piezoelectric Polymer Transducers

The discovery of a strong piezoelectric effect in polyvinylidene fluoride (PVDF) [7.94] recently led to a number of significant applications of this material in electroacoustic and electromechanical transducers [7.17]. The piezoelectric d- and g-constants of PVDF as well as a number of other constants of this polymer are compared in Table 7.1 with data of other polymers, piezoelectric ceramics and crystals (see also Table 1.1 for related data).

It is evident that the piezoelectric d-constant and the pyroelectric coefficient of the polymer materials are considerably lower than the corresponding constants of the ceramic materials. Of great importance for the use in

Table 7.1. Piezoelectric, pyroelectric, and other constants of polymers and other materials [7.19, 95]

Material	Density	Dielectric constant	Piezoelectric constants		Pyroelectric coefficient
			d	g	p
	$[g\,cm^{-3}]$		$[10^{-12}\,C/N]$	$[10^{-3}\,Vm/N]$	$[nC\,cm^{-2}K^{-1}]$
PVDF	1.76	11	$20\,(d_{31})$	$200\,(g_{31})$	4
PVF	1.38	5	$1\,(d_{31})$	$20\,(g_{31})$	1.0
PVDF-TEE Copolymer	1.85	9.6	$12.1\,(d_{31})$	$140\,(g_{31})$	2.7
PZT-5	7.75	1700	$171\,(d_{31})$	$11\,(g_{31})$	6–50
BaTiO$_3$	5.7	1700	$78\,(d_{31})$	$5\,(g_{31})$	20
Quartz	2.66	4.5	$2\,(d_{11})$	$50\,(g_{11})$	—
TGS	1.7	50	—	—	35

transducers is, however, the electromechanical coupling coefficient $k = \sqrt{dg/c}$, where c is the elastic compliance. The two coupling coefficients of interest, k_{31} and k_{33}, equal 0.16 and 0.20 for uniaxially stretched PVDF [7.96], as compared to 0.34 and 0.70 for PZT-5. In spite of their smaller coupling coefficients, the polymer materials are frequently attractive for use in piezoelectric and pyroelectric devices because of their excellent mechanical properties, such as flexibility and ruggedness, their availability as thin films (down to 2 μm thickness), and their low cost.

7.7.1 Transducer Principles

Piezoelectric transducers are either based on the longitudinal or on the transverse piezoelectric effects. While the former is mostly employed in ultrasonic transducers, the latter has found broad applications in the audio-frequency domain. Transducers for airborne sound utilizing the transverse effect require a mechanical transformation to reduce the high mechanical impedance of the piezoelectric materials. In the case of ceramic materials this transformation is often achieved with bimorph arrangements, combining two wafers of the material in a way to produce a bendertype motion.

While this principle has also been applied to polymer transducers (see below) these materials allow, due to their flexibility, the more advantageous arrangement shown schematically in Fig. 7.20 [7.17]. If an electrical signal is applied between the electrodes of the plane polymer film shown in Fig. 7.20a the film vibrates in a transverse direction. Better coupling to the air can be achieved by the pulsator arrangement shown in part b of the figure, where the film is given a cylindrical or convex curvature and clamped at the two straight edges. Due to the shape of the film, the transverse motion is now converted into a pulsating vibration which strongly couples to the air.

The relations between voltage and sound pressure follow from a combination of the piezoelectric relations and the equation of motion of the

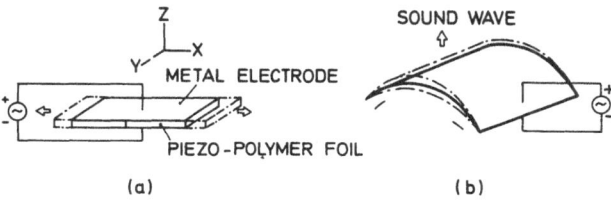

Fig. 7.20a, b. Schematic diagram showing (a) transverse motion of plane piezopolymer element and (b) conversion of transverse movement into pulsating motion by cylindrical element. z axis: direction of voltage application; x axis: direction of elongation [7.17]

membrane. The static (or low-frequency) relations for the transverse mode of a piezoelectric polymer material are approximately given by

$$S_1 = c_{11} T_1 + d_{31} E_3, \tag{7.8}$$

$$D_3 = d_{31} T_1 + \varepsilon_{33} E_3, \tag{7.9}$$

where S_1 is the strain, c_{11} the compliance, T_1 the stress, E_3 the electric field, D_3 the electric displacement, and ε_{33} the dielectric constant. The indices indicate the components of the vectors or tensors. The equation of motion, i.e., the relation between sound pressure and deflection or strain, is dependent upon membrane geometry, material constants, and parameters of the sound field. Discussion of this relation is beyond the scope of this volume. Some results of analytical studies of this kind on microphones will be discussed in Sect. 7.7.2.

7.7.2 Transducers Using Transverse Piezoelectric Effect

A schematic drawing of a commercial headphone operating on the pulsator principle explained above is shown in Fig. 7.21 [7.97]. In this transducer, the 8 μm thick PVDF film is placed over a polyurethane backing which gives the film the desired curvature and also provides sufficient mechanical damping to suppress undesirable resonances. Such transducers have, compared to electrostatic systems, the relatively large capacitance of $1.5\,\mathrm{nF\,cm}^{-2}$ (total capacitance about $0.1\,\mu F$).

The frequency response of a piezoelectric headphone, measured with an artificial ear, is shown in Fig. 7.22. A signal voltage of 3 V without a step-up transformer has been applied to the transducer for these experiments. Nonlinear distortions do not exceed 1 % at sound pressure levels of 110 dB.

The same principle has been used for the design of a commercial omnidirectional tweeter system [7.17, 97], as shown in Fig. 7.23. In this case, a full cylindrical PVDF element is used to obtain an omnidirectional radiation pattern in the horizontal plane. To cover the frequency range from 2 to 20 kHz with variations of less than ± 2 dB, two units of different size (radii 3 and 6 cm), one mounted above the other, are employed. A voltage of 2 V and a step-up transformer are needed to generate a sound-pressure level of 90 dB at 1 m

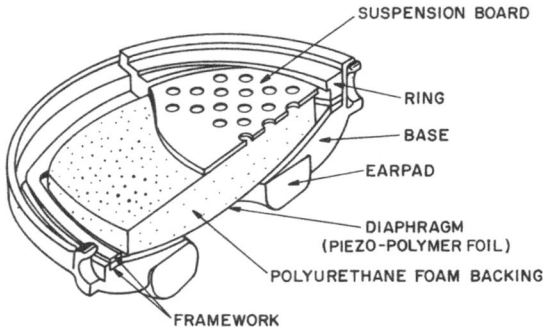

Fig. 7.21. Cross-sectional view of piezopolymer headphone [7.17]

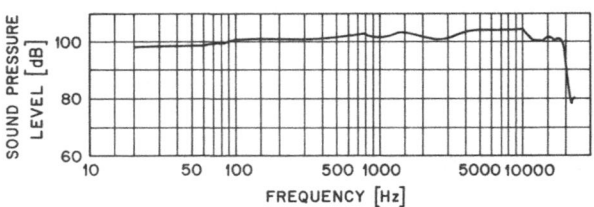

Fig. 7.22. Frequency response of piezopolymer stereophonic headphone measured with B & K artificial ear (type 4153) when input voltage is 3 V without step-up transformer [7.17]

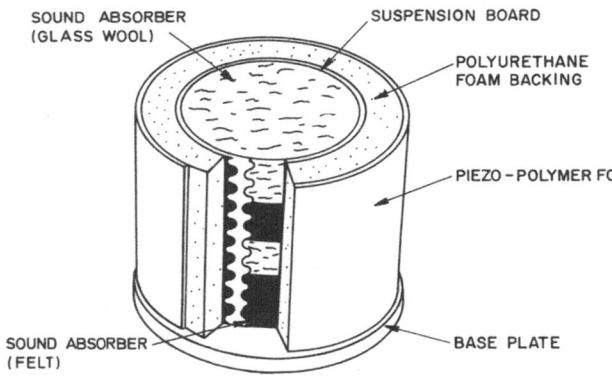

Fig. 7.23. Structure of piezopolymer tweeter [7.17]

distance. The harmonic distortion of such systems is low and an accurate transient response is obtained.

Also based on the curved-diaphragm arrangement is a piezoelectric polymer microphone whose design is similar to that of the earphone shown in Fig. 7.21 but smaller in size [7.17]. Although the sensitivity of such microphones is only in the range of -70 to -80 dB, relative to a sound-pressure level of 1 μbar, the capacitance is relatively high (about $400 \, \text{pF cm}^{-2}$) so that the equivalent noise level is as low as 10 dB SPL. Piezoelectric microphones can be simplified by using a self-supporting dome-shaped structure [7.98].

A number of investigations of piezopolymer microphones with spherical or cylindrical diaphragm geometry were carried out recently [7.99–101], of which two deserve particular attention. In one study [7.99], sensitivity and resonance

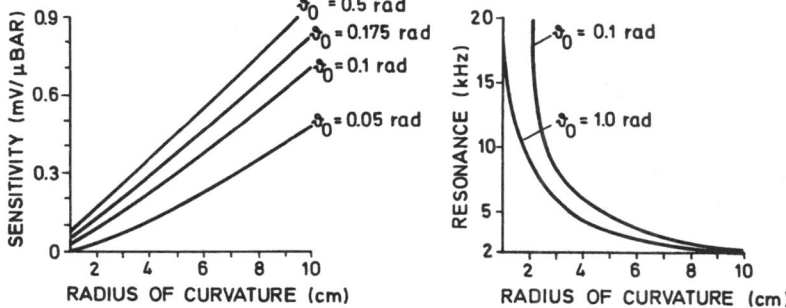

Fig. 7.24. Sensitivity and resonance frequency of spherical piezopolymer microphones as function of radius of curvature with opening angle as parameter [7.99]

frequency of spherical, self-supporting piezopolymer membranes were analyzed numerically by taking the elastic properties of the film fully into account. Some of the results are shown in Fig. 7.24. It appears that the sensitivity rises almost proportionally with the radius of curvature of the membrane, while the first resonance decreases. Both quantities are relatively independent of the opening angle. Studies of this kind allow one to optimize the diaphragm shape.

The same parameters were also investigated under simplifying assumptions for cylindrical membranes [7.100] with results similar to those obtained for spherical geometry. At the same time, a microphone with extremely small vibration sensitivity was suggested. This design, shown in Fig. 7.25, is based on the use of two cylindrical membranes arranged in a way that oppositely polarized surfaces face each other. By connecting the inner electrodes and taking the output between the outer electrodes, the system has an acoustical sensitivity of about $-72\,\mathrm{dB}$ re $1\,\mathrm{V}/\mu\mathrm{bar}$ or about twice that of a single-membrane unit but $20\,\mathrm{dB}$ lower vibration sensitivity.

Two other interesting uses of curved-diaphragm transducers have been suggested recently. In one application [7.102], a circular PVDF film, mounted at its edge, is spherically deformed by a pressure difference between its two surfaces. The deformation generates an acoustical focus which can be scanned axially by variation of the pressure difference. Such variable-focus transducers may be of use in medical ultrasonics. In the other application [7.103], curved-diaphragm transmitters operating in the $10\text{--}20\,\mathrm{kHz}$ range are mounted on the hull of marine vessels. The acoustic radiation emanating from these transducers prevents the deposition and growth of marine life on the hull. To achieve such antifouling action, only about $1\,\%$ of the hull has to be covered with transducers.

Apart from the pulsator designs discussed above, polymer transducers have also been implemented as bimorphs. An example is a microphone consisting of two $25\,\mu\mathrm{m}$ PVDF films polarized in opposite directions and electrically connected in series [7.104]. The $6\,\mathrm{mm}$ diameter films are rolled together and mounted in a self-supporting, edge-clamped configuration, similar to other

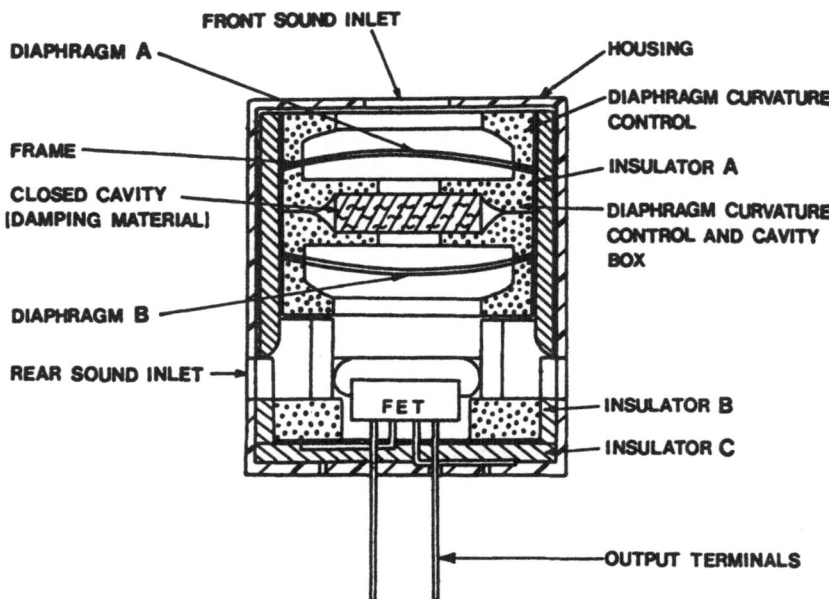

Fig. 7.25. Cross-sectional view of noise-cancelling microphone using two piezopolymer diaphragms [7.100]

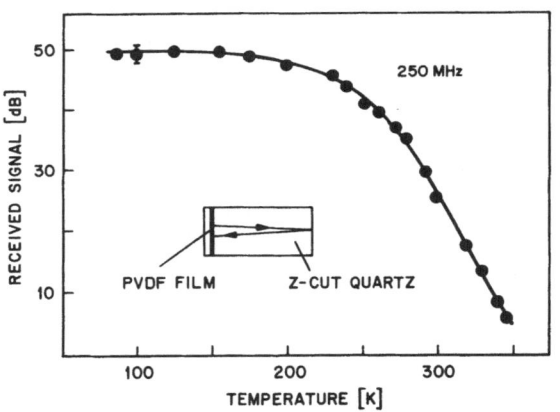

Fig. 7.26. Relative increase of the received signal generated and detected by PVDF transducer, as a function of the temperature. Transducer operated in resonance (250 MHz) [7.107]

bimorph transducers. Since this particular microphone has been designed for telephone use it has a frequency response and overall sensitivity (including amplification) which is close to that of the carbon transmitter.

Another bender-type transducer with piezofilm has been suggested for underwater use [7.105]. In this hydrophone, a PVDF film is mounted on a flexing circular plate which serves as a support and also guarantees linear operation. The sensitivity can be increased by connecting several pieces of polymer film, mounted on two support plates, in series. Combinations of 8 films yield a flat response within the frequency range from 2 Hz to 1 kHz with a sensitivity of about −200 dB re 1 V/μPa.

The transverse piezoeffect in polymers has also been employed in electromechanical transducers, such as phonograph cartridges [7.17]. Utilizing 1 × 2 mm strips of PVDF as the transducing elements, a high-compliance vibrating systems of high resonance frequency is obtained. The 45 kHz frequency response makes such cartridges suitable for playback of CD-4 records. Another electromechanical application of the transverse piezoelectric effect in polymers has been suggested in coin sensors [7.106]. Such sensors have 2–3 times greater sensitivity and allow simpler geometries than corresponding PZT devices.

7.7.3 Transducers Using Longitudinal Piezoelectric Effect

The longitudinal piezoelectric effect is of interest in ultrasonic applications where the wavelength is comparable to the thickness dimensions of the transducer. However, some electromechanical transducers also use the longitudinal effect.

It was shown [7.107, 108] that PVDF exhibits a strong piezoelectric effect in the ultrasonic range and that the effect increases with decreasing temperature down to 80 K, as shown in Fig. 7.26. Pulse-echo measurements with 25 and 50 μm thick films coated and poled in the usual manner and cemented to a quartz rod showed characteristic resonances whenever the film vibrates in a thickness mode. Such experiments were initially performed up to frequencies of 500 MHz.

More recent experiments [7.109] demonstrated, however, that two-dimensionally stretched PVDF can be used for the generation and detection of ultrasonic waves up to frequencies of about 9 GHz. The sound waves were generated and detected by a 12 μm nonmetallized PVDF film positioned in a microwave cavity. The film was bonded to a corundum cylinder whose temperature was kept at 1.3 K and served as the propagation medium.

Because of their relatively low acoustic impedance, which is comparable to that of water, PVDF transducers operated in the thickness mode are ideally suited for transducer applications in water [7.110]. For example, arrays with 200 parallel strip electrodes for underwater imaging have been constructed and evaluated [7.111]. Another application based on the low acoustic impedance is an acousto-optic light deflector which consists of a PVDF transducer bonded to a polymethyl methacrylate (PMMA) parallelepiped [7.112]. Light propagating in the PMMA block can be totally diffracted by the ultrasonic wave at a sufficiently high acoustic-power level.

Devices operating with electromechanical transducers based on the longitudinal piezoeffect in polymers have also been described. Among these are a typewriter [7.113], various touch dials [7.114–118], and an impact-sensitive device [7.119].

7.8 Pyroelectric Polymer Devices

7.8.1 Pyroelectric Response

Apart from their piezoelectric properties, a number of polarized polymers, such as PVDF and PVF, also exhibit a pyroelectric response [7.95] (see Table 7.1). The pyroelectric coefficients of these materials are about an order of magnitude smaller than those of such widely used substances as triglycine sulfate (TGS) and lanthanum modified lead zirconate titanate (PLZT) [7.19, 120]. However, the polymers are attractive for pyroelectric applications because of their excellent mechanical properties, their availability as thin films, and their small thermal conductivity.

As opposed to other thermal detectors, pyroelectric materials exhibit a current response that depends on the rate of change of temperature rather than on temperature directly. Thus, they are sensitive to chopped or modulated electromagnetic (in particular infrared) radiation. They have maximum response at angular frequencies above the inverse of the thermal relaxation time τ_T of the device. The voltage responsivity (output voltage to power input) at $\omega \gg 1/\tau_T$ is given by [7.18]

$$\gamma = Ae|Z|p/C_T, \tag{7.10}$$

where A is the electroded area, e the fraction of the incident power which thermalizes in the detector, $|Z| = R/\sqrt{1+(\omega RC)^2}$ the impedance of the detector-load circuit, p the pyroelectric coefficient, and C_T the thermal capacity of the pyroelectric material. The bandwidth extends from $\omega_1 = 1/\tau_T$ to $\omega_2 = 1/RC$ and thus increases with decreasing R and C while the response within the passband increases with R. The ultimate upper cutoff frequency is given by the thermalization rate of the absorbed radiation within the detector. For absorbing electrodes this is determined by the thermal diffusion time which, depending on materials and geometry, is in the microsecond to subnanosecond range, while for transmitting electrodes the energy is directly absorbed by the pyroelectric material resulting in shorter relaxation times.

Pyroelectric polymers have been suggested for use as detectors of electromagnetic (in particular infrared) radiation [7.121, 122], as sensitive materials in vidicons [7.19, 123] and for application in an electrostatic copying process [7.124] (see also [7.18]). Although few of these applications are presently used commercially, they all appear to be of sufficient interest to warrant a brief discussion.

7.8.2 Applications

A schematic view of a pyroelectric polymer *detector* is shown in Fig. 7.27 [7.122]. The PVDF film, available in thicknesses down to 2 μm, is cemented to

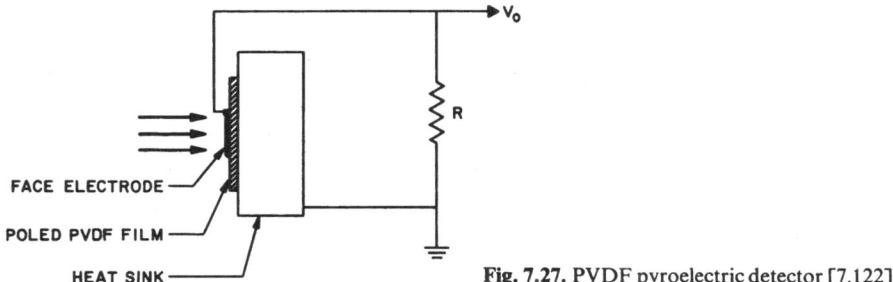

Fig. 7.27. PVDF pyroelectric detector [7.122]

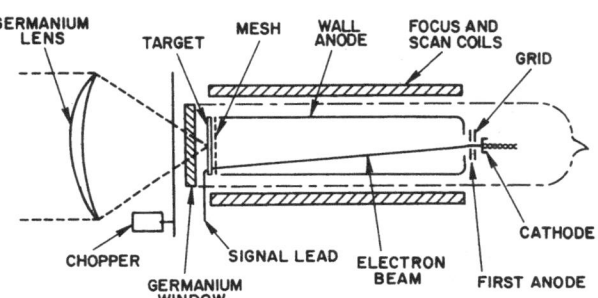

Fig. 7.28. Schematic cross section of pyroelectric vidicon [7.126]

a metal heat sink. The thermal capacity of the film and the thermal conductance to the heat sink determine the thermal relaxation time. Face electrodes which are either highly absorbing or partially transmitting for the incident radiation are used. A number of other configurations are also in practical use [7.120]. Although inferior in signal-to-noise ratio to TGS or other devices, pyroelectric polymer detectors are preferred in many applications because of their simplicity and superior mechanical properties. Such systems are presently marketed as inexpensive detectors of faulty steam traps and building insulation [7.125]. Arrays of pyroelectric detectors consisting of up to 50 elements have also been designed [7.114].

An experimental pyroelectric *vidicon* is schematically shown in Fig. 7.28. Thermal (infrared) radiation from a scene to be recorded is imaged through a chopper onto an electrode covering one side of a pyroelectric substance, e.g., a suitable polymer film [7.126].

A temperature pattern is generated which, due to the pyroelectric response, results in a voltage pattern on the nonmetalized side of the polymer film. This can be interrogated with a scanning electron beam, resulting in the generation of an output signal on the electrode [7.19].

PVDF vidicons have a smaller response than similar TGS devices and the possibility of noise introduced by charging effects by the electron beam exists. However, due to the small thermal diffusivity of PVDF, it yields better thermal resolution than other pyroelectric materials.

In the electrostatic *copying process* [7.124], a visible light image of the object is projected onto a piezoelectric polymer such as PVDF. The exposure generates a temperature distribution which results in a corresponding charge distribution in the sense that the hotter regions generate a negative charge while the cool regions develop the (relative) positive charge. The latent charge image may be developed with a toner and transferred to paper by standard techniques. Since this method is inherently wavelength independent it can be more readily adapted to color copying than photoconductive processes.

Acknowledgement. The authors are grateful to Mrs. E. Kubli for her assistance in compiling the literature list for this chapter.

References

7.1   S. Nishikawa, D. Nukijama: Proc. Imp. Acad. (Tokyo) **4**, 290 (1928)
7.2   A. Gemant: Philos. Mag. **20**, 929 (1935)
7.3   R. T. Rutherford: U.S. Patent 2,024,705 (1935)
7.4   Bogen catalogue (1939) p. 16
        W. Bruno: U.S. Patent 2,284,039 (1942)
7.5   F. Gutmann: Rev. Mod. Phys. **20**, 457 (1948)
7.6   G. M. Sessler, J. E. West: J. Acoust. Soc. Am. **34**, 1787 (1962)
7.7   G. M. Sessler, J. E. West: J. Acoust. Soc. Am. **40**, 1433 (1966)
7.8   G. M. Sessler, J. E. West: J. Acoust. Soc. Am. **53**, 1589 (1973)
7.9   P. Selenyi: U.S. Patent 1,818,760 (1931); J. Appl. Phys. **9**, 637 (1938)
7.10   C. F. Carlson: U.S. Patent 2,221,776 (1940)
7.11   R. M. Schaffert: *Electrophotography* (Wiley, New York 1975)
7.12   J. van Turnhout, C. van Bochove, G. J. van Veldhuizen: Staub. Reinhalt. Luft **36**, 36 (1976)
7.13   O. D. Jefimenko: "Electrostatic Motors", in *Electrostatics and its Applications*, ed. by A. D. Moore (Wiley, New York 1973) pp. 131–142
7.14   V. A. Andryushchemko: Autom. Remote Control (USSR) **21**, 93 (1960)
        G. Dreyfus, J. Lewiner: J. Appl. Phys. **46**, 4357 (1975); Electronics, 55 (1977)
7.15   G. J. Hine, G. L. Brownell (eds.): *Radiation Dosimetry* (Academic Press, New York 1956)
        F. H. Attix, W. C. Roesch (eds.): *Radiation Dosimetry* (Academic Press, New York 1968)
7.16   E. Fukada: In Proc. 6th Int. Cong. Acoust., Tokyo, ed. by Y. Kohasi (Acoustical Society of Japan; Tokyo 1968) paper D-3-1; Ultrason. **6**, 229 (1968)
7.17   M. Tamura, T. Yamaguchi, T. Oyaba, T. Yoshimi: J. Audio Eng. Soc. **23**, 21 (1975)
7.18   M. E. Lines, A. M. Glass: *Principles and Applications of Ferroelectrics and Related Materials* (Clarendon Press, Oxford 1977)
7.19   L. E. Garn, E. J. Sharp: IEEE Trans. PHP-**10**, 208 (1974)
7.20   P. V. Murphy, S. Merchant: In *Electrets, Charge-Storage, and Transport in Dielectrics* (Electrochemical Society, Princeton, N.J. 1973) pp. 627–649
7.21   S. Mascarenhas: In *Electrets, Charge-Storage, and Transport in Dielectrics* (Electrochemical Society, Princeton, N.J. 1973) pp. 650–656/657–660
7.22   E. Fukada, T. Takamatsu, I. Yasuda: Jpn. J. Appl. Phys. **14**, 2079 (1975)
        J. J. Konikoff, J. E. West: 1978 *Annu. Rep. Conf. Electric. Insul. Diel. Phenom.* (NAS, Washington) pp. 304–310
7.23   P. M. Morse: *Vibration and Sound* (McGraw-Hill, New York 1948)
7.24   W. Kuhl, G. R. Schodder, F. K. Schroeder: Acustica **4**, 519 (1954)
        K. Matsuzawa: J. Phys. Soc. Jpn. **13**, 1533 (1958)
7.25   C. W. Reedyk: IEEE Trans. AU-**19**, 1 (1971)
7.26   M. C. Killion, E. V. Carlson: J. Audio Eng. Soc. **22**, 237 (1974)
        M. C. Killion: J. Audio Eng. Soc. **23**, 123 (1975)

7.27   J.O.Hedman: Tele **1**, 49 (1976)
7.28   G.M.Sessler, J.E.West: Bell Lab. Rec. **47**, 244 (1969)
7.29   Y.Kodera: Proc. Conf. Inst. Elec. Jpn. (1971) p. 480
7.30   F.Fraim, P.Murphy: J. Audio Eng. Soc. **18**, 511 (1970)
7.31   J.van Turnhout: J. Electrostat. **1**, 147 (1975)
7.32   S.V.Djuric: J. Acoust. Soc. Am. **51**, 129 (1972)
       A.P.G.Peterson, E.E.Gross, Jr.: *Handbook of Noise Measurement*, 7th ed. (General Radio, Concord, Mass. 1972) p. 314
7.33   H.J.Griese: J. Audio Eng. Soc. **20**, 324 (1972); Radio Mentor Electron. **8**, 377 (1972)
7.34   G.M.Sessler, J.E.West: Proc. 5th Int. Cong. Acoust., Liège (1965), paper J42
7.35   G.M.Sessler, J.E.West: J. Acoust. Soc. Am. **46**, 1081 (1969)
7.36   P.V.Murphy, F.W.Fraim: In *Electrets, Charge Storage, and Transport in Dielectrics*, ed. by M.M.Perlman (Electrochemical Society, Princeton, N.J. 1973) pp. 603–621
7.37   G.R.Hruska, E.B.Magrab, W.B.Penzer: J. Acoust. Soc. Am. **61**, 206 (1977)
7.38   H.J.Griese: Proc. 9th Int. Cong. Acoust., Madrid (1977) paper Q29
       S.V.Djuric: Proc. 9th Int. Cong. Acoust., Madrid (1977) paper Q35
7.39   J.E.Warren, J.F.Hamilton, A.N. Brzezinski: J. Acoust.Soc. Am. **52**, 711 (1972)
7.40   F.W.Fraim, P.V.Murphy, R.J.Ferran: J. Acoust. Soc. Am. **53**, 1601 (1973)
       H.S.Madsen: J. Acoust. Soc. Am. **53**, 1616 (1973)
7.41   E.V.Carlson, M.C.Killion: U.S. Patent 3,740,496 (1973)
7.42   G.M.Sessler, J.E.West: Proc. 7th Int. Cong. Acoust., Budapest (Akademiai Kiado, Budapest 1971) paper 23E1
7.43   D.Legros, J.Lewiner: In *Electrets, Charge Storage, and Transport in Dielectrics* (Electrochemical Society, Princeton, N.J. 1973) pp. 622–626; J. Acoust. Soc. Am. **53**, 1663 (1973)
7.44   G.M.Sessler, J.C.French: Proc. 8th Int. Cong. Acoust., London (1974), Vol. II, p. 695
7.45   L.A.Farrow, R.E.Richton: J. Appl. Phys. **48**, 4962 (1977)
       C.K.N.Patel, R.J.Kerl: Appl. Phys. Lett. **30**, 578 (1977)
       J.Heiserman, J.P.Hulin, J.Maunard, I.Rudnick: Phys. Rev. B **14**, 2862 (1976)
7.46   A.Boeryd, L.Branden, J.A.Hedman, O.Larsson: Ericsson Rev. **3**, 118 (1976)
       N.Gleiss: Tele **1**, 41 (1976)
       R.Ceituli, R.Marion: Elettron. Telecommunic. **25**, 105 (1976)
7.47   C.B.McDowell, III, L.E.O'Boyle: Proc. IEEE Fall Electron. Conf. (1971) pp. 97–102
7.48   G.M.Sessler, J.E.West: IEEE Trans. AU-19, 19 (1971)
7.49   E.V.Carlson, M.C.Killion: J. Audio Eng. Soc. **22**, 92 (1974)
7.50   C.W.Reedyk: J. Acoust. Soc. Am. **53**, 1609 (1973)
       P.H.D'Amico, P.Kuhn: J. Audio Eng. Soc. **24**, 117 (1976)
7.51   G.M.Sessler, J.E.West, M.R.Schroeder: J. Acoust. Soc. Am. **46**, 28 (1969)
       G.M.Sessler, J.E.West: J. Acoust. Soc. Am. **58**, 273 (1975)
7.52   H.Sell: Z. Tech. Phys. **18**, 3 (1937)
       K.Geide: Acustica **10**, 295 (1960)
       W.W.Wright: "High Frequency Electrostatic Transducers for Use in Gases", Tech. Memo. No. 47, Harvard University (1962)
       G.M.Sessler, J.E.West: J. Acoust. Soc. Am. **34**, 1774 (1962)
7.53   Philips: Audio-Info **43**, 1 (1978)
7.54   G.M.Sessler, J.E.West: Proc. 4th Int. Cong. Acoust., Copenhagen ed. by A.K.Nielson (1962) paper N55
7.55   R.E.Collins: Proc. IREE **34**, 381 (1973)
       G.M.Sessler, J.E.West: Polym. Lett. **7**, 367 (1969)
7.56   H.J.Griese, G.Kock: Funkschau **49**, 1251 (1977)
7.57   C.W.Reedyk: North. Electr. Telesis **1**, 22 (1967)
7.58   N.Sakamoto, T.Gotoh, N.Atoji, T.Aoi: J. Audio Eng. Soc. **24**, 368 (1976)
7.59   G.Morgenstern: Acustica **40**, 81 (1978)
7.60   H.Kawakami: Audio Eng. Soc., Preprint No. 693 (B-3) (1969)

380   G. M. Sessler and J. E. West

7.61 Y.Mochizuki, S.Watanabe, M.Kobayashi, N.Yakushiji, H.Imazeki: Toshiba Rev. 35–38 (1972)
7.62 G.M.Sessler, J.E.West, R.L.Wallace,Jr.: IEEE Trans. COM-21, 61 (1973)
      G.M.Sessler, R.L.Wallace,Jr., J.E.West: U.S. Patent 3,668,417 (1972)
7.63 S.F.Demirdjioghlu, R.M.VanDyk: U.S. Patent 3,668,698 (1972)
      J.R.Webb, R.C.Webb: U.S. Patent 3,653,038 (1972)
7.64 J.L.Bruneel, F.Micheron: Appl. Phys. Lett. 30, 382 (1977)
7.65 D.Perino, J.Lewiner, G.Dreyfus: L'onde Electr. 57, 688 (1977)
7.66 M.V.Scherb, G.P.Kazokas, J.A.Zelik, J.R.Mastandrea, D.C.Mackallor: NASI-10840 (1972)
7.67 A.N.Gubkin, V.F.Sergienko, N.M.Torfimenko: Prib. Tekh. Eksp. 2, 166 (1961)
7.68 G.K.Miller: "Electret Tape Transducers", GTE Sylvania F30602-75-6-0075; "High-Pressure Transducers", GTE Sylvania N00014-72-C-0307
7.69 J.D.Zook, S.T.Liu: J. Appl. Phys. 43, 1304 (1972)
7.70 C.Hennion, J.Lewiner: 9th Int. Cong. Acoust., Madrid (1977) paper Q36
7.71 C.Hennion, J.Lewiner: J. Acoust. Soc. Am. 63, 279 (1978)
7.72 C.Hennion, J.Lewiner: J. Acoust. Soc. Am. 63, 1229 (1978)
7.73 R.M.Schaffert: Electrophotography (Wiley, New York 1975)
7.74 J.H.Dessauer, H.E.Clark (eds.): Xerography and Related Processes (Focal Press, London 1965)
7.75 R.B.Comizzoli, G.S.Lozier, D.A.Ross: Proc. IEEE 60, 348 (1972)
7.76 D.R.White (ed.): Electrophotography, Second Internat. Conf. (Society of Photographic Scientists and Engineers, Washington, D.C. 1974)
7.77 V.M.Fridkin, I.S.Zheludev: Photoelectrets and the Electrophotographic Process (Consultants Bureau, N.Y. 1961)
7.78 F.Bestenreiner, J.Freund, D.Giglberger, U.Greis, J.Heinburger, A.Rheude, K.Stadtler, J. van Engeland, W.Simm: In Electrophotography, Second International Conference, ed. by D.R.White (Society of Photographic Scientists and Engineers, Washington, D.C. 1974) pp. 9–28
7.79 T.L.Thourson: IEEE Trans. ED-19, 495 (1972)
7.80 J.W.Boag, A.J.Stacey, R.David: J. Photogr. Sci. 19, 45 (1971)
7.81 C.J.Young, H.G.Craig: RCA Rev. 15, 471 (1954)
7.82 R.L.Jepsen, G.F.Day: In Electrophotography, Second International Conference, ed. by D.R.White (Society of Photographic Scientists and Engineers, Washington, D.C. 1974) pp. 28–36
7.83 U.Rothgordt: Philips Tech. Rundsch. 36, 98 (1976/77)
7.84 J.J.Stone: IEEE Trans. ED-19, 563 (1972)
7.85 G.M.Sessler, J.E.West: In Electrophotography, Second International Conference, ed. by D.R.White (Society Photographic Scientists and Engineers, Washington, D.C. 1974) pp. 162–166
7.86 J.Feder: J. Appl. Phys. 47, 1741 (1976)
7.87 H.F.Frohbach: IEEE Trans. ED-19, 579 (1972)
7.88 D.E.Richardson, J.J.Brophy, H.Seiwatz, J.E.Dickens, R.J.Kerr: IRE Trans. AU-10, 95 (1962)
7.89 H.Seiwatz, D.E.Richardson: IEEE Trans. AU-12, 63 (1964)
      J.J.Brophy, D.E.Richardson, H.Seiwatz: IEEE Trans. AU-12, 111 (1964)
7.90 G.Zebel: J. Colloid Sci. 20, 522 (1965)
      C.N.Davies: Air Filtration (Academic Press, London 1973)
7.91 O.Jefimenko, D.K.Walker: Conf. on Diel. Mater., Meas. Appl. (Inst. of Electrical Eng., London 1970) pp. 146–149
7.92 A.N.Gubkin: Electrets (Academy of Science, Moscow 1961) pp. 130–133
7.93 V.G.Nazarov: Elektrichestvo 7, 60 (1954)
7.94 H.Kawai: Jpn. J. Appl. Phys. 8, 975 (1969)

7.95   G.T.Davis, M.G.Broadhurst: *Intern. Symp. on Electrets and Dielectrics*, ed. by M.S.deCampos (Academia Brasil de Ciencas Rio de Janeiro 1977) pp. 299–319
       M.Latour: J. Electrostat. **2**, 241 (1977)
       J.C.Hicks, T.E.Jones, J.C.Logan: J. Appl. Phys. **49**, 6092 (1978)
7.96   H.Ohigashi: J. Appl. Phys. **47**, 949 (1976)
7.97   M.Tamura, K.Ogasawara, T.Yoshimi: Ferroelectrics **10**, 125 (1976)
       K.Hatakeyama, S.Kinoshita, A.Haeno, T.Asanuma: Audio Eng. Soc. Preprint No. 1056 (L-1) (1975)
7.98   G.M.Sessler: In *Fortschritte der Akustik*, DAGA 76, Heidelberg (VDI Verlag, Düsseldorf 1976) pp. 81–95
7.99   R.Lerch: In *Fortschritte der Akustik*, DAGA '78, Bochum (VDE Verlag, Berlin 1978) pp. 661–664
7.100 H.Naono, T.Gotoh, M.Matsumoto, S.Ibaraki, Audio Eng. Soc. Preprint No. 1271 (0-1) (1977)
7.101 W.D.Cragg, N.W.Tester: Elektr. Nachr. **52**, 359 (1977)
7.102 S.D.Bennett, J.Chambers: Electron. Lett. **13**, 110 (1977)
7.103 M.Latour, O.Guelorget, P.Murphy: In *Proc. Int. Workshop Electrical Charges Dielectrics*, ed. by E.Fukada (Kyoto Japan, 1978) p. 102–103
7.104 R.Carpenter, G.M.Garner, J.F.Sear: "Piezoelectric and Pyroelectric Materials and Applications", IEE Science, Education and Management Division, Digest No. 1975/25, Contrib. No. 10
7.105 T.D.Sullivan, J.M.Powers: J. Acoust. Soc. Am. **63**, 1396 (1978)
7.106 C.R.Crane: IEEE Trans. SU-**25**, 393 (1978)
7.107 H.Sussner, D.Michas, A.Assfalg, S.Hunklinger, K.Dransfeld: Phys. Lett. **45**A, 475 (1973)
7.108 H.Sussner: „Ursache und Anwendung des starken piezoelektrischen Effekts in dem Hochpolymer PVF$_2$"; Dissertation (1976)
7.109 C.Alquie, J.Lewiner, C.Friedman: Appl. Phys. Lett. **29**, 69 (1976)
7.110 B.Woodward: Acustica **38**, 264 (1977)
7.111 B.Woodward: Proc. 9th Int. Cong. Acoust., Madrid (1977) paper K40
7.112 H.Ohigashi, R.Shigenari, M.Yokota: Jpn. J. Appl. Phys. **14**, 1085 (1975)
7.113 H.Ohigashi: Private communication
7.114 H.R.Gallantree, R.M.Quilliam: Marconi Rev. **39**, 189 (1976)
7.115 M.Quilliam: Electron. Ind. **3**, 23 (1977)
7.116 H.Ohigashi: U.S. Patent 3,940,637 (1976)
7.117 M.Yoshida, M.Segawa, H.Obara: U.S. Patent 3,935,485 (1976)
7.118 N.Murayama, K.Nakamura, H.Obara, M.Segawa: Ultrasonics **14**, 15 (1976)
7.119 A.S.DeReggi: "Piezoelectric Polymer Transducer for Impact Pressure Measurement", Rpt. NBS, COM-75-11127 (1975)
7.120 S.T.Liu, D.Long: Proc. IEEE **66**, 14 (1978)
7.121 A.M.Glass, J.M.McFee, J.G.Bergman: J. Appl. Phys. **42**, 5219 (1971)
7.122 J.H.McFee, J.G.Bergman, G.R.Crane: Ferroelectrics **3**, 305 (1972)
7.123 E.H.Putley, R.Watton, J.H.Ludlow: Ferroelectrics **3**, 263 (1972)
7.124 J.G.Bergman, G.R.Crane, A.A.Ballman, H.M.O'Bryan: Appl.Phys. Lett. **21**, 497 (1972)
7.125 N.Murayama: U.S. Patent 3,872,328 (1975)
7.126 R.G.F.Taylor, H.A.H.Best: Contemp. Phys. **14**, 55 (1973)

8. Recent Progress in Electret Research

R. Gerhard-Multhaupt, B. Gross, and G. M. Sessler

With 22 Figures

8.1 Introduction

In this chapter, the progress achieved in the field of electret research since publication of the First Edition is briefly reviewed. The chapter thus covers essentially the time period of 1979 to early 1986. Occasionally, reference is made to older papers, particularly if they are of importance for the understanding of significant new developments.

A large number of publications has appeared in this time period. It is estimated that only in the field of piezoelectric polymers almost 1000 papers were published since 1979. Electret research has also been discussed in a number of recent books. These include monographs on the subject in general [8.1, 2], treatises on electrical properties of polymers [8.3, 4], books on thermally stimulated processes [8.5, 6], the Annual Reports of the Conferences on Electrical Insulation and Dielectric Phenomena [8.7], the proceedings of the 5th International Symposium on Electrets [8.8] and records of conferences on related topics [8.9, 10].

In order to keep this new edition within the size limits established for "Topics" volumes, the authors had to severely limit the size of this chapter. It was therefore necessary to select from the large number of papers relatively few and present only the most important results from these in very concise form. The authors apologize if they have, in the process, eliminated or inadequately discussed many of the relevant publications.

The present chapter is subdivided in seven sections (Sects. 8.1–7) which correspond to, and update, Chaps. 1–7. To facilitate use of the newly included material, reference is frequently made to corresponding parts of preceding chapters.

8.2 Physical Principles

A number of improvements in our basic understanding and control of electrets have been reported in the recent literature. Examples are new developments in relaxation theory, an improved understanding of the charge decay in insulators, the discovery of the superior charge-retention properties of SiO_2, the emergence of methods for measuring charge and polarization distributions with micrometer resolution, and further improvements of electret-charging methods. In the following, a brief review of these and other significant contributions is given.

8.2.1 Charging and Poling

Materials exhibiting dielectric relaxation necessarily show *dielectric absorption* and *electret behavior* if an electric field is applied to them [8.11]. "Universal" relations for the dielectric relaxation function have been established [8.12, 13]. They have led to the development of theories of dielectric relaxation in which the concept of independently relaxing elementary units [8.14] is abandoned in favor of models involving cooperative phenomena, such as many-body interactions [8.12] and cluster formation [8.15]. It has been shown recently that the formalism of isothermal relaxation theory cannot be applied to thermally activated relaxation, since the latter is described by differential equations with time-dependent coefficients [8.16].

Contact electrification (Sect. 2.2.1) has been extensively reviewed [8.17–22]. It is assumed to be due to electronic charge transfer by tunneling from delocalized states in the metal to traps in the dielectric; the charge penetrates up to ∼10 nm into the insulator [8.18]. Electron transfer was confirmed, e.g., by means of photoemission measurements of PE charged by contact with metals with and without applied field [8.23]. Other experiments utilizing repeated contacts and contacts with variable duration indicate an increase of the charge density with number and time of contacts [8.24, 25]. It appears, however, that details of the charging process are critically dependent on material properties. The above model has to be modified, for example, for charge exchange between an insulating polymer and a metal with a thin interfacial oxide layer [8.26]. *Frictional charging* was studied with respect to its dependence on temperature and friction speed [8.27].

A comprehensive study of *room-temperature charging* of two-side metallized PVDF was reported [8.28]. In particular, the time and field dependence of the pyroelectric activity was investigated in detail. Some results are shown in Fig. 8.1. It appears from these results, as from results of corona charging, that room-temperature poling of PVDF is capable of producing piezo- and pyroelectric activities comparable to those achieved with thermal poling (Sect. 8.5). In this method, flashover around the edge of the sample may be avoided by obstructing the flashover path [8.29]. Other experiments showed that the poling of PVDF is a very fast process, taking only 4 μs at 20 °C under a field of 200 MV/m [8.30].

Corona methods (pp. 30–31) were considerably refined and adapted for controlled and stable charging. Particularly useful is the constant-current corona [8.31], utilizing feedback control of the output of the high-voltage source to exactly set the current. The constant-current corona is like an electron beam of very low energy providing either negative or positive charges. The uniformity of the surface potential in corona-charged samples can be improved

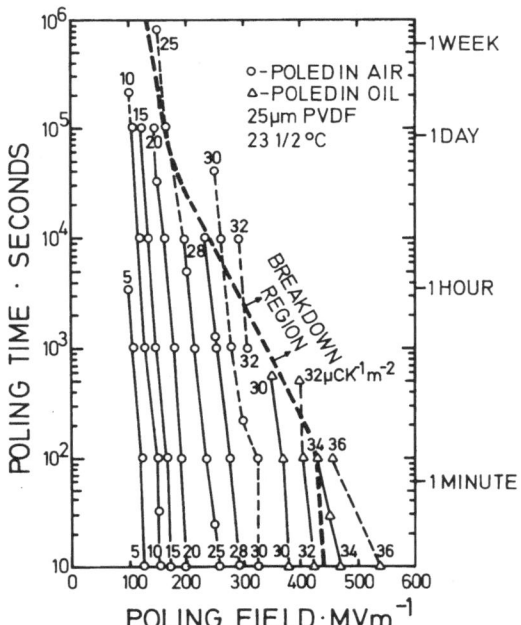

by adding a metal aperture at grid potential between grid and sample [8.32]. By measuring the corona current during charging, it is possible to monitor the total charge or polarization present in a sample at any time as long as leakage currents are absent [8.33, 34]. Another important development concerning the corona method is the use of elevated temperatures during charging to stabilize charge retention by filling deeper traps [8.35, 36] (Sect. 2.6.5).

Breakdown charging, i.e. the method of utilizing a sandwich structure consisting of the sample to be charged and a resistive layer (pp. 31–32) was further developed [8.37, 38]. First, the method was applied to two-side metallized samples of PVDF, allowing application of fields up to 1.1 GV/m without destructive breakdown [8.37]. Then it was demonstrated that on one-side metallized samples the electret charge may be closely controlled by varying the thickness of an air gap between the resistive layer and the nonmetallized side of the sample [8.38].

An important new method for aligning dipoles is *deformation-aided poling*, which is accomplished by exposing the sample to a strong electric field at room temperature while it is being stretched [8.39–42]. In some materials, the stretching causes recrystallization. Thus the ordering by the applied field is more efficient and requires lower field strength [8.42]. Generally, the dipolar mobility increases and reaches for PVC at the yield point a value equal to that at the glass-transition temperature. The process is aided by the line-up of the main chains parallel to the draw direction and by the increase of the poling field due to the thinning of the sample [8.40]. A modification of this method consists of simultaneous field application and rolling of the sample [8.41]. Poling during stretching and rolling has been successfully applied to PVDF and other polymers.

Other charging methods that have been recently introduced or improved are *plasma poling* [8.43], simultaneous plasma polymerization and poling [8.44], positive charging with low-energy electrons (Sect. 8.4), and *ionization-chamber techniques*, where charge is deposited by application of a field generated by biasing the chamber electrodes [8.45, 46] or by use of a biasing electret [8.47].

8.2.2 Methods for the Measurement of Charge and Polarization Distributions

A number of important new developments have taken place with respect to measuring distributions of charge, polarization, and piezo- or pyroelectricity in electrets. Particularly noteworthy is the advent of methods to probe such distributions nondestructively with μm resolution in the thickness direction of thin-film samples. Reviews of this field have been recently given [8.48, 49].

Considerable progress was made with respect to *pressure-pulse methods* (Sect. 2.4.6). Particularly remarkable is the generation of very short pressure pulses by laser excitation or by piezoelectric means. The laser-induced pressure pulses (LIPPs) are generated as the laser light interacts with the sample surface via ablation, thermal stress effects, and radiation pressure. The method was gradually refined by using shorter light pulses and by improving the coupling layer [8.50–58]. For the highest-resolution measurements reported, 70 ps light pulses were directly (without a distorting coupling disk) absorbed by a very thin layer deposited on the sample, as shown in Fig. 8.2, resulting in pressure pulses of below 500 ps duration [8.58–60]. Thus, with sound velocities around 2 km/s in many polymers, the resolution is about 1 μm. If the shape of the pressure pulse is known, the resolution can be further improved by deconvolution [8.51]. Considerably longer (~ 1 μs) acoustic pulses were generated by a diaphragm-type acoustic pulse generator [8.61]. Pressure steps of about 1 ns

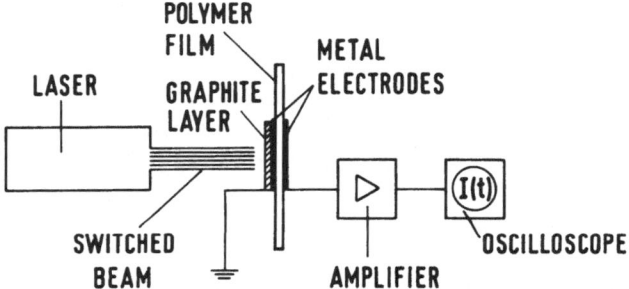

Fig. 8.2. Experimental setup for laser-induced pressure-pulse (LIPP) method [8.56]

rise time were produced piezoelectrically by electric excitation of an X-cut Quartz plate [8.62, 63]. The relations between the electrical signal generated by the pressure disturbance and the charge or polarization density have been worked out by a number of authors [8.50–64] and were recently discussed comprehensively for a variety of conditions [8.65]. The advantages of pressure-pulse techniques, such as direct display of charge or field distributions without need of deconvolution, have made them the methods of choice.

The *thermal-pulse method* (Sect. 2.4.5) was further refined and its potential was explored. The method has been extremely successful for determining the charge centroid in one- and two-side metallized samples [8.66, 67]. There has been some discussion about the usefulness of the method to determine details of charge distributions [8.68, 69]. As mentioned in Sect. 2.4.5, three to ten spatial Fourier coefficients of the charge distribution can be found in samples metallized on one side only, the actual number depending on the location of the charge relative to the metal layer. New evaluations show, however, that by using two-side metallized samples and pulsing from both surfaces, about 12 Fourier coefficients can be determined and a resolution of 0.1 of the sample thickness, independent of charge location, may be achieved [8.69]. The limitations are imposed by experimental accuracy and not by the computational method.

An interesting modification of the thermal-pulse technique consists in the periodic heating of a sample electrode with a laser beam modulated at frequencies between 0.1 and 100 kHz [8.70–73]. The heat produces temperature waves within the sample whose attenuation increases with frequency. The nonuniform heating generates pyroelectric currents which are a function of the distribution of real and polarization charges. It is claimed that both distributions are obtained separately from a least-square analysis.

Electron-beam probing (Sect. 2.4.7) has been further investigated and its applicability outlined [8.74–77]. Among the parameters studied is the efficiency of charge compensation and the location of the virtual electrode as function of electron-beam energy [8.74]. To eliminate the effect of charging due to the beam current, calibration runs on the same sample are necessary [8.74, 75]. A modification of the method consists in using a series of electron pulses at each energy, allowing the dielectric to discharge between pulses [8.76]. A method for use in SiO_2 was also discussed [8.77].

Other methods recently suggested for determining charge distributions are: (i) Analysis of the *depolarization current* and the *dielectric displacement* during isothermal discharge of the sample [8.78]; this method requires the assumption of a constant carrier mobility; (ii) analysis of *field-dependent changes in the Raman spectrum* or of the *photostimulated current* as one illuminates various points over the thickness dimension of the sample [8.79, 80]; the resolution of these methods is relatively coarse ($\sim 100 \, \mu m$); (iii) analysis of currents induced by *exposure* of the dielectric to *solvent vapors* [8.81]; and (iv) generation of *pressure pulses by electrical excitation* [8.82].

While all the methods discussed so far in this subsection refer to probing of charge distributions in the thickness direction, new instrumentation was also described for *probing along the sample surface* [8.83]. By using a small commercial probe in a carefully designed positioning system, a lateral resolution of better than 100 μm was achieved.

8.2.3 Storage and Decay of Real and Dipolar Charges

Retention of real and dipolar charges (Sects. 2.6.1, 2) was further studied with thermally stimulated discharge (TSD) and thermoluminescence (TL) methods. These experiments have revealed important new information which will also be discussed in Sect. 8.3. For charge trapping in polymers, the results point to a broad distribution of activation energies, as shown in Fig. 8.3 [8.84]. Such distributions follow also from calculations of polarization energies of traps in an amorphous matrix [8.85] and from an interpretation of peak-shift data in irradiated Teflon and thermoluminescence data of polystyrene [8.86]. Origins of traps in PE are believed to be defects, unstable or stable oxidation products, cross links and antistatic agents [8.84]. Differences between surface and bulk traps, previously reported for Teflon, have now been studied in other polymers [8.87, 88] (Sect. 8.3.2). Dipole polarization, ionic space charge, and Maxwell-Wagner polarization in ionic solids, also investigated with TSD methods, was reviewed [8.89].

Important new results on the *spatial distribution of real charges* (Sect. 2.6.3), injected by different methods and subject to drift motion have been obtained during the last few years. Electron-beam charged Teflon was studied with a number of methods [8.57, 58, 66, 67]. Results obtained [8.58] with the LIPP method (Sect. 8.2.2), depicted in Fig. 8.4, indicate the presence of relatively narrow charge peaks close to the electron range, as also predicted theoretically. The distribution during isothermal and nonisothermal charge decay of negatively and positively corona-charged Teflon was also studied with the pressure-pulse method [8.90]. The results, to be discussed in more detail under "charge decay", show a small volume charge and a large surface charge for negatively charged samples and a large volume charge with no surface trapping for positively charged films. However, thermal-pulse experiments on positively charged Teflon indicate, for identical conditions, only small charge penetration [8.91]. Measurements on electron-beam charged PETP, PP, and PMMA were also performed with pressure-pulse methods [8.57, 64, 92–94]. Additional information for PETP samples has been obtained with electron-beam probing [8.74, 75]. As in Teflon, the samples show a relatively narrow peak close to the electron range with an additional peak in the vicinity of the electrode of injection. This structure can be attributed to a maximum in the radiation-induced conductivity between the two peaks which prevents permanent charge deposition in this area. Other electrets investigated include PETP and PE charged by electrode injection [8.55, 95], PS charged by breakdown [8.96], aluminum oxide charged by dark injection [8.97], and silicon dioxide charged by electron beam [8.77].

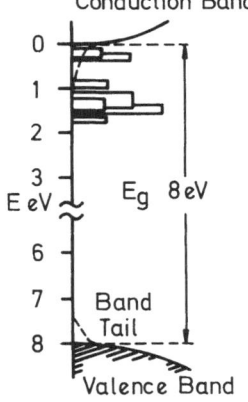

Conduction Band

◀ **Fig. 8.3.** Energy level diagram of carrier traps in PE [8.84]

E eV

E_g 8 eV

Band
Tail

Valence Band

Fig. 8.4. Current responses, corresponding to charge distributions, of 24 μm Teflon FEP samples charged with 10, 20, 30, and 50 keV electron beams, as measured with the LIPP method [8.58]

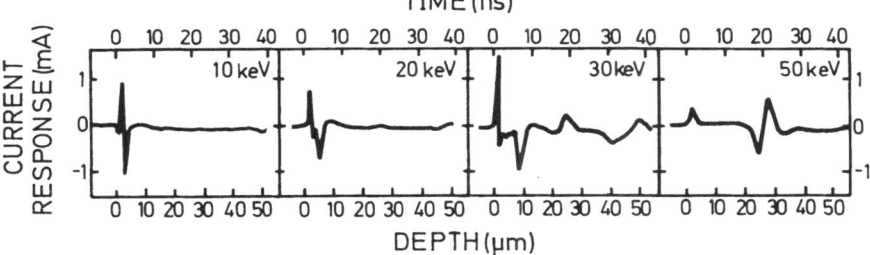

TIME (ns)

CURRENT RESPONSE (mA)

10 keV   20 keV   30 keV   50 keV

DEPTH (μm)

The *spatial distribution of dipolar charges* (Sect. 2.6.3) in PVDF and its copolymers was extensively studied with thermal-pulse, laser-modulation, and pressure-pulse methods [8.69–72, 98–104]. The data show that uniform distributions form only under special experimental conditions, preferably for room-temperature poling [8.98]. Nonuniform distributions are often found in thermally poled samples and may be due to field distortion caused by excessive charge injection at elevated temperatures or by charge migration [8.99, 100, 102]. Preexisting nonuniform polarizations were detected in 25 μm thick PVDF samples taken from a roll [8.101]. Polarization distributions were also measured for PE [8.103, 104] and polyamide [8.102]. Experimentally determined piezoelectricity profiles in PVDF and its copolymers are discussed in Sect. 8.5.5.

Charge- or potential-decay theory of electrets (Sect. 2.6.4) was discussed in terms of space-charge-perturbed currents and field-dependent polarization [8.105–107]. Charge-decay models taking into account surface trapping, injection into the bulk, and bulk trapping were proposed for negatively and positively charged Teflon [8.108, 109], PE [8.110–112], and naphthalene [8.113]. In all cases, the models could be accurately fitted to experimental data.

An important experimental investigation of *charge decay* in negatively and positively charged Teflon was performed by application of the pressure-pulse method [8.90], affording a view of the charge distribution during decay. An

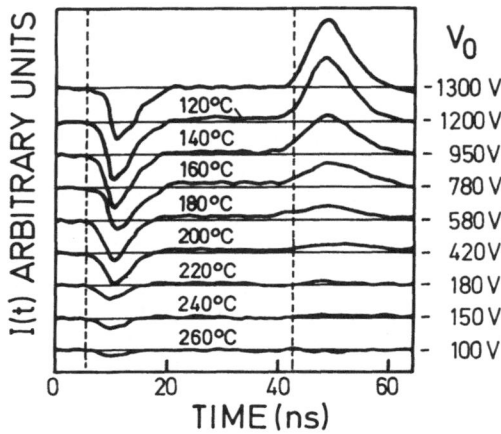

Fig. 8.5. Charge evolution in negatively corona-charged 50 µm Teflon FEP after thermal discharge at temperatures shown. The surface potentials V_0 after heating are also shown [8.90]

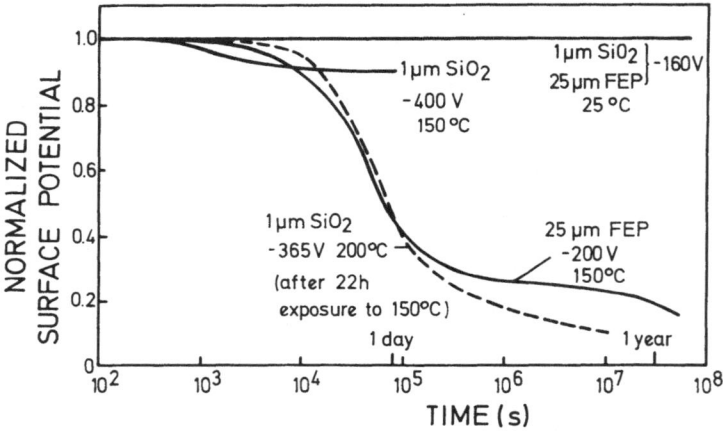

Fig. 8.6. Time dependence of surface potential of 1 µm SiO_2 and 25 µm Teflon electrets at various temperatures. The original surface potentials are indicated [8.114]

example is shown in Fig. 8.5. It can be concluded from this and other data that negative charges, after being released from the surface, are trapped in the volume. When the available traps of density 3×10^{13} cm^{-3} are filled (in Fig. 8.5 at 120 °C), the released surface charges apparently move through the saturated volume without being further affected. Corresponding results for positive charging indicate the absence of surface trapping at 40 °C and above, a higher volume-trap density (10^{14} cm^{-3}) and complete discharge between 230 and 260 °C.

Superior charge-retention properties were found for *silicon-dioxide* electrets. Results for negatively corona-charged, thermally grown SiO_2 are compared in Fig. 8.6 with data for negatively charged Teflon [8.114]. The charge is stable for both materials at room temperature. At 150 °C, the surface potentials decay, but the decrease is less pronounced for SiO_2.

Other recent experimental studies of charge decay have been performed on electrets of such materials as silicone resin [8.115], polyolefins [8.116], PS [8.117], and Teflon [8.118]. Chemical modification of surface and bulk of some polymers was found to affect charge lifetimes [8.119, 120]. In particular, etching of Teflon improves storage of positive charges (Sect. 8.3.2, Fig. 8.10). Other studies refer to the influence of surface impurities on charge storage in Teflon [8.121], the effect of humidity on charge storage in PE [8.122], the effect of vibrationally excited diatomic molecules on charge injection from the surface into the bulk of polymers [8.123], and the influence of additives deposited in the bulk on charge trapping and conductivity in PE [8.124].

New studies of *polarization decay* (Sect. 2.6.1) were conducted for PVDF, PETP, and other polymers. In PVDF, different decay rates were found for uniaxially and biaxially stretched materials; the depolarization rate also depends critically on poling parameters, on crystallinity and on the particular PVDF materials as discussed in Sect. 8.5.5. In PETP and other polymers, the effect of aging on the dipolar peak was studied [8.125, 126]; this shall be discussed in Sect. 8.3.

Charge-dipole interaction, in particular the effect of the field generated by the real charges on the alignment of dipoles, was investigated for PETP and PVC [8.126–129]. TSD measurements suggest a field-induced alignment of dipoles in PETP-Hostaphan and PETP-Mylar at temperatures up to 90° and up to 120 °C, respectively [8.126, 128]. The dipolar alignment was confirmed by thermal-pulse experiments indicating an increase of pyroelectric activity of PETP-Mylar during TSD up to 120 °C [8.128].

8.2.4 Conduction Phenomena

Our knowledge of conduction phenomena (Sect. 2.6.6) in electret-forming materials has been furthered by a number of recent studies on mobility, Schubweg, conductivity, electrode injection, etc. Particularly useful for investigations of the bulk conductivity are corona- or electron-beam-injection experiments. Measurements with positive and negative coronas yield direct information about the contributions of positive and negative carriers to conduction [8.130]. The available methods for measuring *carrier mobility* in low-mobility materials and results obtained in PE by these methods were recently reviewed [8.131]. These and new data for HDPE [8.132] demonstrate the dependence on measuring method and time frame of the experiment typical for materials with a wide range of trapping levels (pp. 72–73). The temperature dependence of the mobility in PE was found to be Arrhenius-like [8.85, 133], a field dependence was not observed up to 30 MV/m [8.132]. In charged materials, the mobility depends on charge density since it is determined by the non-occupied trapping levels. Such a dependence was observed in PETP

[8.134]. Trap-filling explains also relatively large electron mobilities in FEP [8.135]. Mobilities in PVDF and PI were also determined [8.136, 137]. The *hole Schubweg* in Teflon at room temperature was found to be independent of field [8.109, 138].

Space-charge-limited currents (SCLCs) in PVDF and PI follow from current-voltage characteristics obtained by direct charge injection into a nonmetallized sample surface [8.136, 137]. SCLCs were also occasionally seen in two-side metallized samples [8.139] in cases where the currents are not electrode limited. If the flow of current is from a freely injecting electrode across the surface of the insulator to a second electrode, one obtains surface-charge-limited currents. It has been found that these are also proportional to the square of the applied voltage with a transient behavior controlled by the transit time [8.140].

Current limitation by the electrode-dielectric interface is often determined by Schottky emission, frequently modified by space-charge effects [8.18]. Schottky emission is favored as the controlling process over the Poole-Frenkel effect if metallized samples yield much smaller currents than samples subject to direct charge injection into a nonmetallized surface. Such experiments confirmed Schottky limitation in LDPE [8.84] and FEP, and Schottky limitation modified by trapped charge in PI [8.137]. The presence of trapped-charge layers due to electrode injection into PETP metallized with Al was proven with electron-beam probing [8.76].

Dispersive hopping (p. 5), which affects charge transport in disordered insulators [1.53, 8.4], was further discussed. Of particular interest has been the effect of the electric field of an injected space charge on the carrier drift [8.107, 141]. Recently, an analytical solution of the space-charge-perturbed case has been obtained [8.107].

8.3 Thermally Stimulated Discharge of Electrets

Thermally-stimulated-discharge (TSD) methods continued to be useful for the analysis of space-charge and dipolar phenomena in dielectrics. Apart from this, the methods themselves were refined or modified in many respects. The entire field has recently been reviewed in several papers and books [8.5, 6, 142–147]. Since TSD analysis is a widely used tool, results obtained with this method are also discussed in other sections of this chapter. In particular, TSD results pertaining to irradiated dielectrics are dealt with in Sect. 8.4.

For brevity, related methods such as thermoluminescence (pp. 192–194), thermally stimulated conductivity (pp. 194–196), deep-level transient spectroscopy (pp. 190–192), etc. are not discussed in this section; reference is made to recent reviews on these topics [8.5, 148–149].

8.3.1 Analysis and Refinement of TSD Methods

Analysis of TSD data depends critically on the assumptions made about the *distribution of relaxation times* (Sects. 3.6.3 and 3.8). The TSD response has been calculated for a number of distribution functions with an improved Ninomiya-Ferry method based on converting ε'' data into current data ([Ref. 3.11, p. 88] and [8.150]). As an example, a family of TSD curves for a Cole-Cole distribution is given in Fig. 8.7. Such calculations allow one to fit experimental TSD curves to theoretical ones.

The effect of the *temperature dependence of the dielectric constant* on dipolar TSD and TSP (thermally stimulated polarization) currents has been studied in detail [8.151]. If the sample is poled at temperatures decreasing from a poling temperature T_p, the polarization depends on the cooling rate. A subsequent TSD experiment will therefore not yield currents truly characteristic of the relaxation properties of the sample at T_p. It was also found that a current reversal must occur in TSP measurements; more recently, new calculations by the same authors showed that this reversal is not only caused by the temperature dependence of the polarization but also by *thermal-expansion effects*, particularly around the glass-transition temperature [8.152].

Analytical methods for determining *trap depths* (pp. 171–173) have been further developed. For example, an existing method depending on the use of different heating rates [3.176] has been modified to yield the trap depth as a function of temperature from derivatives of two or more voltage curves [8.153]. Results for PE, shown in Fig. 8.8, indicate a constant trap depth up to 95 °C rising to a maximum at higher temperatures. Another method utilizing different heating rates is based on computer-fitting of TSD curves [8.154]. Other examples for the extension of analytical techniques are modifications of the well-

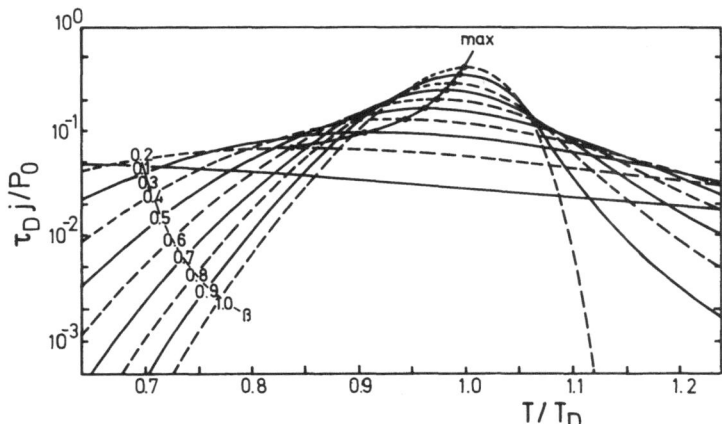

Fig. 8.7. Family of calculated TSD current curves for a Cole-Cole response. The quantity β is the Cole-Cole parameter, the activation energy is $20\,kT_D$ [8.150]

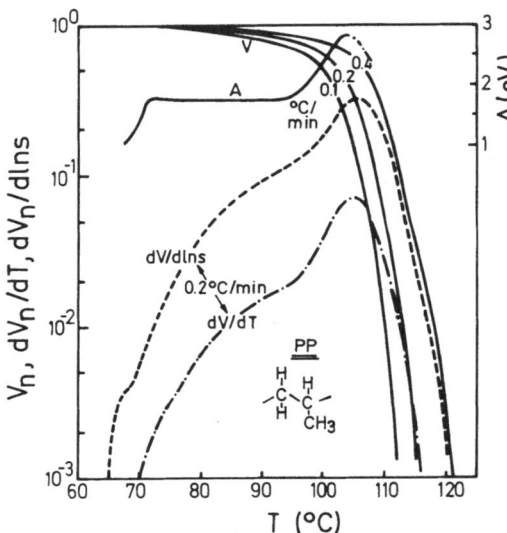

Fig. 8.8. Evaluation of activation energy A for PP from derivatives of the TSD voltage curves V, shown for three heating rates 1/s. Only the derivatives of the central voltage curve are given [8.153]

known initial-rise, Grossweiner, and similar methods to yield more accurate trap depths [8.155, 156].

The detrapping of charge carriers by multi-phonon processes was also treated with *quantum mechanics* [8.157]. This analysis leads to much smaller activation energies than the initial rise or variable-heating-rate methods; this suggests that the high activation energies and pre-exponential factors found by the classical methods are incorrect.

Experimental improvements of TSD methods have also been implemented. An example is a differential technique [8.158]. In this method, the sample to be investigated and an uncharged but otherwise identical sample are arranged electrically in series for the TSD experiment. Thus, parasitic currents due to thermal expansion, temperature-induced changes of permittivity, etc. are cancelled and the reproducibility of the measurements is improved.

8.3.2 Real-Charge Effects

TSD currents in a number of materials are now reasonably well understood by means of *models assuming appropriate trap distributions*. For Teflon FEP, surface and bulk traps have been separated by a combination of open-circuit and short-circuit TSD methods (p. 57) [2.135]. Based an these results, a theoretical model for the TSD process, assuming carrier transport controlled by a surface and a bulk trapping level, was developed [8.159]. For corona-charged samples, the model predicts TSD currents characterized by peaks at 150 and 200 °C, due to the emptying of surface and bulk traps, respectively. As Fig. 8.9 shows, there is excellent agreement between the theoretical and the experimental results. A detailed analysis of the first TSD

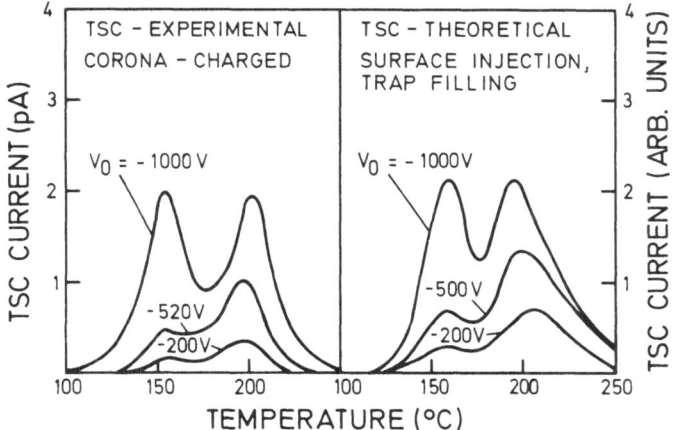

Fig. 8.9. Comparison of open-circuit TSD current measurements with theory for 25 μm Teflon FEP negatively corona-charged to different potentials V_0 [8.159]

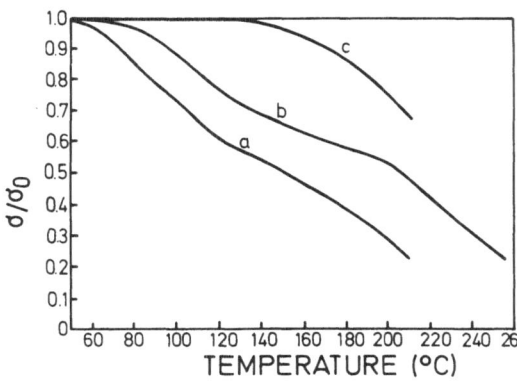

Fig. 8.10. TSD charge (or voltage) decay for 25 μm Teflon PTFE for *(a)* virgin sample, +400 V, *(b)* sample etched with Na–Naph., +400 V, *(c)* sample etched with Na–Naph., +160 V [8.120]

peak of FEP taking into account the initial Schubweg and the fraction of charge injected into the sample was also carried out recently [8.160]. For PTFE, a model with a total of five surface and bulk traps is necessary to explain the experimental TSD data [8.161].

Measurements of TSD currents due to real-charge effects have been reported for a variety of materials. For *Teflon FEP and PTFE* (Sects. 3.10.1 and 3.13.2), the above-mentioned results [8.88, 120, 159, 161] were supplemented by studies of the effect of ion injection [8.162], structural transitions around room temperature [8.163], and electron-beam charging [8.164] on TSD. The TSD characteristics of samples etched with a sodium-naphthalene complex which modifies the PTFE surface were also determined [8.120]. Figure 8.10 shows the improvements obtained for positive charges. Similar results were found earlier for vapor- or liquid-treated surfaces of TFE-copolymer films [8.119]. Other experiments showed that Aclar (PCTFE), a related polyolefin material, improves its TSD and charge-retention properties if charged at elevated temperatures [8.165] (p. 68).

As discussed above (Sect. 8.2.3), the charge-dipole interaction in PETP (pp. 139–140, 170) was extensively studied with TSD methods [8.126–128]. Other TSD measurements showed that in this material no significant differences between surface and bulk traps exist [8.88]. Apart from these and other above-room-temperature studies [8.166], the temperature range down to $-150\,°C$ was also investigated [8.167]. These measurements yielded three β-peaks caused by electron detrapping due to the relaxation of polar groups.

Another widely studied material is PE. Measurements on LDPE in a broad temperature range show, apart from low-temperature dipolar peaks, space-charge peaks at $40–60\,°C$ [8.87, 88, 168–171] and $90\,°C$ [8.87, 122, 171]. Indications are that these peaks are due to bulk and surface traps, respectively [8.87, 88]. In humid samples of LDPE, two other peaks were found [8.122]. HDPE has also been studied [8.172]. Chemical modification of the surface regions of PE with acids and bulk modification with antioxidants can result in a shift of TSD peaks or charge decays to higher temperatures [8.120, 124]. A model to explain the surface-potential crossovers observed in PE (p. 71), based on the action of excited molecules present in corona discharges, has been suggested [8.87].

Other TSD studies on polymers relate to space-charge effects in PP, where a strong influence of the thermal history of the sample on the population of TSD peaks was observed [8.173]. Negatively charged electrets of PP were found to discharge at higher temperatures than positively charged ones [8.174]. TSD due to the presence of space charge, often with dipolar contributions, was also observed in Kapton PI [8.175], PVC [8.176], Polyvinylbutyral [8.177], PS [8.178, 179], Polyvinylalcohol [8.180], Tetracene [8.181], PMMA/PVAc blends [8.182], Chloranil [8.183] and PVDF (Sect. 8.3.3).

Trapping effects in *inorganic materials* (Sects. 3.14, 15) were frequently studied with thermally-stimulated-conductivity experiments, i.e. with an applied bias [8.184, 185]. TSD methods for charged samples have, however, only been used on a few insulating inorganic materials [8.186, 187].

8.3.3 Dipole Effects

TSD currents in *ionic solids* (Sect. 3.11.1) are commonly called ionic thermo-currents (ITCs) [3.7, 8.145] and are used to study impurity-vacancy dipoles in doped crystals [8.188, 189], dielectric relaxations caused by the movement of ionic dipoles in crystals [8.190] and glasses [8.191], the glass transition in disordered crystals [8.192], transitions between metastable and stable phases in crystalline heterocyclic compounds [8.193], non-Debye behavior in single crystals with dipolar interaction [8.194], and similar phenomena. Apart from these purely inorganic materials, hybrid substances such as mica-loaded polyvinyl chloride (PVC) [8.176] and polyethylene oxide (PEO) complexed with potassium thiocyanate (KSCN) [8.195] were investigated by means of TSD techniques; the results obtained were interpreted in terms of dipolar relaxations and space-charge phenomena.

A detailed study of the β and γ *relaxations* (Sect. 3.11.7) in LDPE seems to indicate that the so-called compensation law can be employed [8.196]; thus, the logarithm of the pre-exponential factor, the activation energy, and the activation entropy appear to be linearly related for these two relaxations. For PETP, the above-mentioned investigations of the charge-dipole interactions (Sect. 8.2.3) also revealed an *aging behavior* (pp. 159–160) of the amorphous phase which can be understood in terms of free-volume relaxations and structure changes [8.126–128, 197]. Results of more recent measurements on PETP [8.198] and other polar polymers [8.27, 199–204] should probably be interpreted along the same lines.

This is especially true for TSD currents around the *glass-transition* (p. 153) temperature in PVC; here, similar behavior was found for friction-charged films [8.199, 203], thermally poled plates [8.176, 200], and very thin thermopoled P(VC–VAc) copolymer samples [8.201]. As with pure LDPE [8.196], the compensation law seems to hold for the dipolar relaxations [8.176, 200]. This is shown in Fig. 8.11 where the natural logarithm of the pre-exponential factor is plotted as a function of the apparent activation energy of thermally-stimulated-creep and TSD-current peaks with a compensation temperature of 356.5 K [8.200].

In PS, charge-dipole interaction (pp. 391, 396) again seems to be the key for understanding the TSD behavior of various samples [8.27, 202, 204]; the free-volume concept is particularly useful when dielectric and mechanical behavior are studied together [8.200, 204]. Activation energies and relaxation times were also obtained from TSD measurements on, e.g., polyvinyl formal [8.205], Nylon 6/66/610 terpolymer [8.206], polyetherimide [8.207], and PEBAX block copolymer [8.208].

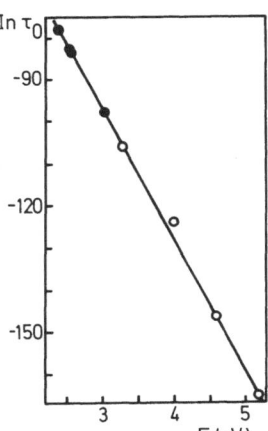

Fig. 8.11. Variation of the pre-exponential factor vs the apparent activation energy in the elementary peaks isolated by thermally stimulated creep (ooo) and TSD (●●●) around the glass transition of PVC. In the TSD case only the dipolar elementary peaks have been considered [8.200]

In relatively thick unpoled PMMA and PTFE samples, TSD currents are observed when *temperature gradients* (p. 194) are applied [8.209]; they are explained by dipole reorientations at phase-transition temperatures. TSD studies of cellulose nitrate [8.210], hydrated cellulose [8.211], and polyampholytes [8.212] reveal the strong influence of *water* and other additives on the electret behavior of biological materials (see also Sect. 8.6). Both, real-charge and dipole contributions are found in the TSD of PMMA [8.213]; the dipolar relaxations are strongly affected by the *tacticity* of the PMMA chains [8.214].

TSD experiments on *polymer blends* (p. 155) can be used to investigate phase transitions and structural changes in these exciting new materials (see also Sect. 8.5.2). For PP/PC blends, the dipoles seem to be so constrained that they do not contribute to the TSD currents [8.215]. In PVDF/PMMA alloys, however, the dipoles of the mixed amorphous phase can be partially aligned by the field of corona-deposited surface charges when they become mobile at the respective glass-transition temperature and thus yield a current peak [8.197, 216]. Figure 8.12 demonstrates that the temperature of this peak at the glass transition depends on the composition of the blend. The following peak is probably caused by a strongly increased conductivity which allows the surface charges to leave the samples [8.216]. The interpretation of these TSD results is again based on charge-dipole interaction [8.127] (pp. 391, 396–397).

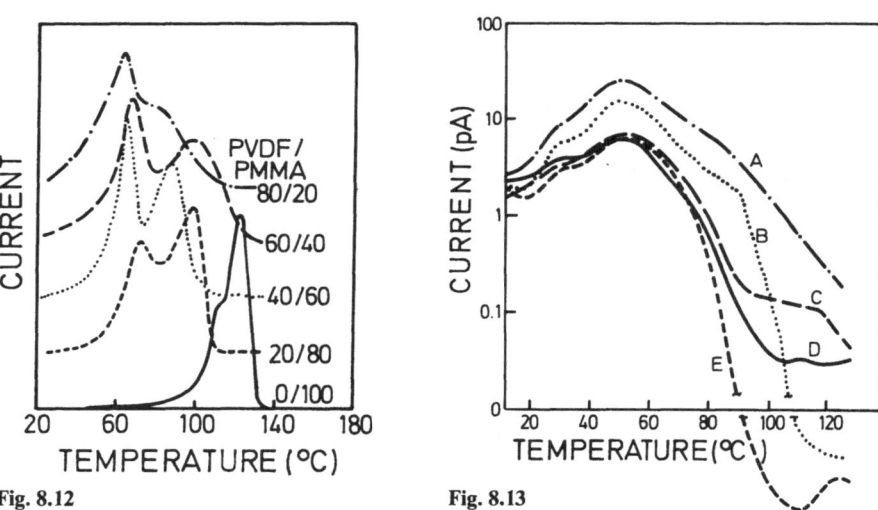

Fig. 8.12   Fig. 8.13

Fig. 8.12. Open-circuit TSD for PVDF/PMMA alloys of different blend composition. Poling temperature: 23 °C; poling field: 100 kV/cm; heating rate: 3 K/min [8.216]

Fig. 8.13. The effects of electrode material and charging environment on TSD in 15 μm thick PVDF charged for 100 s at −2 kV, 18 °C (cooled to 10 °C in presence of field). Curves: *A*: moist N_2, Al electrodes; *B*: moist N_2, Au electrodes; *C*: high vacuum, Al electrodes; *D*: dry N_2, Al electrodes; *E*: dry O_2, Al electrodes. Heating rate 0.5 °C min^{-1} [8.221]

TSD studies of pure PVDF (Sect. 3.11.8) led to current curves which vary significantly from laboratory to laboratory and whose interpretation is by no means conclusive [8.217–221]. Below room temperature, micro-Brownian motion and fluidification seem to cause two dipolar relaxations [8.217]. Only one of these is still found after stretching of the samples [8.218]. Interfacial charge trapping and charge injection are thought to be responsible for two TSD-current peaks above room temperature [8.218, 219]. Again, one of them disappears upon stretching [8.218]. Recently, it was suggested that the above-mentioned differences between experimental results from various laboratories may be explained by the influence of electrode materials and charging environments [8.221]; Fig. 8.13 demonstrates how strong this influence can be. The importance of charge injection and conduction for the poling of PVDF will also be discussed below (Sect. 8.5.5).

8.4 Radiation Effects

Interest in the interaction between ionizing radiation and dielectrics has continued, particularly in view of the increasing use of dielectric materials in high-radiation environments. Several books [8.222, 223] and review articles [8.31] were published. In the following we shall briefly discuss and list some of the papers which have recently appeared in the field of radiation effects in dielectrics.

8.4.1 Conduction and Charge Transport

The theory of charge transport (Sect. 4.2.5) was developed in terms of *rate equations* [8.224] taking into account the carrier-generation rate and its dependence on the absorbed-dose rate of the incident radiation, trapping and detrapping in shallow and deep traps, recombination, mobility-controlled charge transport, formation and injection of excess charge, diffusion, and carrier extraction at the electrodes. Trapping rates depend on the level of occupancy of traps, repulsion of carriers by occupied traps, and generation of trapping states by the incident radiation [8.225]. Deep trapping and recombination coefficients [8.226] are introduced in the presence of shallow traps whose occupants are in thermal equilibrium with free carriers. Radiation-induced currents depend on the behavior of the electrodes: Blocking electrodes prevent carrier injection from the plates, but allow carrier collection. Neutral electrodes assure continuity of the current density at the electrode-dielectric interface. Electrodes which are blocking in the dark might become neutral under irradiation.

8.4.2 Irradiation with X-Rays

Irradiation with fully penetrating X-rays gives a uniform carrier-generation rate. For *neutral electrodes* (Sect. 4.2.1), formation of space charge is avoided and the bulk of the dielectric remains electrically neutral. In this case conduction can be characterized fully by a depth-independent radiation-induced conductivity. Its value is dose-dependent, field-dependent, and time-dependent, as in Teflon FEP where only holes are mobile. The current goes through a build-up region characterized by deep trapping, reaches a maximum and decreases to a final recombination-controlled steady-state value [8.226], as shown in Fig. 8.14. Experimental results are explained by the rate theory of monopolar conduction, assuming deep trapping and trap-modulated recombination [8.226, 227]. A similar behavior has been found in Teflon PTFE [8.228].

For *blocking electrodes* (Sect. 4.2.1), the irradiated dielectric behaves like a solid-state ionization chamber which differs from a gaseous ionization chamber by the presence of trapping. The general rate equations, including diffusion, have been solved numerically [8.229].

Parameters were determined by fitting existing data for the photoconductor PbO (lead oxide). Induced currents decrease with the time of irradiation due to the build-up of space-charge clouds in the electrode regions. The internal electric field shows a minimum in the center region of the sample and increases in absolute value towards the electrodes. An analogous situation is found in the gaseous ionization chamber.

Without diffusion, the rate equations for the steady state can be solved analytically. Results show that diffusion plays a minor role and might be neglected in most cases [8.230].

Charge injection from previously blocking electrodes (*electrode switching*) can be induced by the build-up of high space-charge fields in the electrode regions. When switching occurs, the current starts to increase with time and subsequently becomes constant at an increased level. The saturation effect is

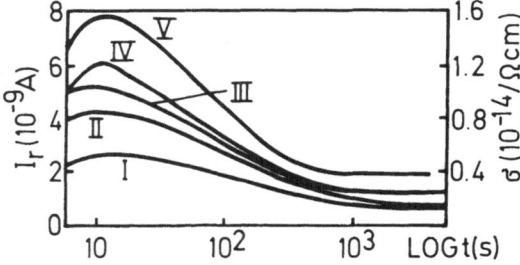

Fig. 8.14. Radiation-induced currents and induced conductivity in Teflon FEP irradiated with X-rays. Applied field 3.6 MV/m. Exposure rates: *I*-55 R/s; *II*-110 R/s; *III*-165 R/s; *IV*-195 R/s; *V*-220 R/s. The current peak indicates transition from trap-controlled to recombination-controlled conduction [8.226]

due to recombination of electrode-injected carriers with carriers of the opposite polarity present in the space-charge clouds. Such an effect has been reported for PETP [8.231].

An effect indicative of blocking electrodes has been found in PVDF. An overshoot of the current is observed when a bias voltage is applied to a sample which has been irradiated previously in short-circuit. Application of the field removes free carriers from the sample and reduces their concentration to the value compatible with the applied field. If the electrodes were neutral, removed carriers would be replaced by carriers provided by the electrodes and the carrier concentration would remain the same [8.232].

The opposite effect is observed in PI (Kapton) and PETP. Application of a voltage to a sample previously irradiated in short-circuit produces a current equal to that which would have been observed if the sample had been irradiated and poled continuously. Reversal of the applied voltage during irradiation changes the polarity of the current, but not its magnitude [8.233, 234].

Interface effects have been reported for irradiation with high-energy (18 MeV) [8.235] and low-energy (10–220 keV) [8.236] X-rays. The last paper gives a theoretical treatment and lists some of the numerous publications in the field.

Charge leakage in irradiated dielectrics due to radiation-induced currents is a disadvantage for many applications. *Radiation-hardened dielectrics* can be obtained by chemical doping which leads to the generation of traps for electrons and holes. In particular, PETP, doped with 2,4,6-trinitro-9-fluorene, has a radiation-induced conductivity of 1% of that of the undoped material [8.237, 238].

8.4.3 Irradiation with Partially Penetrating Electrons

Effects in dielectrics irradiated with non-penetrating or partially penetrating electrons (Sect. 4.3) are complex due to the injection of excess charge by the incident beam, the depth dependence of the absorbed dose, and the formation of space charges associated with the gradient of the beam current. Experiments are frequently carried out by the method of the split Faraday cup (Sect. 4.3.2). If the maximum range of the electron beam is smaller than the sample thickness, the rear electrode can be considered blocking while the irradiated electrode is usually assumed to be neutral.

A simple approach is found in the *box model* in which the rear section of the sample (between the maximum electron range and the rear electrode) is assumed non-conducting and charge-free, while the front section (between the front electrode and the maximum electron range) is assumed conducting and equally charge-free, with a charge layer formed at the interface between these sections (Sect. 4.4.1). The model gives a good description of the trap-controlled build-up of *radiation-induced conductivity* in Teflon irradiated with 40 keV electrons [8.239]. The radiation-induced current goes through a maximum before finally approaching zero. The model has been used for determination of values of radiation-induced prompt and delayed conductivities in Kapton, Teflon, Mylar, and solar-cell cover glass [8.240].

A more general model still divides the dielectric into a non-conducting non-irradiated and a conducting irradiated region, but takes into account *current and dose-deposition profiles* in the irradiated region (Sect. 4.3.6) [8.241]. The model has been further generalized by considering the dark conductivity in the non-irradiated region and back-scattering from the front surface and has been used for calculation of the surface-potential build-up under irradiation [8.242]. Bremsstrahlung and X-ray effects, which were also considered, can be neglected in most cases [8.243].

The previously discussed "conduction" models neglect transfer of carriers from the irradiated into the non-irradiated region through the virtual electrode generated by the incident electron beam (Sect. 4.3.4). Carrier transfer from the irradiated into the non-irradiated region can no longer be neglected for biased samples. It leads to the generation of monopolar currents which involve hole or electron transfer, depending on the polarity of the applied voltage. Measurements of carrier transport through PVDF have shown long-lasting transients leading to a steady state where the current-voltage characteristic is given by *Child*'s law [8.244]. Analysis allowed determination of the carrier mobilities which were found to be the same for electrons as for holes [8.136].

While the previous model still assumes a radiation-induced conductivity in the irradiated region, a very *general model* introduces mobility-controlled carrier transport in the irradiated as well as in the non-irradiated regions and realistic current and dose profiles. This model has been worked out in detail for irradiation of samples in short-circuit, biased by an external voltage, or with the front-surface floating. Comparison between experimental and theoretical results for Teflon gave good agreement [8.245, 246] (Fig. 8.15). Experimental data for PMMA, PETP, PS, ionic crystals of LiF and glasses [8.247, 248] were also analyzed by means of a similar model.

An alternative to the rate theory of conduction is the theory of *dispersive charge transport* (Sect. 4.1.1) [8.107, 236, 249, 250]. A recent paper applies the formalism of this theory to radiation-induced charge transport [8.251].

When materials are irradiated with electron beams, a fraction of the incoming electrons is re-emitted in the backward direction, due to backscattering of primary electrons and to *emission of secondary electrons*. The energies of the secondaries reach 50 eV and those of the back-scattered primaries go up to the primary energy. The total emission yield y increases with decreasing primary energy and exceeds unity between the first cross-over point (electron energy around 100 eV) and the second cross-over point (energy for polymers between 800 and 1600 eV) [8.252]. Data for secondary emission yields of polymeric materials fall on a master curve [8.253]. When y exceeds unity, the net current incident on the material is positive. If the irradiated material is a dielectric, irradiation with a low-energy beam can lead to *positive charging* of the surface of incidence [8.254]. This effect has been used for the measurement

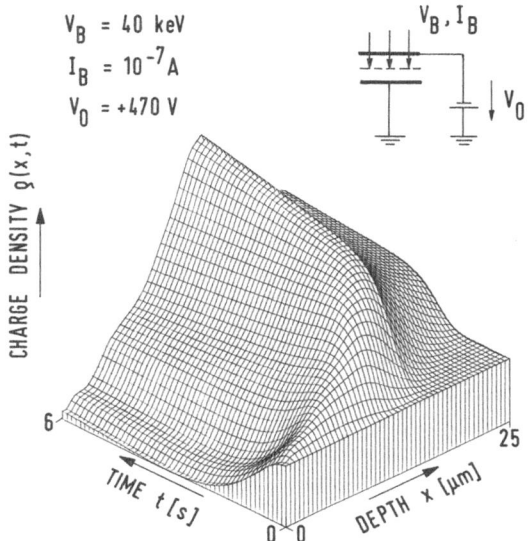

$V_B = 40$ keV

$I_B = 10^{-7}$ A

$V_0 = +470$ V

CHARGE DENSITY $\varrho(x,t)$

V_B, I_B

V_0

TIME t [s]

DEPTH x [μm]

6

25

0 0

Fig. 8.15. Theoretical charge density ϱ as a function of depth x and irradiation time t for sample irradiated through thin metal electrode with monoenergetic electrons. Irradiation set-up and parameters as shown in figure [8.245]

of y between the two cross-over points where the irradiated dielectric becomes positively biased by the beam [8.255]. With this method, emission yields for several polymers (Teflon, Aclar, Mylar, and Kapton) have been determined [8.256]. Methods using two electron beams, one charging the surface of the sample and the other intermittently probing the secondary emission, have also come into use [8.257].

8.4.4 Analysis of Internal-Field and Space-Charge Distributions by Electron Beams

Externally measurable quantities (currents, fields, surface charges) are notoriously insensitive to the spatial distribution of charges and fields. Therefore, methods of internal analysis are indispensable for the confirmation of conclusions drawn from experimental and theoretical studies and for the detailed exploration of bulk properties.

The formation of a *carrier-depletion layer* (Schottky layer) during measurements of conductivity-glow curves (Sect. 4.6.4) for Teflon has been detected by irradiation of sample surfaces with low-energy (5 keV) electron beams. Irradiation generates an increased conductivity in the electrode region and thus neutralizes the charge layer by the generation of discharge currents [8.258].

Important new methods with high resolution are the gradual exploration of the bulk of a charged dielectric by means of pressure pulses and monoenergetic electron beams whose energy is increased in steps (Sect. 8.2.2).

The knowledge of *charge-deposition profiles* is required for the interpretation of effects induced in dielectrics by irradiation with partially penetrating electron beams. Integral and differential profiles can be determined by a method in which samples are cut into sections of different thicknesses (Sect. 4.3.1) which are made conductive by carbon filling or externally metallized. During irradiation the profiles are obtained by measuring currents derived from successive sections of the sample [8.259]. Current transmission, integral charge deposition profiles, values of the charge centroid and of the maximum electron range have been measured for PETP irradiated with electrons between 10 and 50 keV [8.260].

Measurements of charge deposition profiles are completed by information about range profiles of photons and electrons. Extensive tables of stopping powers of a variety of materials [8.261–263] are available.

During electron irradiation of dielectrics with low conductivity and low carrier mobility, charge build-up might occur which leads to a *reduction of the electron range* [4.77]. This effect has been found to lead to local dose enhancement in plastic phantoms [8.264–266]. It occurs also during irradiation with ^{60}Co γ-rays due to Compton electron contamination of the beam [8.267, 268].

8.4.5 Radiation-Induced Breakdown Effects

Injection of electrons into dielectrics containing deep electron traps leads to the formation of space-charge layers whose field might become high enough to induce spontaneous breakdown. This effect has become of importance in connection with the problem of *space-craft charging* [8.269, 270]. Space-craft systems are usually covered by a thermal blanket made of a thin polymer film consisting of, e.g., Teflon or Kapton. In space such blankets are exposed to electron radiation with energies of the order of 20 keV. The range of these electrons is smaller than the thickness of the dielectric films (about 25 μm). If internal leakage does not limit the potential, the dielectric is charged up to the surface potential V_s at which the secondary plus back-scatter emission yield has become unity. This gives $eV_s = E_b - E_{II}$, where E_b is the electron energy and E_{II} the energy of the second cross-over point. Teflon, for which $E_{II} = 1.6$ keV, will charge up to 18.4 keV when $E_b = 20$ keV. Charging curves have been measured in model experiments and found to behave according to theory [8.271] except for Kapton where internal leakage apparently cannot be neglected.

Final surface potentials are high enough to produce breakdown, accompanied by surface flash-over, electron emission, and radio-frequency noise which often interferes with the satellite communication system. For this reason, these effects [8.272, 273], in particular structural damage caused by breakdown as shown in Fig. 8.16 [8.274], have been extensively investigated. A literature survey is given in a review paper [8.275] where further references are found. Discharges were also observed in Teflon foils of 25 μm thickness corona-charged to 3 kV [8.31] and in Mylar at similar voltages [8.276].

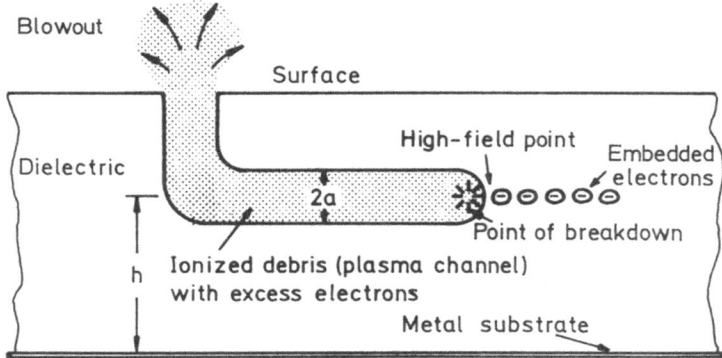

Fig. 8.16. Model for surface breakdown in a sample irradiated in vacuum by an electron beam [8.274]

A method of storage of radioactive waste is vitrification. The end product of this process is what might be called a *radioactive dielectric*, that is a dielectric body containing radioactive material. Corpuscular and γ radiation generated within the material can lead to formation of space charge and internal fields high enough to induce breakdown, as well as affect the kinetics of dissolution, sorption, and possibly evaporation. The dielectric body as a whole is neutral, since surface charges become compensated by conduction in the surrounding medium (air, a liquid, or a solid material). A particular case is that of a solid dielectric containing a β^- radioactive isotope. Properties of radioactive dielectrics have been considered in a series of papers by authors from the USSR reported in a recent survey [8.277]. Charge distributions were explored by pressure-pulse methods (Sect. 8.2.2). Spontaneous internal discharges could be observed in low-conductivity glasses (conductivity 10^{-20} S/m) containing up to 30 mCi/g of radioactive ^{90}Sr in equilibrium with ^{90}Y. Charge accumulation and occurrence of discharges are reduced by a temperature increase of the material. A particular case is the appearance of hot aerosol particles with sizes of 1 μm and activities of 1 mCi, or 10 Ci/g. Such particles acquire a net positive charge.

8.4.6 Dosimetry

Considerable interest has been devoted to the development of electret dosimetry (Sect. 4.7). The system discussed by a majority of researchers is the *self-biased ionization chamber*, where the field otherwise provided by a charged electrode or an external battery, is produced by an electret. Its advantage is the reduction of spontaneous leakage from externally accessible charged electrodes, present in conventional battery-charged meters. General aspects of electret dosimetry have been discussed in survey articles [8.278–280].

The electrets used in these chambers are charged by such methods as thermo-polarization, corona charging, and liquid-contact charging. In a recently proposed method they are charged by ionizing radiation in air. Charge carriers produced in the air by, e.g. X-rays, drift in an externally applied field and are trapped on the polymer surface generating surface-charge densities of up to $10 \, mC/m^2$ [8.45, 46]. In a modification of this method, the externally applied field is replaced by the field of a second electret. No external power supply is needed and after the charging process two electrets with the same surface potential are obtained [8.281].

The configuration of the *electric field* and the potential in such chambers has been investigated [8.282]. Electret chambers usually operate in the saturation region, their accuracy being controlled by the degree of saturation that can be maintained during irradiation. Recharging becomes necessary when the saturation ratio has fallen below a certain limit. *Saturation curves* for electret chambers are not different from those for conventional ionization chambers. Approximate analytical [8.283] and rigorous numerical [8.284–287] calculations were developed for the current as a function of the potential as well as for the space-charge distribution for different chamber currents. An approximate semi-empirical relation was also proposed [8.288]. The energy dependence of the sensitivity of electret ionization chambers for X-rays and γ-rays between 20 keV and 1.25 MeV has also been investigated [8.289, 290].

Several devices were constructed for use as *personal-protection dosimeters*. One of them operates with a moving electrode subjected to two opposing forces, one of them being the field which decreases during irradiation and the other an elastic or magnetic force (caused, respectively, by a spring and a magnet). An acoustic alarm is generated when the displacement of the electrode indicates that a predetermined dose has been reached [8.291]. A cylindrical dosimeter using corona- or radiation-charged Teflon electrets was described which can be used like the conventional ionization chamber pocket dosimeters and for read-out is plugged into a Keithley electrometer with automatic recording [8.292]. A Teflon-electret dosimeter charger was also discussed [8.293]. Several systems were designed for the measurement of Thoron and Radon in air [8.294, 295] and for passive radon-daughter dosimetry [8.296].

The energy of generation of a free carrier pair in argon is lower than in air. Therefore, use of argon as a filling gas gives a higher sensitivity than air. The sensitivity can be further increased by operation with a field strength which is sufficiently high to produce *carrier multiplication*, in analogy to the operation of the conventional proportional counter [8.297]. Other researchers have described systems which are particularly suitable for measurement of low doses in connection with radiation therapy [8.298, 299].

For measurement of *fast neutrons* one can use the ionization generated by recoil protons from highly hydrogenated wall materials, like PE [8.300–302]. *Low-energy neutrons* often generate *nuclear reactions* associated with the emission of energetic particles whose ionization effects can be measured. For this purpose, a chamber has been used whose wall material is covered with a sheet of $Li_2B_4O_7$. Irradiation with neutrons produces an (n, α) reaction with the ^6Li and ^{10}B nuclides [8.303].

X-ray dosimetry by thermally stimulated currents, generated during heating of a previously irradiated dielectric biased by an external voltage, is an alternative method of measurement. Most researchers determine the absorbed dose from the integral over the external current [8.285]. The need for current integration involves, however, some specialized equipment and the length of the measuring period is inconveniently long. These disadvantages might be avoided if the peak value of the thermally stimulated current is taken to represent the dose absorbed by the sample. The method has been successfully applied for PE [8.304]. Changes in the response function of PE have, however, been found after exposure to large doses (10 krad) of X-rays and neutrons. They have been attributed to creation of traps by the radiation [8.305].

8.5 Piezo-, Pyro-, and Ferroelectric Properties

Almost two decades [5.6] after the discovery of a strong piezoelectric effect in suitably prepared polyvinylidene fluoride (PVDF) films, piezo- and pyroelectric polymers are still a major focus of attention in electret research. The development in the understanding and the application of these fascinating materials is covered in, e.g., an extensive bibliography [8.306], three kook articles (Chap. 5) [8.307, 308], a journal volume [8.309], and two conference reports devoted specifically to piezo- and pyroelectric polymers [8.310, 311]. In view of the vast amount of original literature on the subject, only a number of highlights can be mentioned in the following. Most relevant review articles are listed in recent surveys [8.312]; in addition, a large number of early papers on the piezo- and pyroelectricity of polymers can be found in [8.313].

8.5.1 Molecular Conformation, Crystalline Structure and Morphology of Polyvinylidene Fluoride (PVDF)

The macroscopic properties of a PVDF sample depend upon the internal organisation of the material at three levels (Sects. 5.3 and 5.4.3): (1) The sequence of molecular groups in the chain, its length, and the amount of branching determine the behavior of a polymer molecule. (2) The ordered packing of several segments of chain molecules leads to crystalline lamellae with peculiar mechanical, dielectric and (sometimes) ferroelectric properties. (3) The shape, size, number, and distribution of crystallites in the amorphous polymer matrix strongly influences the observable material behavior. These properties are more or less interdependent and should be known altogether for a complete characterization of a given sample. New results about the three levels promise a much better understanding of the intriguing polymer electret PVDF and its copolymers.

Crystallization and melting in PVDF are influenced by so-called head-to-head ($-CF_2-CF_2-$) and tail-to-tail ($-CH_2-CH_2-$) defects (p. 298) which make up between 3.5 and 6% of the chain molecules in most commercial samples [8.314, 315]. A better regularity of the polymer chains and a greatly reduced frequency of chain branching was achieved in deuterated PVDF [8.315] where all hydrogen atoms are replaced by deuterium atoms. Here, nuclear magnetic resonance (NMR) of fluorine allowed for a detailed examination of chain regularity and branching in PVDF [8.315]; it may also be used to study the orientation changes of molecular dipoles upon poling [8.316].

Today, at least five *crystalline modifications of* PVDF are known [8.307]: The *β phase* with its parallel stacking of planar zig-zag (TTTT) chains is obtained from α-phase material mainly by stretching [8.317] or by application of high electric fields [8.318]. The non-polar *α phase* with alternating layers of helix molecules (TGTG'), whose dipole moments are oriented in opposite directions, is the form usually produced with extruders; details of chain packing in the α phase were derived from model calculations [8.319] and X-ray measurements [8.320]. The polar *δ phase* differs from the α phase insofar as the helical chains (TGTG') are arranged with their dipole moments parallel [8.321]; it is obtained, for example, by application of electric fields above 100 MV/m to α-phase samples [5.107, 77, 8.322, 323]; several mechanisms were suggested for this field-induced α to δ transformation [8.324–326]. The polar *γ phase* represents an intermediate between the *β* (zig-zag) and the *δ* (helix) conformations at the molecular level; its chains (TTTGTTTG') are packed with parallel dipole moments [8.327, 328]. Antiparallel packing of the same chains (TTTGTTTG') leads to the non-polar *ε phase* [8.328, 329]. The crystal structures of the three basic PVDF polymorphs (α, β, and γ) and the structural disorder in PVDF were discussed recently [8.330].

The highest piezo- and pyroelectric coefficients are found in uniaxially oriented PVDF films which consist of lamellar *β* crystals (Fig. 5.18) whose polar axes are preferentially perpendicular to the film surface [8.331, 332]. Controlled zone-drawing of PVDF can be used to obtain films with regular sequences (long periods of about 11 nm) of crystalline lamellae and amorphous layers [8.333]. In these samples, the crystallites are linked by taut tie-molecule regions which probably account for about 1.2–1.4% of the sample volume and which influence the mechanical and electrical behavior of the amorphous phase [8.333]. Biaxially oriented PVDF films prepared by simultaneous rolling and poling [8.41] exhibit long periods of about 8.5 nm and crystalline lamellae whose normals form an angle of between 20° and 33° with the film plane [8.334].

8.5.2 Copolymers and Blends (Alloys) with Polyvinylidene Fluoride (PVDF)

In order to obtain the planar zig-zag conformation of the *β*-phase molecules without stretching, monomer units of either trifluoroethylene (TrFE), tetrafluoroethylene (TFE), or chlorotrifluoroethylene (CTFE) may be copolymerized together with vinylidene fluoride (VDF) monomer. The resulting copolymers with varying portions of the respective monomers offer an increased spectrum of properties in comparison to pure PVDF and thus became a major focus of attention in recent years [8.310].

P(VDF–TrFE) copolymer forms basically three crystalline phases: regular all-trans conformation is found in a polar ferroelectric phase similar to the PVDF β phase; disordered trans molecules make up a disordered crystal structure as in pure PTrFE; a statistical combination of TTTG, TG, TTTG', and TG' conformations probably forms a paraelectric phase which is related to the α and γ phases of pure PVDF [8.335]. The examination of transitions between these phases led to an improved understanding of ferroelectric phenomena in polymer electrets [8.336–338]. Double hysteresis loops on P(VDF–TrFE) can be interpreted in terms of field-induced phase changes [8.339]; other measurements revealed a pressure effect on the Curie transition in P(VDF–TrFE) [8.340].

The existence of a basically similar behavior in *P(VDF–TFE) copolymer* demonstrates that VDF is the essential monomer for the observed ferro-, pyro-, and piezoelectricity in these copolymers [8.341, 342]. Very recently, it was found that the transition from the ferroelectric to the paraelectric phase in both copolymers can be induced not only by heating [8.336], but also by electron irradiation [8.343]. *P(VDF–CTFE) copolymer* seems to contain crystalline phases similar to the PVDF α and β phases as well [8.344].

While these copolymers differ mainly in the behavior of their crystalline phases, *compatible blends of PVDF* and a few other polymers allow one to vary the properties of the (mixed) amorphous phase systematically. Complete miscibility with PVDF (dependent on volume ratio and temperature) is found for polyvinyl acetate (PVAc), polymethyl methacrylate (PMMA), polymethyl acrylate (PMA), polyethyl methacrylate (PEMA), polyethyl acrylate (PEA), and a few other polymers [8.345]. The blends often contain pure PVDF crystallites so that their ferroelectric properties are mainly determined by the PVDF content; the dielectric behavior of, e.g., PVDF/PMMA blends is, however, strongly influenced by the PMMA content of the amorphous phase(s) [8.216, 346, 347].

8.5.3 Ferroelectricity and Switching in PVDF and Its Copolymers

While *hysteresis behavior* and switching of the polarization in β and δ PVDF were known for some time (Sect. 5.5.2) [8.348, 349], the existence of a clear transition from the ferroelectric to the paraelectric phase at the *Curie temperature* could not be demonstrated for PVDF. Only recently, the intensive study of P(VDF–TrFE) copolymers with different VDF contents led to the conclusion that the Curie transition in PVDF is masked by the onset of melting [8.336, 350, 351]; for β PVDF, a Curie temperature of about 205 °C (approximately 20 °C above the melting temperature) was extrapolated from the copolymer data as shown in Fig. 8.17 [8.336]. In P(VDF–TrFE) and

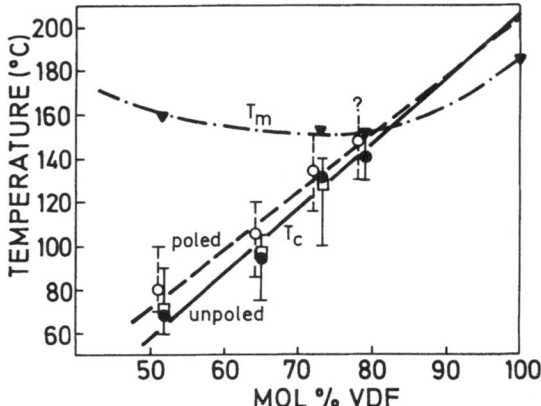

Fig. 8.17. Curie temperatures T_C and melting temperatures T_m for P(VDF–TrFE) copolymers as functions of the VDF content. Bars denote the entire ranges of the transitions as observed by means of X-ray diffraction. ○ maximum transformation rates for unpoled samples; ● maximum transformation rates for poled samples; □ dielectric peaks [8.336]

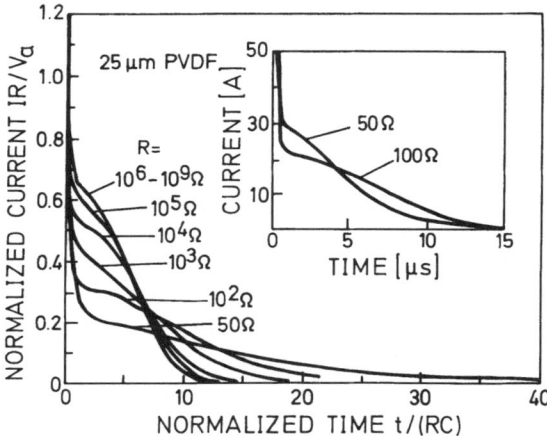

Fig. 8.18. Experimental poling currents for PVDF samples biased with $V_a = 7\,kV$. Samples with a capacitance C of approximately 7 pF and resistors of values as indicated were employed [8.360]

P(VDF–TFE) copolymers, the Curie transition is clearly established at temperatures well below the melting point [8.336, 341, 351–353]. Its detection is, however, complicated by the influence of space charges which may also lead to highly non-uniform polarization profiles in P(VDF–TrFE) [8.337, 354], P(VDF–TFE) [8.355], and PVDF [8.59] (see below).

Characteristic *switching currents* are found in PVDF [5.93] and P(VDF–TrFE) [8.338] when, at room temperature, electric fields in excess of about 100 and 50 MV/m, respectively, are applied to unpoled β samples or to samples with a polarization parallel to the field (in which case the polarization is reversed leading to a twice as high switching current). The observed currents result from the charging of the sample capacitance, which is determined by the RC time constant of the poling circuit, and from the switching itself, which is controlled by the total current available [5.93, 356–360]. Thus, the apparent switching times are, in most cases, given by the circuit parameters and not by the microscopic switching processes; this is demonstrated by the series of switching currents in Fig. 8.18 which were obtained with RC time constants ranging from 0.35 μs to 7 s [8.360]. Similar currents are also found on a different time scale (on the order of minutes) in corona-poling experiments [8.34]. Recently, high-field switching times as low as 100 ns were measured on P(VDF–TrFE) films produced by spin coating [8.361].

8.5.4 Other Piezo-, Pyro- or Ferroelectric Polymers

A survey of several piezo-, pyro- and ferroelectric polymers including a list of their piezo- and pyroelectric constants can be found in [8.308]. This review also covers some biopolymers and polymer-ceramic composites. More recently, ferroelectric hysteresis effects in uniaxially stretched films of polyamide 11 (PA 11) were discussed and compared to the respective behavior of PVDF [8.362]; the piezoelectric constants of such PA 11 films are about one order of magnitude smaller than their PVDF counterparts [8.308, 362]. Pyroelectric behavior was investigated on polyvinyl fluoride (PVF) [8.363] and PA 5, 7 [8.364]. Intrinsic shear piezoelectricity was found in a copolymer of β-hydroxybutyrate and β-hydroxyvalerate [8.365]; it is explained in terms of a spherical-dispersion model.

   Amorphous copolymers of tributyltin methacrylate (TBTM) and methyl methacrylate (MMA) [8.366] and of vinylidene cyanide (VDCN) and vinyl acetate (VAc) [8.308, 367] as well as other VDCN copolymers [8.367] were also studied for their piezo- and pyroelectric properties; the most promising of these materials is P(VDCN–VAc) which seems to surpass PVDF with respect to the piezoelectric activity in the thickness direction [8.367].

8.5.5 Preparation, Modification, and Control of Piezo- and Pyroelectric Electrets

The preparation of piezo- and pyroelectric polymer electrets includes polymerization, extrusion or casting, sometimes stretching, and usually poling [8.307]. Improved poling methods as well as the simultaneous combination of stretching and poling are discussed in Sect. 8.2.1 above. At high temperatures and high final drawing stresses, PVDF films can be drawn to ratios of about 7 [8.368, 369]; this ultra-drawn material possesses enhanced stiffness and increased piezoelectric response. Mainly for underwater applications, thick voided PVDF films with lower mechanical thickness compliance are produced by a suitable stretching procedure [8.370]. Similarly, increased piezo- and pyroelectric activity was obtained by adding small amounts of plasticizer to PVDF [8.371, 372]. Ultrathin PVDF films can be prepared by drawing from a melt on a heated glass slide [8.373] or by using a centrifuge to spread a PVDF solution evenly on a substrate [8.374].

VOLTAGE ON

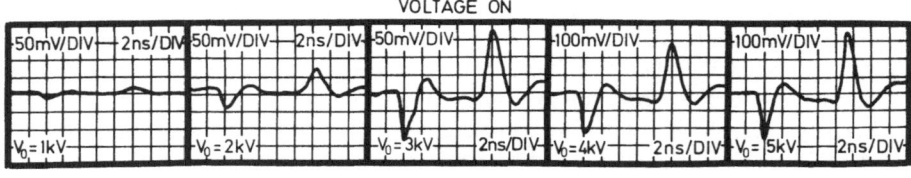

VOLTAGE OFF

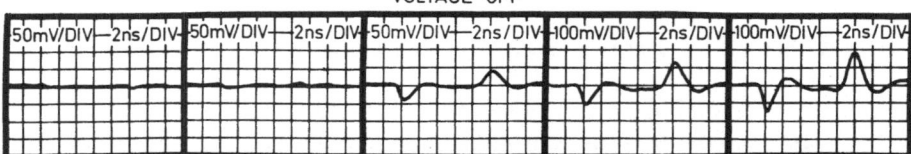

Fig. 8.19. Piezoelectricity gradients (oscilloscope traces of LIPP measurements [8.99, 100]) in 25 μm thick PVDF films during (*upper row*) and after (*lower row*) application of voltages as indicated. A permanent piezoelectric activity is only obtained for poling fields above 100 MV/m [8.100]

In addition to measuring *piezo- and pyroelectric coefficients* by means of various techniques, it is now possible to scan their *profiles* in the thickness direction of a polymer film with a resolution of about 1 μm (Sect. 8.2.2). Room-temperature poling of PVDF at fields around 100 MV/m often leads to an almost uniform profile or to a profile with a broad maximum near the center of the film [8.100, 375]. In contrast, thermal or corona poling at lower fields yields strongly non-uniform distributions (pp. 309–310) which peak near the positive electrode [8.62, 63, 99, 100, 375]. Similar profiles were apparently produced by subjecting multiple-layer samples to thermal gradients during poling [8.376, 377]; it was found that the cooler side was always more active, which is in agreement with the assumption that space-charge injection and movement is enhanced at higher temperatures. Such conduction effects may distort not only the piezoelectricity profile, but also the ferroelectric-hysteresis curves [8.378]. The field-dependent build-up of a uniform piezoelectricity profile in PVDF as a result of room-temperature poling is shown in Fig. 8.19 [8.99, 100]. The upper row shows piezoelectricity gradients during direct application of different poling fields, the lower row contains the results obtained after short-circuiting the sample electrodes. Asymmetrically polarized films can, however, be useful in flexure-mode or high-frequency applications [8.376, 377]. The so-called bending piezoelectricity (change of the electric displacement D upon a change of the sample curvature) was investigated on PP, PVDF and P(VDF–TrFE) electrets [8.379].

The *stability of the piezo- and pyroelectric activity* under various conditions is of utmost importance for industrial applications of such materials. In general, both, the piezo- and the pyroelectric coefficients decrease with increasing pressure and increase with temperature, as long as thermal depolarization is avoided [8.380–382]. The decay of the polarization seems to begin at temperatures around 60 °C [8.383–385]; it may, however, be slowed by annealing the samples at elevated temperature under a fixed strain [8.386], which reduces the activity [8.387], but enhances its long-term stability [8.386]. Thermally activated processes do not seem to reduce the reliability of pyroelectric PVDF [8.388]. γ-irradiation of PVDF films increases crosslinking in the amorphous phase, thus leading to improved dimensional and piezo-electric stability [8.389, 390]. However, the tensile strength decreases with increasing radiation dose; from the experimental data, an optimal dose of about 40 Mrad is suggested [8.390].

8.5.6 Theoretical Descriptions of the Piezo-, Pyro-, and Ferroelectricity in Polymer Electrets

A systematic classification of piezoelectric effects in different types of polymers can be found in a recent review [8.391]. For PVDF and its copolymers, there is now wide-spread agreement that the main contribution (about 2/3) to their piezo- and pyroelectricity stems from dimensional changes of the amorphous phase which lead to changes of the electrode charges induced by the permanent polarization of the crystalline phase (Sect. 5.7) [8.308, 392]. Very recently, the importance of this dimensional effect in some of the P(VDF–TrFE) copolymers was questioned [8.393]; instead, a significant contribution from electrostriction in the crystalline phase was proposed. About a dozen effects were suggested in order to explain the remaining part of the activity.

One of the main problems is the assumption of a physically realistic, but mathematically treatable *phase geometry*; most model calculations are based either on thin crystalline lamellae [5.145, 8.394] or on spherical crystallites [8.395, 396] embedded in a piezoelectrically inactive amorphous matrix. Recently, ellipsoidal crystallites were suggested as a better approximation of the real phase geometry [8.397]. The assumption of invariant crystallite shapes is, however, challenged by experimental evidence for reversible changes in the crystallinity of PVDF [8.398, 399]; therefore, models for the piezo- and pyroelectricity of PVDF, which consider the contribution of a temperature and electric-field dependence of the crystallinity, were introduced [8.400, 401]. Furthermore, a piezoelectric contribution from extended chain segments in the amorphous phase cannot be ruled out [8.402]. Both, the crystallinity changes and the contribution from extended-chain dipoles involve the amorphous phase which was examined by means of far-infrared experiments [8.403]. The temperature dependence of the piezo- and pyroelectric constants of PVDF was modelled on the basis of the simple assumption of crystalline spheres [8.404]; qualitative agreement with experimental data was achieved.

Recent calculations of the *piezoelectric response from the PVDF crystallites* started either with a point-charge model [8.396], in which the atoms of the macromolecular chain are viewed as point charges [8.405–407], or with a dipole model [8.394], in which the fields of molecular point dipoles [8.408, 409] or of extended molecular dipoles [8.410, 411] and their interactions are considered. A *quantitative comparison* [8.412] of three modelling approaches [5.145, 8.394, 396] led to the conclusion that piezoelectric and pyroelectric coefficients are predicted within the correct order of magnitude; more accurate predictions, however, are only possible with a better knowledge of fundamental material parameters [8.412].

A *phenomenological approach* to the calculation of the piezoelectric activity in PVDF [8.413, 414] led to estimates which are in reasonable agreement with the experimental data; these estimates include the pyroelectric contribution to the piezoelectric effect which is caused by adiabatic heating of the sample and thus depends on the respective time scale or frequency domain [8.414]. On the other hand, the pyroelectric activity stems in part from a piezoelectric response brought about by the thermal expansion of the material (secondary pyroelectricity). The ferroelectric phase transition of P(VDF–TrFE) copolymers was the subject of two theoretical studies involving a statistical model with two order parameters [8.415] and a phenomenological model based on the free energy [8.416].

8.6 Bioelectrets

Research on *conduction processes and electret effects in biological materials* (Chap. 6) is covered in a recent book [8.417] which contains chapters on electronic conduction in biopolymers, biological and organic superconduction at physiological temperatures, piezoelectricity in biological materials, pyroelectric effects in biological materials, and mechano-electrical effects in biological systems. Piezoelectric properties of natural biomaterials and biological polymers were also discussed in two recent reviews [8.418, 419]. The term "electro-biorheology" is used for the description of piezoelectric phenomena in biopolymers which involve also rheological processes [8.420].

Polarization phenomena in collagens from various tissues show characteristic differences related to the respective intra- and intermolecular mobilities [8.421]. Dielectric and electret effects in violuric-acid monohydrate [8.422], desoxyribonucleic acid (DNA) [8.423], lysozyme [8.424], and cellulose [8.425, 426] have the common feature of being strongly dependent upon the respective *water content*. The same is true for the piezoelectric properties of human skin [8.427] which, however, decrease with increasing moisture content; thus, piezoelectricity does not seem to be a major effect in moist skin [8.427]. All these bioelectret studies [8.422–427] are mainly based on thermal stimulation (Sect. 8.3) of the samples under investigation. A calculation of such thermally stimulated depolarization currents for the case of α-keratin was based on a correlated-walk model [8.428]. The connection between bioelectret properties and the successful application (Sect. 8.7) of piezoelectric or space-charge electrets for the stimulation of bone formation [8.429, 430] or pseudoarthrosis [8.431] is still far from being clear, since the bioelectret effect itself and its relation to other biophysical processes is not yet well understood.

8.7 Applications

Of the large number of electret applications suggested in the recent literature, only a very small selection can be discussed in the following. Particularly noteworthy are the proposed use of inorganic electrets in electroacoustic transducers and the numerous piezo- and pyroelectric devices finally reaching the commercial arena. Most applications require only small amounts of electret material; however, the devices fabricated represent a significant commerical value.

8.7.1 Capacitive Transducers

New devices now under development are electret microphones (Sect. 7.1.1) made by batch-processing techniques on silicon wafers. These transducers consist of an ultrathin diaphragm and a perforated backplate, both etched from the silicon by directional etching techniques. A charged silicon dioxide layer on the diaphragm or backplate serves as electret [8.432]. In a prototype design, charged SiO_2 layers were successfully used as backplate electrets in otherwise conventional transducers [8.433]. Very recently, silicon-etching and processing techniques were employed to construct a microminiature capacitive microphone, shown in Fig. 8.20 [8.434, 435]; it is intended to include electret biasing in later models.

Other novel electret transducers operating in the audio-frequency range or below include an integrated electret-microphone MOSFET device that has small leakage current and therefore exhibits improved infrasonic frequency response [8.436], a slit-effect transducer consisting of two pairs of electrodes separated by slits with an electret between them [8.437], an earphone that directly converts a digital input signal into an analog acoustic output signal by use of a subdivided backplate [8.438], a directional antenna with a toroidal sensitivity pattern consisting of an oblong backplate covered with a foil electret [8.439], and a toroidal microphone consisting of four electret transducers

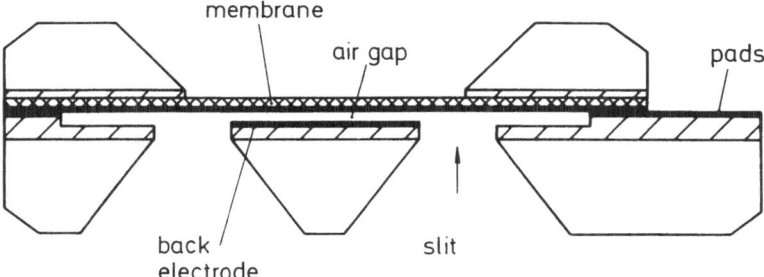

Fig. 8.20. Micro-miniature capacitive microphone made by etching technique [8.435] (Membrane area: $800 \times 800 \ \mu m^2$, thickness: $0.15 \ \mu m$)

placed in the wall of a hollow cylinder [8.440]. A transducer for the ultrasonic frequency range, made from a silicon substrate carrying a charged aluminum-oxide layer covered by silicone rubber and an aluminum electrode was also described [8.441]. At frequencies up to about 1 MHz, this transducer has a sensitivity considerably higher than conventional ZnO sensors. A foil-electret microphone optimized for picking up heart sounds by contact with the human skin was developed for blood-pressure monitoring [8.442].

Considerable progress has been made in the *analysis and optimization of existing transducer* structures. One example is an optimization scheme for electret earphones for use in telephony which are subject to material, geometric, and performance constraints [8.443]. Numerical evaluations indicate that the optimum design incorporates a backplate electret and a PETP diaphragm supported at the edge and along an inner, circular ridge. At 1 kHz, this system can produce a sound-pressure level of about 90 dB in a 6 cm^3 coupler in response to a 1 V RMS signal. Purely circular systems have smaller sensitivities, since the electret bias has to be lower to assure stability [8.444, 445]. Also of interest is a detailed analysis of the effect of gas flow damping in the air gap on the sensitivity of electret microphones [8.446]. This study shows the predominance of flow damping over radiation losses in such devices.

Applications of mass-produced electret transducers (pp. 354–355) have recently expanded into new fields. An example is the use of electret microphones in many American and other telephone handsets [8.447, 448]. For hearing-aid applications, the miniaturization potential and the low vibration sensitivity of electret microphones are of particular importance [8.449]. Recent designs have dimensions of only about 3×4 mm^2 and diaphragm thicknesses of 1–2 µm. The total annual production of all electret microphones is estimated to be about 200 million in 1985.

8.7.2 Air Filters

Electret filters (Sect. 7.4) are now available with different kinds of electret fibers [8.450, 451]: The most important ones are split fibers made by splitting of corona-charged polypropylene film [8.452] and electrostatically spun fibers ("ES" fibers) produced by atomising a polycarbonate solution and solidifying the forming fluid fibers in a strong, inhomogeneous electric field [8.453]. One of the differences between the two kinds of fibers is size, with diameters for split fibers ranging up to 20 µm or more and for ES fibers up to 1 µm. The split fibers have bipolar charge while the ES fibers are unipolarly charged with polarity slowly and irregularly changing along the fiber length. Split fibers produce larger electrostatic fields than ES fibers.

Because of these differences, the filtration characteristics for the two kinds of fibers are also different. Split-fiber filters allow higher gas velocity and offer less flow resistance than ES-fiber filters. While both kinds of filters show high initial fractional efficiency (low penetration), the efficiency decreases more for

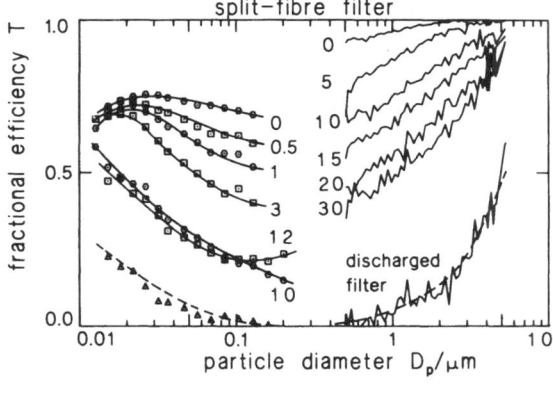

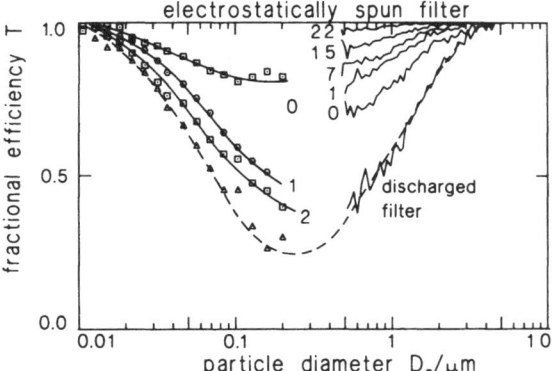

Fig. 8.21. Time-dependent fractional separation functions of two electret filters. Parameter is filtration time in hours. Note that the ES filter increases its efficiency when loaded with larger particles, due to obstruction effects [8.451]

discharged split-fiber filters than for ES-fiber filters, as shown in Fig. 8.21. This has to be attributed to the mechanical filter action of the denser ES fibers.

Air filters of these kinds are used in large numbers as face masks designed for dusty environments, as exhaust filters in vacuum cleaners, and as dust collectors in air circulators and air conditioners. Related granular-bed filters consisting of electrified thermally stable materials such as alumina are being used for the filtration of flue gases.

8.7.3 Piezoelectric Polymer Devices

A large number of suggested piezo-polymer applications were discussed in several recent articles [8.454–459]; many application-oriented papers are also listed in the bibliography by *Lang* [8.306]. An interesting example is an *audio-frequency transducer* with two PVDF diaphragms bonded in the center, as schematically shown in Fig. 8.22 [8.460]. Earphones of this design generate a sound-pressure level of 100 dB/V at 1 kHz if the diaphragms are operated in a push-pull arrangement. Such transducers may also be used as microphones or

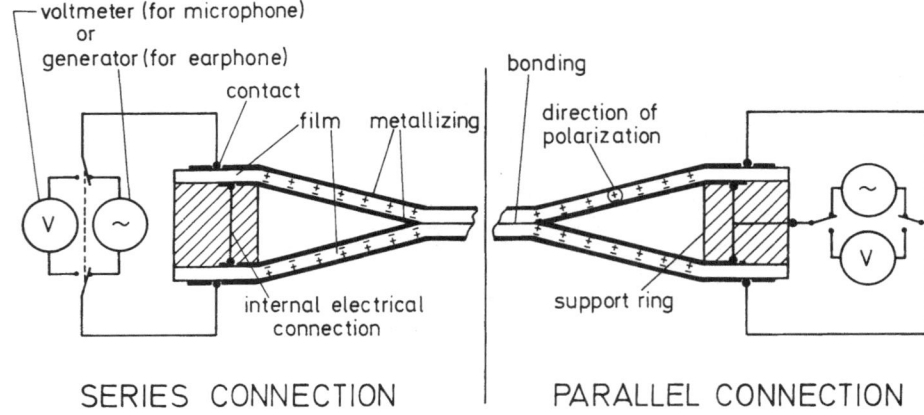

Fig. 8.22. Schematic diagram of double-diaphragm PVDF transducer [8.460]

tone ringers for telephone sets. Other novel audio-frequency transducers include designs with rigidly supported membranes [8.461] and improved systems with self-supporting, dome-shaped piezo-polymer structures [8.462]. Piezoelectric PVDF is also employed in a family of newly designed string instruments whose weak acoustic output can be electronically amplified to almost any desired level [8.463].

Presently, *ultrasonic transducers* are probably the most important application of piezoelectric polymer electrets [8.374, 454–458, 464–473]: PVDF membrane hydrophones are now routinely employed for the calibration of ultrasonic equipment [8.466]. Needle-tip hydrophones achieve spatial resolutions of better than 500 μm in complex ultrasonic fields [8.467, 468]. Multilayer transducers with Barker-coded polarity sequences show high sensitivities when used as ultrasonic-pulse receivers or transmitters [8.468, 469]. A dynamically focussed PVDF transducer composed of several annular elements was constructed for the non-destructive testing of turbine shafts [8.470]. A linear transducer array made from a single 95 μm thick P(VDF–TrFE) film produces a well-focussed ultrasonic field [8.471]. High-resolution ultrasonic imaging is achieved with an electron-beam-scanned PVDF layer which rests on the inside surface of the acoustically transparent faceplate in an image-converter tube ("ultrasonic camera") [8.472]. Solution-cast PVDF layers of about 1 μm thickness were operated as ultrasonic transducers for phonon research at frequencies of up to 24 GHz [8.374].

Medical applications of piezoelectric polymer electrets include not only ultrasonic hydrophones [8.466–471, 474], but also a wide variety of transducers for diagnostic measurements and prosthetic devices [8.474]. The very same sensors may also be employed in robotics. PVDF catheter-tip transducers were suggested for pressure, sound, and flow measurements inside the cardiovascular system [8.475]. So-called pedobarography (analysis of the pressure distribution under the foot) is possible by means of a PVDF insole with several separate electrodes [8.476].

The combination of an *optical fiber* and a piezoelectric-polymer coating permits electrically induced *phase modulation* of the transmitted light. This mechanism can be used, e.g., to measure voltages with a fiberoptic interferometer [8.477]. Recently, the frequency response of a fiber-optic phase modulator with a piezoelectric-polymer (P(VDF–TrFE)) jacket was analyzed [8.478]. Light modulation is achieved in an integrated fiber-optic Mach-Zehnder device with a PVDF film as phase shifter [8.479].

Other suggested applications of piezoelectric polymer electrets as electromechanical transducers include *vibration devices and generators:* Ring-shaped transducer elements are employed to transmit speech signals to deaf people by means of a vibrational stimulation of finger phalanges [8.480]. The vibration of boat or ship hulls by means of piezoelectric polymer films was found to greatly reduce marine fouling on the underwater surfaces of the vessels [8.481]. Implanted PVDF films were proposed as physiological power supplies for pacemakers and other artificial implants [8.482]. The use of piezoelectric electrets as mechano-electrical energy converters in sea [8.483] or wind [8.484] power plants was also suggested. The wind generators are supposed to operate on the basis of flexure-mode devices such as bimorphs whose basic properties and potential applications were extensively described in recent surveys [8.485, 486].

8.7.4 Pyroelectric Polymer Sensors

Pyroelectric polymer electrets find wide-spread use as infrared sensors in a variety of devices; only a few examples will be given here: Analogous to its utilization as target layer in an ultrasonic camera [8.472], PVDF may be employed in infrared-vidicon tubes [8.487]. In an extension of this idea, pyroelectric P(VDF–TrFE) films are deposited directly onto a silicon charge-transfer device to form a hybrid solid-state image sensor [8.488]. The fast response of thin pyroelectric PVDF films is exploited for the detection of thermal transients occuring in thermal analysis and spectroscopy [8.489]. An interesting recent application of pyroelectric polymer materials is their employment in detectors for ultraheavy nuclei and comet-dust particles [8.490]. Here, the loss of polarization caused by the impinging particle is used together with the pyroelectric response in order to determine the deposited energy and thus to identify the particle.

The high temperature sensitivity of PVDF and similar electret materials is of great advantage in pyroelectric applications, but may cause problems in piezoelectric applications. Electronic compensation of the temperature sensitivity of PVDF pressure sensors was recently demonstrated [8.491].

8.7.5 Other Electret Applications

An interesting potential application of polymer electrets is in *solar cells*. It has been demonstrated that charged polymers can deplete or invert the near-surface region in a semiconductor. This effect has been used to make experimental solar cells [8.492]. *Electret relays* (p. 360) are still receiving attention due to their low power consumption [8.493].

Recently, *slit-effect devices* (p. 368) were again studied theoretically and experimentally [8.437, 494, 495]. Two different theoretical response equations are given in the literature [8.437, 494–496]; therefore, equations relating to the slit effect should not be used without taking notice of the relevant papers. Apart from these basic studies of electret slit-effect devices, electrometers [8.497] and generators [8.498] were investigated in more detail.

Electret dosimeters (Sects. 4.7.1, 2) are an interesting application from the viewpoint of the large number of engineering implementations and because of their significant commercial use. New developments in this active field are discussed in Sect. 8.4.6 above.

In the field of *biomedical applications* (Sect. 6.9), a few important new developments have occured: Particularly significant was the demonstration that bone formation is induced by PVDF implants [8.499]. Related new studies with bimorph piezoelectric PVDF films confirmed this finding [8.429]. Recent work comparing the effects of piezoelectric (PVDF) and real-charge (FEP) implants has shown that both types of electrets are effective with respect to stimulation of bone growth [8.429].

Other recent biomedical applications are in wound treatment, where electret bandages placed on skin incisions considerably improved the tensile strength of the wound over a given period of time and thus accelerated the wound-healing process [8.500] and in endodontics, where charged Teflon was shown to enhance the formation of esteodontin when applied to core-filling material [8.501].

References

8.1   A.N.Gubkin: *Electrets* (Nauka, Moscow 1978) (In Russian);
       V.V.Gromov: *Electricheskit zarriad v obluch mater* (Energiya, Moscow 1982)
8.2   B.Hilczer, J.Małecki: *Electrets* (Elsevier, Amsterdam 1986)
8.3   D.A.Seanor (ed.): *Electrical Properties of Polymers* (Academic, New York 1982)
8.4   J.Mort, G.Pfister (eds.): *Electronic Properties of Polymers* (Wiley, New York 1982)
8.5   R.Chen, Y.Kirsh (eds.): *Analysis of Thermally Stimulated Processes* (Pergamon, Oxford 1981)
8.6   P.Bräunlich (ed.): *Thermally Stimulated Relaxation in Solids*, Topics Appl. Phys., Vol. 37
       (Springer, Berlin, Heidelberg 1979)
8.7   Annual Reports, Conference on Electrical Insulation and Dielectric Phenomena (National
       Academy of Science, Washington, DC 1979, 1980); (IEEE Service Center, Piscataway, NJ
       1981–1986)
8.8   G.M.Sessler, R.Gerhard-Multhaupt (eds.): Proc. 5th Int. Symp. on Electrets, Heidelberg
       (IEEE Service Center, Piscataway, NJ 1985)

8.9   C.H.L.Goodman (ed.): *Physics of Dielectric Solids*, 1980, Conf. Ser. 58 (Inst. of Physics, Bristol 1980)

8.10   Proc. 1st and 2nd Int. Conf. on Conduction and Breakdown in Solid Dielectrics (IEEE Service Center, Piscataway, NJ 1983 and 1986)

8.11   B.Gross: In [Ref. 8.8, pp. 9–46]

8.12   A.K.Jonscher: *Dielectric Relaxation in Solids* (Chelsea Dielectrics Press, London 1984)

8.13   G.Williams, D.C.Watt: Trans. Faraday Soc. **66**, 80 (1971)
       G.Williams, D.C.Watt, S.B.Down, A.N.North: Trans. Faraday Soc. **67**, 1323 (1972)

8.14   H.Froehlich: *Theory of Dielectrics* (Oxford Univ. Press, Oxford 1940)

8.15   K.L.Ngai, A.K.Rajagopal, R.W.Rendell, S.Teitler: In [Ref. 8.8, pp. 265–270]

8.16   B.Gross, M.T. de Figueiredo: J. Phys. D **18**, 617 (1985)

8.17   D.K.Davies: Static Electricity, in *Methods of Experimental Physics*, Vol. 16. Pt. C, ed. by R. Fava (Academic, London 1980) pp. 422–442;
       D.K.Davies: Contact Electrification of Polymers and Its Elimination, in [Ref. 8.3, pp. 285–318]

8.18   T.J.Lewis: IEEE Trans. EI-**19**, 210 (1984)

8.19   R.Brück: Kunststoffe **71**, 234 (1981)

8.20   C.B.Duke, T.J.Fabish: J. Appl. Phys. **49**, 315 (1978)

8.21   J.J.Ritsko: Electronic States and Triboelectricity, in [Ref. 8.4, pp. 13–57]

8.22   J.Fuhrmann, J.Kuerschner: J. Electrostatics **10**, 115 (1981)

8.23   Y.Murata, T.Hodoshima, S.Kittaka: Jpn. J. Appl. Phys. **18**, 2215 (1979)

8.24   C.Barnes, P.G.Lederer, T.J.Lewis, R.Toomer: J. Electrostatics **10**, 107 (1981)

8.25   D.K.Davies: IEE Proc. A **128**, 153 (1981)

8.26   W.Pong, D.Brandt, Z.X.He, W.Imaino: J. Appl. Phys. **58**, 896 (1985)

8.27   K.Ohara: J. Electrostatics **15**, 249 (1984)

8.28   J.M.Kenney, S.C. Roth: J. Res. NBS **84**, 447 (1979)

8.29   T.T.Wang, H. von Seggern: J. Appl. Phys. **54**, 4602 (1983)

8.30   T.Furukawa, G.E.Johnson: Appl. Phys. Lett. **38**, 1027 (1981)

8.31   B.Gross, J.A.Giacometti, G.F.Leal Ferreira: IEEE Trans. NS-**28**, 4513 (1981)

8.32   H. von Seggern: IEEE Trans. IA-**20**, 1623 (1984)

8.33   R.Gerhard-Multhaupt: 1983 Annual Report, CEIDP, pp. 454–459

8.34   R.Gerhard-Multhaupt: Proc. 4th Int. Conf. Diel. Materials, Measurements and Applications (1984) pp. 53–56

8.35   S.S.Bamji, K.J.Kao, M.M.Perlman: J. Electrostatics **6**, 373 (1979)

8.36   H. von Seggern, J.E.West: J. Appl. Phys. **55**, 2754 (1984)

8.37   T.T.Wang, J.E.West: J. Appl. Phys. **53**, 6552 (1982)

8.38   W.Stark, W.Kuenstler, R.Danz: Exp. Techn. Phys. **29**, 223 (1981)

8.39   T.Furukawa, T.Goho, M.Date, T.Takamatsu, E.Fukada: Konbunshi Ronbunshu **36**, 685 (1979)

8.40   P.T.A.Klaase, J. van Turnhout: Europhysics Conf. Macromolecular Physics: *Structure and Motion in Polymer Glasses*. Europhysics Abstracts **4A** (1980) pp. 46–51

8.41   B.Servet, S.Ries, D.Broussoux, F.Micheron: J. Appl. Phys. **55**, 2763 (1984)

8.42   P.Pantelis: Pyro- and Piezoelectric Polymers, in *Trends in Physics* 1984 (Proc. 6th General Conf. European Physical Society, Prag 1984) Vol. 2, pp. 537–542

8.43   J.E.McKinney, G.T.Davis, M.G.Broadhurst: J. Appl. Phys. **51**, 1676 (1980)

8.44   S.Sapieha, J.E.Klemberg-Sapieha, L.Ouellet, M.R.Wertheimer: In [Ref. 8.8, pp. 571–577]

8.45   B.G.Fallone, E.B.Podgorsak: Phys. Rev. B **27**, 2615 (1983)

8.46   B.G.Fallone, E.B.Podgorsak: Phys. Rev. B **27**, 5062 (1983)

8.47   G.Pretzsch: Nucl. Instrum. Meth. **227**, 350 (1984)

8.48   G.M.Sessler, R.Gerhard-Multhaupt: Rad. Phys. Chem. **23**, 363 (1984)

8.49   P.E.Secker, J.N.Chubb: J. Electrostatics **16**, 1 (1984)

8.50   A.G.Rozno, V.V.Gromov: Sov. Tech. Phys. Lett. **5**, 266 (1979)

8.51   C.Alquié, G.Dreyfus, J.Lewiner: Phys. Rev. Lett. **47**, 1483 (1981)

8.52   C.Alquié, J.Lewiner, G.Dreyfus: J. Phys.-Lett. **44**, L-171 (1983)

8.53   J.Lewiner: In [Ref. 8.8, pp. 429–443]

8.54 A. Migliori, T. Hofler: Rev. Sci. Instr. **53**, 662 (1982)
8.55 R. A. Anderson, S. R. Kurtz: J. Appl. Phys. **56**, 2856 (1984);
      S. R. Kurtz, R. A. Anderson: J. Appl. Phys. **60**, 681 (1986)
8.56 G. M. Sessler, J. E. West, R. Gerhard: Polym. Bull. **6**, 109 (1981)
8.57 G. M. Sessler, J. E. West, R. Gerhard: Phys. Rev. Lett. **48**, 563 (1982)
8.58 G. M. Sessler, J. E. West, R. Gerhard-Multhaupt, H. von Seggern: IEEE Trans. NS-29, 1644 (1982)
8.59 R. Gerhard-Multhaupt, G. M. Sessler, J. E. West, K. Holdik, M. Haardt, W. Eisenmenger: J. Appl. Phys. **55**, 2769 (1984)
8.60 G. M. Sessler, R. Gerhard-Multhaupt, J. E. West, H. von Seggern: J. Appl. Phys. **58**, 119 (1985)
8.61 A. Migliori, J. D. Thompson: J. Appl. Phys. **51**, 479 (1980)
8.62 W. Eisenmenger, M. Haardt: Solid State Commun. **41**, 917 (1982)
8.63 M. Haardt, W. Eisenmenger: 1982 Annual Report, CEIDP, pp. 46–51
8.64 A. G. Rozno, V. V. Gromov: Rad. Phys. Chem. **23**, 295 (1984)
8.65 R. Gerhard-Multhaupt: Phys. Rev. B**27**, 2494 (1983)
8.66 R. E. Collins: Ferroelectrics **33**, 65 (1981)
8.67 H. von Seggern, J. E. West, R. A. Kubli: Rev. Sci. Instrum. **55**, 964 (1984)
8.68 R. E. Collins: J. Appl. Phys. **51**, 2973 (1980)
8.69 F. I. Mopsik, A. S. DeReggi: J. Appl. Phys. **53**, 4333 (1982)
8.70 S. B. Lang, D. K. Das-Gupta: Ferroelectrics **39**, 1249 (1981)
8.71 S. B. Lang, D. K. Das-Gupta: Ferroelectrics **55**, 151 (1984); **60**, 23 (1984)
8.72 S. B. Lang, D. K. Das-Gupta: J. Appl. Phys. **59**, 2151 (1986)
8.73 S. B. Lang, D. K. Das-Gupta: In [Ref. 8.8, pp. 444–449]
8.74 G. M. Sessler, J. E. West, H. von Seggern: J. Appl. Phys. **53**, 4320 (1982)
8.75 D. W. Tong: IEEE Trans. EI-17, 377 (1982)
8.76 J. Hirsch, A. Y. Ko, A. Y. Irfan: IEEE Trans. EI-19, 190 (1984)
8.77 D. M. Taylor, A. A. Al-Jassar: J. Phys. D**17**, 1493 (1984)
8.78 L. Badian, J. Klocek: J. Electrostatics **8**, 69 (1979)
8.79 M. Latour, G. Donnet: J. Electrostatics **8**, 81 (1979)
8.80 S. G. Boev, G. I. Sigaev: Instrum. Exper. Techniques **24**, 1054 (1981)
8.81 M. Falck, G. Dreyfus, J. Lewiner: Phys. Rev. B**25**, 5499 (1982); 5509 (1982)
8.82 T. Takada, T. Maeno, H. Kushibe: In [Ref. 8.8, pp. 450–455]
8.83 R. Gerhard-Multhaupt, W. Petry: J. Phys. E**16**, 418 (1983)
8.84 M. Ieda: IEEE Trans. EI-19, 162 (1984)
8.85 H. J. Wintle, W. Scribuzy: J. Phys. D**17**, 2267 (1984)
8.86 J. Hagekyriakou, R. J. Fleming: J. Polym. Sci.: Polym. Phys. Ed. **21**, 1691 (1983);
      R. J. Fleming, A. Markiewicz: In [Ref. 8.8, pp. 471–475]
8.87 K. J. Kao, S. S. Bamji, M. M. Perlman: J. Appl. Phys. **50**, 8181 (1979)
8.88 H. von Seggern: J. Appl. Phys. **52**, 4086 (1981)
8.89 R. Capelletti, R. Fieschi, G. Lenzi, M. Manfredi, C. Mora, R. Reverberi: In [Ref. 8.8, pp. 463–470]
8.90 C. Alquié, F. Chapeau, J. Lewiner: 1984 Annual Report, CEIDP, pp. 488–494
8.91 H. von Seggern: J. Appl. Phys. **52**, 4081 (1981)
8.92 R. Gerhard-Multhaupt, M. Haardt, W. Eisenmenger, G. M. Sessler: J. Phys. D**16**, 2247 (1983)
8.93 A. G. Rozno, V. V. Gromov: Russ. J. Phys. Chem. **54**, 1482 (1980)
8.94 A. G. Rozno, V. V. Gromov: Rad. Phys. Chem. **22**, 555 (1983)
8.95 Y. Suzuoki, H. Muto, T. Mizutani, M. Ieda: Jpn. J. Appl. Phys. **24**, 604 (1985)
8.96 R. Singh, S. C. Datt: J. Electrostatics **6**, 247 (1979)
8.97 H. Kliem, R. Schmidt, G. Arlt: J. Appl. Phys. **53**, 1590 (1982)
8.98 F. I. Mopsik, A. S. DeReggi: Appl. Phys. Lett. **44**, 65 (1984)
8.99 G. M. Sessler, R. Gerhard-Multhaupt, H. von Seggern, J. E. West: 1984 Annual Report, CEIDP, pp. 393–398
8.100 G. M. Sessler, R. Gerhard-Multhaupt, H. von Seggern, J. E. West: In [Ref. 8.8, pp. 565–570]
8.101 S. B. Lang, A. S. DeReggi, F. I. Mopsik, M. G. Broadhurst: J. Appl. Phys. **54**, 5598 (1983)
8.102 K. Holdik, W. Eisenmenger: In [Ref. 8.8, pp. 553–558]

8.103 S.B.Lang, D.K.Das-Gupta: In [Ref. 8.8, pp. 547–552]
8.104 F.Chapeau, C.Alquié, J.Lewiner: In [Ref. 8.8, pp. 559–564]
8.105 A.I.Rudenko, V.I.Arkhipov: J. Electrostatics **4**, 309 (1978); **9**, 97 (1980)
8.106 B.Gross: J. Electrostatics **9**, 91 (1980)
8.107 V.I.Arkhipov, J.A.Popova, A.I.Rudenko: J. Electrostatics **18**, 23 (1986)
8.108 H. von Seggern: J. Appl. Phys. **50**, 7039 (1979)
8.109 B.Gross, J.A.Giacometti, G.F.Leal Ferreira: Appl. Phys. A **37**, 89 (1985)
8.110 R.Toomer, T.J.Lewis: J. Phys. D **13**, 1343 (1980)
8.111 H. von Berlepsch, L.Brehmer, D.Geiss: In [Ref. 8.8, pp. 316–321]
8.112 H. von Berlepsch: J. Phys. D **18**, 1155 (1985)
8.113 M.Campos, J.A.Giacometti: J. Appl. Phys. **52**, 4546 (1981)
8.114 D.Hohm: Ph.D. Thesis, Technical University of Darmstadt (1986)
8.115 M.M.Perlman, K.J.Kao, S.S.Bamji: J. Appl. Phys. **50**, 3622 (1979)
8.116 A.Mishra: J. Appl. Polym. Sci. **27**, 381 (1982)
8.117 P.K.Watson: 1983 Annual Report, CEIDP, pp. 421–426
8.118 J.A.Giacometti, G.F.Leal Ferreira, B.Gross: 1983 Annual Report, CEIDP, pp. 421–426, pp. 427–432;
       S.Tamura, H.Yoshioha, S.Watanabe, M.Kobayashi: In *Charge Storage, Charge Transport and Electrostatics with their Applications*, ed. by Y.Wada, M.M.Perlman, H.Kokaido (Elsevier, Amsterdam 1979) pp. 128–132
8.119 V.G.Boitsov, A.A.Rychkov: Sov. Tech. Phys. Lett. **6**, 42 (1981)
8.120 S.Haridoss, M.M.Perlman: J. Appl. Phys. **55**, 1332 (1984)
8.121 W.Stark, R.Danz: Acta Polymerica **33**, 9 (1982)
8.122 M.Beyer, K.-D.Eckhardt, Q.-Q.Lei: etz-Archiv **7**, 41 (1985)
8.123 S.Haridoss, M.M.Perlman, C.Carlone: J. Appl. Phys. **53**, 6106 (1982)
8.124 M.M.Perlman, S.Haridoss: In [Ref. 8.8, pp. 88–93]
8.125 A.J.Kovacs, J.J.Aklonis, J.M.Hutchinson, A.R.Ramos: J. Polym. Sci.: Polym. Phys. Ed. **17**, 1097 (1979)
8.126 W.A.Schneider, J.H.Wendorff, R.Gerhard-Multhaupt: 1983 Annual Report, CEIDP, pp. 441–446
8.127 D.Broemme, R.Gerhard, G.M.Sessler: 1981 Annual Report, CEIDP, pp. 129–135
8.128 H. von Seggern: 1984 Annual Report, CEIDP, pp. 468–473
8.129 A.S.DeReggi, A.J.Bur, F.I.Mopsik: Bull. Am. Phys. Soc. (II) **28**, 394 (1983)
8.130 M.Ieda, T.Mizutani, S.Ikeda: In [Ref. 8.8, pp. 106–111]
8.131 L.Brehmer: Acta Polymerica **32**, 415 (1981)
8.132 L.Brehmer, M.Pinnow, M.Kornelson, H. von Berlepsch, J.Hanspach: J. Electrostatics **14**, 19 (1983)
8.133 D.K.Davies: J. Phys. D **14**, 261 (1981)
8.134 B.Medycki, B.Hilczer: Mater. Sci. **10**, 479 (1984)
8.135 J.A.Giacometti, G.F.Leal Ferreira, B.Gross: phys. stat. sol. (a) **88**, 297 (1985)
8.136 B.Gross, H. von Seggern, R.Gerhard-Multhaupt: J. Phys. D **18**, 2497 (1985)
8.137 G.M.Sessler, B.Hahn, D.Y.Yoon: J. Appl. Phys. **60**, 318 (1986)
8.138 H. von Seggern, B.Gross, D.A.Berkley: Appl. Phys. A **34**, 163 (1984)
8.139 R.Singh, S.C.Datt: J. Electrostatics **6**, 95 (1979)
8.140 T.C.Chapman, H.J.Wintle: J. Appl. Phys. **53**, 7425 (1982)
8.141 P.Röhl: Siemens Forsch.- u. Entwickl.-Ber. **14**, 104 (1985)
8.142 M.Ieda, T.Mizutani, Y.Suzuoki: Memoirs Faculty Engin. Nagoya Univers. **32**, 173 (1980)
8.143 S.H.Carr: Thermally-Stimulated Discharge Current Analysis, in [Ref. 8.3, pp. 215–239]
8.144 D.A.Seanor: Electrothermal Analysis in *Treatise on Analytical Chemistry*, Part I, Vol. 12, Sect. J (Wiley, New York 1983) pp. 501–550
8.145 R.Capelleti, R.Fieschi, G.Lenzi, M.Manfredi, C.Mora, R.Reverberi: In [Ref. 8.8, pp. 463–470]
8.146 P.Müller: phys. stat. sol. (a) **67**, 11 (1981)
8.147 A.Kessler: phys. stat. sol. (a) **90**, 715 (1985)

8.148 H.Fritzsche, N.Ibaraki: Phil. Mag. B **52**, 299 (1985)
8.149 P.J.Dean, A.G.Cullis, A.M.White: In *Handbook on Semiconductors*, Vol. 3, ed. by S.P.Keller (North-Holland, Amsterdam 1980) pp. 113–215
8.150 J. van Turnhout: Europhys. Conf. Abstr. 7G, 24–29 (EPS, Geneva 1983); Adv. Polym. Sci. (to be published)
8.151 J.Vanderschueren, A.Linkens, J.Gasiot, J.P.Filliard, P.Parot: J. Appl. Phys. **51**, 4967 (1980); J.Vanderschueren, M.Ladang, J.Niezette: IEEE Trans. EI-**17**, 189 (1982)
8.152 J.Vanderschueren, M.Ladang, J.Niezette, M.Corapci: J. Appl. Phys. **58**, 4654 (1985)
8.153 J. van Turnhout, K.E.D.Wapenaar: Europhys. Conf. Abstr. 7 G, 128–130 (EPS, Geneva 1983)
8.154 Y.D.Dekhtyar, G.L.Sagalovich: phys. stat. sol. (a) **73**, K 35 (1982); G.Rudlof, H.Glaefeke: phys. stat. sol. (a) **85**, K 149 (1984)
8.155 S.G.Elkomoss, M.Samimi, M.Hage-Ali, P.Siffert: J. Appl. Phys. **57**, 5313 (1985)
8.156 S.Maeta, K.Sakaguchi: Jpn. J. Appl. Phys. **19**, 519 (1980); **19**, 597 (1980); A.Opanowicz: Acta Phys. Polon. A **67**, 653 (1985)
8.157 M.Böhm, O.Erb, A.Scharmann: Appl. Phys. A **37**, 165 (1985)
8.158 J.-P.Reboul, A.Toureille: J. Polym. Sci.: Polym. Phys. Ed. **22**, 21 (1984)
8.159 H. von Seggern: 1980 Annual Report, CEIDP, pp. 345–352
8.160 R.A.Moreno, M.T.Figueiredo: In [Ref. 8.8, pp. 283–287]
8.161 R.L.Remke, H. von Seggern: J. Appl. Phys. **54**, 5262 (1983)
8.162 K.Ikezaki, M.Miki, J.Tamura: Jpn. J. Appl. Phys. **20**, 1741 (1981)
8.163 S.Ikeda, K.Matsuda: Jpn. J. Appl. Phys. **21**, 359 (1982)
8.164 P.J.Atkinson, R.J.Fleming: 1980 Annual Report, CEIDP, pp. 337–344
8.165 S.S.Bamji: J. Phys. D **15**, 911 (1982)
8.166 K.Shindo: In [Ref. 8.8, pp. 169–174]
8.167 D.Ito, T.Nakakita: J. Appl. Phys. **51**, 3273 (1980)
8.168 T.Mizutani, T.Tsukahara, M.Ieda: Jpn. J. Appl. Phys. **19**, 2095 (1980)
8.169 S.Kobayashi, K.Yahagi: Jpn. J. Appl. Phys. **18**, 261 (1979)
8.170 D.K.Das-Gupta, J.S.Duffy, D.E.Cooper: J. Electrostat. **14**, 99 (1983)
8.171 S.C.Datt, R.Singh: In [Ref. 8.8, pp. 196–201]
8.172 K.Ikezaki: phys. stat. sol. (a) **90**, 383 (1985)
8.173 A.Baba, K.Ikezaki: Appl. Phys. Lett. **40**, 1027 (1982); J. Appl. Phys. **57**, 359 (1985)
8.174 J.M.Keller, S.C.Datt: phys. stat. sol. (a) **91**, 205 (1985)
8.175 R.P.Bhardwaj, J.K.Quamara, B.L.Sharma, K.K.Nagpaul: J. Phys. D **17**, 1013 (1984)
8.176 J.J.Del Val, A.Alegria, J.Colmenero, C.Lacabanne: In [Ref. 8.8, pp. 175–180]; P.K.C.Pillai, A.Tripathi, A.K.Tripathi, T.C.Goel: In [Ref. 8.8, pp. 117–125]; see also [8.200]
8.177 P.C.Mehendru, S.Chand, N.Kumar: Thin Solid Films **105**, 103 (1983)
8.178 I.M.Talwar, H.C.Sinha, A.P.Srivastava: Thin Solid Films **113**, 251 (1984)
8.179 R.Singh, S.C.Datt: In [Ref. 8.8, pp. 191–195]
8.180 R.Sharma, L.V.Sud, P.K.C.Pillai: Polymer **21**, 925 (1980)
8.181 T.Sukurai: Jpn. J. Appl. Phys. **24**, 79 (1985)
8.182 R.Sekar, A.K.Tripathi, T.C.Goel, P.K.C.Pillai: In [Ref. 8.8, pp. 181–190]
8.183 P.K.C.Pillai, Rashmi: J. Appl. Phys. **51**, 3425 (1980)
8.184 G.Micocci, S.Mongelli, A.Rizzo, A.Tepore: Solid State Commun. **53**, 873 (1985)
8.185 D.S.Misra, V.A.Singh, S.C.Agarwal: Solid State Commun. **55**, 147 (1985); Phys. Rev. B **32**, 4052 (1985)
8.186 E.B.Yakimov, N.A.Yarykin, V.I.Nikitenko: phys. stat. sol. (a) **84**, 443 (1984)
8.187 R.J.Cava, R.M.Fleming, E.A.Rietman, R.G.Dunn, L.F.Schneemeyer: Phys. Rev. Lett. **53**, 1677 (1984)
8.188 R.Capelletti, M.G.Bridelli, M.Friggeri, G.Ruani, R.Földvari, L.Kovacs, A.Watterich: In [Ref. 8.8, pp. 294–298]
8.189 J.Prakash, Rahul, A.K.Nishad: J. Appl. Phys. **59**, 2129 (1986)
8.190 M.Böhm, G.Peter, A.Scharmann, M.Thomae: In [Ref. 8.8, pp. 229–234]
8.191 M.B.Dutt, D.E.Day: J. Non-Cryst. Sol. **71**, 125 (1985)
8.192 F.Fischer, J.Prakash, W.Schober: phys. stat. sol. (a) **89**, K 215 (1985)
8.193 D.Andre, A.Dworkin, P.Figuiere, A.H.Fuchs, H.Szwarc: J. Phys. Chem. Sol. **46**, 505 (1985)

8.194 A.Torres, J.Jiminez, J.C.Merino, J.A.DeSaja, F.Sobron: J. Phys. Chem. Sol. **46**, 665 (1985); A.Torres, J.Jiminez, B.Vega, J.A.DeSaja: In [Ref. 8.8, pp. 530–535]
8.195 J.P.Calame, J.J.Fontanella, M.C.Wintersgill, C.G.Andeen: J. Appl. Phys. **58**, 2811 (1985)
8.196 D.Ronarc'h, P.Audren, S.Haridoss, J.Herrou: J. Appl. Phys. **54**, 4439 (1983)
8.197 W.A.Schneider, J.H.Wendorff: In *Electronic Properties of Polymers and Related Compounds*, ed. by H.Kuzmany, M.Mehring, S.Roth, Springer Ser. Solid-State Sci., Vol. 63 (Springer, Berlin, Heidelberg 1985) pp. 309–316
8.198 J.Belana, P.Colomer, M.Pujal, S.Montserrat: J. Macromol. Sci.-Phys. B **23**, 467 (1985); Y.Zhou, W.Li, G.Tong, Z.Qi, R.Hayakawa: In [Ref. 8.8, pp. 277–282]
8.199 K.Ohara: J. Electrostatics **8**, 299 (1980)
8.200 J.J. del Val, A.Alegria, J.Colmenero, C.Lacabanne: J. Appl. Phys. **59**, 3829 (1986)
8.201 P.C.Mehendru, K.Jain, N.P.Gupta, S.Chand: J. Macromol. Sci.-Phys. B **20**, 581 (1981)
8.202 S.K.Shrivastava, J.D.Ranadi, A.D.Srivastava: Thin Solid Films **67**, 201 (1980)
8.203 K.Ohara: J. Electrostatics **18**, 179 (1986)
8.204 A.Bernes, D.Chatain, J.-P.Ibar, C.Lacabanne: In [Ref. 8.8, pp. 399–403]
8.205 V.V.R.Narasimha Rao, N.Narsang Das: Polymer **27**, 523 (1986)
8.206 J.L.Gil-Zambrano, C.Juhasz: In [Ref. 8.8, pp. 235–240]
8.207 B.L.Sharma, J.K.Quamara: In [Ref. 8.8, pp. 241–246]
8.208 H.S.Faruque, C.Lacabanne: Polymer **27**, 527 (1986)
8.209 J.Sworakowski, M.T.Figueiredo, G.F.Leal Ferreira, M.Campos: J. Appl. Phys. **54**, 4523 (1983); **56**, 1149 (1984)
8.210 K.K.Saini, I.M.Talwar, N.Lal, K.Mahash, K.K.Nagpaul: Acta Polymerica **34**, 16 (1983)
8.211 P.Pissis: J. Phys. D **18**, 1897 (1985)
8.212 E.Marchal, V.M.Monroy Soto, J.-C.Galin: In [Ref. 8.8, pp. 247–252]
8.213 K.Mazur: In [Ref. 8.8, pp. 271–276]
8.214 A.Gourari, M.Bendaoud, C.Lacabanne, R.F.Boyer: J. Polym. Sci.: Polym. Phys. Ed. **23**, 889 (1985)
8.215 P.K.C.Pillai, G.K.Narula, A.K.Tripathi, R.G.Mendiratta: Phys. Rev. B **27**, 2508 (1983)
8.216 H.Frensch, J.H.Wendorff: In [Ref. 8.8, pp. 132–137]
8.217 C.Lacabanne, D.Chatain, T.El Sayed, D.Broussoux, F.Micheron: Ferroelectrics **30**, 307 (1980)
8.218 T.Mizutani, T.Yamada, M.Ieda: J. Phys. D **14**, 1139 (1981); T.Yamada, T.Mizutani, M.Ieda: J. Phys. D **15**, 289 (1982)
8.219 T.Mizutani, T.Nagata, M.Ieda: J. Phys. D **17**, 1883 (1984); T.Yamada, T.Mizutani, M.Ieda: Jpn. J. Appl. Phys. **23**, 738 (1984)
8.220 S.Eliasson: J. Phys. D **18**, 275 (1985)
8.221 K.Doughty, D.K.Das-Gupta: J. Phys. D **19**, L 29 (1986)
8.222 W.Ehrenberg, D.J.Gibbons: *Electron Bombardment Induced Conductivity* (Academic, London 1981)
8.223 V.A.J.van Lint, T.M.Flanagan, R.E.Leadon, J.A.Naber, V.C.Rogers: *Mechanism of Radiation Effects in Electronic Materials*, Vol. 1 (Wiley, New York 1980)
8.224 L.Weaver, J.K.Shultis, R.E.Faw: J. Appl. Phys. **48**, 2762 (1977)
8.225 D.R.Wolters, J.J.van der Schoot: J. Appl. Phys. **58**, 831 (1985)
8.226 B.Gross, R.M.Faria, G.F.Leal Ferreira: J. Appl. Phys. **52**, 571 (1981)
8.227 B.Gross, H. von Seggern, D.A.Berkley: phys. stat. sol. (a) **79**, 607 (1983)
8.228 A.Frederickson: 1981 Annual Report, CEIDP, pp. 45–51
8.229 R.C.Hughes, R.J.Sockel: J. Appl. Phys. **52**, 6743 (1981); **53**, 7414 (1982)
8.230 M.de Freitas: The solid state ionization chamber, Thesis, University of São Paulo, Brazil (1984)
8.231 R.C.Hughes: J. Appl. Phys. **51**, 5933 (1980)
8.232 R.M.Faria, B.Gross, G.F.Leal Ferreira: In [Ref. 8.8, pp. 636–641]
8.233 R.Gregorio Filho, B.Gross, R.M.Faria: In [Ref. 8.8, pp. 629–635]
8.234 S.Mochizuki, N.Tamura, K.Yahagi: J. Appl. Phys. **54**, 4433 (1983)
8.235 B.Gross: Z. Angew. Physik **30**, 323 (1971)
8.236 J.C.Garth: IEEE Trans. NS-**25**, 1598 (1978)

8.237 S.R.Kurtz, C.Arnold Jr., R.C.Hughes: IEEE Trans. NS-30, 4077 (1983); Appl. Phys. Lett. 43, 1132 (1985)

8.238 S.R.Kurtz, C.Arnold Jr.: J. Appl. Phys. 57, 2532 (1985); IEEE Trans. NS-31, 1284 (1984)

8.239 B.Gross, J.E.West, H. von Seggern, D.A.Berkley: J. Appl. Phys. 51, 4875 (1980)

8.240 B.C.Passenheim, J.A.Riddell, V.A.J. van Lint, R.Kitterer: IEEE Trans. NS-29, 1594 (1982)

8.241 K.Labonte: 1981 Annual Report, CEIDP, pp. 52–57

8.242 V.W.Pine, B.L.Beers, S.T.Ives: IEEE Trans. NS-30, 429 (1983)

8.243 A.R.Frederickson, S.Woolf: IEEE Trans. NS-29, 2004 (1982)

8.244 C.D.Child: Phys. Rev. 32, 498 (1911)

8.245 K.Labonte: IEEE Trans. NS-29, 1650 (1982)

8.246 K.Labonte: Rad. Phys. Chem. 23, 359 (1984)

8.247 V.A.Dyrkov, O.B.Evdokimov, N.I.Yagushkin: Rad. Phys. Chem. 23, 331 (1984)

8.248 O.B.Evdokimov, Yu.A.Solovyov: Rad. Phys. Chem. 23, 341 (1984)

8.249 J.Noolandi: Phys. Rev. B 16, 4466, 4474 (1977)

8.250 F.W.Schmidlin: Solid State Commun. 22, 451 (1977)

8.251 I.Arkhipov, A.I.Rudenko, S.D.Shutov: In [Ref. 8.8, pp. 592–596]

8.252 A.J.Dekker: Secondary electron emission in Solid State Physics 6, 251 (Academic, New York 1958)

8.253 E.A.Burke: IEEE Trans. NS-27, 1760 (1980)

8.254 B.Gross, H.von Seggern, J.E.West: J. Appl. Phys. 56, 2333 (1984)

8.255 H.von Seggern: IEEE Trans. NS-32, 1503 (1985)

8.256 B.Gross, H.von Seggern, A.Berraissoul: In [Ref. 8.8, pp. 608–615]

8.257 M.S.Leung, M.B.Tueling, E.R.Schnauss: Effects of Secondary Electron Emission on Charging, in: Space Craft Charging Technology 1980, NASA Conf. Publ. 2128, AFGL-TR-0270

8.258 B.Gross, J.E.West, D.A.Berkley: 1978 Annual Report, CEIDP, pp. 163–168

8.259 A.R.Frederickson, S.Woolf: IEEE Trans. NS-28, 4186 (1981)

8.260 B.Gross, R.Gerhard-Multhaupt, K.Labonte, A.Berraissoul: Coll. Polym. Sci. 262, 93 (1984)

8.261 J.H.Hubbell: Int. J. Appl. Radiat. Isot. 33, 1269 (1982)

8.262 N.J.Berger, S.M.Seltzer: Stopping powers and ranges of electrons and positrons, NBSIR 82-2550-A (1983)

8.263 "Stopping powers, ranges, and radiation yields for electrons", ICRU report 37 (1984)

8.264 D.M.Galbraith, J.A.Rawlinson, P.Munro: Med. Phys. 11, 197 (1984)

8.265 J.A.Rawlinson, A.F.Bielajew, P.Munro, D.M.Galbraith: Med. Phys. 11, 814 (1984)

8.266 L.O.Mattson, H.Svensson: Act. Rad. Oncologica 23, 393 (1984)

8.267 D.M.Galbraith, J.A.Rawlinson: Med. Phys. 11, 336 (1984)

8.268 D.M.Galbraith, J.A.Rawlinson: Med. Phys. 12, 273 (1985)

8.269 A.Rosen (ed.): Spacecraft charging by magnetospheric plasma, Progr. Astronautics and Aeronautics, Vol. 47, AIAA 1976

8.270 H.B.Garrett, C.P.Pike (eds.): Space systems and their interactions with earth's space environment, Progr. Astronautics and Aeronautics, Vol. 71, AIAA 1980

8.271 R.C.Hazelton, E.J.Yadlowski, R.C.Churchill, L.W.Parker: IEEE Trans. NS-28, 4541 (1981)

8.272 A.I.Akishin, E.A.Vitoshkin, Y.I.Tyutrin: Rad. Phys. Chem. 23, 305 (1984)

8.273 A.I.Akishin, Yu.S.Goncharov, L.S.Novikov, Yu.I.Tyutrin, L.I.Tseplyaev: Rad. Phys. Chem. 23, 319 (1984)

8.274 K.G.Balmain, G.R.Dubois: IEEE Trans. NS-26, 5146 (1979)

8.275 K.G.Balmain: In [Ref. 8.8, pp. 597–601]

8.276 P.C.Coackley, B.Kitterer, M.Treadway: IEEE Trans. NS-29, 1639 (1982)

8.277 V.V.Gromov, A.G.Sakharov: Rad. Phys. Chem. 23, 307 (1984)

8.278 G.Pretzsch: Kernenergie 27, 178 (1984)

8.279 S.Mascarenhas: In [Ref. 8.8, pp. 602–607]

8.280 Y.Kirsh, R.Chen: Rad. Eff. 83, 161 (1984)

8.281 G.Pretzsch: Nucl. Instrum. Meth. 227, 350 (1984)

8.282 E.G.Fallone, E.B.Podgorsak: J. Appl. Phys. 54, 4739 (1983)

8.283 G.Mie: Ann. Phys. Lpz. 13, 957 (1904)

8.284 J.W.Boag, T.Wilson: Brit. J. Appl. Phys. **3**, 222 (1952)
8.285 W.Armstrong, P.A.Tate: Phys. Med. Biol. **9**, 143 (1965)
8.286 P.B.Scott, J.R.Greening: Brit. J. Radiol. **34**, 791 (1961)
8.287 G.F.Leal Ferreira, L.Nunes de Oliveira, B.Gross: J. Res. NBS **79 B**, 65 (1975)
8.288 B.G.Fallone, E.B.Podgorsak: Med. Phys. **10**, 191 (1983)
8.289 P.Kotrappa, P.C.Gupta, S.K.Dua, S.D.Soman: Rad. Prot. Dosim. **2**, 175 (1982)
8.290 G.Pretzsch, U.Koestel: Rad. Prot. Dosim. **5**, 179 (1984)
8.291 G.Dreyfus, J.Lewiner, D.Périno, W.Buttler, J.C.Magne: Proc. 5th IRPA, Jerusalem (1980)
       Vol. II, pp. 153–156
8.292 J.Cameron, S.Mascarenhas: Nucl. Instr. Meth. **175**, 117 (1980)
8.293 P.Kotrappa, P.C.Gupta, S.K.Dua: Health Phys. **39**, 566 (1980)
8.294 P.Kotrappa,   S.K.Dua,   N.S.Pimpale,   P.C.Gupta,   K.S.V.Nambl,   A.M.Bhagwat,
       S.D.Soman: Health Phys. **43**, 399 (1982)
8.295 P.Kotrappa, S.K.Dua, P.C.Gupta, Y.S.Mayya: Health Phys. **41**, 35 (1981)
8.296 A.Khan, C.R.Philips: Health Phys. **46**, 141 (1984)
8.297 M.Ikeya, T.Miki: Health Phys. **37**, 797 (1980)
8.298 M.Ricard, A.Lisbona, J.P.Morucci, J.Roulleau: Nucl. Instr. Meth. **175**, 129 (1980)
8.299 J.Roulleau, A.Bui, A.Lisbona, J.P.Morucci, M.Ricard, J.Segui: Acta Radiol. Onc. Radiat.
       Ther. Phys. Biol. **20**, 143 (1981)
8.300 L.L.Campos, A.A.Suarez, S.Mascarenhas: Health Phys. **43**, 731 (1982)
8.301 G.Pretzsch, U.Koestel: Nucl. Instr. Meth. **223**, 155 (1984)
8.302 G.Pretzsch: Nucl. Instr. Meth. **243**, 183 (1985)
8.303 M.Ikeya: Jpn. J. Appl. Phys. **20**, 1615 (1981)
8.304 D.Okkalides: J. Phys. D **18**, 129 (1985)
8.305 D.Okkalides: J. Phys. D **17**, 563 (1984)
8.306 S.B.Lang: Ferroelectrics **32**, 191 (1981); **34**, 239 (1981); **45**, 283 (1982); **46**, 51 (1982); **47**, 259
       (1983); **61**, 157 (1984); **62**, 259 (1985); **67**, 223 (1986)
8.307 A.J.Lovinger: Poly(vinylidene fluoride), in *Developments in Crystalline Polymers*, ed. by
       D.C.Bassett (Applied Science, London 1982) pp. 195–273
8.308 Y.Wada: Piezoelectricity and Pyroelectricity, in [Ref. 8.4, pp. 109–160]
8.309 M.G.Broadhurst, F.Micheron, Y.Wada (eds.): PVDF and Associated Piezoelectric Polymers
       (Special Issue), Ferroelectrics **32** (1981)
8.310 K.D.Pae, Y.Wada (eds.): Piezoelectric Polymers (Joint U.S.-Japan Cooperative Science
       Seminar, Honolulu, Hawaii, 17–23 July 1983), Ferroelectrics **57** (1984)
8.311 P.M.Galletti, M.G.Broadhurst, D.E.DeRossi (eds.): Piezoelectricity in Biomaterials and
       Biomedical Devices (Intern. Symp., Pisa, Italy, 20–22 June 1983), Ferroelectrics **60** (1984)
8.312 R.Gerhard-Multhaupt: Dissertation, Technical University of Darmstadt (1984); ntz-Archiv
       7, 133 (1985); Ferroelectrics **77**, in press (1987)
8.313 W.T.Chen, E.Sacher, D.H.Strope, J.J.Woods: J. Macromol. Sci.-Phys. B **21**, 397 (1982)
8.314 L.T.Chen, C.W.Frank: Ferroelectrics **57**, 51 (1984)
8.315 R.E.Cais, J.M.Kometani: Macromolecules **17**, 1887 (1984)
8.316 R.E.Cais: Personal communication
8.317 K.Matsushige, K.Nagata, S.Imada, T.Takemura: Polymer **21**, 1391 (1980)
8.318 R.Danz: Polym. Bull. **7**, 497 (1982)
8.319 M.A.Bachmann, J.B.Lando: Macromolecules **14**, 40 (1981)
8.320 Y.Takahashi, Y.Matsubara, H.Tadokoro: Macromolecules **16**, 1588 (1983)
8.321 M.Bachmann, W.L.Gordon, S.Weinhold, J.B.Lando: J. Appl. Phys. **51**, 5095 (1980)
8.322 D.Naegele, D.Y.Yoon, M.G.Braodhurst: Macromolecules **11**, 1297 (1978)
8.323 G.R.Davies, H.Singh: Polymer **20**, 772 (1979)
8.324 H.Dvey-Aharon, P.L.Taylor, A.J.Hopfinger: J. Appl. Phys. **51**, 5184 (1980)
8.325 A.J.Lovinger: Macromolecules **14**, 225 (1981)
8.326 S.Weinhold, M.Litt, J.B.Lando: Ferroelectrics **57**, 277 (1984)
8.327 M.A.Bachmann, W.L.Gordon, J.L.Koenig, J.B.Lando: J. Appl. Phys. **50**, 6106 (1979)
8.328 A.J.Lovinger: Macromolecules **14**, 322 (1981)
8.329 A.J.Lovinger: Macromolecules **15**, 40 (1982)

8.330 Y.Takahashi, H.Tadokoro: Ferroelectrics **57**, 187 (1984)
8.331 T.T.Wang: J. Appl. Phys. **50**, 6091 (1979)
8.332 V.J.McBrierty, D.C.Douglas, T.T.Wang: Appl. Phys. Lett. **41**, 1051 (1982)
8.333 K.Yamada, M.Oie, M.Takanayagi: J. Polym. Sci.: Polym. Phys. Ed. **21**, 1063 (1983)
8.334 J.F.Legrand, J.Lajzerowicz: Ferroelectris **51**, 129 (1983)
8.335 K.Tashiro, K.Takano, M.Kobayashi, Y.Chatani, H.Tadokoro: Ferroelectrics **57**, 297 (1984)
8.336 A.J.Lovinger, T.Furukawa, G.T.Davis, M.G.Broadhurst: Ferroelectrics **50**, 227 (1983)
8.337 G.T.Davis, M.G.Broadhurst, A.J.Lovinger, T.Furukawa: Ferroelectrics **57**, 73 (1984)
8.338 T.Furukawa, M.Date, M.Ohuchi, A.Chiba: J. Appl. Phys. **56**, 1481 (1984)
8.339 N.Koizuimi, Y.Murata, H.Tsununashima: In [Ref. 8.8, pp. 936–941]
8.340 K.Matsushige, T.Horiuchi, S.Taki, T.Takemura: Jpn. J. Appl. Phys. **24**, L203 (1985)
8.341 A.J.Lovinger, G.E.Johnson, H.E.Bair, E.W.Anderson: J. Appl. Phys. **56**, 2412 (1984)
8.342 S.Tasaka, S.Miyata: J. Appl. Phys. **57**, 906 (1985)
8.343 A.J.Lovinger: Bull. Am. Phys. Soc. (II) **30**, 385 (1985); Macromolecules **18**, 910 (1985)
8.344 M.Latour: Ferroelectrics **60**, 71 (1984)
8.345 D.R.Paul, J.W.Barlow: J. Macromol. Sci.-Rev. Macromol. Chem. C **18**, 109 (1980)
8.346 J.K.Krüger, A.Marx, R.Roberts, H.-G. Unruh, J.H.Wendorff: Ferroelectrics **55**, 147 (1984);
       B.R.Hahn, J.H.Wendorff: Polymer **26**, 1611 (1985)
8.347 C.Domenici, D.DeRossi, A.Nannini, R.Verni: Ferroelectrics **60**, 61 (1984)
8.348 T.Furukawa, M.Date, E.Fukada: J. Appl. Phys. **51**, 1135 (1980)
8.349 J.I.Scheinbeim, C.H.Yoon, K.D.Pae, B.A.Newman: J. Appl. Phys. **51**, 5156 (1980)
8.350 Y.Higashihata, J.Sako, T.Yagi: Ferroelectrics **32**, 85 (1981)
8.351 T.Tashiro, K.Takano, M.Kobayashi, Y.Chatani, H.Tadokoro: Polymer **24**, 199 (1983)
8.352 T.Furukawa, G.E.Johnson, H.E.Bair, Y.Tajitsu, A.Chiba, E.Fukada: Ferroelectrics **32**, 61 (1981)
8.353 T.Yamada, T.Kitayama: J. Appl. Phys. **52**, 6859 (1981)
8.354 J.Ohwaki, H.Yamazaki, T.Kitayama: J. Appl. Phys. **52**, 6856 (1981)
8.355 M.G.Broadhurst, G.T.Davis, A.S.DeReggi, S.C.Roth, R.E.Collins: Polymer **23**, 22 (1982)
8.356 K.Matsushige, S.Imada, T.Takemura: Polym. J. **13**, 493 (1981)
8.357 T.Furukawa, G.E.Johnson: Appl. Phys. Lett. **38**, 1027 (1981)
8.358 Y.Takase, A.Odajima: Jpn. J. Appl. Phys. **21**, L707 (1982); **24**, 87 (1985)
8.359 K.Matsushige, T.Takemura: Crystal Transformation, Piezoelectricity, and Ferroelectric
       Polarization Reversal in Poly(vinylidene Fluoride), in *Structure-Property Relationships of
       Polymeric Solids*, ed. by A.Hiltner (Plenum, New York 1983) pp. 115–137
8.360 H.von Seggern, T.T.Wang: J. Appl. Phys. **56**, 2448 (1984)
8.361 T.Furukawa, H.Matsuzaki, M.Shiina, Y.Tajitsu: Jpn. J. Appl. Phys. **24**, L661 (1985)
8.362 S.C.Mathur, J.I.Scheinbeim, B.A.Newman: J. Appl. Phys. **56**, 2419 (1984)
8.363 S.B.Lang, A.S.DeReggi, M.G.Broadhurst, G.T.Davis: Ferroelectrics **33**, 119 (1981)
8.364 M.H.Litt, Ju-Chui Lin: Ferroelectrics **57**, 171 (1984)
8.365 Y.Ando, M.Minato, K.Nishida, E.Fukada: In [Ref. 8.8, pp. 871–876]
8.366 R.Liepins, M.L.Timmons, N.Morosoff: J. Polym. Sci.: Polym. Phys. Ed. **21**, 751 (1983)
8.367 S.Tasaka, K.Miyasato, M.Yoshikawa, S.Miyata, M.Ko: Ferroelectrics **57**, 267 (1984)
8.368 J.C.McGrath, I.M.Ward: Polymer **21**, 855 (1980)
8.369 E.L.Nix, L.Holt, J.C.McGrath: Ferroelectrics **32**, 103 (1981)
8.370 J.C.McGrath, L.Holt, D.M.Jones, I.M.Ward: Ferroelectrics **50**, 13 (1983)
8.371 B.A.Newman, J.I.Scheinbeim, A.Sen: Ferroelectrics **57**, 229 (1984)
8.372 A.Sen, J.I.Scheinbeim, B.A.Newman: J. Appl. Phys. **56**, 2433 (1984)
8.373 E.Häusler, W.Kaufmann, J.Petermann, L.Stein: Ferroelectrics **60**, 55 (1984)
8.374 A.Ambrosy, K.Holdik: J. Phys. E **17**, 856 (1984)
8.375 W.Eisenmenger, M.Haardt, K.Holdik: 1982 Annual Report, CEIDP, pp. 52–57
8.376 M.A.Marcus: Ferroelectrics **32**, 149 (1981)
8.377 M.A.Marcus: J. Appl. Phys. **62**, 6273 (1981)
8.378 S.Ikeda, S.Kobayashi, Y.Wada: J. Polym. Sci.: Polym. Phys. Ed. **23**, 1513 (1985)
8.379 T.Takamatsu, R.W.Tian, H.Sasabe: In [Ref. 8.8, pp. 942–946]
8.380 J.I.Scheinbeim, K.T.Chung, K.D.Pae, B.A.Newman: J. Appl. Phys. **51**, 5106 (1980)

8.381 K.T.Chung, B.A.Newman, J.I.Scheinbeim, K.D.Pae: J. Appl. Phys. **53**, 6557 (1982)
8.382 P.Destruel, F.Soto Rojas, D.Tougne, Hoang-The-Giam: J. Appl. Phys. **56**, 3298 (1984)
8.383 L.L.Blyler, Jr., G.E.Johnson, N.M.Hylton: Ferroelectrics **28**, 303 (1980)
8.384 A.G.Kolbeck: J. Polym. Sci.: Polym. Phys. Ed. **20**, 1987 (1982)
8.385 D.K.Das-Gupta, K.Doughty: Ferroelectrics **60**, 51 (1984)
8.386 T.T.Wang: J. Appl. Phys. **53**, 1828 (1982)
8.387 R.C.DeMattei, R.K.Route, R.S.Feigelson: J. Appl. Phys. **53**, 7615 (1982)
8.388 G.E.Johnson, L.L.Blyler, Jr., G.R.Crane, C.Gieniewski: Ferroelectrics **32**, 43 (1981)
8.389 T.T.Wang: J. Polym. Sci.: Polym. Lett. Ed. **19**, 289 (1981)
8.390 T.T.Wang: Ferroelectrics **41**, 213 (1982)
8.391 Y.Wada: In [Ref. 8.8, pp. 851–856]
8.392 M.G.Broadhurst, G.T.Davis: Ferroelectrics **60**, 3 (1984)
8.393 T.Furukawa: In [Ref. 8.8, pp. 883–888]
8.394 C.K.Purvis, P.L.Taylor: J. Appl. Phys. **54**, 1021 (1983)
8.395 Y.Wada, R.Hayakawa: Ferroelectrics **32**, 115 (1981)
8.396 K.Tashiro, H.Tadokoro, M.Kobayashi: Ferroelectrics **32**, 167 (1981)
8.397 G.R.Davies: "The mechanisms of piezoelectricity and pyroelectricity in poly(vinylidene fluoride)". The Dielectric Society 1984 Meeting, Abstracts of Invited Lectures
8.398 J.M.Schultz, J.S.Lin, R.W.Hendricks, R.R.Lagasse, R.G.Kepler: J. Appl. Phys. **51**, 5508 (1980)
8.399 R.G.Kepler, R.A.Anderson, R.R.Lagasse: Phys. Rev. Lett. **48**, 1274 (1982)
8.400 H.Dvey-Aharon, T.J.Sluckin, P.L.Taylor: Ferroelectrics **32**, 25 (1981)
8.401 R.G.Kepler, R.A.Anderson, R.R.Lagasse: Ferroelectrics **57**, 151 (1984)
8.402 J.DeJonge: Personal communication
8.403 M.Latour, H.Abo Dorra, J.L.Galigné: J. Polym. Sci.: Polym. Phys. Ed. **22**, 345 (1984)
8.404 Y.Wada: Ferroelectrics **57**, 343 (1984)
8.405 K.Tashiro, M.Kobayashi, H.Tadokoro, E.Fukada: Macromolecules **13**, 691 (1980)
8.406 K.Tashiro, M.Kobayashi, H.Tadokoro: Polym. Bull. **2**, 397 (1980)
8.407 K.Tashiro, H.Tadokoro: Macromolecules **16**, 961 (1983)
8.408 C.K.Purvis, P.L.Taylor: Phys. Rev. B **26**, 4547 (1982)
8.409 C.K.Purvis, P.L.Taylor: Phys. Rev. B **26**, 4564 (1982)
8.410 R.Al-Jishi, P.L.Taylor: J. Appl. Phys. **57**, 897 (1985)
8.411 R.Al-Jishi, P.L.Taylor: J. Appl. Phys. **57**, 902 (1985)
8.412 B.A.Capron, D.W.Hess: IEEE Trans. UFFC-33, 33 (1986)
8.413 A.S.DeReggi: Ferroelectrics **50**, 21 (1983)
8.414 A.S.DeReggi: Ferroelectrics **60**, 83 (1983)
8.415 A.Odajima: Ferroelectrics **57**, 159 (1984)
8.416 T.Furukawa: Ferroelectrics **57**, 63 (1984)
8.417 B.Lipinski (ed.): *Electronic Conduction and Mechanoelectrical Transduction in Biological Materials* (Dekker, New York 1982)
8.418 E.Fukada: Quart. Rev. Biophys. **16**, 59 (1983)
8.419 E.Fukada: Ferroelectrics **60**, 285 (1984)
8.420 E.Fukada: Biorheology **21**, 75 (1984)
8.421 A.Lemure, N.Hitmi, C.Lacabanne, M.-F.Harmand, D.Herbage: In [Ref. 8.8, pp. 738–743]
8.422 S.Mascarenhas, J.N.Onuchic: An. Acad. Bras. Cienc. **52**, 165 (1983)
8.423 M.Neubert, R.Bakule, J.Nedbal: In [Ref. 8.8, pp. 825–830]
8.424 M.G.Bridelli, R.Capelletti, G.Ruani, A.Vecli: In [Ref. 8.8, pp. 831–835]
8.425 S.Aouadi, M.Bendaoud, C.Lacabanne: In [Ref. 8.8, pp. 836–841]
8.426 P.Pissis, A.Anagnostopoulou-Konsta: In [Ref. 8.8, pp. 842–847]
8.427 D.DeRossi, C.Domenici, P.Pastacaldi: In [Ref. 8.8, pp. 877–882]
8.428 E.Blaisten-Barojas, S.Mascarenhas: J. Chem. Phys. **76**, 5643 (1982)
8.429 J.J.Ficat, G.Escourrou, M.J.Fauran, R.Durraux, P.Ficat, C.Lacabanne, F.Micheron: Ferroelectrics **51**, 121 (1983);
       J.J.Ficat, T.Thiechart, P.Ficat, C.Lacabanne, F.Micheron, I.Bab: Ferroelectrics **60**, 313 (1984);
       H.-L.Graf, W.Stark, R.Danz, D.Geiss: In [Ref. 8.8, pp. 813–818]

8.430 E.Fukada: In *Mechanisms of Growth Control*, ed. by R.O.Becker (C.C.Thomas, Springfield, Illinois 1981) pp. 192–210

8.431 A.E.Rodrigues Fuentes, J.P.M.Souza, V.Valeri, S.Mascarenhas: Clin. Orth. Rel. Res. **183**, 267 (1984)

8.432 D.Hohm, G.M.Sessler: Proc. 11th Intern. Congr. Acoust., Paris, Vol. **6**, pp. 29–32 (1983)

8.433 D.Hohm, R.Gerhard-Multhaupt: J. Acoust. Soc. Am. **75**, 1297 (1984)

8.434 D.Hohm: In „Fortschritte der Akustik – DAGA 85" (DPG, Bad Honnef 1985) pp. 847–850; ibid. (1986) pp. 813–816

8.435 D.Hohm: Proc. 12th Intern. Congr. Acoust., Toronto (1986) paper L3–3

8.436 J.A.Voorthuyzen, P.Bergveld: IEEE Trans. ED-32, 1185 (1985)

8.437 R.Gerhard-Multhaupt: J. Phys. D 17, 649 (1984)

8.438 J.L.Flanagan: Bell Syst. Techn. J. **59**, 1693 (1980)

8.439 I.J.Busch-Vishniac, J.E.West, R.L.Wallace: J. Acoust. Soc. Am. **76**, 1609 (1984)

8.440 G.M.Sessler, J.E.West: Acustica **57**, 193 (1985)

8.441 J.J.Bernstein, R.M.White: Proc. IEEE 1982 Ultrasonics Symp., pp. 525–528

8.442 J.E.West, I.J.Busch-Vishniac, G.A.Harshfield, T.G.Pickering: J. Acoust. Soc. Am. **74**, 680 (1983)

8.443 I.J.Busch-Vishiniac: J. Acoust. Soc. Am. **78**, 398 (1985)

8.444 I.J.Busch-Vishniac: J. Acoust. Soc. Am. **75**, 977, 990 (1984)

8.445 R.Zahn: Acustica **55**, 175 (1984)

8.446 R.Zahn: Acustica **54**, 69 (1983); **57**, 200 (1985)

8.447 J.C.Baumhauer, A.M.Brzezinski: Bell Syst. Techn. J. **58**, 1557 (1979)

8.448 S.P.Khanna, R.L.Remke: Bell Syst. Techn. J. **59**, 745 (1980)

8.449 P.V.Murphy, K.Huebschi: In [Ref. 8.8, pp. 732–737]

8.450 P.H.de Haan, J.van Turnhout, K.E.D.Wapenaar: In [Ref. 8.8, pp. 756–765]

8.451 H.Baumgartner, F.Loeffler, H.Umhauer: In [Ref. 8.8, pp. 772–777]

8.452 J.van Turnhout, W.J.Hoeneveld, J.W.C.Adams, M.van Rossen: IEEE Trans. IA-17, 240 (1981)

8.453 W.Simm: Chem. Ing. Techn. **41**, 8 (1969); K.Schmidt: Melliand Text. Ber. **61**, 495 (1980)

8.454 G.M.Sessler: J. Acoust. Soc. Am. **70**, 1596 (1981)

8.455 M.A.Marcus: Ferroelectrics **40**, 29 (1982)

8.456 N.Murayama, H.Obara: Jnp. J. Appl. Phys. **22**, Suppl, **22–3**, 3 (1983)

8.457 A.G.Holmes-Siedle, P.D.Wilson, A.P.Verrall: Mater. & Design **4**, 910 (1984)

8.458 M.A.Marcus: In [Ref. 8.8, pp. 724–731]

8.459 R.Lerch: ntz-Archiv **7**, 145 (1985)

8.460 P.V.Murphy, G.C.Maurer: In [Ref. 8.8, pp. 783–788]

8.461 R.Lerch, G.M.Sessler: J. Acoust. Soc. Am. **67**, 1379 (1980)

8.462 F.Micheron, P.Ravinet, D.Guillon, C.Claudepierre: Ferroelectrics **51**, 143 (1983)

8.463 T.S.Perry, Ch.Whiting: IEEE Spectrum **22** (11), 26 (Nov. 1985)

8.464 H.R.Gallantree: Marconi Rev. **45**, 49 (1982)

8.465 H.R.Gallantree: IEE Proc. I **130**, 219 (1983)

8.466 R.C.Preston, D.R.Bacon, A.J.Livett, K.Rajendran: J. Phys. E **16**, 786 (1983)

8.467 M.Mueller, M.Platte: Acustica **58**, 215 (1985)

8.468 M.Platte: Acustica **54**, 23 (1983)

8.469 M.Platte: Acustica **56**, 29 (1984)

8.470 W.E.Glenn, J.B.O'Maley: Mater. Eval. **41**, 1412 (1983)

8.471 K.Kimura, N.Hashimoto, H.Ohigashi: IEEE Trans. SU-32, 566 (1985)

8.472 P.H.Brown: Ferroelectrics **60**, 251 (1984)

8.473 P.A.Lewin: Ferroelectrics **60**, 127 (1984)

8.474 D.DeRossi, P.Dario: Ferroelectrics **49**, 49 (1983)

8.475 P.Dario, D.DeRossi, R.Bedini, R.Francesconi, M.G.Trivella: Ferroelectrics **60**, 149 (1984)

8.476 A.Pedotti, R.Assente, G.Fusi, D.DeRossi, P.Dario, C.Domenici: Ferroelectrics **60**, 163 (1984)

8.477 L.J.Donalds, W.G.French, W.C.Mitchell, R.M.Swinehart, T.Wei: Electron. Lett. **18**, 327 (1982)

8.478 J.Jarzynski: J. Appl. Phys. **55**, 3243 (1984)
8.479 M.Kimura, H.Takahashi, T.Miyamoto: Electron. Lett. **20**, 772 (1984)
8.480 H.Leysieffer: Acustica **58**, 196 (1985)
8.481 M.Latour, P.V.Murphy: Ferroelectrics **32**, 33 (1981)
8.482 E.Haeusler, L.Stein, G.Harbauer: Ferroelectrics **60**, 277 (1984)
8.483 E.Haeusler, L.Stein: Proc. MELECON '83 (IEEE, New York 1983) pp. D 7.12/1–3
8.484 V.H.Schmidt, M.Klakken, H.Darejeh: Ferroelectrics **51**, 105 (1983)
8.485 M.A.Marcus: Ferroelectrics **57**, 203 (1984)
8.486 P.Th.A.Klaase: Ferroelectrics **60**, 215 (1984)
8.487 J.E.Jacobs, S.A.Remily: Infrared Phys. **19**, 1 (1979)
8.488 E.Yamaka: Ferroelectrics **57**, 337 (1984)
8.489 H.Coufal: In [Ref. 8.8, pp. 801–807]
8.490 J.A.Simpson, A.J.Tuzzolino: Phys. Rev. Lett. **52**, 601 (1984); Physics Today **38** (3), 113 (March 1985)
8.491 A.J.Bur, S.C.Roth: In [Ref. 8.8, pp. 712–717]
8.492 A.Filion, M.M.Perlman: Conf. Rec. 16th IEEE Photovolt. Special. Conf. (IEEE, New York 1982) pp. 973–977;
       A.Filion, B.Noirhomme, V.Gelfandbein, A.Rambo: Conf. Rec. 17th IEEE Photovolt. Special. Conf. (IEEE, New York 1984) pp. 965–970;
       M.A.Green, T.Szpitalak, M.R.Willison, A.W.Blakers, Y.W.Lam: Conf. Rec. 15th IEEE Photovolt. Special. Conf. (IEEE, New York 1981) pp. 1418–1421;
       E.L.Heasell: Solid-State Electron. **27**, 475 (1984)
8.493 R.Sato, T.Takamatsu: In [Ref. 8.8, pp. 744–749]
8.494 A.Abazi, O.D.Jefimenko: J. Appl. Phys. **54**, 4076 (1983)
8.495 A.Prechtl: in *The Mechanical Behavior of Electromagnetic Solid Continua*, ed. by G.A.Maugin (Elsevier, Amsterdam 1984) pp. 17–21
8.496 O.D.Jefimenko: Am. J. Phys. **51**, 988 (1983);
       R.Gerhard-Multhaupt: Am. J. Phys. **53**, 375 (1985);
       O.D.Jefimenko: Am. J. Phys. **53**, 376 (1985)
8.497 O.D.Jefimenko, A.Abazi: Rev. Sci. Instrum. **53**, 1746 (1982)
8.498 Y.Tada: In [Ref. 8.8, pp. 750–755]
8.499 H.Suzuki: Abs. Jap. Comm. Electr. Frac. Bone Healing 11 (1969)
8.500 J.J.Konikoff, J.E.West: 1978 Ann. Rep., CEIDP, pp. 304–310
8.501 N.M.West, J.E.West, J.H.Revere, M.C.England: J. Endodont. **5**, 208 (1979)

Partial List of Symbols

(Additional symbols defined in text)

A	Activation energy for dipole relaxation (also U)	$\bar{r}$	Mean depth of stored charge; average range
C	Capacitance	s	Energy loss per cm; electret thickness
D	Electric displacement; dose		
$\dot{D}$	Dose rate	t	Time
E	Electric field in electret	$t_\lambda, t_{0\lambda}$	transit time
G	Energy for production of carrier pair	$\left.\begin{array}{l} x \\ y \\ z \end{array}\right\}$	Coordinates
I	Current		
I_c	Conduction current	α	Relaxation frequency of dipoles; coefficient of expansion; factor
P	Polarization		
R	Resistance		
T	Absolute temperature	β	Heating rate; compressibility
U	Activation energy		
V	Voltage, potential	∂	Density
a	Area	ε_0	Permittivity of free space
b	Recombination coefficient	$\left.\begin{array}{l} \varepsilon \\ \kappa \end{array}\right\}$	Dielectric constant
c	Velocity of light		
d_{ij}	Piezoelectric constant	μ_0	Mobility of free carriers
e	Electronic charge	μ	Trap-modulated mobility
f	Frequency; fraction	ν	Escape frequency
g	Conductivity (intrinsic, radiation-induced and delayed)	ϱ	Volume-charge density (charge per unit volume)
i	Current density (current per unit area)	σ	Planar charge density (charge per unit area)
i_c	Conduction current density	$\hat{\sigma}$	$= \int \varrho(x)dx$ integrated volume charge density
j	$\sqrt{-1}$		
k	Boltzmann's constant	$\left.\begin{array}{l} \sigma_{i1} \\ \sigma_{i2} \end{array}\right\}$	Induction-charge densities on metal electrodes
p	Carrier production rate; pyroelectric coefficient	τ	Time constant; relaxation time
r	Particle range	$\omega = 2\pi f$	Angular frequency

Additional References with Titles

Chapter 2

Adamec, V., Calderwood, J.H.: Charging phenomena and SCL regime in polymeric dielectrics. IEEE Trans. EI-21, 389 (1986)

Alquie, C., Lewiner, J.: A new method for studying piezoelectric materials. Rev. Phys. Appl. 20, 395 (1985)

Attard, A.E., Kuehls, J.F.: Polarization hysteresis in dielectric films using light diffraction by acoustic waves. Appl. Phys. Lett. 44, 522 (1984)

Bernes, A., Chatain, D., Ibar, J.P., Lacabanne, C.: Polarization phenomena in rheomolded polymers. IEEE Trans. EI-21, 347 (1986)

Bernstein, J.J., White, R.M.: Surface potential difference of anodized Al_2O_3 Electrets. J. Electrochem. Soc. 132, 1140 (1985)

Bondarenko, E.I., Zagoruiko, V.A., Kuz'minov, Yu.S., Pavlov, A.N., Panchenko, E.M., Prokopalo, O.I.: Model of the electret state in oxygen-octahedral materials. Sov. Phys. Solid State 27, 629 (1985)

Bouillier, G., Alquie, C., Dreyfus, G.: New nondestructive method for the investigation of insulator-semiconductor structures. Appl. Phys. Lett. 47, 506 (1985)

Chand, S., Mehendru, P.C.: Electrical conduction in PVF_2 films: the effect of iodine. J. Phys. D 19, 857 (1986)

Chapeau, F., Alquie, C., Lewiner, J.: The pressure-wave propagation method for the analysis of insulating materials: Applications to LPDE used in HV cables. IEEE Trans. EI-21, 405 (1986)

Coelho, R., Levy, L., Sarrail, D.: On the natural decay of corona charged Teflon sheet. Conf. Rec. Industr. Appl. Soc. IEEE-IAS-1984 Annual Meeting. (New York, IEEE 1984) pp. 1033–1037

Coelho, R.: The electrostatic characterization of insulating materials. J. Electrostatics 17, 13 (1985)

Doi, A.: Possible protonic conduction in poly(vinylidene fluoride). J. Appl. Phys. 59, 2068 (1986)

Faria, R.M., Leal Ferreira, G.F.: Vacuum-induced depolarization in Mylar under an electric field. IEEE Trans. EI-21, 339 (1986)

Fruth, B., Richter, H.J., Meurer, D.: Electrical conduction and space-charge formation in semi-crystalline polymers. IEEE Trans. EI-21, 327 (1986)

Furukawa, T., Date, M., Ishida, K., Ikeda, Y.: Computer-controlled apparatus for measuring complex elastic, dielectric, and piezoelectric constants of polymer films. Rev. Sci. Instrum. 57, 285 (1986)

Garcia, A., Grosse, C., Brito, P.: On the effect of volume charge distribution on the Maxwell-Wagner relaxation. J. Phys. D 18, 739 (1985)

George, E.P., Hanscomb, J.R., Ho, J.: Discharge currents in space-charge limited charge distributions. J. Phys. D 17, 1423 (1984)

Gross, B.: Electret research: Stages in its development. IEEE Trans. EI-21, 249 (1986)

Henson, B.L.: The integro-differential equations predicting transient decay of space-charge currents in media. J. Appl. Phys. 60, 1689 (1986)

Hibma, T., Pfluger, P., Zeller, H.R.: Electronic Processes in Polymeric Dielectrics under High Electrical Fields: in *Electronic Properties of Polymers and Related Compounds*, ed. by H. Kuzmany, M. Mehring, and S. Roth, Springer Ser. Solid-State Sci., Vol. 63 (Springer, Berlin, Heidelberg 1985) pp. 317–326

Ieda, M.: In pursuit of better electrical insulating solid polymers: Present status and future trends. IEEE Trans. EI-21, 793 (1986)

Ieda, M., Mizutani, T., Ikeda, S.: Electrical conduction and chemical structure of insulating polymers. IEEE Trans. EI-21, 301 (1986)

Jarman, R. H.: A note on the analysis of space-charge-limited current data. J. Appl. Phys. 60, 1210 (1986)

Jerzyniak, S., Hilczer, B., Szlaferek, A.: Effect of stretching on the stability of Teflon-FEP electrets. J. Electrostatics 18, 305 (1986)

Kegel, G. H R.: Fast neutrons: generation, dosimetry, and applications. IEEE Trans. EI-21, 271 (1986)

Kerimov, M. K.: Electron paramagnetic resonance of stabilized electrons in PTFE coronoelectrets. Phys. Stat. Sol. (a) 90, K 71 (1985)

Kitani, I., Arii, K.: Study on the polarity of residual voltage in polymeric insulating films. Jpn. J. Appl. Phys. 25, 1332 (1986)

Kuz'min, Yu. I., Tairov, V. N.: Percolation model for charge relaxation in electrets. Sov. Phys. Tech. Phys. 29, 575 (1984)

Lang, S. B., Das-Gupta, D. K.: Polarization and space charge distribution in thermally-poled polyethylene and a comparison with polyvinylidene fluoride. IEEE Trans. EI-21, 399 (1986)

Leal Ferreira, G. F., Oliveira, L. N., Oliveira, O. N., Giacometti, J. A.: An experimentally verified current-conservation relation. IEEE Trans. EI-21, 275 (1986)

Leal Ferreira, G. F., Oliveira, O. N., Jr., Giocometti, J. A.: Point-to-plane corona: Current-voltage characteristics for positive and negative polarity with evidence of an electronic component. J. Appl. Phys. 59, 3054 (1986)

Lewiner, J.: Evolution of experimental techniques for the study of electrical properties of insulating materials. IEEE Trans. EI-21, 351 (1986)

Lewis, T. J.: Electrical effects at interfaces and surfaces. IEEE Trans. EI-21, 289 (1986)

Maji, M. L., Chatterjee, S. D.: Magneto-Electret. Proc. Indian natn. Sci. Acad. 51, A316 (1985)

Marchal, E.: Factors affecting the location of the peak above Tg in TSD currents in polystyrene. IEEE Trans. EI-21, 323 (1986)

Matsuoka, S.: Dielectric Relaxation, in *Encyclopedia of Polymer Science and Engineering*, Vol. 5, ed. by H. F. Mark, N. M. Bikales, Ch. G. Overberger, G. Menges, and J. I. Kroschwitz (Wiley, New York 1986) pp. 23–36

Mehendru, P. C., Singh, R., Panwar, V. S., Gupta, N. P.: Hopping transport and relaxation phenomena in poly(vinylchloride: vinylacetate) terpolymer. IEEE Trans. EI-21, 297 (1986)

Moreno, R. A., de Figueiredo, M. T., Leal Ferreira, G. F.: Injection of charge from surface traps into films with deep bulk traps. IEEE Trans. EI-21, 319 (1986)

Nath, R., Kumar, A.: Polarity-reversal current transient in cellulose acetate films. IEEE Trans. EI-21, 333 (1986)

Ngai, K. L., Rajagopal, A. K., Rendell, R. W., Teitler, S.: Models of Kohlrausch relaxations. IEEE Trans. EI-21, 313 (1986)

Oda, T., Ueno, K.: Surface charge density measurements of dielectric films using ultrasonic vibration. IEEE Trans. EI-21, 375 (1986)

Pohl, H. A.: Superdielectrics polymers. IEEE Trans. EI-21, 683 (1986)

Rysiakiewicz-Pasek, E.: Electret effect in silica glass containing two types of alkaline ions. J. Electrostatics 16, 123 (1984)

Sessler, G. M., Gerhard-Multhaupt, R., von Seggern, H., West, J. E.: Charge and polarization profiles in polymer electrets. IEEE Trans. EI-21, 411 (1986)

Singh, S., Hearn, G. L.: Development and application of an electrostatic microprobe. J. Electrostatics 16, 353 (1985)

Smycz, E.: Boundary problems in the SCM theory – Characterization of contact surfaces. J. Electrostatics 15, 173 (1984)

Sworakowski, J., Leal Ferreira, G. F.: Space-charge-limited currents and trap-filled limit in one-dimensional insulators. J. Phys. D 17, 135 (1984)

Torres, A., Jimenez, J., Vega, B., de Saja, J. A.: Non-Debye behavior of dipolar relaxation in systems with dipolar interactions. IEEE Trans. EI-21, 395 (1984)

Totterdell, D.H.J.: Electron injection effects in polymer dielectrics. J. Phys. D **19**, L111 (1986)

Toureille, A., Reboul, J.P.: HV charging technique and TSD studies of polyethylene thermo-electrets. IEEE Trans. EI-**21**, 343 (1986)

Tsutsumi, N., Yamamoto, M., Nishijima, Y.: Hole transport behavior of N-phenylcarbazole-doped poly(bisphenol A carbonate) films in a glass transition region. J. Appl. Phys. **59**, 1557 (1986)

van Roggen, A., Meijer, P.H.E.: The effect of electrode-polymer interfacial layers on polymer conduction. IEEE Trans. EI-**21**, 307 (1986)

Von Seggern, H.: New developments in charging and discharging of polymers. IEEE Trans. EI-**21**, 281 (1986)

Wintle, H.J.: Capacitor edge corrections. IEEE Trans. EI-**21**, 361 (1986)

Wintle, H.J.: Point-plane and edge-plane space charge limited flows. IEEE Trans. EI-**21**, 365 (1986)

Wintle, H.J.: Reversals in Electrical Current and other Anomalies in Insulating Polymers. IEEE Trans. EI-**21**, 747 (1986)

Wintle, H.J.: Hole schubweg in FEP (Fluorinated Ethylene Propylene Copolymer). Radiat. Phys. Chem. **28**, 315 (1986)

Yamanaka, S., Fukuda, T., Sawa, G., Ieda, M.: Residual voltage in low-density Polyethylene film containing antioxidant. Jpn. J. Appl. Phys. **23**, 741 (1984)

Zahn, M.: Electro-optic field and space-charge mapping measurements in high-voltage stressed dielectrics. Phys. Technol. **16**, 288 (1985)

Chapter 3

Audren P., Ronarc'h, D.: Determination of the chain segment lengths involved in γ relaxation of low density polyethylene by thermostimulated depolarization measurements. J. Appl. Phys. **60**, 946 (1986)

Christodoulides C.: Determination of activation energies by using the widths of peaks of thermoluminescence and thermally stimulated depolarization currents. J. Phys. D **18**, 1501 (1985)

Das-Gupta, D.K., Noel, S., Cooper, D.E.: Field assisted thermally stimulated current in low-density polyethylene. J. Electrostatics **18**, 233 (1986)

Datta, T., Noufi, R., Deb, S.K.: Thermally stimulated current in P-type $CuInSe_2$ thin films. J. Appl. Phys. **59**, 1548 (1986)

Eliasson, S.: On TSD in PVDF in the temperature range – 60 to 165 c⁰: II. J. Phys. D **19**, 1965 (1986)

Felix-Vandorpe, M.C., Maitrot, M., Ongaro, R.: Electrical properties at very low frequencies of αPVDF subjected to thermal cycling. J. Phys. D **18**, 1385 (1985)

Giacometti, J.A., Malmonge, J.A.: Open-circuit TSD method and anomalous air-gap current in Teflon FEP. IEEE Trans. EI-**21**, 383 (1986)

Kalley, R., Singh, R., Datt, S.C.: Study of open circuit thermally stimulated charge decay in polypropylene. Ind. J. Pure and Appl. Phys. **23**, 107 (1985)

Lazaridou, M.: On the connection between the melting temperature and the parameters obtained from thermally stimulated depolarization experiments. Phys. Stat. Sol. (a) **88**, K19 (1985)

Sharma, S.K., Gupta, A.K., Pillai, P.K.C.: Thermally stimulated discharge current studies in poly(acrylonitrile butadien styrene) sensitized with leucomalachite green for application in electrothermography. J. Mat. Sci. Lett. **5**, 244 (1986)

Stark, W., Pinnow, M.: „Zu Problemen der Messung des thermisch stimulierten Depolarisationsstromes an Elektreten". Experim. Techn. Physik **33**, 315 (1985)

Tagyik, Y., Talwar, I.M., Lal, N., Nagpaul, K.K.: Effect of ageing on cellulose electrets. J. Mat. Sci. Lett. **5**, 331 (1986)

Takeda, Y., Kinjo, N., Narahara, T.: Thermally stimulated current in aluminium phthalocyanine chloride films. J. Mat. Sci. Lett. **4**, 80 (1985)

Chapter 4

Balmain, K.G., Battagin, A., Dubois, G.R.: Thickness scaling for arc discharges on electron-beam-charged dielectrics. IEEE Trans. NS-**32**, 4073 (1985)

Balmain, K. G.: Surface arc discharges on spacecraft dielectrics. IEEE Trans. EI-21, 427 (1986)

Berraissoul, A., Gerhard-Multhaupt, R., Gross, B.: Radiation-induced conductivity in poly-(ethylene terephthalate) irradiated with 10–40 keV electrons. Appl. Phys. A39, 203 (1986)

Boesch, H. E., Jr., Taylor, T. L.: Charge and interface state generation in field oxides. IEEE Trans. NS-32, 1273 (1984)

Boev, S. G.: Charge storage by dielectrics irradiated by accelerated electrons. Sov. Phys. Tech. Phys. 29, 1419 (1984)

Danchenko, V., Brashears, S. S., Fang, P. H.: Electron trapping in rad-hard RCA IC's irradiated with electrons and gamma rays. IEEE Trans. NS-31, 1492 (1984)

Gregorio Filho, R., Gross, B., Faria, R. M.: Induced conductivity of Mylar and Kapton irradiated by x-rays. IEEE Trans. EI-21, 431 (1986)

Kurtz, S. R., Anderson, R. A.: Radiation-induced space charge in polymer film capacitors. Appl. Phys. Lett. 49, 1484 (1986)

Miki, T., Ikeya, M., Matsuyama, M., Watanabe, K.: Tritium detection using electret dosimeter. Jpn. J. Appl. Phys. 23, L931 (1984)

Miki, T., Ikeya, M.: Theoretical response of electret dosimeter to ionizing radiation. Jpn. J. Appl. Phys. 24, 496 (1985)

Prasad, A., Kaushik, B. K., Chakarvarti, S. K., Sukheeja, B. D.: Effect of thermal annealing on alpha irradiated cellulose nitrate electrets. Acta Polymer. 36, 341 (1985)

Pretzsch, G., Dorschel, B., Schonmuth, T.: Dosimeter properties of electret ionization chambers. IEEE Trans. EI-21, 437 (1986)

Rozno, A. G., Gromov, V. V.: Electric charge distribution and radiation effects in irradiated dielectrics. IEEE Trans. EI-21, 417 (1986)

Stettner, R., De Wald, A. B.: A surface discharge model for spacecraft dielectrics. IEEE Trans. NS-32, 4079 (1985)

Chapter 5

Ando, Y., Fukada, E.: Piezoelectric properties and molecular motion of poly(β-hydroxybutyrate) films. J. Polym. Sci.: Polym Phys. Ed. 22, 1821 (1984)

Ando, Y., Minato, M., Nishida, K., Fukada, E.: Primary piezoelectric relaxation in a copolymer of beta-hydroxybutyrate and beta-hydroxyvalerate. IEEE Trans. EI-21, 505 (1986)

Bloomfield, P. E., Preis, S.: Piezoelectric and dielectric properties of heat-treated and polarized VF2/VF3 copolymers. IEEE Trans. EI-21, 533 (1986)

Bur, A. J., Barnes, J. D., Wahlstrand, K. J.: A study of thermal depolarization of polyvinylidene fluoride using x-ray pole-figure observations. J. Appl. Phys. 59, 2345 (1986)

Bur, A. J., Roth, S. C.: Measurements of a piezoelectric d constant for poly(vinylidene fluoride) transducers using pressure pulses. J. Appl. Phys. 57, 113 (1985)

Date, M.: Effect of electric field on the phase transition in vinylidene fluoride-trifluoroethylene copolymers. IEEE Trans. EI-21, 539 (1986)

de Rossi, D., Domenici, C., Pastacaldi, P.: Piezoelectric properties of dry human skin. IEEE Trans. EI-21, 511 (1986)

Elling, B., Danz, R., Weigel, P.: Reversible pyroelectricity in the melting and crystallization region of polyvinylidene fluoride. Ferroelectrics 56, 179 (1984)

Fuhrmann, J., Hofmann, R., Streibel, H. J., Jahn, U.: Electron injection at PVDF/metal interfaces. IEEE Trans. EI-21, 529 (1986)

Furukawa, T., Wen, J. X.: Electrostriction and piezoelectricity in ferroelectric polymers. Jpn. J. Appl. Phys. 23, L677 (1984)

Furukawa, T., Date, M., Nakajima, K., Kosaka, T., Seo, I.: Large dielectric relaxations in an alternate copolymer of vinylidene cyanide and vinyl acetate. Jpn. J. Appl. Phys. 25, 1178 (1986)

Furukawa, T., Wen, J. X., Suzuki, K., Takashina, Y., Date, M.: Piezoelectricity and pyro-electricity in vinylidene fluoride/trifluoroethylene copolymers. J. Appl. Phys. 56, 829 (1984)

Green, J. S., Farmer, B. L., Rabolt, J. F.: Effect of thermal and solution history on the Curie point of VF$_2$-TrFE random copolymers. J. Appl. Phys. 60, 2690 (1986)

Green, J. S., Rabe, J. P., Rabolt, J. F.: Studies of chain conformation above the Curie point in a vinylidene fluoride/trifluoroethylene random copolymer. Macromolecules **19**, 1725 (1986)

Hahn, B. R.: Studies on the nonlinear piezoelectric response of polyvinylidene fluoride. J. Appl. Phys. **57**, 1294 (1985)

Hahn, B., Wendorff, J., Yoon, D. Y.: Dielectric relaxation of the crystal-amorphous interphase in poly(vinylidene fluoride) and its blend with poly(methyl-methacrylate). Macromolecules **18**, 718 (1985)

Hsu, C. C., Geil, P. H.: Morphology-structure-property relationships in ultraquenched poly-(vinylidene fluoride). J. Appl. Phys. **56**, 2404 (1984)

Hughes, S. T., Piercy, A. R.: The simultaneous measurement of pyroelectric and relaxation currents in poly(vinylidene fluoride). J. Phys. E **19**, 976 (1986)

Humphreys, J., Ward, I. M., McGrath, J. C., Nix, E. L.: The measurement of the piezoelectric coefficients d_{31} and d_{3h} for uniaxially oriented polyvinylidene fluoride. Ferroelectrics **67**, 131 (1986)

Itoh, T., Kunimura, S., Sakaoku, K.: Transmission electron microscopic observation on ferroelectric phase transitions in polyvinylidene fluoride induced by high electric field. Ferroelectrics **65**, 95 (1985)

Jimbo, M., Fukada, T., Takeda, H., Suzuki, F., Horino, K., Koyama, K., Ikeda, S., Wada, Y.: Ferroelectric switching characteristics of 73/27 copolymer of vinylidene fluoride and trifluoroethylene. J. Polym. Sci.: Polym. Phys. Edit. **24**, 909 (1986)

Jimbo, M., Kobayashi, S., Horino, K., Ikeda, S., Wada, Y.: Analysis of the hysteresis curve of ferroelectric polymers by a phenomenological relaxation theory. J. Polym. Sci.: Polym. Phys. Edit. **22**, 2139 (1984)

Kimura, K., Ohigashi, H.: Polarization behavior in vinylidene fluoride-trifluoroethylene copolymer thin films. Jpn. J. Appl. Phys. **25**, 383 (1986)

Kobayashi, M.: Rotational isomerism and physical properties of long-chain molecules in solid states. J. Molec. Struct. **126**, 193 (1985)

Koga, K., Ohigashi, H.: Piezoelectricity and related properties of vinylidene fluoride and trifluoroethylene copolymers. J. Appl. Phys. **59**, 2142 (1986)

Koizumi, N., Murata, Y., Tsunashima, H.: Polarization reversal and double hysteresis loop in copolymers of vinylidene fluoride and trifluoroethylene. IEEE Trans. EI-**21**, 543 (1986)

Latour, M., Dorra, H. A.: Study of phase transitions in piezoelectric polyvinylidene fluoride. Revue Phys. Appl. **20**, 137 (1985)

Latour, M., Moreira, R. L.: A spectroscopic study in the mm region of poly(vinylidene fluoride) and copolymers. IEEE Trans. EI-**21**, 525 (1986)

Lu, F. J., Waldman, D. A., Hsu, S. L.: A spectroscopic study to interpret the increased piezoelectric effect at high temperature in poly(vinylidene fluoride). J. Polym. Sci.: Polym. Phys. Edit. **22**, 827 (1984)

Marcus, M. A.: Orientation effects on dielectric and piezoelectric properties of poly(vinylidene fluoride) films. IEEE Trans. EI-**21**, 519 (1986)

Micheron, F.: Comment on "Bulk photovoltaic effect in polyvinylidene fluoride". Appl. Phys. Lett. **47**, 67 (1985)

Morra, B. S., Stein, R. S.: The crystalline morphology of poly(vinylidene fluoride)/poly(methyl-methacrylate) blends. Polym. Engineer. and Sci. **24**, 311 (1984)

Muralidhar, C., Pillai, P. K. C.: Pyroelectric behavior in barium titanate/polyvinylidene fluoride composites. IEEE Trans. EI-**21**, 501 (1986)

Nix, E. L.: A direct method for measurement of the film-thickness piezoelectric coefficient of polyvinylidene fluoride. Ferroelectrics **67**, 125 (1986)

Nix, E. L., Ward, I. M.: The measurement of the shear piezoelectric coefficients of polyvinylidene fluoride. Ferroelectrics **67**, 137 (1986)

Ogden, T. R., Gookin, D. M.: Bulk photovoltaic effect in polyvinylidene fluoride. Appl. Phys. Lett. **45**, 995 (1984)

Oka, Y., Koizumi, N.: Pyroelectricity in oriented polytrifluoroethylene. Jpn. J. Appl. Phys. **24**, 669 (1985)

Olsen, R. B., Bruno, D. A., Briscoe, J. M., Jacobs, E. W.: High electric field resistivity and pyroelectric properties of vinylidene fluoride-trifluoroethylene copolymers. J. Appl. Phys. **58**, 2854 (1985)

Olsen, R. B., Bruno, D. A., Briscoe, J. M., Jacobs, E. W.: Pyroelectric conversion cycle of vinylidene fluoride-trifluoroethylene copolymer. J. Appl. Phys. **57**, 5036 (1985)

Reneker, D. H., Mazur, J.: Modelling of chain twist boundaries in poly(vinylidene fluoride) as a mechanism for ferroelectric polarization. Polymer **26**, 821 (1985)

Takase, Y., Odajima, A., Wang, T. T.: A modified nucleation and growth model for ferroelectric switching in form I poly(vinylidene fluoride). J. Appl. Phys. **60**, 2920 (1986)

Ueda, H., Fukada, E., Karasz, F. E.: Piezoelectricity in three-phase systems: effect of the boundary phase. J. Appl. Phys. **60**, 2672 (1986)

Wang, T. T., von Seggern, H., West, J. E., Keith, H. D.: High field poling of poly(vinylidene fluoride) films using a current limiting circuit. Ferroelectrics **61**, 249 (1984)

Wen, J. X.: Piezoelectric properties in uniaxially drawn copolymers of vinylidene fluoride and trifluoroethylene. Jpn. J. Appl. Phys. **23**, 1434 (1984)

Yamada, K., Oie, M., Takayanagi M.: Piezoelectricity and viscoelastic crystalline dispersion in highly oriented poly(vinylidene fluoride). J. Polym. Sci.: Polym. Phys. Edit. **22**, 245 (1984)

Yamauchi, N., Kato, K., Wada, T.: Observation of ferroelectricity in very thin vinylidene fluoride trifluoroethylene copolymer [P(VDF · TrFE)] films by high frequency C-V measurements of Al–SiO$_2$–P(VDF · TrFE)–SiO$_2$–Si capacitors. Jpn. J. Appl. Phys. **23**, L671 (1984)

Chapter 6

Lamure, A., Hitmi, N., Lacabanne, C., Harmand, M. F., Herbage, D.:Polarization phenomena in collagens from various tissues. IEEE Trans. EI-21, 443 (1986)

Takamatsu, T., Sasabe, H., Okada, K.: Callus Formation by the Piled Polymer Electrets, in *Bioelectrical Repair and Growth*, ed. by Fukada, Inoue, Sakou, Takahashi, Tsuyama (Nishimura, Kyoto 1985) pp. 159–165

Chapter 7

Baumgartner, H., Löffler, F., Umhauer, H.: Deep-bed electret filters: The determination of single fiber charge and collection efficiency. IEEE Trans. EI-21, 477 (1986)

Becker, C., Lenz, D.: PVDF transducers for MHz pulse-echo attenuation measurements in metals at low temperature. IEEE Trans. EI-21, 487 (1986)

Blackford, D. B., Brown, R. C.: An air filter made from an electret and a conductor. IEEE Trans. EI-21, 471 (1986)

Brigham, G. A.: Theoretical study of a plane, uniform piezoelectric polymer hydrophone element. J. Acoust. Soc. Am. **79**, 2067 (1986)

Choi, P. K., Takagi, K.: An attempt at ultrasonic resonator with piezoelectric polymer film. J. Acoust. Soc. Jpn. (E) **6**, 15 (1985)

Coufal, H.: Highly time-resolved calorimetry using pyroelectric thin-film electrets. IEEE Trans. EI-21, 495 (1986)

Coufal, H.: Pyroelectric detection of radiation-induced thermal wave phenomena. IEEE Trans. UFFC-33, 507 (1986)

Das-Gupta, D. K., Doughty, K.: Electro-active polymers in non-destructive dielectric evaluation. IEEE Trans. EI-20, 20 (1985)

de Haan, P. H., Van Turnhout, J., Wapenaar, K. E. D.: Fibrous and granular filters with electrically enhanced dust-capturing efficiency. IEEE Trans. EI-21, 465 (1986)

Franz, J., Hohm, D.: Elektroakustische Polymerwandler. Der Fernmelde-Ingenieur, **40**/11, 1 (1986)

Fridkin, V. M., Shlensky, A. L., Verchovskaya, K. A., Bilke, W. D., Markiewitz, N. N., Pietsch, H., Sydow, M.: Electrophotographic process based on the pyroelectric and photovoltaic effects in the ferroelectric polymer PVF$_2$. Disp. Technol. **1**, 49 (1985)

Haeusler, E., Stein, L.: Hydromechanic-electric power converter, in "Ocean Engineering and the Environment", Conf. Record. 1985 (sponsored by: Marine Techn. Soc. and IEEE Ocean Engin. Soc.) pp. 1313–1316

Hohm, D., Franz, J., Hoffmann, R.: Piezopolymerhörer als elektroakustische Digital-Analog-Wandler für linear und logarithmisch quantisierte Signale. Acustica 57, 218 (1985)

Imai, M., Tanizawa, H., Ohtsuka, Y., Takase, Y., Odajima, A.: Piezoelectric copolymer jacketed single-mode fibers for electric-field sensor application. J. Appl. Phys. 60, 1916 (1986)

Kashyap, R., Pantelis, P.: Optical fibre absorption loss measurements using a pyro-electric poly(vinylidene fluoride) tube. J. Phys. D 18, 1709 (1985)

Lancee, C. T., Souquet, J., Ohigashi, H., Bom, N.: Ferro-electric ceramics versus polymer piezo-electric materials. Ultrasonics, 23, 138 (1985)

Lee, T. M., Anderson, A. P., Benson, F. A.: Microwave field-detecting element based on pyroelectric effect in PVDF. Electron. Lett. 22, 200 (1986)

Lerch, R.: Simulation von Ultraschall-Wandlern Acustica 57, 205 (1985)

Meeks, S. W., Ting, R. Y.: The evaluation of PVF_2 for underwater shock-wave sensor application. J. Acoust. Soc. Am. 75, 1010 (1984)

Meixner, H., Kleinschmidt, P.: Plastik spürt Wärme. Der Elektroniker 10, 81 (1985)

Ogden, T. R., Gookin, D. M.: Ferroelectric polymers as an optical memory material. Mater. Lett. 3, 127 (1985)

Platte, M.: A polyvinylidene fluoride needle hydrophone for ultrasonic applications. Ultrasonics, 23, 113 (1985)

Ricketts, D.: The frequency of flexural vibration of completely free composite piezoelectric polymer plates. J. Acoust. Soc. Am. 80, 723 (1986)

Ricketts, D.: Model for a compliant tube polymer hydrophone. J. Acoust. Soc. Am. 79, 1603 (1986)

Sato, R., Takamatsu, T.: An electret switch. IEEE Trans. EI-21, 449 (1986)

Shombert, D. G., Harris, G. H.: Use of miniature hydrophones to determine peak intensities typical of medical ultrasound devices. IEEE Trans. UFFC-33, 287 (1986)

Tada, Y.: Theoretical characteristics of generalized electret generator, using polymer film electrets. IEEE Trans. EI-21, 457 (1986)

Veit, I.: The efficiency of sound transmitters with cylindrically formed PVDF films as active element. Acustica 60, 159 (1986)

Voorthuyzen, J. A.: The pressfet – an integrated electret-mosfet structure for application as a catheter tip blood-pressure sensor. Dissertation Technische Hogeschool Twente (1986)

Subject Index

Topics in Applied Physics Founded by Helmut K. V. Lotsch